Student Solutions Manual for Goodman/Hirsch's

Precalculus
Understanding Functions
A Graphing Approach

D1537529

Ross Rueger
College of the Sequoias

Thomson Learning™

Australia • Canada • Mexico • Singapore • Spain • United Kingdom • United States

Assistant Editor: Stephanie Schmidt
Marketing Manager: Karin Sandberg
Marketing Assistant: Beth Kroenke
Editorial Assistant: Emily Davidson
Production Coordinator: Stephanie Andersen

Cover Design: Roy Neuhaus
Cover Illustration: Gregory Ochocki/Digital Stock
Print Buyer: Micky Lawler
Printing and Binding: Webcom Limited

For more information about this or any other Brooks/Cole product, contact:
BROOKS/COLE
511 Forest Lodge Road
Pacific Grove, CA 93950 USA
www.brookscole.com
1-800-423-0563 (Thomson Learning Academic Resource Center)

For permission to use material from this work, contact us by
Web: www.thomsonrights.com
fax: 1-800-730-2215
phone: 1-800-730-2214

Printed in Canada

5   4   3   2   1

ISBN: 0-534-37182-5

Contents

In memory of Tom Riddle and John Schaeffer.

Chapter 1
Algebra: The Fundamentals

1.1 The Real Numbers

1. Let $x = 0.22\overline{2}$, so $10x = 2.22\overline{2}$. Subtracting yields $9x = 2$, so $x = \frac{2}{9}$. Thus $0.22\overline{2} = \frac{2}{9}$.

3. Let $x = 4.55\overline{5}$, so $10x = 45.55\overline{5}$. Subtracting yields $9x = 41$, so $x = \frac{41}{9}$. Thus $4.55\overline{5} = \frac{41}{9}$.

5. Let $x = 8.238238\overline{238}$, so $1000x = 8238.238238\overline{238}$. Subtracting yields $999x = 8230$, so $x = \frac{8230}{999}$.
Thus $8.238238\overline{238} = \frac{8230}{999}$.

7. This statement is true: Associative property of addition

9. This statement is false: Let $x = 1$ and $y = 1$, now evaluate each side:
$$3 + (xy) = 3 + (1 \cdot 1) = 3 + 1 = 4$$
$$(3 + x)(3 + y) = (3 + 1)(3 + 1) = 4 \cdot 4 = 16$$

11. This statement is true: Associative property of multiplication

13. This statement is true: Distributive property

15. This statement is true: Multiplicative inverse property

17. Let $\dfrac{a}{b}$ and $\dfrac{c}{d}$ represent the two rational numbers (a, b, c, and d are integers). Their product is given
by: $\dfrac{a}{b} \cdot \dfrac{c}{d} = \dfrac{a \cdot c}{b \cdot d} = \dfrac{ac}{bd}$
Since ac and bd are both integers, this product is a rational number.

19. No. If the two irrational numbers are $\sqrt{2}$ and $-\sqrt{2}$, their sum is $\sqrt{2} + \left(-\sqrt{2}\right) = 0$, which is a rational number.

21. Graphing the interval $\{x \mid x < 4\}$:

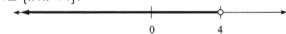

23. Graphing the interval $\{x \mid x > 5\}$:

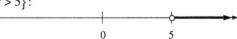

25. Graphing the interval $\{x \mid -8 < x < -2\}$:

27. The interval notation is $(5, \infty)$. Graphing the interval:

29. The interval notation is $[-5,\infty)$. Graphing the interval:

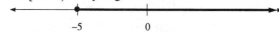

                            −5   0

31. The interval notation is $[-8,-5)$. Graphing the interval:

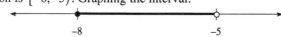

                            −8   −5

33. The interval notation is $(-\infty,-2]$. Graphing the interval:

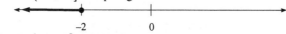

                            −2   0

35. The interval notation is $(-9,-2]$. Graphing the interval:

                            −9   2

37. The value is $|3-5|=|-2|=2$.

39. Since $\sqrt{2}>1$, the value is $\left|\sqrt{2}-1\right|=\sqrt{2}-1$.

41. If $x\ge 5$, $x-5\ge 0$ and if $x<5$, $x-5<0$. The expression is:
$$|x-5|=\begin{cases}x-5 & \text{if } x\ge 5\\ 5-x & \text{if } x<5\end{cases}$$

43. Since $x^2+1>0$, the expression is $\left|x^2+1\right|=x^2+1$.

45. The distance is $|2-3|=|-1|=1$.

47. The distance is $|-3-8|=|-11|=11$.

49. If $a<b$, then $b-a$ is a positive number. If $b<c$, then $c-b$ is a positive number. Thus $(b-a)+(c-b)=c-a$ is a positive number, so $a<c$.

51. Let c represent the diagonal of the rectangle with base $\sqrt{2}$ and height 1. Using the Pythagorean Theorem:
$$\left(\sqrt{2}\right)^2+1^2=c^2$$
$$c^2=3$$
$$c=\sqrt{3}$$

53. If n is even, then $n=2k$ for some integer k. Thus $n^2=4k^2$, and so n^2 is even. If n^2 is even, then $n^2=2l$ for some integer l. Then $n=\sqrt{2l}$, and for n to be an integer, l must have a factor of 2. Let $l=2p$, and thus $n=\sqrt{2\cdot 2p}=2\sqrt{p}$. Thus n is an even integer. Thus n^2 is even if and only if n is even. Now let $\sqrt{2}=\dfrac{m}{n}$, therefore:
$$\sqrt{2}=\frac{m}{n}$$
$$n\sqrt{2}=m$$
$$\left(n\sqrt{2}\right)^2=m^2$$
$$2n^2=m^2$$
Thus m^2 is even, and thus m is even. Let $m=2k$, thus:
$$\sqrt{2}=\frac{2k}{n}$$
$$n\sqrt{2}=2k$$
$$n=\frac{2k}{\sqrt{2}}$$

Thus n is an even number. If m and n are both even, then $\dfrac{m}{n}$ cannot be a fraction reduced to lowest terms, which contradicts our original assumption that $\sqrt{2}$ is rational. Thus $\sqrt{2}$ is an irrational number.

1.2 Operations with Real Numbers

1. Evaluating the expression: $-3+(-6)-(+4)-(-8)=-3+(-6)+(-4)+8=-5$
3. Evaluating the expression: $(-6)(-2)(-3)=(12)(-3)=-36$
5. Evaluating the expression: $-2-3.552=-2+(-3.552)=-5.552$
7. Evaluating the expression: $-4+7.29=3.29$
9. Evaluating the expression: $-2[3-(2-5)]=-2[3-(-3)]=-2(3+3)=-2(6)=-12$
11. Evaluating the expression: $2-(-3)^2=2-9=2+(-9)=-7$
13. Evaluating the expression: $6-\left[4-(5-8)^2\right]=6-\left[4-(-3)^2\right]=6-(4-9)=6-(-5)=6+5=11$
15. Evaluating the expression: $\left|-6-5\right|-\left|6-5\right|=\left|-11\right|-\left|1\right|=11-1=10$
17. Evaluating the expression: $\dfrac{\left|-9-5\right|}{\left|-9\right|-\left|5\right|}=\dfrac{\left|-14\right|}{9-5}=\dfrac{14}{4}=\dfrac{7}{2}$
19. Evaluating the expression: $\dfrac{3}{4}-\dfrac{2}{3}+\dfrac{1}{2}=\dfrac{9}{12}-\dfrac{8}{12}+\dfrac{6}{12}=\dfrac{7}{12}$
21. Evaluating the expression: $\left(-\dfrac{2}{5}\right)^2-\dfrac{3}{4}=\dfrac{4}{25}-\dfrac{3}{4}=\dfrac{16}{100}-\dfrac{75}{100}=-\dfrac{59}{100}$
23. Evaluating the expression: $6\left(-\dfrac{2}{3}\right)^2+\left(-\dfrac{2}{3}\right)-2=6\left(\dfrac{4}{9}\right)-\dfrac{2}{3}-2=\dfrac{8}{3}-\dfrac{2}{3}-2=2-2=0$
25. Evaluating the expression: $3\left(-\dfrac{1}{5}\right)^2+2\left(-\dfrac{1}{5}\right)-3=3\left(\dfrac{1}{25}\right)-\dfrac{2}{5}-3=\dfrac{3}{25}-\dfrac{10}{25}-\dfrac{75}{25}=-\dfrac{82}{25}$
27. Simplifying the complex fraction: $\dfrac{3+\frac{3}{5}}{5-\frac{1}{8}}\cdot\dfrac{40}{40}=\dfrac{120+24}{200-5}=\dfrac{144}{195}=\dfrac{48}{65}$
29. Simplifying the complex fraction: $\dfrac{\frac{2}{3}-\frac{1}{2}}{\frac{1}{8}+\frac{2}{5}}\cdot\dfrac{120}{120}=\dfrac{80-60}{15+48}=\dfrac{20}{63}$
31. Evaluating when $x=-1$ and $y=-2$: $2x^2-4y^2=2(-1)^2-4(-2)^2=2(1)-4(4)=2-16=-14$
33. Evaluating when $x=-1$ and $y=-2$: $\dfrac{x^2-2xy+y^2}{x-y}=\dfrac{(-1)^2-2(-1)(-2)+(-2)^2}{-1-(-2)}=\dfrac{1-4+4}{-1+2}=1$
35. Finding the geometric mean: $g=\sqrt[7]{5\cdot8\cdot7\cdot9\cdot7\cdot8\cdot6}=\sqrt[7]{846720}\approx7.0278$
37. Finding the value of s_e: $s_e=s_y\sqrt{1-r_{xy}^2}=1.25\sqrt{1-0.4^2}\approx1.15$
39. Computing the value of Z: $Z=\dfrac{Z_{r_1}-Z_{r_2}}{\sqrt{\dfrac{1}{n_1-3}+\dfrac{1}{n_2-3}}}=\dfrac{0.50-0.32}{\sqrt{\frac{1}{62}+\frac{1}{80}}}\approx1.06$
41. Computing the value of σ_r: $\sigma_r=\sqrt{\dfrac{1-\rho^2}{n-1}}=\sqrt{\dfrac{1-0.5^2}{99}}\approx0.087$
43. Note that $\dfrac{a}{c}+\dfrac{b}{c}=\dfrac{1}{c}\cdot a+\dfrac{1}{c}\cdot b=\dfrac{1}{c}(a+b)=\dfrac{a+b}{c}$.

45. **a.**   Since $\dfrac{1}{\sqrt{2}} \cdot \dfrac{\sqrt{2}}{\sqrt{2}} = \dfrac{\sqrt{2}}{\sqrt{4}} = \dfrac{\sqrt{2}}{2}$, the two fractions are equal.

b.   Since $\dfrac{5}{\sqrt{3}} \cdot \dfrac{\sqrt{3}}{\sqrt{3}} = \dfrac{5\sqrt{3}}{\sqrt{9}} = \dfrac{5\sqrt{3}}{3}$, the two fractions are equal.

47.   For $x = 1, y = 1 + 1 = 2$. For $x = 3, y = 3^2 = 9$.

1.3 Polynomials and Rational Expressions

1.   Performing the operations: $\left(3x^2 y\right)\left(-2xy^2\right) = -6x^3 y^3$

3.   Performing the operations: $(-3xy)^2 (5xy) = \left(9x^2 y^2\right)(5xy) = 45x^3 y^3$

5.   Performing the operations: $(2x - 3)(3x + 2) = 6x^2 - 9x + 4x - 6 = 6x^2 - 5x - 6$

7.   Performing the operations: $(2x - 7)(3x + 1) = 6x^2 - 21x + 2x - 7 = 6x^2 - 19x - 7$

9.   Performing the operations:
$$(3x - 2)\left(2x^2 - 3x + 1\right) = 3x\left(2x^2 - 3x + 1\right) - 2\left(2x^2 - 3x + 1\right)$$
$$= 6x^3 - 9x^2 + 3x - 4x^2 + 6x - 2$$
$$= 6x^3 - 13x^2 + 9x - 2$$

11.   Performing the operations:
$$(5x + 3)\left(25x^2 - 15x + 9\right) = 5x\left(25x^2 - 15x + 9\right) + 3\left(25x^2 - 15x + 9\right)$$
$$= 125x^3 - 75x^2 + 45x + 75x^2 - 45x + 27$$
$$= 125x^3 + 27$$

13.   Performing the operations:
$$(x - 3)(x - 2)(x + 1) = (x - 3)\left(x^2 - x - 2\right)$$
$$= x\left(x^2 - x - 2\right) - 3\left(x^2 - x - 2\right)$$
$$= x^3 - x^2 - 2x - 3x^2 + 3x + 6$$
$$= x^3 - 4x^2 + x + 6$$

15.   Performing the operations: $(x - 3y)^2 = x^2 - 2(x)(3y) + (3y)^2 = x^2 - 6xy + 9y^2$

17.   Performing the operations: $\left(x^2 - 7\right)\left(x^2 + 7\right) = \left(x^2\right)^2 - (7)^2 = x^4 - 49$

19.   Performing the operations: $(2x - 3y)^2 = (2x)^2 - 2(2x)(3y) + (3y)^2 = 4x^2 - 12xy + 9y^2$

21.   Performing the operations: $(x - y + 7)(x - y - 7) = (x - y)^2 - (7)^2 = x^2 - 2xy + y^2 - 49$

23.   Performing the operations: $(x - y + 2)^2 = (x - y)^2 + 4(x - y) + 4 = x^2 - 2xy + y^2 + 4x - 4y + 4$

25.   Performing the operations:
$$(x - 2)^2 - (x - 2)(x + 2) = \left(x^2 - 4x + 4\right) - \left(x^2 - 4\right) = x^2 - 4x + 4 - x^2 + 4 = -4x + 8$$

27.   Performing the operations:
$$5y(y - 5)^2 - (y - 5)(y + 5) = 5y\left(y^2 - 10y + 25\right) - \left(y^2 - 25\right)$$
$$= 5y^3 - 50y^2 + 125y - y^2 + 25$$
$$= 5y^3 - 51y^2 + 125y + 25$$

29.   Performing the operations:
$$2(x - 3)^2 - 2x^2 = 2\left(x^2 - 6x + 9\right) - 2x^2 = 2x^2 - 12x + 18 - 2x^2 = -12x + 18$$

31. Performing the operations:
$$2(x+h)^2 + 1 - \left(2x^2 + 1\right) = 2\left(x^2 + 2xh + h^2\right) + 1 - 2x^2 - 1$$
$$= 2x^2 + 4xh + 2h^2 + 1 - 2x^2 - 1$$
$$= 4xh + 2h^2$$

33. The area is given by: $\pi(x+3)^2 - \pi(3)^2 = \pi\left(x^2 + 6x + 9\right) - 9\pi = \pi x^2 + 6\pi x + 9\pi - 9\pi = \pi x^2 + 6\pi x$

35. The dimensions of the box are $2 - 2x$ by $3 - 2x$ by x, so the surface area is given by:
$$A = (2-2x)(3-2x) + 2x(2-2x) + 2x(3-2x)$$
$$= 6 - 10x + 4x^2 + 4x - 4x^2 + 6x - 4x^2$$
$$= 6 - 4x^2 \ \text{ft}^2$$

37. Factoring the expression: $x^2 + x - 20 = (x+5)(x-4)$

39. Factoring the expression: $12x^2 + 10x - 12 = 2\left(6x^2 + 5x - 6\right) = 2(3x-2)(2x+3)$

41. Factoring the expression: $3a(b-2) - (b-2) = (b-2)(3a-1)$

43. Factoring the expression: $5x(x-1)^2 + 10(x-1)^3 = 5(x-1)^2\left[x + 2(x-1)\right] = 5(x-1)^2(3x-2)$

45. Factoring the expression:
$$3x(x-2)\left(2x^2+1\right) - 6(x+3)\left(2x^2+1\right) = 3\left(2x^2+1\right)\left[x(x-2) - 2(x+3)\right]$$
$$= 3\left(2x^2+1\right)\left(x^2 - 2x - 2x - 6\right)$$
$$= 3\left(2x^2+1\right)\left(x^2 - 4x - 6\right)$$

47. Factoring the expression: $ax + bx - 2a - 2b = x(a+b) - 2(a+b) = (a+b)(x-2)$

49. Factoring the expression: $x^3 - 5x - 3x^2 + 15 = x\left(x^2 - 5\right) - 3\left(x^2 - 5\right) = \left(x^2 - 5\right)(x-3)$

51. Factoring the expression: $x^2 - 9 = (x+3)(x-3)$

53. Factoring the expression: $x^2 - 8x + 16 = (x-4)(x-4) = (x-4)^2$

55. Factoring the expression: $8x^3 - 1 = (2x-1)\left(4x^2 + 2x + 1\right)$

57. Factoring the expression: $81x^3 - 24 = 3\left(27x^3 - 8\right) = 3(3x-2)\left(9x^2 + 6x + 4\right)$

59. Factoring the expression:
$$x^3 + 3x^2 - 16x - 48 = x^2(x+3) - 16(x+3) = (x+3)\left(x^2 - 16\right) = (x+3)(x+4)(x-4)$$

61. Factoring the expression: $x^4 - 10x^2 + 24 = \left(x^2 - 6\right)\left(x^2 - 4\right) = \left(x^2 - 6\right)(x+2)(x-2)$

63. Factoring the expression: $x^2 - 2xy + y^2 - 16 = (x-y)^2 - 16 = (x-y+4)(x-y-4)$

65. Factoring the expression:
$$a^5 - a^3 - a^2 + 1 = a^3\left(a^2 - 1\right) - 1\left(a^2 - 1\right)$$
$$= \left(a^2 - 1\right)\left(a^3 - 1\right)$$
$$= (a+1)(a-1)(a-1)\left(a^2 + a + 1\right)$$
$$= (a+1)(a-1)^2\left(a^2 + a + 1\right)$$

67. Simplifying the fraction:
$$\frac{2x^3 - 2xy^2}{4x^4 - 8x^3y + 4x^2y^2} = \frac{2x\left(x^2 - y^2\right)}{4x^2\left(x^2 - 2xy + y^2\right)} = \frac{2x(x+y)(x-y)}{4x^2(x-y)^2} = \frac{x+y}{2x(x-y)}$$

69. Simplifying the fraction: $\dfrac{3(x+h)+1-(3x+1)}{h} = \dfrac{3x+3h+1-3x-1}{h} = \dfrac{3h}{h} = 3$

71. Simplifying the fraction:

$$\dfrac{\left(x^2-4\right)^2(-4)-(-4x)(4x)\left(x^2-4\right)}{\left(x^2-4\right)^4} = \dfrac{-4\left(x^2-4\right)\left[x^2-4-4x^2\right]}{\left(x^2-4\right)^4}$$

$$= \dfrac{-4\left(x^2-4\right)\left(-3x^2-4\right)}{\left(x^2-4\right)^4}$$

$$= \dfrac{4\left(3x^2+4\right)}{\left(x^2-4\right)^3}$$

73. Performing the operations:

$$\dfrac{6x^2-7x-3}{4x^2-12x+9}\cdot\dfrac{2x-3}{3x^2-5x-2} = \dfrac{(3x+1)(2x-3)}{(2x-3)^2}\cdot\dfrac{2x-3}{(3x+1)(x-2)} = \dfrac{1}{x-2}$$

75. Performing the operations:

$$\dfrac{27x^3-8}{9x^2-4}\div\left(9x^2+6x+4\right) = \dfrac{(3x-2)\left(9x^2+6x+4\right)}{(3x+2)(3x-2)}\cdot\dfrac{1}{9x^2+6x+4} = \dfrac{1}{3x+2}$$

77. Performing the operations:

$$\dfrac{x}{x-y}-\dfrac{x}{y}-\dfrac{xy}{y(x-y)} = \dfrac{x(x-y)}{y(x-y)}-\dfrac{xy-x^2+xy}{y(x-y)} = \dfrac{2xy-x^2}{y(x-y)} = \dfrac{x(2y-x)}{y(x-y)}$$

79. Performing the operations: $\dfrac{2a+6}{a^2+5a+6}+\dfrac{5a+1}{a+2} = \dfrac{2(a+3)}{(a+3)(a+2)}+\dfrac{5a+1}{a+2} = \dfrac{2}{a+2}+\dfrac{5a+1}{a+2} = \dfrac{5a+3}{a+2}$

81. Performing the operations:

$$\dfrac{3x}{x-1}+\dfrac{x+3}{x-2}+2x-3 = \dfrac{3x(x-2)}{(x-1)(x-2)}+\dfrac{(x+3)(x-1)}{(x-1)(x-2)}+\dfrac{(2x-3)(x-1)(x-2)}{(x-1)(x-2)}$$

$$= \dfrac{3x^2-6x+x^2+2x-3+2x^3-9x^2+13x-6}{(x-1)(x-2)}$$

$$= \dfrac{2x^3-5x^2+9x-9}{(x-1)(x-2)}$$

83. Performing the operations:

$$-\dfrac{x^2+1}{(x-2)^2}+\dfrac{2x}{x-2} = \dfrac{-x^2-1}{(x-2)^2}+\dfrac{2x(x-2)}{(x-2)^2} = \dfrac{-x^2-1+2x^2-4x}{(x-2)^2} = \dfrac{x^2-4x-1}{(x-2)^2}$$

85. Simplifying the fraction: $\dfrac{1-\dfrac{1}{x^2}}{\dfrac{x-1}{x}}\cdot\dfrac{x^2}{x^2} = \dfrac{x^2-1}{x(x-1)} = \dfrac{(x+1)(x-1)}{x(x-1)} = \dfrac{x+1}{x}$

87. Simplifying the fraction:

$$\dfrac{\dfrac{1}{x+3}-\dfrac{1}{x+1}}{2}\cdot\dfrac{(x+3)(x+1)}{(x+3)(x+1)} = \dfrac{x+1-x-3}{2(x+3)(x+1)} = \dfrac{-2}{2(x+3)(x+1)} = \dfrac{-1}{(x+3)(x+1)}$$

89. Simplifying the fraction:

$$\frac{\dfrac{5}{x+h}-\dfrac{5}{x}}{h}\cdot\frac{x(x+h)}{x(x+h)}=\frac{5x-5(x+h)}{hx(x+h)}=\frac{5x-5x-5h}{hx(x+h)}=\frac{-5h}{hx(x+h)}=\frac{-5}{x(x+h)}$$

91. Simplifying the fraction:

$$\frac{\dfrac{3x+1}{x-4}}{\dfrac{2x+1}{x+2}-\dfrac{x}{x-4}}\cdot\frac{(x+2)(x-4)}{(x+2)(x-4)}=\frac{(3x+1)(x+2)}{(2x+1)(x-4)-x(x+2)}$$

$$=\frac{3x^2+7x+2}{2x^2-7x-4-x^2-2x}$$

$$=\frac{3x^2+7x+2}{x^2-9x-4}$$

93. As x gets larger and larger, the polynomial x^2+2x+3 will become larger.

1.4 Exponents and Radicals

1. Simplifying the fraction: $\dfrac{\left(x^2y\right)^2\left(x^3y\right)^2}{(xy)^4}=\dfrac{x^4y^2\cdot x^6y^2}{x^4y^4}=\dfrac{x^{10}y^4}{x^4y^4}=x^6$

3. Simplifying the fraction:

$$\frac{\left(3x^{-1}y^{-2}\right)^{-3}\left(x^2y^{-1}\right)^3}{\left(9x^{-2}y\right)^{-2}}=\frac{3^{-3}x^3y^6\cdot x^6y^{-3}}{9^{-2}x^4y^{-2}}=\frac{3^{-3}x^9y^3}{9^{-2}x^4y^{-2}}=\frac{9^2x^9y^5}{3^3x^4}=\frac{81}{27}x^5y^5=3x^5y^5$$

5. Simplifying the fraction: $\dfrac{x^{-1}+2}{x+2}=\dfrac{\dfrac{1}{x}+2}{x+2}\cdot\dfrac{x}{x}=\dfrac{1+2x}{x(x+2)}=\dfrac{2x+1}{x^2+2x}$

7. Simplifying the fraction: $\dfrac{x^{-1}+y^{-1}}{xy^{-1}}=\dfrac{\dfrac{1}{x}+\dfrac{1}{y}}{\dfrac{x}{y}}\cdot\dfrac{xy}{xy}=\dfrac{y+x}{x^2}=\dfrac{x+y}{x^2}$

9. Writing in scientific notation: $7,500,000,000=7.5\times10^9$

11. Dividing: $\dfrac{93,000,000\text{ miles}}{186,000\text{ miles / second}}=\dfrac{9.3\times10^7}{1.86\times10^5}$ seconds $=500$ seconds $=8$ min 20 sec.

13. Evaluating the expression: $32^{-3/5}=\left(32^{1/5}\right)^{-3}=2^{-3}=\frac{1}{8}$

15. Evaluating the expression: $\left(-\frac{8}{27}\right)^{-2/3}=\left(\left(-\frac{8}{27}\right)^{1/3}\right)^{-2}=\left(-\frac{2}{3}\right)^{-2}=\left(-\frac{3}{2}\right)^2=\frac{9}{4}$

17. Simplifying the expression: $\dfrac{x^{1/2}y^{2/3}}{x^{-1/3}}=x^{1/2+1/3}y^{2/3}=x^{5/6}y^{2/3}$

19. Simplifying the expression:

$$\left(x^2+1\right)^{-1/2}\left(x^2+1\right)^{3/5}=\left(x^2+1\right)^{-1/2+3/5}=\left(x^2+1\right)^{-5/10+6/10}=\left(x^2+1\right)^{1/10}$$

21. Simplifying the expression: $\left(x^{1/2}+2\right)^2 = \left(x^{1/2}\right)^2 + 2\left(x^{1/2}\right)(2)+(2)^2 = x+4x^{1/2}+4$

23. Simplifying the expression: $\left(x^{1/2}+2\right)\left(x^{1/2}-2\right) = \left(x^{1/2}\right)^2 - (2)^2 = x-4$

25. Writing as a single fraction:

$$3(x+1)^{-1}-3x(x+1)^{-2} = \frac{3}{x+1}-\frac{3x}{(x+1)^2} = \frac{3(x+1)}{(x+1)^2}-\frac{3x}{(x+1)^2} = \frac{3x+3-3x}{(x+1)^2} = \frac{3}{(x+1)^2}$$

27. Writing as a single fraction:

$$4x^3\left(x^2+1\right)^{-1/2}+4x\left(x^2+1\right)^{1/2} = 4x\left(x^2+1\right)^{-1/2}\left(x^2+x^2+1\right)$$

$$= 4x\left(x^2+1\right)^{-1/2}\left(2x^2+1\right)$$

$$= \frac{4x\left(2x^2+1\right)}{\left(x^2+1\right)^{1/2}}$$

29. Writing as a single fraction:

$$\frac{3x^3}{2}\left(x^3-3\right)^{-1/2}+\left(x^3-3\right)^{1/2} = \frac{1}{2}\left(x^3-3\right)^{-1/2}\left[3x^3+2\left(x^3-3\right)\right]$$

$$= \frac{1}{2}\left(x^3-3\right)^{-1/2}\left[5x^3-6\right]$$

$$= \frac{5x^3-6}{2\left(x^3-3\right)^{1/2}}$$

31. Rewriting without radicals: $\dfrac{3}{x\sqrt{x^3-3}} = \dfrac{3}{x\left(x^3-3\right)^{1/2}} = 3x^{-1}\left(x^3-3\right)^{-1/2}$

33. Simplifying the radical: $\sqrt{24x^4y^6} = \sqrt{4x^4y^6}\sqrt{6} = 2x^2y^3\sqrt{6}$

35. Simplifying the radical: $\sqrt{\dfrac{16x^8}{9}} = \dfrac{4x^4}{3}$

37. Simplifying the radical:

$$\sqrt{3x^3}\sqrt[3]{81x^5} = \left(3x^3\right)^{1/2}\left(81x^5\right)^{1/3}$$

$$= 3^{1/2}x^{3/2}\cdot 3^{4/3}x^{5/3}$$

$$= 3^{1/2+4/3}x^{3/2+5/3}$$

$$= 3^{11/6}x^{19/6}$$

$$= 3x^3\sqrt[6]{3^5 x}$$

39. Rationalizing the denominator: $\dfrac{5}{\sqrt{3x^5}}\cdot\dfrac{\sqrt{3x}}{\sqrt{3x}} = \dfrac{5\sqrt{3x}}{\sqrt{9x^6}} = \dfrac{5\sqrt{3x}}{3x^3}$

41. Rationalizing the denominator: $\dfrac{1}{\sqrt{x}-4}\cdot\dfrac{\sqrt{x}+4}{\sqrt{x}+4} = \dfrac{\sqrt{x}+4}{x-16}$

43. Rationalizing the numerator:

$$\frac{\sqrt{(x-2)(x-1)}}{x-1}\cdot\frac{\sqrt{(x-2)(x-1)}}{\sqrt{(x-2)(x-1)}} = \frac{(x-2)(x-1)}{(x-1)\sqrt{(x-2)(x-1)}} = \frac{x-2}{\sqrt{(x-2)(x-1)}}$$

45. Rationalizing the numerator: $\dfrac{\sqrt{x}-3}{x-9} \cdot \dfrac{\sqrt{x}+3}{\sqrt{x}+3} = \dfrac{x-9}{(x-9)(\sqrt{x}+3)} = \dfrac{1}{\sqrt{x}+3}$

47. Rationalizing the numerator:

$$\dfrac{\sqrt{x+h}-\sqrt{x}}{h} \cdot \dfrac{\sqrt{x+h}+\sqrt{x}}{\sqrt{x+h}+\sqrt{x}} = \dfrac{x+h-x}{h\left(\sqrt{x+h}+\sqrt{x}\right)} = \dfrac{h}{h\left(\sqrt{x+h}+\sqrt{x}\right)} = \dfrac{1}{\sqrt{x+h}+\sqrt{x}}$$

49. Simplifying the radicals:

$$\left(3xy^2\sqrt{x^2y}\right)\left(2x\sqrt{18xy^2}\right) = 6x^2y^2\sqrt{18x^3y^3} = 6x^2y^2 \cdot 3xy\sqrt{2xy} = 18x^3y^3\sqrt{2xy}$$

51. Simplifying the radicals: $\dfrac{\sqrt{x^2-2x+1}}{\sqrt{x-1}} = \dfrac{\sqrt{(x-1)^2}}{\sqrt{x-1}} = \sqrt{x-1}$

53. Simplifying the radicals:

$$\sqrt{27}-3\sqrt{18}+6\sqrt{12} = 3\sqrt{3}-3 \cdot 3\sqrt{2}+6 \cdot 2\sqrt{3} = 3\sqrt{3}-9\sqrt{2}+12\sqrt{3} = 15\sqrt{3}-9\sqrt{2}$$

55. Simplifying the radicals: $\left(2\sqrt{3}-5\right)\left(2\sqrt{3}+2\right) = 4\sqrt{9}-10\sqrt{3}+4\sqrt{3}-10 = 12-6\sqrt{3}-10 = 2-6\sqrt{3}$

57. Simplifying the radicals: $\left(\sqrt{a}-5\right)\left(\sqrt{a}+5\right) = \left(\sqrt{a}\right)^2 - (5)^2 = a-25$

59. Simplifying the radicals:

$$\left(\sqrt{x}-3\right)^2 - \left(\sqrt{x-3}\right)^2 = x-6\sqrt{x}+9-(x-3) = x-6\sqrt{x}+9-x+3 = 12-6\sqrt{x}$$

61. Simplifying the radicals: $\dfrac{\sqrt{2}}{\sqrt{7}-\sqrt{5}} \cdot \dfrac{\sqrt{7}+\sqrt{5}}{\sqrt{7}+\sqrt{5}} = \dfrac{\sqrt{14}+\sqrt{10}}{7-5} = \dfrac{\sqrt{14}+\sqrt{10}}{2}$

63. Simplifying the radicals:

$$\dfrac{12}{\sqrt{6}-2} - \dfrac{36}{\sqrt{6}} = \dfrac{12}{\sqrt{6}-2} \cdot \dfrac{\sqrt{6}+2}{\sqrt{6}+2} - \dfrac{36}{\sqrt{6}} \cdot \dfrac{\sqrt{6}}{\sqrt{6}} = \dfrac{12\left(\sqrt{6}+2\right)}{6-4} - \dfrac{36\sqrt{6}}{6} = 6\left(\sqrt{6}+2\right)-6\sqrt{6} = 12$$

65. Expressing as a single fraction: $2x \cdot \dfrac{\sqrt{x^2-2}}{\sqrt{x^2-2}} - \dfrac{1}{\sqrt{x^2-2}} = \dfrac{2x\sqrt{x^2-2}-1}{\sqrt{x^2-2}}$

67. Expressing as a single fraction:

$$3\sqrt{x^2-2} \cdot \dfrac{\sqrt{x^2-2}}{\sqrt{x^2-2}} - \dfrac{2}{\sqrt{x^2-2}} = \dfrac{3\left(x^2-2\right)-2}{\sqrt{x^2-2}} = \dfrac{3x^2-6-2}{\sqrt{x^2-2}} = \dfrac{3x^2-8}{\sqrt{x^2-2}}$$

69. Expressing as a single fraction: $\dfrac{\sqrt{\dfrac{2}{x+h}}-\sqrt{\dfrac{2}{x}}}{h} \cdot \dfrac{\sqrt{x}\sqrt{x+h}}{\sqrt{x}\sqrt{x+h}} = \dfrac{\sqrt{2x}-\sqrt{2(x+h)}}{h\sqrt{x(x+h)}}$

1.5 The Complex Numbers

1. Simplifying: $i^{16} = \left(i^4\right)^4 = (1)^4 = 1$

3. Simplifying: $i^{100} = \left(i^4\right)^{25} = (1)^{25} = 1$

5. Simplifying: $i^{4n+3} = i^{4n} \cdot i^3 = \left(i^4\right)^n \cdot i^2 \cdot i = (1)^n(-1)i = -i$

7. Writing in the form $a+bi$: $3+\sqrt{-16} = 3+4i$

9. Writing in the form $a+bi$: $5-\sqrt{-\dfrac{1}{12}}=5-\dfrac{i}{\sqrt{12}}\cdot\dfrac{\sqrt{3}}{\sqrt{3}}=5-\dfrac{\sqrt{3}}{6}i$

11. Performing the operations: $(3+2i)+(6-8i)=9-6i$

13. Performing the operations: $(8-2i)+(9+2i)=17$

15. Performing the operations: $(2-3i)(5+i)=10-15i+2i-3i^2=13-13i$

17. Performing the operations: $(5-6i)(5+6i)=25-36i^2=61$

19. Performing the operations: $(2-3i)^2=4-12i+9i^2=-5-12i$

21. Performing the operations: $\dfrac{2+8i}{2}=1+4i$

23. Performing the operations: $\dfrac{2+3i}{i}\cdot\dfrac{i}{i}=\dfrac{2i+3i^2}{i^2}=\dfrac{2i-3}{-1}=3-2i$

25. Performing the operations: $\dfrac{3}{2+i}\cdot\dfrac{2-i}{2-i}=\dfrac{6-3i}{4-i^2}=\dfrac{6-3i}{5}=\dfrac{6}{5}-\dfrac{3}{5}i$

27. Performing the operations: $\dfrac{i}{2-i}\cdot\dfrac{2+i}{2+i}=\dfrac{2i+i^2}{4-i^2}=\dfrac{2i-1}{5}=-\dfrac{1}{5}+\dfrac{2}{5}i$

29. Performing the operations: $\dfrac{i+1}{i-1}\cdot\dfrac{i+1}{i+1}=\dfrac{i^2+2i+1}{i^2-1}=\dfrac{-1+2i+1}{-1-1}=\dfrac{2i}{-2}=-i=0-1i$

31. Performing the operations: $\dfrac{6-2i}{5+2i}\cdot\dfrac{5-2i}{5-2i}=\dfrac{30-10i-12i+4i^2}{25-4i^2}=\dfrac{26-22i}{29}=\dfrac{26}{29}-\dfrac{22}{29}i$

33. Performing the operations:
$(2+i)^2-4(2+i)=4+4i+i^2-8-4i=4+4i-1-8-4i=-5=-5+0i$

35. Performing the operations: $(2-2i)^2+2(2+i)=4-8i+4i^2+4+2i=4-8i-4+4+2i=4-6i$

37. Substituting $x=3+2i$ into the equation:
$$x^2-6x+13=(3+2i)^2-6(3+2i)+13$$
$$=9+12i+4i^2-18-12i+13$$
$$=9+12i-4-18-12i+13$$
$$=0$$
Thus $x=3+2i$ is a solution to the equation.

39. Expressing in $a+bi$ form: $i^{-1}=\dfrac{1}{i}\cdot\dfrac{i}{i}=\dfrac{i}{i^2}=-i=0-1i$

41. Using the hint:
$$\sqrt[6]{-1}=\sqrt{\sqrt[3]{-1}}=\sqrt{-1}=i$$
$$\sqrt[10]{-1}=\sqrt{\sqrt[5]{-1}}=\sqrt{-1}=i$$
$$\sqrt[14]{-1}=\sqrt{\sqrt[7]{-1}}=\sqrt{-1}=i$$

1.6 First-Degree Equations and Inequalities in One Variable

1. Solving the equation:
$$2-3(x-4)=2(x-1)$$
$$2-3x+12=2x-2$$
$$-3x+14=2x-2$$
$$-5x=-16$$
$$x=\tfrac{16}{5}$$

3. Solving the inequality:

$$4x + \tfrac{2}{3} \le 2x - (3x + 1)$$

$$4x + \tfrac{2}{3} \le 2x - 3x - 1$$

$$4x + \tfrac{2}{3} \le -x - 1$$

$$5x \le -\tfrac{5}{3}$$

$$x \le -\tfrac{1}{3}$$

The solution is $\left(-\infty, -\tfrac{1}{3}\right]$.

5. Solving the inequality:

$$5\{y - [2 - (y - 3)]\} > y - 2$$

$$5[y - (2 - y + 3)] > y - 2$$

$$5[y - (5 - y)] > y - 2$$

$$5(y - 5 + y) > y - 2$$

$$5(2y - 5) > y - 2$$

$$10y - 25 > y - 2$$

$$9y > 23$$

$$y > \tfrac{23}{9}$$

The solution is $\left(\tfrac{23}{9}, \infty\right)$.

7. Solving the equation:

$$\tfrac{2}{3}x - 5 = \tfrac{3}{2}x + 4(x - 1)$$

$$6\left(\tfrac{2}{3}x - 5\right) = 6 \cdot \tfrac{3}{2}x + 6 \cdot 4(x - 1)$$

$$4x - 30 = 9x + 24x - 24$$

$$4x - 30 = 33x - 24$$

$$-29x = 6$$

$$x = -\tfrac{6}{29}$$

9. Solving the equation:

$$-\frac{7}{y} + 1 = -13$$

$$y\left(-\frac{7}{y} + 1\right) = -13y$$

$$-7 + y = -13y$$

$$14y = 7$$

$$y = \tfrac{1}{2}$$

11. Solving the inequality:

$$\frac{3x + 1}{2} - \frac{1}{3} < 1$$

$$6 \cdot \frac{3x + 1}{2} - 6 \cdot \frac{1}{3} < 1 \cdot 6$$

$$3(3x + 1) - 2 < 6$$

$$9x + 3 - 2 < 6$$

$$9x < 5$$

$$x < \tfrac{5}{9}$$

The solution is $\left(-\infty, \tfrac{5}{9}\right)$.

13.   Solving the equation:
$$-\frac{7}{y}+3=3$$
$$y\left(-\frac{7}{y}+3\right)=3y$$
$$-7+3y=3y$$
$$-7=0$$

Since this statement is false, there is no solution to the equation.

15.   Solving the equation:
$$\frac{2}{x+3}+4=\frac{5-x}{x+3}$$
$$(x+3)\left(\frac{2}{x+3}+4\right)=(x+3)\left(\frac{5-x}{x+3}\right)$$
$$2+4(x+3)=5-x$$
$$2+4x+12=5-x$$
$$4x+14=5-x$$
$$5x=-9$$
$$x=-\frac{9}{5}$$

17.   Solving the equation:
$$\frac{4a+1}{(a-3)(a+2)}=\frac{2}{a-3}+\frac{5}{a+2}$$
$$(a-3)(a+2)\left(\frac{4a+1}{(a-3)(a+2)}\right)=(a-3)(a+2)\left(\frac{2}{a-3}+\frac{5}{a+2}\right)$$
$$4a+1=2(a+2)+5(a-3)$$
$$4a+1=2a+4+5a-15$$
$$4a+1=7a-11$$
$$-3a=-12$$
$$a=4$$

19.   Solving the equation:
$$\frac{5}{2x+1}+\frac{3}{2x-1}=\frac{22}{(2x+1)(2x-1)}$$
$$(2x+1)(2x-1)\left(\frac{5}{2x+1}+\frac{3}{2x-1}\right)=(2x+1)(2x-1)\left(\frac{22}{(2x+1)(2x-1)}\right)$$
$$5(2x-1)+3(2x+1)=22$$
$$10x-5+6x+3=22$$
$$16x-2=22$$
$$16x=24$$
$$x=\frac{3}{2}$$

21. Solving the equation:

$$\frac{6}{3x+5} - \frac{2}{x-4} = \frac{10}{(3x+5)(x-4)}$$

$$(3x+5)(x-4)\left(\frac{6}{3x+5} - \frac{2}{x-4}\right) = (3x+5)(x-4)\left(\frac{10}{(3x+5)(x-4)}\right)$$

$$6(x-4) - 2(3x+5) = 10$$
$$6x - 24 - 6x - 10 = 10$$
$$-34 = 10$$

Since this statement is false, there is no solution to the equation.

23. Solving for y:
$$3x + 2y - 4 = 5x - 3y + 2$$
$$5y = 2x + 6$$
$$y = \frac{2x+6}{5}$$

25. Solving for W:
$$S = 2LH + 2LW + 2WH$$
$$S - 2LH = W(2L + 2H)$$
$$W = \frac{S - 2LH}{2L + 2H}$$

27. Solving for h:
$$A = \frac{1}{2}h\left(b_1 + b_2\right)$$
$$2A = h\left(b_1 + b_2\right)$$
$$h = \frac{2A}{b_1 + b_2}$$

29. Setting each factor equal to zero results in $x = \frac{2}{3}$ or $y = \frac{1}{2}$. So $y = \frac{1}{2}$ as long as $x \neq \frac{2}{3}$.

31. Since $s > 0$, each side of the inequality can be multiplied by s:
$$\frac{x - \mu}{s} < 1.96$$
$$x - \mu < 1.96s$$
$$x > \frac{x - \mu}{1.96}$$

33. Solving for f:
$$\frac{1}{f} = \frac{1}{f_1} + \frac{1}{f_2}$$
$$ff_1f_2 \cdot \frac{1}{f} = ff_1f_2\left(\frac{1}{f_1} + \frac{1}{f_2}\right)$$
$$f_1f_2 = ff_2 + ff_1$$
$$f_1f_2 = f\left(f_2 + f_1\right)$$
$$f = \frac{f_1f_2}{f_1 + f_2}$$

35. Solving for x:

$$y = \frac{3x - 2}{x}$$
$$xy = 3x - 2$$
$$xy - 3x = -2$$
$$x(y - 3) = -2$$
$$x = \frac{-2}{y - 3} = \frac{2}{3 - y}$$

37. Solving the inequality:

$$-4 < 3x - 2 < 5$$
$$-2 < 3x < 7$$
$$-\tfrac{2}{3} < x < \tfrac{7}{3}$$

Graphing the interval $\left(-\tfrac{2}{3}, \tfrac{7}{3}\right)$:

$-2/3$ $\qquad\qquad$ $7/3$

39. Solving the inequality:

$$0 < 5 - 2x \le 4$$
$$-5 < -2x \le -1$$
$$\tfrac{5}{2} > x \ge \tfrac{1}{2}$$

Graphing the interval $\left[\tfrac{1}{2}, \tfrac{5}{2}\right)$:

$1/2$ $\qquad\qquad$ $5/2$

41. Solving the inequality:

$$-\tfrac{3}{2} < \tfrac{1}{3} - x < 2$$
$$6\left(-\tfrac{3}{2}\right) < 6\left(\tfrac{1}{3} - x\right) < 6(2)$$
$$-9 < 2 - 6x < 12$$
$$-11 < -6x < 10$$
$$\tfrac{11}{6} > x > -\tfrac{5}{3}$$

Graphing the interval $\left(-\tfrac{5}{3}, \tfrac{11}{6}\right)$:

$-5/3$ $\qquad\qquad$ $11/6$

43. Solving each inequality:

$$2x - 3 < 4 \qquad\qquad 4 < 3x - 1$$
$$2x < 7 \qquad\qquad\qquad 5 < 3x$$
$$x < \tfrac{7}{2} \qquad\qquad\qquad \tfrac{5}{3} < x$$

Graphing the interval $\left(\tfrac{5}{3}, \tfrac{7}{2}\right)$:

$5/3$ $\qquad\qquad$ $7/2$

45. The new value of s is given by: $s' = k(3t)^2 = 9kt^2 = 9s$

So s is multiplied by 9.

47. The new value of E is given by: $E' = \dfrac{k}{(3d)^2} = \dfrac{k}{9d^2} = \tfrac{1}{9}E$

So E is divided by 9.

49. Let x represent the number of 20 lb boxes and $50 - x$ represent the number of 25 lb boxes. The equation is:
$$20(x) + 25(50 - x) = 1175$$
$$20x + 1250 - 25x = 1175$$
$$-5x + 1250 = 1175$$
$$-5x = -75$$
$$x = 15$$
There are 15 20 lb boxes and 35 25 lb boxes.

51. Let x represent the number of orchestra seats and $56 - x$ represent the number of balcony seats. The equation is:
$$48(x) + 28(56 - x) = 2328$$
$$48x + 1568 - 28x = 2328$$
$$20x + 1568 = 2328$$
$$20x = 760$$
$$x = 38$$
The club bought 38 orchestra seats.

53. Let t represent the time for the older machine and $110 - t$ represent the time for the newer machine. The equation is:
$$50(t) + 70(110 - t) = 7200$$
$$50t + 7700 - 70t = 7200$$
$$7700 - 20t = 7200$$
$$-20t = -500$$
$$t = 25$$
The older machine made $50(25) = 1250$ copies.

55. Let x represent the ounces of 20% solution. The equation is:
$$0.20(x) + 0.50(5) = 0.30(x + 5)$$
$$0.2x + 2.5 = 0.3x + 1.5$$
$$-0.1x = -1.0$$
$$x = 10$$
Thus 10 ounces of 20% solution should be mixed.

57. Let x represent the amount Cindy invests in the 14% certificates. The inequality is:
$$0.09(8000 - x) + 0.14(x) \geq 890$$
$$720 - 0.09x + 0.14x \geq 890$$
$$720 + 0.05x \geq 890$$
$$0.05x \geq 170$$
$$x \geq 3400$$
She should invest at least $3400 in the higher-risk certificate.

59. The rate for pipe A is $\frac{1}{3}$, and that for pipe B is $\frac{1}{2}$. Let t represent the time for both pipes (together) to fill the pool. The equation is:
$$\tfrac{1}{3}t + \tfrac{1}{2}t = 1$$
$$6\left(\tfrac{1}{3}t + \tfrac{1}{2}t\right) = 6(1)$$
$$2t + 3t = 6$$
$$5t = 6$$
$$t = \tfrac{6}{5} = 1\tfrac{1}{5}$$
It will take $1\tfrac{1}{5}$ days to fill the pool.

61.   The rate to fill the tub is $\frac{1}{10}$, and the rate to drain the tub is $\frac{1}{15}$. Let t represent the time to fill the tub with the faucet on and the drain open. The equation is:

$$\frac{1}{10}t - \frac{1}{15}t = 1$$
$$30\left(\frac{1}{10}t - \frac{1}{15}t\right) = 30(1)$$
$$3t - 2t = 30$$
$$t = 30$$

It will take 30 minutes to fill the tub.

63.   If $\frac{1}{a} < 0$, it must be the case that $a < 0$.

65.   The student's step of $3 < 4(x-2)$ is assuming that $x < 2$, since the inequality was reversed. The other case is for $3 > 4(x-2)$, which occurs when $x > 2$.

1.7 Absolute Value Equations and Inequalities

1.   Solving the equation:
$$|x| = 12$$
$$x = -12, 12$$

3.   Solving the inequality:
$$|x| < 9$$
$$-9 < x < 9$$
The solution is $(-9, 9)$.

5.   Solving the inequality:
$$|x - 8| \leq 2$$
$$-2 \leq x - 8 \leq 2$$
$$6 \leq x \leq 10$$
The solution is $[6, 10]$.

7.   Solving the equation:
$$|2x - 4| = 7$$

$$2x - 4 = 7 \qquad \text{or} \qquad 2x - 4 = -7$$
$$2x = 11 \qquad\qquad\qquad 2x = -3$$
$$x = \tfrac{11}{2} \qquad\qquad\qquad x = -\tfrac{3}{2}$$

9.   Solving the inequality:
$$|5 + 2x| < 3$$
$$-3 < 5 + 2x < 3$$
$$-8 < 2x < -2$$
$$-4 < x < -1$$
The solution is $(-4, -1)$.

11.   Solving the inequality:
$$\left|x - \tfrac{2}{3}\right| > \tfrac{2}{3}$$

$$x - \tfrac{2}{3} < -\tfrac{2}{3} \qquad \text{or} \qquad x - \tfrac{2}{3} > \tfrac{2}{3}$$
$$x < 0 \qquad\qquad\qquad x > \tfrac{4}{3}$$
The solution is $(-\infty, 0) \cup \left(\tfrac{4}{3}, \infty\right)$.

13.   Solving the inequality:
$$|3-2x|<3$$
$$-3<3-2x<3$$
$$-6<-2x<0$$
$$3>x>0$$
The solution is $(0,3)$.

15.   Solving the equation:
$$\left|\frac{x}{3}+1\right|=\frac{3}{4}$$

$$\frac{x}{3}+1=-\frac{3}{4} \qquad \text{or} \qquad \frac{x}{3}+1=\frac{3}{4}$$

$$\frac{x}{3}=-\frac{7}{4} \qquad\qquad\qquad \frac{x}{3}=-\frac{1}{4}$$

$$x=-\frac{21}{4} \qquad\qquad\qquad x=-\frac{3}{4}$$

17.   Solving the inequality:
$$\left|\frac{x}{2}-\frac{2}{3}\right|>\frac{1}{2}$$

$$\frac{x}{2}-\frac{2}{3}<-\frac{1}{2} \qquad \text{or} \qquad \frac{x}{2}-\frac{2}{3}>\frac{1}{2}$$

$$6\left(\frac{x}{2}-\frac{2}{3}\right)<6\left(-\frac{1}{2}\right) \qquad 6\left(\frac{x}{2}-\frac{2}{3}\right)>6\left(\frac{1}{2}\right)$$

$$3x-4<-3 \qquad\qquad 3x-4>3$$

$$3x<1 \qquad\qquad\qquad 3x>7$$

$$x<\frac{1}{3} \qquad\qquad\qquad x>\frac{7}{3}$$

The solution is $\left(-\infty,\frac{1}{3}\right)\cup\left(\frac{7}{3},\infty\right)$.

19.   Solving the inequality:
$$\left|\frac{x}{2}-\frac{x}{3}\right|\geq\frac{1}{2}$$

$$\frac{x}{2}-\frac{x}{3}\leq-\frac{1}{2} \qquad \text{or} \qquad \frac{x}{2}-\frac{x}{3}\geq\frac{1}{2}$$

$$6\left(\frac{x}{2}-\frac{x}{3}\right)\leq6\left(-\frac{1}{2}\right) \qquad 6\left(\frac{x}{2}-\frac{x}{3}\right)\geq6\left(\frac{1}{2}\right)$$

$$3x-2x\leq-3 \qquad\qquad 3x-2x\geq3$$

$$x\leq-3 \qquad\qquad\qquad x\geq3$$

The solution is $(-\infty,-3]\cup[3,\infty)$.

21.   Solving the equation:
$$\left|\frac{2x-3}{2}\right|=4$$

$$\frac{2x-3}{2}=-4 \qquad \text{or} \qquad \frac{2x-3}{2}=4$$

$$2x-3=-8 \qquad\qquad 2x-3=8$$

$$2x=-5 \qquad\qquad\qquad 2x=11$$

$$x=-\frac{5}{2} \qquad\qquad\qquad x=\frac{11}{2}$$

23.   Solving the equation:

$$\left|\frac{5-x}{5}\right| = \tfrac{1}{2}$$

$$\frac{5-x}{5} = -\tfrac{1}{2} \qquad \text{or} \qquad \frac{5-x}{5} = \tfrac{1}{2}$$

$$10 - 2x = -5 \qquad\qquad\qquad 10 - 2x = 5$$

$$-2x = -15 \qquad\qquad\qquad -2x = -5$$

$$x = \tfrac{15}{2} \qquad\qquad\qquad\quad x = \tfrac{5}{2}$$

25.   Solving the equation:

$$|3x - 2| = |-5|$$

$$3x - 2 = -5 \qquad \text{or} \qquad 3x - 2 = 5$$

$$3x = -3 \qquad\qquad\qquad 3x = 7$$

$$x = -1 \qquad\qquad\qquad x = \tfrac{7}{3}$$

27.   Solving the equation:

$$|3x - 4| = |x|$$

$$3x - 4 = -x \qquad \text{or} \qquad 3x - 4 = x$$

$$-4 = -4x \qquad\qquad\qquad -4 = -2x$$

$$x = 1 \qquad\qquad\qquad\quad x = 2$$

29.   Solving the equation:

$$|3x - 2| + 3 = 6$$

$$|3x - 2| = 3$$

$$3x - 2 = -3 \qquad \text{or} \qquad 3x - 2 = 3$$

$$3x = -1 \qquad\qquad\qquad 3x = 5$$

$$x = -\tfrac{1}{3} \qquad\qquad\qquad x = \tfrac{5}{3}$$

31.   Solving the inequality:

$$|5 - 3x| - 4 < 8$$

$$|5 - 3x| < 12$$

$$-12 < 5 - 3x < 12$$

$$-17 < -3x < 7$$

$$\tfrac{17}{3} > x > -\tfrac{7}{3}$$

The solution is $\left(-\tfrac{7}{3}, \tfrac{17}{3}\right)$.

33.   Solving the inequality:

$$|3x - 2| + 3 \le 1$$

$$|3x - 2| \le -2$$

This is impossible, since $|3x - 2| \ge 0$ for all reals x. There is no solution.

35.   The equivalent expression is: $\left|x^2 + 8\right| = x^2 + 8$

37.   The equivalent expression is: $\left|-4x^2 - 9\right| = |-1|\left|4x^2 + 9\right| = 4x^2 + 9$

39.   Since $\left|3x^3 + 3\right| = 3\left|x^3 + 1\right|$, and $x^3 + 1$ can be positive or negative (depending on the value of x), it is not possible to rewrite the expression without absolute values.

41.   The equivalent expression is: $\left|\dfrac{-5 - x^6}{4}\right| = \left|-\tfrac{1}{4}\right|\left|x^6 + 5\right| = \dfrac{x^6 + 5}{4}$

43. Multiplyiing each side of the inequality by $|ab| > 0$:

$$\left|\frac{1}{a}\right| < \left|\frac{1}{b}\right|$$

$$|ab|\left|\frac{1}{a}\right| < |ab|\left|\frac{1}{b}\right|$$

$$|b| < |a|$$

1.8 Quadratic Equations and Equations in Quadratic Form

1. Solving the equation:

$$y^2 - 65 = 0$$

$$y^2 = 65$$

$$y = \pm\sqrt{65}$$

3. Solving the equation:

$$2x^2 - 7x = 15$$

$$2x^2 - 7x - 15 = 0$$

$$(2x + 3)(x - 5) = 0$$

$$x = -\tfrac{3}{2}, 5$$

5. Solving the equation:

$$x - 3 = \frac{1}{x + 3}$$

$$(x + 3)(x - 3) = (x + 3) \cdot \frac{1}{x + 3}$$

$$x^2 - 9 = 1$$

$$x^2 = 10$$

$$x = \pm\sqrt{10}$$

7. Solving by completing the square:

$$x^2 + 2x - 4 = 0$$

$$x^2 + 2x = 4$$

$$x^2 + 2x + 1 = 4 + 1$$

$$(x + 1)^2 = 5$$

$$x + 1 = \pm\sqrt{5}$$

$$x = -1 \pm \sqrt{5}$$

9.   Solving by completing the square:

$$\frac{1}{a-5}+\frac{3}{a+2}=4$$

$$(a-5)(a+2)\left(\frac{1}{a-5}+\frac{3}{a+2}\right)=4(a-5)(a+2)$$

$$a+2+3(a-5)=4\left(a^2-3a-10\right)$$

$$a+2+3a-15=4a^2-12a-40$$

$$4a-13=4a^2-12a-40$$

$$4a^2-16a-27=0$$

$$a^2-4a=\frac{27}{4}$$

$$a^2-4a+4=\frac{27}{4}+4$$

$$(a-2)^2=\frac{43}{4}$$

$$a-2=\pm\frac{\sqrt{43}}{\sqrt{4}}$$

$$a-2=\pm\frac{\sqrt{43}}{2}$$

$$a=\frac{4\pm\sqrt{43}}{2}$$

11.   Solving the equation:

$$(x+5)(x-7)=3$$

$$x^2-2x-35=3$$

$$x^2-2x-38=0$$

Using the quadratic formula with $a=1$, $b=-2$, and $c=-38$:

$$x=\frac{-(-2)\pm\sqrt{(-2)^2-4(1)(-38)}}{2(1)}=\frac{2\pm\sqrt{156}}{2}=\frac{2\pm2\sqrt{39}}{2}=1\pm\sqrt{39}$$

13.   Solving the equation:

$$\frac{3}{x-2}=x$$

$$3=x^2-2x$$

$$x^2-2x-3=0$$

$$(x-3)(x+1)=0$$

$$x=-1,3$$

15.   Solving the equation:

$$\frac{x}{x+2}-\frac{3}{x}=\frac{x+1}{x}$$

$$x(x+2)\left(\frac{x}{x+2}-\frac{3}{x}\right)=x(x+2)\left(\frac{x+1}{x}\right)$$

$$x^2-3(x+2)=(x+1)(x+2)$$

$$x^2-3x-6=x^2+3x+2$$

$$-3x-6=3x+2$$

$$-6x=8$$

$$x=-\frac{4}{3}$$

17. Solving for s_y:

$$1.24 = s_y \sqrt{1 - 0.63^2}$$
$$1.24 = s_y \sqrt{0.6031}$$
$$s_y = \frac{1.24}{\sqrt{0.6031}} \approx 1.60$$

19. Solving for n_1:

$$0.8 = \frac{0.62 - 0.52}{\sqrt{\dfrac{1}{n_1 - 3} + \dfrac{1}{83 - 3}}}$$

$$0.8 = \frac{0.1}{\sqrt{\dfrac{1}{n_1 - 3} + \dfrac{1}{80}}}$$

$$\sqrt{\frac{1}{n_1 - 3} + \frac{1}{80}} = \frac{1}{8}$$

$$\frac{1}{n_1 - 3} + \frac{1}{80} = \frac{1}{64}$$

$$\frac{1}{n_1 - 3} = \frac{1}{320}$$

$$n_1 - 3 = 320$$

$$n_1 = 323$$

21. Solving for n:

$$3.2 = \frac{70 - 68}{3.4 \Big/ \sqrt{n}}$$

$$3.2 = \frac{2\sqrt{n}}{3.4}$$

$$2\sqrt{n} = 10.88$$

$$\sqrt{n} = 5.44$$

$$n \approx 30$$

23. Solving for s:

$$K = \frac{2gm}{s^2}$$

$$Ks^2 = 2gm$$

$$s^2 = \frac{2gm}{K}$$

$$s = \sqrt{\frac{2gm}{K}} \cdot \frac{\sqrt{K}}{\sqrt{K}}$$

$$s = \frac{\sqrt{2gmK}}{K}$$

25. Solving for a:

$$3a^2 + 2b = 5b - 2a^2x - 2$$
$$3a^2 + 2a^2x = 3b - 2$$
$$a^2(3 + 2x) = 3b - 2$$
$$a^2 = \frac{3b - 2}{2x + 3}$$
$$a = \pm\sqrt{\frac{3b - 2}{2x + 3} \cdot \frac{\sqrt{2x + 3}}{\sqrt{2x + 3}}}$$
$$a = \pm\frac{\sqrt{(3b - 2)(2x + 3)}}{2x + 3}$$

Note that $x \neq -\frac{3}{2}$ for this solution to be valid.

27. Solving for s_y:

$$s_e = s_y\sqrt{1 - r_{xy}^2}$$
$$s_y = \frac{s_e}{\sqrt{1 - r_{xy}^2}} \cdot \frac{\sqrt{1 - r_{xy}^2}}{\sqrt{1 - r_{xy}^2}}$$
$$s_y = \frac{s_e\sqrt{1 - r_{xy}^2}}{1 - r_{xy}^2}$$

29. Let x represent the number. The equation is:

$$x + \frac{1}{x} = \frac{13}{6}$$
$$6x\left(x + \frac{1}{x}\right) = 6x\left(\frac{13}{6}\right)$$
$$6x^2 + 6 = 13x$$
$$6x^2 - 13x + 6 = 0$$
$$(3x - 2)(2x - 3) = 0$$
$$x = \frac{2}{3}, \frac{3}{2}$$

The numbers are $\frac{2}{3}$ and $\frac{3}{2}$.

31. Let w represent the width and $2w + 5$ represent the length. The equation is:

$$w(2w + 5) = 75$$
$$2w^2 + 5w = 75$$
$$2w^2 + 5w - 75 = 0$$
$$(2x + 15)(w - 5) = 0$$
$$w = 5 \quad \left(w = -\frac{15}{2} \text{ is impossible}\right)$$
$$2w + 5 = 15$$

The width is 5 ft and the length is 15 ft.

33. Let w represent the width of the path. The dimensions of the pool together with the path are $21 + 2w$ by $21 + 2w$, thus the equation is:

$$(21 + 2w)(21 + 2w) - (21)(21) = 184$$
$$441 + 84w + 4w^2 - 441 = 184$$
$$4w^2 + 84w - 184 = 0$$
$$w^2 + 21w - 46 = 0$$
$$(w + 23)(w - 2) = 0$$
$$w = 2 \quad (w = -23 \text{ is impossible})$$

The width of the path is 2 feet.

35. Solving the equation:

$$\sqrt{x} - 2 = x - 4$$
$$\sqrt{x} = x - 2$$
$$x = x^2 - 4x + 4$$
$$x^2 - 5x + 4 = 0$$
$$(x - 4)(x - 1) = 0$$
$$x = 1, 4$$

Note that $x = 1$ does not check, so the solution is $x = 4$.

37. Solving the equation:

$$\sqrt{7x + 1} - 2\sqrt{x} = 2$$
$$\sqrt{7x + 1} = 2\sqrt{x} + 2$$
$$7x + 1 = 4x + 8\sqrt{x} + 4$$
$$3x - 3 = 8\sqrt{x}$$
$$9x^2 - 18x + 9 = 64x$$
$$9x^2 - 82x + 9 = 0$$
$$(9x - 1)(x - 9) = 0$$
$$x = \tfrac{1}{9}, 9$$

Note that $x = \tfrac{1}{9}$ does not check, so the solution is $x = 9$.

39. Solving the equation:

$$x^4 - 17x^2 + 16 = 0$$
$$\left(x^2 - 16\right)\left(x^2 - 1\right) = 0$$
$$x^2 = 16, 1$$
$$x = \pm 4, \pm 1$$

41. Solving the equation:

$$x^{1/2} + 8x^{1/4} + 7 = 0$$
$$\left(x^{1/4} + 7\right)\left(x^{1/4} + 1\right) = 0$$
$$x^{1/4} = -7 \text{ or } x^{1/4} = -1$$

Since neither of these equations is possible $\left(x^{1/4} \geq 0\right)$, there is no solution to the equation.

43.   Solving the equation:
$$\sqrt{a} - \sqrt[4]{a} - 6 = 0$$
$$\left(\sqrt[4]{a} - 3\right)\left(\sqrt[4]{a} + 2\right) = 0$$
$$\sqrt[4]{a} = 3 \text{ or } \sqrt[4]{a} = -2 \quad \text{(impossible)}$$
$$a = 3^4 = 81$$

45.   Solving the equation:
$$x^3 - 9x + 4x^2 - 36 = 0$$
$$x\left(x^2 - 9\right) + 4\left(x^2 - 9\right) = 0$$
$$\left(x^2 - 9\right)(x + 4) = 0$$
$$(x + 3)(x - 3)(x + 4) = 0$$
$$x = -4, -3, 3$$

47.   Solving the equation:
$$\frac{(x-3)^{1/3}}{(x+7)^{2/3}} = 0$$
$$(x-3)^{1/3} = 0$$
$$x = 3$$

49.   Substituting $x = 2 + i$ into the equation:
$$2(2+i)^2 - 8(2+i) + 10 = 2(4 + 4i - 1) - 16 - 8i + 10 = 6 + 8i - 16 - 8i + 10 = 0$$
So $x = 2 + i$ is a solution to the equation.

51.   Completing the square:
$$Ax^2 + Bx + C = 0$$
$$x^2 + \frac{B}{A}x = -\frac{C}{A}$$
$$x^2 + \frac{B}{A}x + \frac{B^2}{4A^2} = -\frac{C}{A} + \frac{B^2}{4A^2}$$
$$\left(x + \frac{B}{2A}\right)^2 = \frac{B^2 - 4AC}{4A^2}$$
$$x + \frac{B}{2A} = \frac{\pm\sqrt{B^2 - 4AC}}{2A}$$
$$x = \frac{-B \pm \sqrt{B^2 - 4AC}}{2A}$$

1.9 Quadratic and Rational Inequalities

1.   Factoring the inequality:
$$x^2 + 2x - 24 > 0$$
$$(x + 6)(x - 4) > 0$$
The cut points are –6 and 4. Forming a sign chart:

$$\text{For} \quad x = -8 \quad\quad -6 \quad\quad\quad x = 0 \quad\quad\quad 4 \quad\quad x = 6$$

The solution is $(-\infty, -6) \cup (4, \infty)$.

3. Factoring the inequality:
$$2x^2 - 3x - 2 < 0$$
$$(2x+1)(x-2) < 0$$
The cut points are −1/2 and 2. Forming a sign chart:

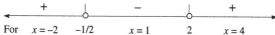

$$\begin{array}{ccccc} + & & - & & + \end{array}$$
For $x = -2$ −1/2 $x = 1$ 2 $x = 4$

The solution is $\left(-\frac{1}{2}, 2\right)$.

5. Factoring the inequality:
$$x^2 - 5x \leq 24$$
$$x^2 - 5x - 24 \leq 0$$
$$(x-8)(x+3) \leq 0$$
The cut points are −3 and 8. Forming a sign chart:

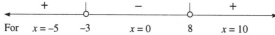

For $x = -5$ −3 $x = 0$ 8 $x = 10$

The solution is $[-3, 8]$.

7. Factoring the inequality:
$$2x^2 - 3x \geq 5$$
$$2x^2 - 3x - 5 \geq 0$$
$$(2x-5)(x+1) \geq 0$$
The cut points are −1 and 5/2. Forming a sign chart:

For $x = -3$ −1 $x = 1$ 5/2 $x = 5$

The solution is $\left(-\infty, -1\right] \cup \left[\frac{5}{2}, \infty\right)$.

9. Factoring the inequality:
$$15x^2 - 2 \leq 7x$$
$$15x^2 - 7x - 2 \leq 0$$
$$(5x+1)(3x-2) \leq 0$$
The cut points are −1/5 and 2/3. Forming a sign chart:

For $x = -2$ −1/5 $x = 0$ 2/3 $x = 4$

The solution is $\left[-\frac{1}{5}, \frac{2}{3}\right]$.

11. Factoring the inequality:
$$8x^2 - 6 > -8x$$
$$8x^2 + 8x - 6 > 0$$
$$(2x-1)(2x+3) > 0$$
The cut points are −3/2 and 1/2. Forming a sign chart:

For $x = -3$ −3/2 $x = 0$ 1/2 $x = 4$

The solution is $\left(-\infty, -\frac{3}{2}\right) \cup \left(\frac{1}{2}, \infty\right)$.

13.   Factoring the inequality:
$$9x^2 - 6x + 1 \geq 0$$
$$(3x-1)^2 \geq 0$$
Since this statement is always true, the solution is all real numbers, or $(-\infty, \infty)$.

15.   Factoring the inequality:
$$x^2 - 2x > -1$$
$$x^2 - 2x + 1 > 0$$
$$(x-1)^2 > 0$$
This statement is true for any value of $x \neq 1$, so the solution is $(-\infty, 1) \cup (1, \infty)$.

17.   Factoring the inequality:
$$y^2 > 1$$
$$y^2 - 1 > 0$$
$$(y+1)(y-1) > 0$$
The cut points are -1 and 1. Forming a sign chart:

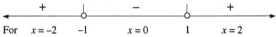

The solution is $(-\infty, -1) \cup (1, \infty)$.

19.   Factoring the inequality:
$$x^3 + 4x^2 - x - 4 < 0$$
$$x^2(x+4) - 1(x+4) < 0$$
$$(x+4)\left(x^2 - 1\right) < 0$$
$$(x+4)(x+1)(x-1) < 0$$
The cut points are -4, -1 and 1. Forming a sign chart:

$$
\begin{array}{ccccccc}
 & - & | & + & | & - & | & + \\
\text{For} & x=-6 & -4 & x=-2 & -1 & x=0 & 1 & x=3
\end{array}
$$

The solution is $(-\infty, -4) \cup (-1, 1)$.

21.   Factoring the inequality:
$$x^3 + 2x^2 - 4x - 8 \geq 0$$
$$x^2(x+2) - 4(x+2) \geq 0$$
$$(x+2)\left(x^2 - 4\right) \geq 0$$
$$(x+2)^2(x-2) \geq 0$$
The cut points are -2 and 2. Forming a sign chart:

$$
\begin{array}{ccccccc}
 & - & | & - & | & + \\
\text{For} & x=-4 & -2 & x=0 & 2 & x=4
\end{array}
$$

The solution is $\{-2\} \cup [2, \infty)$.

23. Use the quadratic formula to find the cut points:

$$x = \frac{-(-3) \pm \sqrt{(-3)^2 - 4(1)(-3)}}{2(1)} = \frac{3 \pm \sqrt{9+12}}{2} = \frac{3 \pm \sqrt{21}}{2}$$

Forming a sign chart:

The solution is $\left(\dfrac{3-\sqrt{21}}{2}, \dfrac{3+\sqrt{21}}{2} \right)$.

25. Use the quadratic formula to find the cut points:

$$x = \frac{-(-1) \pm \sqrt{(-1)^2 - 4(2)(-2)}}{2(2)} = \frac{1 \pm \sqrt{1+16}}{4} = \frac{1 \pm \sqrt{17}}{4}$$

Forming a sign chart:

The solution is $\left(-\infty, \dfrac{1-\sqrt{17}}{4} \right] \cup \left[\dfrac{1+\sqrt{17}}{4}, \infty \right)$.

27. The cut point is -1. Forming a sign chart:

The solution is $(-\infty, -1)$.

29. The cut points are -5 and $1/2$. Forming a sign chart:

The solution is $(-\infty, -5) \cup \left[\frac{1}{2}, \infty \right)$.

31. The cut points are -1 and 2. Forming a sign chart:

The solution is $(-1, 2)$.

33. The cut points are $1/5$ and 2. Forming a sign chart:

The solution is $\left(-\infty, \frac{1}{5} \right] \cup (2, \infty)$.

35. The cut points are 0 and 1. Forming a sign chart:

The solution is $(-\infty, 0] \cup (1, \infty)$.

37.   Solving the inequality:
$$\frac{3x}{x-1} < 1$$
$$\frac{3x}{x-1} - 1 < 0$$
$$\frac{3x - x + 1}{x-1} < 0$$
$$\frac{2x+1}{x-1} < 0$$

The cut points are –1/2 and 1. Forming a sign chart:

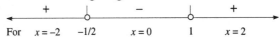

For $x=-2$ –1/2 $x=0$ 1 $x=2$

The solution is $\left(-\frac{1}{2}, 1\right)$.

39.   Solving the inequality:
$$\frac{x-1}{x+1} \geq 2$$
$$\frac{x-1}{x+1} - 2 \geq 0$$
$$\frac{x-1-2x-2}{x+1} \geq 0$$
$$\frac{-x-3}{x+1} \geq 0$$

The cut points are –3 and –1. Forming a sign chart:

For $x=-4$ –3 $x=-2$ –1 $x=0$

The solution is $[-3, -1)$.

41.   The cut points are –1, 2 and 3. Forming a sign chart:

For $x=-2$ –1 $x=1$ 2 $x=5/2$ 3 $x=4$

The solution is $(-1, 2] \cup (3, \infty)$.

43.   Factoring the inequality:
$$\frac{2x}{x^2-1} \leq 0$$
$$\frac{2x}{(x+1)(x-1)} \leq 0$$

The cut points are –1, 0 and 1. Forming a sign chart:

For $x=-2$ –1 $x=-1/2$ 0 $x=1/2$ 1 $x=2$

The solution is $(-\infty, -1) \cup [0, 1)$.

45.   Factoring the inequality:
$$\frac{x^2-9}{x+5} < 0$$
$$\frac{(x+3)(x-3)}{x+5} < 0$$

The cut points are –5, –3 and 3. Forming a sign chart:

For $x = -7$ –5 $x = -4$ –3 $x = 0$ 3 $x = 4$

The solution is $(-\infty, -5) \cup (-3, 3)$.

47.   The only cut point is 0, since $x^2 + 1 > 0$ for all values of x . Forming a sign chart:

For $x = -2$ 0 $x = 2$

The solution is $[0, \infty)$.

49.   Factoring the inequality:

$$\frac{a^2 - 1}{a^2 - 16} > 0$$

$$\frac{(a+1)(a-1)}{(a+4)(a-4)} > 0$$

The cut points are –4, –1, 1 and 4. Forming a sign chart:

For $x = -6$ –4 $x = -2$ –1 $x = 0$ 1 $x = 2$ 4 $x = 6$

The solution is $(-\infty, -4) \cup (-1, 1) \cup (4, \infty)$.

51.   Solving the inequality:

$$|2x - 1| < |x|$$

$$(2x - 1)^2 < x^2$$

$$4x^2 - 4x + 1 < x^2$$

$$3x^2 - 4x + 1 < 0$$

$$(3x - 1)(x - 1) < 0$$

The cut points are 1/3 and 1. Forming a sign chart:

For $x = 0$ 1/3 $x = 2/3$ 1 $x = 2$

The solution is $\left(\frac{1}{3}, 1\right)$.

53.   Solving the inequality:

$$|x - 3| < |x + 2|$$

$$(x - 3)^2 < (x + 2)^2$$

$$x^2 - 6x + 9 < x^2 + 4x + 4$$

$$-6x + 9 < 4x + 4$$

$$-10x < -5$$

$$x > \frac{1}{2}$$

The solution is $\left(\frac{1}{2}, \infty\right)$.

55.   Solving the inequality:

$$1000\left(-T^2 + 42T - 320\right) \geq 96000$$

$$-T^2 + 42T - 320 \geq 96$$

$$T^2 - 42T + 320 \leq -96$$

$$T^2 - 42T + 416 \leq 0$$

$$(T - 16)(T - 26) \leq 0$$

The cut points are 16 and 26. Forming a sign chart:

The solution is $[16,26]$. The water temperature range must be between $16°$ C and $26°$ C, including the endpoints.

1.10 Substitution

1. **a.** Let s represent a side of the square. Then $4s = 84$, so $s = 21$. So the area is given by:
$$A = (21)^2 = 441 \text{ sq. inches}$$

    b. Since $4s = x$, $s = \dfrac{x}{4}$. So the area is given by: $A = s^2 = \left(\dfrac{x}{4}\right)^2 = \dfrac{x^2}{16}$ sq. inches

3. Let s represent a side of the square. Using the Pythagorean theorem:
$$s^2 + s^2 = 5^2$$
$$2s^2 = 25$$
$$s^2 = \tfrac{25}{2}$$

So the area is given by: $A = s^2 = \tfrac{25}{2}$ sq. feet

5. Let s represent a side of the square, and d represent its diagonal. Since $s^2 = 84$, using the Pythagorean theorem:
$$d^2 = s^2 + s^2$$
$$d^2 = 84 + 84$$
$$d^2 = 168$$
$$d = \sqrt{168} = 2\sqrt{42} \approx 12.96 \text{ cm}$$

7. Let s represent a side of the square, and d represent its diagonal. Since $P = 4s$, $s = \dfrac{P}{4}$. Using the Pythagorean theorem:
$$d^2 = s^2 + s^2$$
$$d^2 = 2s^2$$
$$d^2 = 2\left(\dfrac{P}{4}\right)^2$$
$$d^2 = \dfrac{2P^2}{16}$$
$$d = \dfrac{P\sqrt{2}}{4}$$

9. Let s represent a side of the square. Using the Pythagorean theorem:
$$s^2 + s^2 = d^2$$
$$2s^2 = d^2$$
$$s^2 = \dfrac{d^2}{2}$$
$$s = \dfrac{d}{\sqrt{2}} \cdot \dfrac{\sqrt{2}}{\sqrt{2}} = \dfrac{d\sqrt{2}}{2}$$

So the perimeter is given by: $P = 4s = 4\left(\dfrac{d\sqrt{2}}{2}\right) = 2d\sqrt{2}$

11. Since the diameter is 8 feet, the radius is $r = \dfrac{8}{2} = 4$ feet. So the volume is given by:

$$V = \tfrac{4}{3}\pi r^3 = \tfrac{4}{3}\pi(4)^3 = \tfrac{256}{3}\pi \approx 268.08 \text{ ft}^3$$

13. The radius is $r = \dfrac{d}{2}$. So the surface area is given by: $S = 4\pi r^2 = 4\pi\left(\dfrac{d}{2}\right)^2 = \pi d^2$

15. Let s represent a side of the square. Since $r = 5$ in., $s = 2r = 10$ in. So the area of the square is given by: $A = s^2 = (10)^2 = 100$ sq. in.

17. Let s represent a side of the square, so $s = 2r$. So the area of the square is given by:
$$A = s^2 = (2r)^2 = 4r^2 \text{ sq. in.}$$

19.   **a.**   After 2 seconds the radius is $r = 6$ in., so the area is given by:
$$A = \pi r^2 = \pi(6)^2 = 36\pi \approx 113.10 \text{ sq. in.}$$

      b.   After t seconds the radius is $r = 3t$ in., so the area is given by:
$$A = \pi r^2 = \pi(3t)^2 = 9\pi t^2 \text{ sq. in.}$$

21. After 5 minutes the radius is 15 in. The volume is given by:
$$V = \tfrac{4}{3}\pi(15)^3 = 4500\pi \approx 14137.17 \text{ cu. in.}$$

23. After t minutes the radius is $3t$ in. The volume is given by: $V = \tfrac{4}{3}\pi(3t)^3 = 36\pi t^3$ cu. in.

25. The total cost is $8(500) = \$4000$.

27. Let s represent a side of the square. Then:
$$s^2 = 200$$
$$s = \sqrt{200} = 10\sqrt{2}$$
The total cost is given by: $C = 8(4s) = 8\left(40\sqrt{2}\right) = 320\sqrt{2} \approx \$452.55$

29. The circumference of the garden is $2\pi(5) = 10\pi$, so the total cost is given by:
$$C = 2(10\pi) = 20\pi \approx \$62.83$$

31. Finding the radius:
$$\pi r^2 = 90$$
$$r^2 = \frac{90}{\pi}$$
$$r = \sqrt{\frac{90}{\pi}} \cdot \frac{\sqrt{\pi}}{\sqrt{\pi}} = \frac{3\sqrt{10\pi}}{\pi}$$

The circumference of the garden is $2\pi\left(\dfrac{3\sqrt{10\pi}}{\pi}\right) = 6\sqrt{10\pi}$, so the total cost is given by:

$$C = 3\left(6\sqrt{10\pi}\right) = 18\sqrt{10\pi} \approx \$100.89$$

33. After 2 hours the boats have traveled 60 mi and 80 mi, respectively. Let d represent the distance between the boats. Using the Pythagorean theorem:
$$d^2 = 60^2 + 80^2$$
$$d^2 = 10000$$
$$d = \sqrt{10000} = 100$$
The boats are 100 miles apart.

Chapter 1 Review Exercises

1. Let $x = 8.25252\overline{5}$, so $100x = 825.25252\overline{5}$. Subtracting yields $99x = 817$, so $x = \frac{817}{99}$. Thus $8.25252\overline{5} = \frac{817}{99}$.

2. Let $x = 72.4724\overline{7247}$, so $1000x = 72472.4724\overline{7247}$. Subtracting yields $999x = 72400$, so $x = \frac{72400}{999}$. Thus $72.4724\overline{7247} = \frac{72400}{999}$.

3. This is the commutative property of addition.

4. This is the distributive property.

5. The interval notation is $[-4, \infty)$. Graphing the interval:

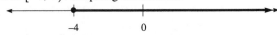

6. The interval notation is $(-8, -1]$. Graphing the interval:

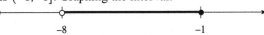

7. The interval notation is $(-\infty, 3)$. Graphing the interval:

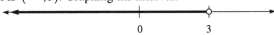

8. The interval notation is $[0, \infty)$. Graphing the interval:

9. Since $\sqrt{5} > 1$, the value is $\left|1 - \sqrt{5}\right| = \sqrt{5} - 1$.

10. If $x \geq 8$, $x - 8 \geq 0$ and if $x < 8$, $x + 8 < 0$. The expression is:
$$|x - 8| = \begin{cases} x - 8 & \text{if } x \geq 8 \\ 8 - x & \text{if } x < 8 \end{cases}$$

11. Evaluating the expression:
$$\begin{aligned} 6 - \{4 - [5 - 2(3 - 9)]\} &= 6 - \{4 - [5 - 2(-6)]\} \\ &= 6 - \{4 - (5 + 12)\} \\ &= 6 - (4 - 17) \\ &= 6 - (-13) \\ &= 19 \end{aligned}$$

12. Evaluating the expression: $|-5| - |-9| - |-5 - 9| = 5 - 9 - 14 = -18$

13. Evaluating the expression: $2\left(\frac{2}{3}\right)^2 - 4\left(\frac{2}{3}\right) + 6 = 2\left(\frac{4}{9}\right) - \frac{8}{3} + 6 = \frac{8}{9} - \frac{24}{9} + \frac{54}{9} = \frac{38}{9}$

14. Evaluating the expression: $5\left(-\frac{1}{3}\right)^2 - 2\left(-\frac{1}{3}\right) - 4 = 5\left(\frac{1}{9}\right) + \frac{2}{3} - 4 = \frac{5}{9} + \frac{6}{9} - \frac{36}{9} = -\frac{25}{9}$

15. Simplifying the complex fraction: $\dfrac{3 - \frac{1}{2}}{\frac{2}{5} - 1} \cdot \dfrac{10}{10} = \dfrac{30 - 5}{4 - 10} = \dfrac{25}{-6} = -\dfrac{25}{6}$

16. Simplifying the complex fraction: $\dfrac{\frac{1}{5} - \frac{2}{3}}{1 - \frac{3}{4}} \cdot \dfrac{60}{60} = \dfrac{12 - 40}{60 - 45} = \dfrac{-28}{15} = -\dfrac{28}{15}$

17. Performing the operations: $(-2xy)^2 (3xy)^3 = (4x^2y^2)(27x^3y^3) = 108x^5y^5$

18. Performing the operations: $3xy^2(2xy^2 - 3y^2) = 6x^2y^4 - 9xy^4$

19. Performing the operations: $(x - 3)(2x + 7) = 2x^2 - 6x + 7x - 21 = 2x^2 + x - 21$

20. Performing the operations: $(3x-4)(2x-1) = 6x^2 - 8x - 3x + 4 = 6x^2 - 11x + 4$

21. Performing the operations: $\left(5a^3 - b\right)\left(5a^3 + b\right) = \left(5a^3\right)^2 - b^2 = 25a^6 - b^2$

22. Performing the operations: $\left(5a^3 - b\right)^2 = \left(5a^3\right)^2 - 2\left(5a^3\right)(b) + b^2 = 25a^6 - 10a^3 b + b^2$

23. Performing the operations: $(3x-1)\left(9x^2 + 3x + 1\right) = (3x)^3 - (1)^3 = 27x^3 - 1$

24. Performing the operations:
$$
\begin{aligned}
(1+2x)\left(1 - 4x + 16x^2\right) &= 1\left(1 - 4x + 16x^2\right) + 2x\left(1 - 4x + 16x^2\right) \\
&= 1 - 4x + 16x^2 + 2x - 8x^2 + 32x^3 \\
&= 32x^3 + 8x^2 - 2x + 1
\end{aligned}
$$

25. Performing the operations:
$$
\begin{aligned}
2x(x-2)^2 - (x+2)(x-2) &= 2x\left(x^2 - 4x + 4\right) - \left(x^2 - 4\right) \\
&= 2x^3 - 8x^2 + 8x - x^2 + 4 \\
&= 2x^3 - 9x^2 + 8x + 4
\end{aligned}
$$

26. Performing the operations:
$$
\begin{aligned}
(x-2)^2 (x-3)^2 &= \left(x^2 - 4x + 4\right)\left(x^2 - 6x + 9\right) \\
&= x^2\left(x^2 - 6x + 9\right) - 4x\left(x^2 - 6x + 9\right) + 4\left(x^2 - 6x + 9\right) \\
&= x^4 - 6x^3 + 9x^2 - 4x^3 + 24x^2 - 36x + 4x^2 - 24x + 36 \\
&= x^4 - 10x^3 + 37x^2 - 60x + 36
\end{aligned}
$$

27. Factoring the expression: $x^2 - 2x - 15 = (x-5)(x+3)$

28. Factoring the expression: $6x^2 - 11x - 10 = (3x+2)(2x-5)$

29. Factoring the expression: $3y(x-2)^2 + 5(x-2) = (x-2)\left[3y(x-2) + 5\right] = (x-2)(3xy - 6y + 5)$

30. Factoring the expression:
$$2x(x+8)^2 - 2x^2(x+8)^3 = 2x(x+8)^2\left[1 - x(x+8)\right] = 2x(x+8)^2\left(1 - 8x - x^2\right)$$

31. Factoring the expression: $25x^2 - 30x + 9 = (5x-3)(5x-3) = (5x-3)^2$

32. Factoring the expression: $a^4 - 81 = \left(a^2 + 9\right)\left(a^2 - 9\right) = \left(a^2 + 9\right)(a+3)(a-3)$

33. Factoring the expression:
$$a^3 + a^2 - 16a - 16 = a^2(a+1) - 16(a+1) = (a+1)\left(a^2 - 16\right) = (a+1)(a+4)(a-4)$$

34. Factoring the expression:
$$
\begin{aligned}
2x^3 + 6x^2 - 18x - 54 &= 2x^2(x+3) - 18(x+3) \\
&= 2(x+3)\left(x^2 - 9\right) \\
&= 2(x+3)(x+3)(x-3) \\
&= 2(x+3)^2(x-3)
\end{aligned}
$$

35. Factoring the expression:
$$
\begin{aligned}
a^5 - a^3 + a^2 - 1 &= a^3\left(a^2 - 1\right) + 1\left(a^2 - 1\right) \\
&= \left(a^2 - 1\right)\left(a^3 + 1\right) \\
&= (a+1)(a-1)(a+1)\left(a^2 - a + 1\right) \\
&= (a+1)^2(a-1)\left(a^2 - a + 1\right)
\end{aligned}
$$

36.   Factoring the expression: $a^2 + 4a + 4 - x^2 = (a+2)^2 - x^2 = (a+2+x)(a+2-x)$

37.   Performing the operations:

$$\frac{4x^2 - 12xy + 9y^2}{2x^2 - xy - 3y^2} \cdot \frac{3x + y}{6x^2 - 7xy - 3y^2} = \frac{(2x - 3y)^2}{(2x - 3y)(x + y)} \cdot \frac{3x + y}{(3x + y)(2x - 3y)} = \frac{1}{x + y}$$

38.   Performing the operations: $\dfrac{2x^2}{x^2 + 3x + 9} \div \dfrac{2x^3 - 18x}{x^3 - 27} = \dfrac{2x^2}{x^2 + 3x + 9} \cdot \dfrac{(x - 3)(x^2 + 3x + 9)}{2x(x + 3)(x - 3)} = \dfrac{x}{x + 3}$

39.   Performing the operations: $\dfrac{3x}{x - 2} + \dfrac{1}{x} = \dfrac{3x^2}{x(x - 2)} + \dfrac{x - 2}{x(x - 2)} = \dfrac{3x^2 + x - 2}{x(x - 2)} = \dfrac{(3x - 2)(x + 1)}{x(x - 2)}$

40.   Performing the operations:

$$\frac{2x - 3}{x + 3} - \frac{x - 1}{x - 3} = \frac{(2x - 3)(x - 3)}{(x + 3)(x - 3)} - \frac{(x - 1)(x + 3)}{(x + 3)(x - 3)}$$

$$= \frac{2x^2 - 9x + 9}{(x + 3)(x - 3)} - \frac{x^2 + 2x - 3}{(x + 3)(x - 3)}$$

$$= \frac{x^2 - 11x + 12}{(x + 3)(x - 3)}$$

41.   Performing the operations: $\dfrac{3x}{(x - 1)^2} - \dfrac{2}{x - 1} = \dfrac{3x}{(x - 1)^2} - \dfrac{2(x - 1)}{(x - 1)^2} = \dfrac{3x - 2x + 2}{(x - 1)^2} = \dfrac{x + 2}{(x - 1)^2}$

42.   Performing the operations:

$$\frac{2x}{x^2 - 1} + \frac{2}{x^2 - 2x + 1} = \frac{2x}{(x + 1)(x - 1)} + \frac{2}{(x - 1)^2}$$

$$= \frac{2x(x - 1)}{(x + 1)(x - 1)^2} + \frac{2(x + 1)}{(x + 1)(x - 1)^2}$$

$$= \frac{2x^2 - 2x + 2x + 2}{(x + 1)(x - 1)^2}$$

$$= \frac{2x^2 + 2}{(x + 1)(x - 1)^2}$$

43.   Simplifying the fraction: $\dfrac{\dfrac{1}{x} + \dfrac{6}{y}}{\dfrac{1}{x^2} - \dfrac{36}{y^2}} \cdot \dfrac{x^2 y^2}{x^2 y^2} = \dfrac{xy^2 + 6x^2 y}{y^2 - 36x^2} = \dfrac{xy(y + 6x)}{(y + 6x)(y - 6x)} = \dfrac{xy}{y - 6x}$

44.   Simplifying the fraction:

$$\frac{\dfrac{3y - 1}{2y + 1} + \dfrac{2y}{4y^2 - 1}}{\dfrac{3}{2y - 1} + 3} \cdot \frac{(2y + 1)(2y - 1)}{(2y + 1)(2y - 1)} = \frac{(3y - 1)(2y - 1) + 2y}{3(2y + 1) + 3(2y + 1)(2y - 1)}$$

$$= \frac{6y^2 - 5y + 1 + 2y}{6y + 3 + 12y^2 - 3}$$

$$= \frac{6y^2 - 3y + 1}{12y^2 + 6y}$$

$$= \frac{6y^2 - 3y + 1}{6y(2y + 1)}$$

45. Simplifying the expression: $\left(xy^2\right)^{-2}\left(xy^{-2}\right)^{-3} = x^{-2}y^{-4}x^{-3}y^6 = x^{-5}y^2 = \dfrac{y^2}{x^5}$

46. Simplifying the expression: $\left(3x^{-1}y^{-2}\right)^{-1}\left(2x^{-2}y\right)^{-3} = 3^{-1}xy^2 \bullet 2^{-3}x^6y^{-3} = \frac{1}{3}x^7 \bullet \frac{1}{8}y^{-1} = \dfrac{x^7}{24y}$

47. Simplifying the expression: $\dfrac{x^{-1}y^{-2}}{xy^{-2}} = x^{-1-1}y^{-2+2} = x^{-2} = \dfrac{1}{x^2}$

48. Simplifying the expression: $\dfrac{x^{-1}+y^{-2}}{xy^{-2}} = \dfrac{\dfrac{1}{x}+\dfrac{1}{y^2}}{\dfrac{x}{y^2}} \bullet \dfrac{xy^2}{xy^2} = \dfrac{y^2+x}{x^2}$

49. Simplifying the expression: $\dfrac{3x^{-2}y^{-2}}{x^{-1}y} = 3x^{-2+1}y^{-2-1} = 3x^{-1}y^{-3} = \dfrac{3}{xy^3}$

50. Simplifying the expression: $\dfrac{3x^{-1}-y^{-2}}{x^{-1}+y} = \dfrac{\dfrac{3}{x}-\dfrac{1}{y^2}}{\dfrac{1}{x}+y} \bullet \dfrac{xy^2}{xy^2} = \dfrac{3y^2-x}{y^2+xy^3}$

51. Simplifying the expression: $\dfrac{x^{2/3}x^{-1/4}}{x^{-2/5}} = x^{2/3-1/4+2/5} = x^{40/60-15/60+24/60} = x^{49/60}$

52. Simplifying the expression: $\dfrac{2x\left(x^2+4\right)^{-3/2}}{\left(x^2+4\right)^{-1/2}} = 2x\left(x^2+4\right)^{-3/2+1/2} = 2x\left(x^2+4\right)^{-1} = \dfrac{2x}{x^2+4}$

53. Evaluating: $(-64)^{-2/3} = \left[(-64)^{1/3}\right]^{-2} = (-4)^{-2} = \dfrac{1}{(-4)^2} = \frac{1}{16}$

54. Evaluating: $81^{-3/4} = \left[81^{1/4}\right]^{-3} = (3)^{-3} = \dfrac{1}{3^3} = \frac{1}{27}$

55. Simplifying:

$$2x\left(x^2+1\right)^{1/2} + x^2\left(\tfrac{1}{2}\right)\left(x^2+1\right)^{-3/2} = \dfrac{2x\left(x^2+1\right)^{1/2}}{1} + \dfrac{x^2}{2\left(x^2+1\right)^{3/2}}$$

$$= \dfrac{4x\left(x^2+1\right)^2}{2\left(x^2+1\right)^{3/2}} + \dfrac{x^2}{2\left(x^2+1\right)^{3/2}}$$

$$= \dfrac{4x\left(x^4+2x^2+1\right)+x^2}{2\left(x^2+1\right)^{3/2}}$$

$$= \dfrac{4x^5+8x^3+x^2+4x}{2\left(x^2+1\right)^{3/2}}$$

56.   Simplifying:

$$\left(x^2-1\right)^{1/3}+\frac{2x^2}{3}\left(x^2-1\right)^{-2/3}=\left(x^2-1\right)^{1/3}+\frac{2x^2}{3\left(x^2-1\right)^{2/3}}$$

$$=\frac{3\left(x^2-1\right)}{3\left(x^2-1\right)^{2/3}}+\frac{2x^2}{3\left(x^2-1\right)^{2/3}}$$

$$=\frac{3x^2-3+2x^2}{3\left(x^2-1\right)^{2/3}}$$

$$=\frac{5x^2-3}{3\left(x^2-1\right)^{2/3}}$$

57.   Rewriting using rational exponents: $3x^2\sqrt{x^2+1}=3x^2\left(x^2+1\right)^{1/2}$

58.   Rewriting using rational exponents: $4x^5\sqrt[3]{x^2-1}=4x^5\left(x^2-1\right)^{1/3}$

59.   Rewriting using radicals: $2x(x+1)^{1/2}+x^2\left(x^2+1\right)^{-3/2}=2x\sqrt{x+1}+\dfrac{x^2}{\sqrt{\left(x^2+1\right)^3}}$

60.   Rewriting using radicals:

$$\left(x^2-1\right)^{1/3}+\frac{2x^2}{3}\left(x^2-1\right)^{-2/3}=\sqrt[3]{x^2-1}+\frac{2x^2}{3\sqrt[3]{\left(x^2-1\right)^2}}$$

$$=\frac{3\left(x^2-1\right)}{3\sqrt[3]{\left(x^2-1\right)^2}}+\frac{2x^2}{3\sqrt[3]{\left(x^2-1\right)^2}}$$

$$=\frac{3x^2-3+2x^2}{3\sqrt[3]{\left(x^2-1\right)^2}}$$

$$=\frac{5x^2-3}{3\sqrt[3]{\left(x^2-1\right)^2}}$$

61.   Simplifying the radical: $\sqrt{48x^{12}y^{15}}=\sqrt{16x^{12}y^{14}\bullet 3y}=4x^6y^7\sqrt{3y}$

62.   Simplifying the radical: $\sqrt{\dfrac{48x^{12}y^{15}}{16x^4y^8}}=\sqrt{3x^8y^7}=\sqrt{x^8y^6\bullet 3y}=x^4y^3\sqrt{3y}$

63.   Rationalizing the denominator: $\dfrac{5x}{\sqrt{x^5}}\bullet\dfrac{\sqrt{x}}{\sqrt{x}}=\dfrac{5x\sqrt{x}}{\sqrt{x^6}}=\dfrac{5x\sqrt{x}}{x^3}=\dfrac{5\sqrt{x}}{x^2}$

64.   Rationalizing the denominator: $\dfrac{\sqrt{8}}{\sqrt{6}-\sqrt{2}}\bullet\dfrac{\sqrt{6}+\sqrt{2}}{\sqrt{6}+\sqrt{2}}=\dfrac{\sqrt{48}+\sqrt{16}}{6-2}=\dfrac{4\sqrt{3}+4}{4}=\sqrt{3}+1$

65.   Simplifying the radicals: $\left(\sqrt{2x}-2\right)^2=\left(\sqrt{2x}\right)^2-4\sqrt{2x}+(2)^2=2x-4\sqrt{2x}+4$

66.   Simplifying the radicals: $\left(\sqrt{2x}-2\right)\left(\sqrt{2x}+2\right)=\left(\sqrt{2x}\right)^2-(2)^2=2x-4$

67. Simplifying the radicals:
$$\left(\sqrt{a-4}\right)^2 - \left(\sqrt{a}-4\right)^2 = a - 4 - \left(a - 8\sqrt{a} + 16\right) = 1 - 4 - a + 8\sqrt{a} - 16 = 8\sqrt{a} - 20$$

68. Simplifying the radicals: $\dfrac{\sqrt{x^2+2x-3}}{\sqrt{x+3}} = \dfrac{\sqrt{(x+3)(x-1)}}{\sqrt{x+3}} = \dfrac{\sqrt{x+3}\sqrt{x-1}}{\sqrt{x+3}} = \sqrt{x-1}$

69. Simplifying the radicals: $\dfrac{\sqrt{8}}{\sqrt{11}-\sqrt{3}} \cdot \dfrac{\sqrt{11}+\sqrt{3}}{\sqrt{11}+\sqrt{3}} = \dfrac{\sqrt{88}+\sqrt{24}}{11-3} = \dfrac{2\sqrt{22}+2\sqrt{6}}{8} = \dfrac{\sqrt{22}+\sqrt{6}}{4}$

70. Simplifying the radicals:
$$\frac{x^2-7x-18}{\sqrt{x}+3} \cdot \frac{\sqrt{x}-3}{\sqrt{x}-3} = \frac{(x-9)(x+2)\left(\sqrt{x}-3\right)}{x-9} = (x+2)\left(\sqrt{x}-3\right) = x\sqrt{x} + 2\sqrt{x} - 3x - 6$$

71. Simplifying the radicals: $2 - \dfrac{2}{\sqrt{x^2-4}} = \dfrac{2\sqrt{x^2-4}}{\sqrt{x^2-4}} - \dfrac{2}{\sqrt{x^2-4}} = \dfrac{2\sqrt{x^2-4}-2}{\sqrt{x^2-4}}$

72. Simplifying the radicals: $5\sqrt{3x^2-2} - \dfrac{5}{\sqrt{3x^2-2}} = \dfrac{5\left(3x^2-2\right)}{\sqrt{3x^2-2}} - \dfrac{5}{\sqrt{3x^2-2}} = \dfrac{15x^2-15}{\sqrt{3x^2-2}}$

73. Performing the operations: $(3+4i)+(2-5i) = 5 - i$

74. Performing the operations: $(6-3i)-(2+5i) = 6 - 3i - 2 - 5i = 4 - 8i$

75. Performing the operations: $(5+i)^2 = 25 + 10i + i^2 = 25 + 10i - 1 = 24 + 10i$

76. Performing the operations: $(3-2i)(3+2i) = 9 - 4i^2 = 9 + 4 = 13$

77. Performing the operations: $\dfrac{2+3i}{2i} \cdot \dfrac{i}{i} = \dfrac{2i+3i^2}{2i^2} = \dfrac{2i-3}{-2} = \dfrac{3}{2} - i$

78. Performing the operations: $\dfrac{6+3i}{2-3i} \cdot \dfrac{2+3i}{2+3i} = \dfrac{12+6i+18i+9i^2}{4-9i^2} = \dfrac{12+24i-9}{4+9} = \dfrac{3+24i}{13} = \dfrac{3}{13} + \dfrac{24}{13}i$

79. Solving the equation:
$$2 - \left\{3 - \left[2(1-x)\right]\right\} = 5x + 3$$
$$2 - (3 - 2 + 2x) = 5x + 3$$
$$2 - 1 - 2x = 5x + 3$$
$$-2x + 1 = 5x + 3$$
$$-7x = 2$$
$$x = -\tfrac{2}{7}$$

80. Solving the equation:
$$3 - 2\left[2 - (x-6)\right] = 2x + 8$$
$$3 - 2(2 - x + 6) = 2x + 8$$
$$3 - 16 + 2x = 2x + 8$$
$$2x - 13 = 2x + 8$$
$$-13 = 8$$
Since this statement is false, there is no solution to the equation.

81.   Solving the inequality:
$$3x - \tfrac{2}{5}x \geq -2x + 3(1-x)$$
$$3x - \tfrac{2}{5}x \geq -2x + 3 - 3x$$
$$3x - \tfrac{2}{5}x \geq -5x + 3$$
$$5\left(3x - \tfrac{2}{5}x\right) \geq 5(-5x + 3)$$
$$15x - 2x \geq -25x + 15$$
$$38x \geq 15$$
$$x \geq \tfrac{15}{38}$$
The solution set is $\left[\tfrac{15}{38}, \infty\right)$.

82.   Solving the inequality:
$$\frac{3x-2}{5} - 2 < \frac{2x-3}{2}$$
$$10\left(\frac{3x-2}{5} - 2\right) < 10\left(\frac{2x-3}{2}\right)$$
$$2(3x-2) - 20 < 5(2x-3)$$
$$6x - 4 - 20 < 10x - 15$$
$$6x - 24 < 10x - 15$$
$$-4x < 9$$
$$x > -\tfrac{9}{4}$$
The solution is $\left(-\tfrac{9}{4}, \infty\right)$.

83.   Solving the equation:
$$\frac{4}{x(x-2)} - \frac{3}{2x} = \frac{17}{6x}$$
$$6x(x-2) \cdot \left(\frac{4}{x(x-2)} - \frac{3}{2x}\right) = 6x(x-2) \cdot \frac{17}{6x}$$
$$24 - 9(x-2) = 17(x-2)$$
$$24 - 9x + 18 = 17x - 34$$
$$-9x + 42 = 17x - 34$$
$$-26x = -76$$
$$x = \tfrac{38}{13}$$

84.   Solving the equation:
$$\frac{3x}{x-1} - 2 = \frac{3}{x-1}$$
$$(x-1) \cdot \left(\frac{3x}{x-1} - 2\right) = (x-1) \cdot \frac{3}{x-1}$$
$$3x - 2(x-1) = 3$$
$$3x - 2x + 2 = 3$$
$$x + 2 = 3$$
$$x = 1$$
Note that $x = 1$ results in a zero denominator, so there is no solution to the equation.

85. Solving the inequality:
$$2 < 5 - 3x \le \tfrac{7}{3}$$
$$6 < 15 - 9x \le 7$$
$$-9 < -9x \le -8$$
$$1 > x \ge \tfrac{8}{9}$$
The solution is $\left[\tfrac{8}{9}, 1\right)$.

86. Solving the inequality:
$$-2 \le \tfrac{5}{2}x + 3 \le 4$$
$$-4 \le 5x + 6 \le 8$$
$$-10 \le 5x \le 2$$
$$-2 \le x \le \tfrac{2}{5}$$
The solution is $\left[-2, \tfrac{2}{5}\right]$.

87. Solving for F:
$$C = \tfrac{5}{9}(F - 32)$$
$$9C = 5F - 160$$
$$5F = 9C + 160$$
$$F = \frac{9C + 160}{5}$$

88. Solving for x:
$$y = \frac{x - 9}{3x + 1}$$
$$3xy + y = x - 9$$
$$3xy - x = -y - 9$$
$$x(3y - 1) = -y - 9$$
$$x = \frac{-y - 9}{3y - 1} = \frac{y + 9}{1 - 3y}$$

89. Let x represent the liters of water to add. The inequality is:
$$0.60(2) + 0(x) < 0.40(x + 2)$$
$$1.2 < 0.4x + 0.8$$
$$0.4x > 0.4$$
$$x > 1$$
Raju should add more than 1 liter.

90. The rate for Collin is $\tfrac{1}{3}$, and that for Mary is $\dfrac{1}{5/2} = \tfrac{2}{5}$. Let t represent the time for both people (together) to paint the room. The equation is:
$$\tfrac{1}{3}t + \tfrac{2}{5}t = 1$$
$$15\left(\tfrac{1}{3}t + \tfrac{2}{5}t\right) = 15(1)$$
$$5t + 6t = 15$$
$$11t = 15$$
$$t = \tfrac{15}{11} = 1\tfrac{4}{11}$$
It will take them $1\tfrac{4}{11}$ hours (together) to paint the room.

91.   Solving the equation:
$$|7x-2|=9$$

$7x-2=9$	or	$7x-2=-9$
$7x=11$		$7x=-7$
$x=\frac{11}{7}$		$x=-1$

92.   Solving the equation:
$$|2x-3|=4$$

$2x-3=4$	or	$2x-3=-4$
$2x=7$		$2x=-1$
$x=\frac{7}{2}$		$x=-\frac{1}{2}$

93.   Solving the inequality:
$$|5-3x|<6$$
$$-6<5-3x<6$$
$$-11<-3x<1$$
$$\frac{11}{3}>x>-\frac{1}{3}$$
The solution is $\left(-\frac{1}{3},\frac{11}{3}\right)$.

94.   Solving the inequality:
$$|5-3x|\geq6$$

$5-3x\leq-6$	or	$5-3x\geq6$
$-3x\leq-11$		$-3x\geq1$
$x\geq\frac{11}{3}$		$x\leq-\frac{1}{3}$

The solution is $\left(-\infty,-\frac{1}{3}\right]\cup\left[\frac{11}{3},\infty\right)$.

95.   Solving the inequality:
$$\left|\frac{1-4x}{3}\right|\geq5$$

$\frac{1-4x}{3}\leq-5$	or	$\frac{1-4x}{3}\geq5$
$1-4x\leq-15$		$1-4x\geq15$
$-4x\leq-16$		$-4x\geq14$
$x\geq4$		$x\leq-\frac{7}{2}$

The solution is $\left(-\infty,-\frac{7}{2}\right]\cup[4,\infty)$.

96.   Solving the inequality:
$$\left|1-\frac{4}{3}x\right|<3$$
$$-3<1-\frac{4}{3}x<3$$
$$-9<3-4x<9$$
$$-12<-4x<6$$
$$3>x>-\frac{3}{2}$$
The solution is $\left(-\frac{3}{2},3\right)$.

97. Solving the equation:
$$|x-1| = |5x+3|$$

$$x-1 = -5x-3 \quad \text{or} \quad x-1 = 5x+3$$
$$6x = -2 \qquad\qquad -4x = 4$$
$$x = -\tfrac{1}{3} \qquad\qquad x = -1$$

98. Solving the equation:
$$|x-2| = |x+2|$$

$$x-2 = -x-2 \quad \text{or} \quad x-2 = x+2$$
$$2x = 0 \qquad\qquad -2 = 2$$
$$x = 0 \qquad\qquad \text{(not possible)}$$

99. The equivalent expression is: $\left|-7x^2 - 3\right| = |-1|\left|7x^2 + 3\right| = 7x^2 + 3$

100. The equivalent expression is: $\left|-6x^4 - 3x^2 - 4\right| = |-1|\left|6x^4 + 3x^2 + 4\right| = 6x^4 + 3x^2 + 4$

101. Solving by completing the square:
$$x^2 - x = 8$$
$$x^2 - x + \tfrac{1}{4} = 8 + \tfrac{1}{4}$$
$$\left(x - \tfrac{1}{2}\right)^2 = \tfrac{33}{4}$$
$$x - \tfrac{1}{2} = \pm\sqrt{\tfrac{33}{4}}$$
$$x - \tfrac{1}{2} = \frac{\pm\sqrt{33}}{2}$$
$$x = \frac{1 \pm \sqrt{33}}{2}$$

102. Solving by completing the square:
$$2t^2 = 8t + 10$$
$$t^2 - 4t = 5$$
$$t^2 - 4t + 4 = 5 + 4$$
$$(t-2)^2 = 9$$
$$t - 2 = \pm 3$$
$$t = 2 \pm 3$$
$$t = -1, 5$$

103. Solving the equation:
$$3x^2 + 8 = 23$$
$$3x^2 = 15$$
$$x^2 = 5$$
$$x = \pm\sqrt{5}$$

104. Solving the equation:
$$6x^2 = 2 - x$$
$$6x^2 + x - 2 = 0$$
$$(3x + 2)(2x - 1) = 0$$
$$x = -\tfrac{2}{3}, \tfrac{1}{2}$$

105. Solving the equation:

$$\frac{3}{x-2}+\frac{2}{x-1}=\frac{3}{2}$$

$$2(x-2)(x-1)\left(\frac{3}{x-2}+\frac{2}{x-1}\right)=2(x-2)(x-1)\cdot\frac{3}{2}$$

$$6(x-1)+4(x-2)=3(x-2)(x-1)$$

$$6x-6+4x-8=3\left(x^2-3x+2\right)$$

$$10x-14=3x^2-9x+6$$

$$3x^2-19x+20=0$$

$$(3x-4)(x-5)=0$$

$$x=\tfrac{4}{3},5$$

106. Solving the equation:

$$\frac{1}{x}+x=2$$

$$x\left(\frac{1}{x}+x\right)=2x$$

$$1+x^2=2x$$

$$x^2-2x+1=0$$

$$(x-1)^2=0$$

$$x=1$$

107. Solving the equation:

$$\frac{6}{(x-3)(x+1)}+\frac{x}{x-3}=3$$

$$(x-3)(x+1)\left(\frac{6}{(x-3)(x+1)}+\frac{x}{x-3}\right)=3(x-3)(x+1)$$

$$6+x(x+1)=3\left(x^2-2x-3\right)$$

$$6+x^2+x=3x^2-6x-9$$

$$2x^2-7x-15=0$$

$$(2x+3)(x-5)=0$$

$$x=-\tfrac{3}{2},5$$

108. Solving the equation:

$$\frac{(x-3)^{1/3}(x+1)^{1/2}}{(x-2)^{1/5}}=0$$

$$(x-3)^{1/3}(x+1)^{1/2}=0$$

$$x=-1,3$$

109. Solving the equation:

$$\sqrt{3x+1}+1=x$$

$$\sqrt{3x+1}=x-1$$

$$3x+1=x^2-2x+1$$

$$x^2-5x=0$$

$$x(x-5)=0$$

$$x=0,5$$

Note that $x=0$ does not check, so the solution is $x=5$.

110. Solving the equation:
$$(x-3)^{-1/3} = 4$$
$$\frac{1}{(x-3)^{1/3}} = 4$$
$$(x-3)^{1/3} = \tfrac{1}{4}$$
$$x-3 = \left(\tfrac{1}{4}\right)^3$$
$$x-3 = \tfrac{1}{64}$$
$$x = \tfrac{193}{64}$$

111. Solving the equation:
$$2x^{2/3} - x^{1/3} = 3$$
$$2x^{2/3} - x^{1/3} - 3 = 0$$
$$\left(2x^{1/3} - 3\right)\left(x^{1/3} + 1\right) = 0$$
$$x^{1/3} = -1, \tfrac{3}{2}$$
$$x = -1, \tfrac{27}{8}$$

112. Solving the equation:
$$\sqrt{5x+1} - 1 = \sqrt{3x}$$
$$\sqrt{5x+1} = \sqrt{3x} + 1$$
$$5x+1 = 3x + 2\sqrt{3x} + 1$$
$$2x = 2\sqrt{3x}$$
$$x = \sqrt{3x}$$
$$x^2 = 3x$$
$$x^2 - 3x = 0$$
$$x(x-3) = 0$$
$$x = 0, 3$$
Both solutions check in the original equation.

113. Solving for x:
$$5x^2 + 7y^2 = 9$$
$$5x^2 = 9 - 7y^2$$
$$x^2 = \frac{9-7y^2}{5}$$
$$x = \pm\sqrt{\frac{9-7y^2}{5}} \cdot \frac{\sqrt{5}}{\sqrt{5}} = \frac{\pm\sqrt{45-35y^2}}{5}$$

114. Solving for a:
$$15a^2 + 7ab - 2b^2 = 0$$
$$(5a-b)(3a+2b) = 0$$
$$a = \frac{b}{5}, -\frac{2b}{3}$$

115. Factoring the inequality:
$$3x^2 - 14x - 5 > 0$$
$$(3x+1)(x-5) > 0$$
The cut points are $-1/3$ and 5. Forming a sign chart:

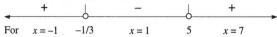

The solution is $\left(-\infty, -\frac{1}{3}\right) \cup (5, \infty)$.

116. Factoring the inequality:
$$2x^2 \geq -5x - 3$$
$$2x^2 + 5x + 3 \geq 0$$
$$(2x+3)(x+1) \geq 0$$
The cut points are $-3/2$ and -1. Forming a sign chart:

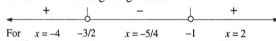

The solution is $\left(-\infty, -\frac{3}{2}\right] \cup [-1, \infty)$.

117. Factoring the inequality:
$$x^2 - 4x + 4 < 0$$
$$(x-2)^2 < 0$$
Since this inequality is never true, there is no solution.

118. Factoring the inequality:
$$x^3 + 2x^2 - x - 2 \leq 0$$
$$x^2(x+2) - 1(x+2) \leq 0$$
$$(x+2)\left(x^2 - 1\right) \leq 0$$
$$(x+2)(x+1)(x-1) \leq 0$$
The cut points are -2, -1 and 1. Forming a sign chart:

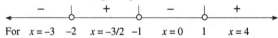

The solution is $(-\infty, -2] \cup [-1, 1]$.

119. The cut points are -1 and $2/3$. Forming a sign chart:

The solution is $(-\infty, -1) \cup \left(\frac{2}{3}, \infty\right)$.

120. The cut points are $3/2$ and 5. Forming a sign chart:

The solution is $\left(-\infty, \frac{3}{2}\right] \cup (5, \infty)$.

121. Solving the inequality:
$$\frac{3x-1}{x} \le 2$$
$$\frac{3x-1}{x} - 2 \le 0$$
$$\frac{3x-1-2x}{x} \le 0$$
$$\frac{x-1}{x} \le 0$$

The cut points are 0 and 1. Forming a sign chart:

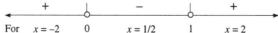

The solution is $(0,1]$.

122. Solving the inequality:
$$\frac{x}{3x+1} < 5$$
$$\frac{x}{3x+1} - 5 < 0$$
$$\frac{x - 15x - 5}{3x+1} < 0$$
$$\frac{-14x - 5}{3x+1} < 0$$

The cut points are $-1/3$ and $-5/14$. Forming a sign chart:

$$\begin{array}{ccccccc} & - & & + & & - & \\ \leftarrow & \quad & \circ & \quad & \circ & \quad & \rightarrow \\ \text{For} & x=-2 & -5/14 & x=-27/84 & -1/3 & x=0 & \end{array}$$

The solution is $\left(-\infty, -\frac{5}{14}\right) \cup \left(-\frac{1}{3}, \infty\right)$.

123. Since $\dfrac{10^{-4}}{10^{-8}} = 10^4$, there are 10^4 angstroms in 1 micron.

124. **a.** Substituting $t = 2$: $s = -16(2)^2 + 80(2) + 44 = -64 + 160 + 44 = 140$
The height of the ball is 140 feet.

b. Finding when $s = 0$:
$$-16t^2 + 80t + 44 = 0$$
$$4t^2 - 20t - 11 = 0$$
$$(2t+1)(2t-11) = 0$$
$$t = -\tfrac{1}{2}, \tfrac{11}{2}$$

It will take $5\frac{1}{2}$ seconds for the ball to hit the ground.

125. Let w represent the width of the path. The equation is:
$$\pi(10+w)^2 - \pi(10)^2 = 44\pi$$
$$(10+w)^2 - 100 = 44$$
$$(10+w)^2 = 144$$
$$10+w = \pm 12$$
$$w = 2 \quad (w = -22 \text{ is impossible})$$

The width of the path is 2 feet.

126. After t minutes the radius is $5t$ inches. The area is given by: $A = \pi(5t)^2 = 25\pi t^2$ sq. inches

127. The circumference of the garden is $2\pi r$ feet, so the cost is given by: $C = 3(2\pi r) = 6\pi r$ dollars

Chapter 1 Practice Test

1. This is the multiplicative inverse property.

2. **a.** The interval notation is $(-\infty, -2]$. Graphing the interval:

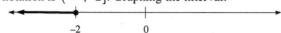

$$-2 \qquad\qquad 0$$

b. The interval notation is $(-6, -2]$. Graphing the interval:

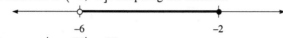

$$-6 \qquad\qquad\qquad -2$$

3. Since $\sqrt{5} > 1$, the value is $\left|1 - \sqrt{5}\right| = \sqrt{5} - 1$.

4. Evaluating the expression: $\dfrac{\frac{2}{5} - 3}{1 - \frac{3}{2}} \cdot \dfrac{10}{10} = \dfrac{4 - 30}{10 - 15} = \dfrac{-26}{-5} = \dfrac{26}{5}$

5. **a.** Performing the operations: $(3a^3 - 2b)(3a^3 + 2b) = (3a^3)^2 - (2b)^2 = 9a^6 - 4b^2$

b. Performing the operations:
$$3x(x-5)^2 - (x+5)(x-5) = 3x(x^2 - 10x + 25) - (x^2 - 25)$$
$$= 3x^3 - 30x^2 + 75x - x^2 + 25$$
$$= 3x^3 - 31x^2 + 75x + 25$$

6. **a.** Factoring the expression: $10x^2 + x - 2 = (5x - 2)(2x + 1)$

b. Factoring the expression:
$$3y(y+3)^2 + 15(y+3) = 3(y+3)[y(y+3) + 5] = 3(y+3)(y^2 + 3y + 5)$$

c. Factoring the expression:
$$x^3 + 2x^2 - 9x - 18 = x^2(x+2) - 9(x+2) = (x+2)(x^2 - 9) = (x+2)(x+3)(x-3)$$

7. **a.** Performing the operations: $\dfrac{3}{x+1} - \dfrac{2x}{(x+1)^2} = \dfrac{3(x+1)}{(x+1)^2} - \dfrac{2x}{(x+1)^2} = \dfrac{3x+3-2x}{(x+1)^2} = \dfrac{x+3}{(x+1)^2}$

b. Performing the operations: $\dfrac{\frac{1}{b} - \frac{5}{a}}{\frac{1}{b^2} - \frac{25}{a^2}} \cdot \dfrac{a^2 b^2}{a^2 b^2} = \dfrac{a^2 b - 5ab^2}{a^2 - 25b^2} = \dfrac{ab(a - 5b)}{(a+5b)(a-5b)} = \dfrac{ab}{a+5b}$

c Performing the operations:
$$\dfrac{x^{-2} + y^{-2}}{(x+y)^{-2}} = \dfrac{\frac{1}{x^2} + \frac{1}{y^2}}{\frac{1}{(x+y)^2}} \cdot \dfrac{x^2 y^2 (x+y)^2}{x^2 y^2 (x+y)^2}$$
$$= \dfrac{y^2(x+y)^2 + x^2(x+y)^2}{x^2 y^2}$$
$$= \dfrac{(x+y)^2(x^2 + y^2)}{x^2 y^2}$$

d.   Performing the operations:

$$\frac{3x^2\left(x^3-2\right)^{-1/3}}{\left(x^3-2\right)^{2/3}}=3x^2\left(x^3-2\right)^{-1/3-2/3}=3x^2\left(x^3-2\right)^{-1}=\frac{3x^2}{x^3-2}$$

8.   **a.**   Simplifying the radicals: $\sqrt{\dfrac{32x^9y^7}{18x^4y^8}}=\sqrt{\dfrac{16x^5}{9y}}\cdot\dfrac{\sqrt{y}}{\sqrt{y}}=\sqrt{\dfrac{16x^5y}{9y^2}}=\dfrac{4x^2\sqrt{xy}}{3y}$

b.   Simplifying the radicals: $\dfrac{5x^2}{\sqrt{2x}}\cdot\dfrac{\sqrt{2x}}{\sqrt{2x}}=\dfrac{5x^2\sqrt{2x}}{2x}=\dfrac{5x\sqrt{2x}}{2}$

c.   Simplifying the radicals: $\left(3x-\sqrt{2}\right)^2=9x^2-6x\sqrt{2}+2$

d.   Simplifying the radicals:

$$\frac{\sqrt{9}}{\sqrt{10}-\sqrt{7}}\cdot\frac{\sqrt{10}+\sqrt{7}}{\sqrt{10}+\sqrt{7}}=\frac{3\left(\sqrt{10}+\sqrt{7}\right)}{10-7}=\frac{3\left(\sqrt{10}+\sqrt{7}\right)}{3}=\sqrt{10}+\sqrt{7}$$

9.   **a.**   Performing the operations: $(3-5i)^2=9-30i+25i^2=9-30i-25=-16-30i$

b.   Performing the operations: $\dfrac{2-3i}{3+i}\cdot\dfrac{3-i}{3-i}=\dfrac{6-11i+3i^2}{9-i^2}=\dfrac{6-11i-3}{9+1}=\dfrac{3-11i}{10}=\dfrac{3}{10}-\dfrac{11}{10}i$

10.   Solving the equation:
$$4-\left[3(2-x)\right]=x+3$$
$$4-6+3x=x+3$$
$$3x-2=x+3$$
$$2x=5$$
$$x=\tfrac{5}{2}$$

11.   Solving the inequality:
$$\frac{x-1}{3}-1<\frac{x+3}{2}$$
$$6\left(\frac{x-1}{3}-1\right)<6\left(\frac{x+3}{2}\right)$$
$$2x-2-6<3x+9$$
$$2x-8<3x+9$$
$$-17<x$$
The solution is $(-17,\infty)$.

12.   Solving the equation:
$$|7+2x|=7$$

$$7+2x=7 \qquad \text{or} \qquad 7+2x=-7$$
$$2x=0 \qquad\qquad\qquad\quad 2x=-14$$
$$x=0 \qquad\qquad\qquad\quad\ x=-7$$

13.   Solving the inequality:
$$\left|\tfrac{2}{3}x-2\right|\le 5$$
$$-5\le\tfrac{2}{3}x-2\le 5$$
$$-3\le\tfrac{2}{3}x\le 7$$
$$-\tfrac{9}{2}\le x\le\tfrac{21}{2}$$
The solution is $\left[-\tfrac{9}{2},\tfrac{21}{2}\right]$.

14. Solving the inequality:

$$|7-2x|>4$$

$$\begin{array}{ccc} 7-2x<-4 & \text{or} & 7-2x>4 \\ -2x<-11 & & -2x>-3 \\ x>\tfrac{11}{2} & & x<\tfrac{3}{2} \end{array}$$

The solution is $\left(-\infty,\tfrac{3}{2}\right)\cup\left(\tfrac{11}{2},\infty\right)$.

15. Solving the equation:

$$|2-3x|=|3x-2|$$

$$\begin{array}{ccc} 2-3x=3x-2 & \text{or} & 2-3x=-3x+2 \\ -6x=-4 & & 2=2 \\ x=\tfrac{2}{3} & & \text{(true)} \end{array}$$

The solution is any real number, or $(-\infty,\infty)$.

16. Solving the equation:

$$(2x-1)(x+2)=5x+2$$

$$2x^2+3x-2=5x+2$$

$$2x^2-2x-4=0$$

$$x^2-x-2=0$$

$$(x-2)(x+1)=0$$

$$x=-1,2$$

17. Solving the equation:

$$3x^2-2x=5-2x$$

$$3x^2=5$$

$$x^2=\tfrac{5}{3}$$

$$x=\pm\sqrt{\tfrac{5}{3}}\cdot\dfrac{\sqrt{3}}{\sqrt{3}}$$

$$x=\pm\dfrac{\sqrt{15}}{3}$$

18. Solving the equation:

$$\frac{2}{x-2}+\frac{3}{x+2}=\frac{5}{(x-2)(x+2)}$$

$$(x-2)(x+2)\left(\frac{2}{x-2}+\frac{3}{x+2}\right)=(x-2)(x+2)\cdot\frac{5}{(x-2)(x+2)}$$

$$2(x+2)+3(x-2)=5$$

$$2x+4+3x-6=5$$

$$5x-2=5$$

$$5x=7$$

$$x=\tfrac{7}{5}$$

19. Solving the equation:
$$\sqrt{x-5} + 4 = x - 1$$
$$\sqrt{x-5} = x - 5$$
$$x - 5 = x^2 - 10x + 25$$
$$x^2 - 11x + 30 = 0$$
$$(x-5)(x-6) = 0$$
$$x = 5, 6$$
Both solutions check in the original equation.

20. Factoring the inequality:
$$2x^2 \le x + 3$$
$$2x^2 - x - 3 \le 0$$
$$(2x-3)(x+1) \le 0$$
The cut points are −1 and 3/2. Forming a sign chart:

The solution is $\left[-1, \frac{3}{2}\right]$.

21. Factoring the inequality:
$$(x-1)(x-2) > 2x - 2$$
$$x^2 - 3x + 2 > 2x - 2$$
$$x^2 - 5x + 4 > 0$$
$$(x-4)(x-1) > 0$$
The cut points are 1 and 4. Forming a sign chart:

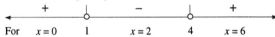

The solution is $(-\infty, 1) \cup (4, \infty)$.

22. The cut points are −1/2 and 2. Forming a sign chart:

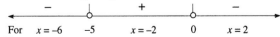

The solution is $\left(-\infty, -\frac{1}{2}\right) \cup (2, \infty)$.

23. Solving the inequality:
$$\frac{x-5}{x} \ge 2$$
$$\frac{x-5}{x} - 2 \ge 0$$
$$\frac{x-5-2x}{x} \ge 0$$
$$\frac{-x-5}{x} \ge 0$$
The cut points are −5 and 0. Forming a sign chart:

The solution is $[-5, 0)$.

24. The rate for Ken is $\frac{1}{3}$, and that for Kim is $\dfrac{1}{7/3} = \frac{3}{7}$. Let t represent the time for both people (together) to process the forms. The equation is:

$$\tfrac{1}{3}t + \tfrac{3}{7}t = 1$$
$$21\left(\tfrac{1}{3}t + \tfrac{3}{7}t\right) = 21(1)$$
$$7t + 9t = 21$$
$$16t = 21$$
$$t = \tfrac{21}{16} = 1\tfrac{5}{16}$$

It will take them $1\frac{5}{16}$ hours (together) to process the 200 forms.

25. Finding when $s = 0$:

$$-16t^2 + 40t + 96 = 0$$
$$2t^2 - 5t - 12 = 0$$
$$(2t + 3)(t - 4) = 0$$
$$t = -\tfrac{3}{2}, 4$$

It will take 4 seconds for the ball to hit the ground.

26. Let r represent the radius of the field. Therefore:

$$\pi r^2 = A$$
$$r^2 = \frac{A}{\pi}$$
$$r = \sqrt{\frac{A}{\pi}} \cdot \frac{\sqrt{\pi}}{\sqrt{\pi}}$$
$$r = \frac{\sqrt{\pi A}}{\pi}$$

The circumference is given by: $2\pi r = 2\pi\left(\dfrac{\sqrt{\pi A}}{\pi}\right) = 2\sqrt{\pi A}$

Therefore the cost of fencing is given by: $C = 6 \cdot 2\sqrt{\pi A} = 12\sqrt{\pi A}$ dollars

Chapter 2
Functions and Graphs: Part I

2.1 The Cartesian Coordinate System: Graphing Straight Lines and Circles

1. To find the x-intercept, let $y = 0$:
$$5(0) - 4x = 20$$
$$-4x = 20$$
$$x = -5$$
To find the y-intercept, let $x = 0$:
$$5y - 4(0) = 20$$
$$5y = 20$$
$$y = 4$$

3. To find the x-intercept, let $y = 0$:
$$x + 3(0) = 6$$
$$x = 6$$
To find the y-intercept, let $x = 0$:
$$0 + 3y = 6$$
$$3y = 6$$
$$y = 2$$

5. To find the x-intercept, let $y = 0$:
$$3x - 8(0) = 16$$
$$3x = 16$$
$$x = \frac{16}{3}$$
To find the y-intercept, let $x = 0$:
$$3(0) - 8y = 16$$
$$-8y = 16$$
$$y = -2$$

7. To find the x-intercept, let $y = 0$:
$$2x + 9(0) + 6 = 0$$
$$2x + 6 = 0$$
$$2x = -6$$
$$x = -3$$
To find the y-intercept, let $x = 0$:
$$2(0) + 9y + 6 = 0$$
$$9y + 6 = 0$$
$$9y = -6$$
$$y = -\frac{2}{3}$$

9.   The x-intercept is –3, and there is no y-intercept.

11.   To find the x-intercept, let $y = 0$:
$$6(0) = 5x + 8$$
$$5x = -8$$
$$x = -\frac{8}{5}$$
To find the y-intercept, let $x = 0$:
$$6y = 5(0) + 8$$
$$6y = 8$$
$$y = \frac{4}{3}$$

13.   To find the x-intercept, let $y = 0$:
$$2x = 3(0)$$
$$2x = 0$$
$$x = 0$$
To find the y-intercept, let $x = 0$:
$$2(0) = 3y$$
$$3y = 0$$
$$y = 0$$

15.   The x-intercept is –7 and the y-intercept is 3. Sketching the graph:

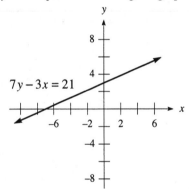

17.   The x-intercept is 4 and the y-intercept is –8. Sketching the graph:

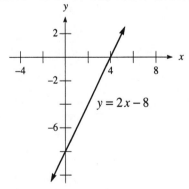

19. The *x*-intercept is 6 and the *y*-intercept is 4. Sketching the graph:

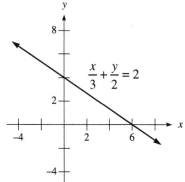

$$\frac{x}{3} + \frac{y}{2} = 2$$

21. The *x*-intercept is −6 and the *y*-intercept is 4. Sketching the graph:

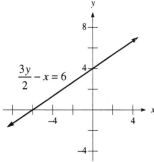

$$\frac{3y}{2} - x = 6$$

23. The *x*-intercept is 9/5 and the *y*-intercept is 9/2. Sketching the graph:

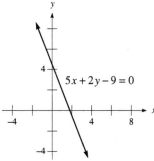

$$5x + 2y - 9 = 0$$

25. There is no *x*-intercept and the *y*-intercept is 3/2. Sketching the graph:

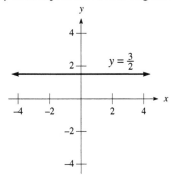

$$y = \frac{3}{2}$$

27. The x-intercept is 0 and the y-intercept is 0. Sketching the graph:

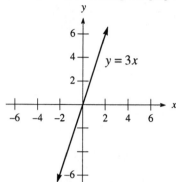

29. The x-intercept is 4/5 and there is no y-intercept. Sketching the graph:

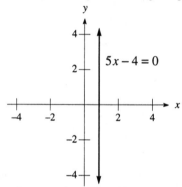

31. The x-intercept is -10 and the y-intercept is 4. Sketching the graph:

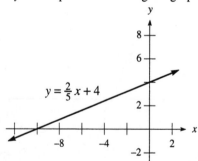

33. To find the x-intercept, let $y = 0$:
$$0 = -2x + 7$$
$$2x = 7$$
$$x = \frac{7}{2}$$
To find the y-intercept, let $x = 0$: $y = -2(0) + 7 = 7$
These values are confirmed on a graphing calculator.

35. To find the *x*-intercept, let *y* = 0:
$$0 = \frac{0.02x + 5}{3}$$
$$0.02x + 5 = 0$$
$$0.02x = -5$$
$$x = -250$$

To find the *y*-intercept, let *x* = 0: $y = \dfrac{0.02(0) + 5}{3} = \dfrac{5}{3} \approx 1.67$

These values are confirmed on a graphing calculator.

37. To find the *x*-intercept, let *y* = 0:
$$2.45x = 0.5(0) - 1$$
$$2.45x = -1$$
$$x \approx -0.41$$

To find the *y*-intercept, let *x* = 0:
$$2.45(0) = 0.5y - 1$$
$$0.5y = 1$$
$$y = 2$$

These values are confirmed on a graphing calculator.

39. To find the *x*-intercept, let *y* = 0:
$$\frac{4.4x + 3(0)}{3} = 2.8$$
$$4.4x = 8.4$$
$$x \approx 1.91$$

To find the *y*-intercept, let *x* = 0:
$$\frac{4.4(0) + 3y}{3} = 2.8$$
$$3y = 8.4$$
$$y = 2.8$$

These values are confirmed on a graphing calculator.

41. Graphing the equation:

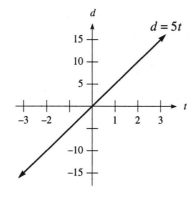

43. Graphing the equation:

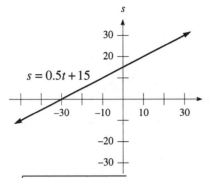

$$s = 0.5t + 15$$

45. The length is given by: $d = \sqrt{(-3-1)^2 + (4+1)^2} = \sqrt{16+25} = \sqrt{41}$

The midpoint is given by: $M = \left(\dfrac{-3+1}{2}, \dfrac{4-1}{2} \right) = \left(-1, \tfrac{3}{2} \right)$

47. The length is given by: $d = \sqrt{(-1-1)^2 + (-4-5)^2} = \sqrt{4+81} = \sqrt{85}$

The midpoint is given by: $M = \left(\dfrac{1-1}{2}, \dfrac{5-4}{2} \right) = \left(0, \tfrac{1}{2} \right)$

49. The length is given by: $d = \sqrt{(a-0)^2 + (a-0)^2} = \sqrt{a^2 + a^2} = \sqrt{2a^2} = |a|\sqrt{2}$

The midpoint is given by: $M = \left(\dfrac{0+a}{2}, \dfrac{0+a}{2} \right) = \left(\dfrac{a}{2}, \dfrac{a}{2} \right)$

51. Drawing the triangle:

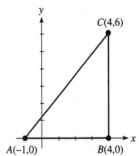

The area is given by: $A = \tfrac{1}{2}(5)(6) = 15$

53. Drawing the triangle:

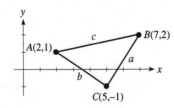

First find each of the lengths:

$$a = \sqrt{(5-2)^2 + (-1-1)^2} = \sqrt{9+4} = \sqrt{13}$$

$$b = \sqrt{(7-5)^2 + (2+1)^2} = \sqrt{4+9} = \sqrt{13}$$

$$c = \sqrt{(7-2)^2 + (2-1)^2} = \sqrt{25+1} = \sqrt{26}$$

Now checking the converse of the Pythagorean theorem:
$$a^2 + b^2 = c^2$$
$$\left(\sqrt{13}\right)^2 + \left(\sqrt{13}\right)^2 = \left(\sqrt{26}\right)^2$$
$$13 + 13 = 26$$
The area is given by: $A = \frac{1}{2} ab = \frac{1}{2}\left(\sqrt{13}\right)\left(\sqrt{13}\right) = \frac{13}{2}$

55. Using the distance formula:
$$\sqrt{(6-0)^2 + (w-3)^2} = 10$$
$$36 + (w-3)^2 = 100$$
$$(w-3)^2 = 64$$
$$w - 3 = -8, 8$$
$$w = -5, 11$$

57. Using the distance formula:
$$\sqrt{(x-5)^2 + (4-1)^2} = 2$$
$$(x-5)^2 + 9 = 4$$
$$(x-5)^2 = -5$$

Since this equation has no real solution, no such point exists.

59. The equation is $(x-2)^2 + (y-3)^2 = 3^2$, or $(x-2)^2 + (y-3)^2 = 9$.

61. The equation is $\left(x-\frac{1}{2}\right)^2 + (y-4)^2 = 6^2$, or $\left(x-\frac{1}{2}\right)^2 + (y-4)^2 = 36$.

63. The center is (3,2) and the radius is $\sqrt{16} = 4$.

65. The center is (0,0) and the radius is $\sqrt{16} = 4$.

67. Completing the square:
$$x^2 + y^2 - 6x - 10y = -9$$
$$\left(x^2 - 6x + 9\right) + \left(y^2 - 10y + 25\right) = -9 + 9 + 25$$
$$(x-3)^2 + (y-5)^2 = 25$$
The center is (3,5) and the radius is $\sqrt{25} = 5$.

69. Completing the square:
$$3x^2 + 3y^2 + 18y = 0$$
$$x^2 + y^2 + 6y = 0$$
$$x^2 + \left(y^2 + 6y + 9\right) = 0 + 9$$
$$x^2 + (y+3)^2 = 9$$
The center is (0,–3) and the radius is $\sqrt{9} = 3$.

71. The center is (2,–3) and the radius is 2. Graphing the circle:

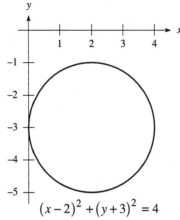

$$(x-2)^2 + (y+3)^2 = 4$$

73. The center of the circle is given by: $\left(\frac{-2+4}{2}, \frac{8-5}{2}\right) = \left(1, \frac{3}{2}\right)$

The radius of the circle is given by: $\sqrt{(4-1)^2 + \left(-5-\frac{3}{2}\right)^2} = \sqrt{9+\frac{169}{4}} = \sqrt{\frac{205}{4}}$

The equation of the circle is: $(x-1)^2 + \left(y-\frac{3}{2}\right)^2 = \frac{205}{4}$

75. The radius of the circle is given by: $\sqrt{(3-2)^2 + (-5-6)^2} = \sqrt{1+121} = \sqrt{122}$

The equation of the circle is: $(x-3)^2 + (y+5)^2 = 122$

77. The radius of the circle is given by: $r = \sqrt{(-3-5)^2 + (4-2)^2} = \sqrt{64+4} = \sqrt{68} = 2\sqrt{17}$

The circumference is given by: $C = 2\pi\left(2\sqrt{17}\right) = 4\pi\sqrt{17}$

79. Its radius must be 2, so the equation is $(x-3)^2 + (y+2)^2 = 4$.

81. The center must be (3,–3) and its radius must be 3, so the equation is $(x-3)^2 + (y+3)^2 = 9$.

83. **a.** The equation of the circle is $x^2 + y^2 = 1$.

b. Checking each point:

$$\left(\frac{3}{5}\right)^2 + \left(-\frac{4}{5}\right)^2 = \frac{9}{25} + \frac{16}{25} = 1$$

$$\left(-\frac{\sqrt{3}}{2}\right)^2 + \left(\frac{1}{2}\right)^2 = \frac{3}{4} + \frac{1}{4} = 1$$

$$\left(\frac{2}{3}\right)^2 + \left(\frac{3}{5}\right)^2 = \frac{4}{9} + \frac{9}{25} = \frac{181}{225} \neq 1$$

Thus $\left(\frac{3}{5}, -\frac{4}{5}\right)$ and $\left(-\frac{\sqrt{3}}{2}, \frac{1}{2}\right)$ are points on the unit circle.

85. Using the distance formula:

$$MP = \sqrt{\left(\frac{x_1+x_2}{2} - x_1\right)^2 + \left(\frac{y_1+y_2}{2} - y_1\right)^2} \qquad MQ = \sqrt{\left(\frac{x_1+x_2}{2} - x_2\right)^2 + \left(\frac{y_1+y_2}{2} - y_2\right)^2}$$

$$= \sqrt{\left(\frac{x_2-x_1}{2}\right)^2 + \left(\frac{y_2-y_1}{2}\right)^2} \qquad\qquad = \sqrt{\left(\frac{x_1-x_2}{2}\right)^2 + \left(\frac{y_1-y_2}{2}\right)^2}$$

$$= \frac{\sqrt{\left(x_2-x_1\right)^2 + \left(y_2-y_1\right)^2}}{2} \qquad\qquad = \frac{\sqrt{\left(x_1-x_2\right)^2 + \left(y_1-y_2\right)^2}}{2}$$

Since $MP = MQ$, M is equidistant from both P and Q.

a. No. We still need to show that M lies on the line passing through P and Q.

b. Yes. Since $PQ = \sqrt{\left(x_2-x_1\right)^2 + \left(y_2-y_1\right)^2}$, note that $MP + MQ = PQ$. We have already shown that $MP = MQ$, thus M is the midpoint of line segment PQ.

c. This is harder than the proof offered in the text.

87. Yes. Finding the distances:

$$PQ = \sqrt{(0+3)^2 + (1-4)^2} = \sqrt{9+9} = \sqrt{18} = 3\sqrt{2}$$

$$QR = \sqrt{(3-0)^2 + (-2-1)^2} = \sqrt{9+9} = \sqrt{18} = 3\sqrt{2}$$

$$PR = \sqrt{(3+3)^2 + (-2-4)^2} = \sqrt{36+36} = \sqrt{72} = 6\sqrt{2}$$

Since $PQ + QR = PR$, the points P, Q, and R lie on the same line. Note also that Q is the midpoint of PR, since $PQ = QR$.

89. The value is: $x_1 + \frac{1}{2}\left(x_2 - x_1\right) = x_1 + \frac{1}{2}x_2 - \frac{1}{2}x_1 = \frac{1}{2}x_1 + \frac{1}{2}x_2 = \frac{x_1+x_2}{2}$

2.2 Slope

1. Sketching the line:

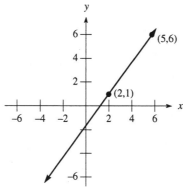

The slope is given by: $m = \dfrac{6-1}{5-2} = \dfrac{5}{3}$

3.   Sketching the line:

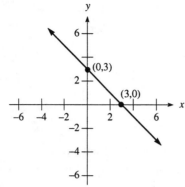

The slope is given by: $m = \dfrac{0-3}{3-0} = \dfrac{-3}{3} = -1$

5.   Sketching the line:

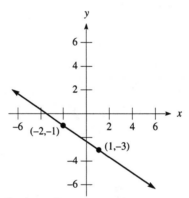

The slope is given by: $m = \dfrac{-3+1}{1+2} = -\dfrac{2}{3}$

7.   Finding the slope: $m = \dfrac{10-7}{4-2} = \dfrac{3}{2}$

9.   Finding the slope: $m = \dfrac{-3-1}{-4-2} = \dfrac{-4}{-6} = \dfrac{2}{3}$

11.   Finding the slope: $m = \dfrac{2-2}{-4-6} = \dfrac{0}{-10} = 0$

13.   Finding the slope: $m = \dfrac{\frac{2}{3}-\frac{3}{5}}{\frac{3}{4}-\frac{1}{2}} = \dfrac{\frac{1}{15}}{\frac{1}{4}} = \dfrac{4}{15}$

15.   Finding the slope: $m = \dfrac{\sqrt{27}-\sqrt{3}}{4-2} = \dfrac{3\sqrt{3}-\sqrt{3}}{2} = \dfrac{2\sqrt{3}}{2} = \sqrt{3}$

17.   Finding the slope: $m = \dfrac{b^2-a^2}{b-a} = \dfrac{(b+a)(b-a)}{b-a} = b+a$

19.   Finding the slope: $m = \dfrac{2s-s}{r+s-r} = \dfrac{s}{s} = 1$

21. Sketching the line:

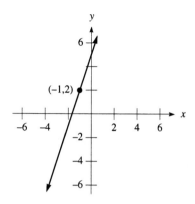

23. Sketching the line:

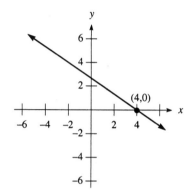

25. Sketching the line:

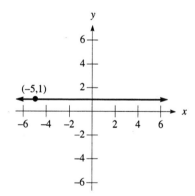

27. Finding the slopes:

$$m_{P_1 P_2} = \frac{3-1}{4-2} = \frac{2}{2} = 1$$

$$m_{P_3 P_4} = \frac{-3+1}{-4+2} = \frac{-2}{-2} = 1$$

Since these slopes are the same, the two lines are parallel.

29. Finding the slopes:
$$m_{P_1P_2} = \frac{1+2}{-1+3} = \frac{3}{2}$$
$$m_{P_3P_4} = \frac{1-4}{7-5} = -\frac{3}{2}$$
Since these slopes are neither the same nor are negative reciprocals, the two lines are neither parallel nor perpendicular.

31. Finding the slopes:
$$m_{P_1P_2} = \frac{4-4}{-3-1} = \frac{0}{-4} = 0$$
$$m_{P_3P_4} = \frac{-3-7}{-2+2} = \frac{-10}{0} = \text{undefined}$$
Since one line is horizontal and the other line is vertical, the two lines are perpendicular.

33. Solving the equation:
$$\frac{c-5}{-4-2} = -\frac{1}{2}$$
$$\frac{c-5}{-6} = -\frac{1}{2}$$
$$c - 5 = 3$$
$$c = 8$$

35. Solving the equation:
$$\frac{-1-t}{t-0} = \frac{-3-2}{2-1}$$
$$\frac{-1-t}{t} = -5$$
$$-1 - t = -5t$$
$$4t = 1$$
$$t = \tfrac{1}{4}$$

37. Solving the equation:
$$\frac{h-1}{-1-h} = -\frac{h-7}{2-h}$$
$$\frac{h-1}{-1-h} = \frac{h-7}{h-2}$$
$$h^2 - 3h + 2 = -h^2 + 6h + 7$$
$$2h^2 - 9h - 5 = 0$$
$$(2h+1)(h-5) = 0$$
$$h = -\tfrac{1}{2}, 5$$

39. Finding the slopes:
$$m_{PQ} = \frac{2-0}{1+3} = \frac{2}{4} = \frac{1}{2}$$
$$m_{QR} = \frac{-2-2}{3-1} = \frac{-4}{2} = -2$$
Since these slopes are negative reciprocals, PQ and QR are perpendicular, and thus ΔPQR is a right triangle.

41. Finding the slopes:
$$m_{PQ} = \frac{4+2}{1+3} = \frac{6}{4} = \frac{3}{2}$$
$$m_{RS} = \frac{1-4}{-4+2} = \frac{-3}{-2} = \frac{3}{2}$$

Since PQ and RS are parallel, then $PQRS$ are the vertices of a trapezoid.

43. Finding each midpoint:
$$M_{AB} = \left(\frac{0+4}{2}, \frac{0+0}{2}\right) = (2,0)$$
$$M_{BC} = \left(\frac{4+5}{2}, \frac{0+2}{2}\right) = \left(\frac{9}{2}, 1\right)$$
$$M_{CD} = \left(\frac{5+1}{2}, \frac{2+2}{2}\right) = (3,2)$$
$$M_{AD} = \left(\frac{0+1}{2}, \frac{0+2}{2}\right) = \left(\frac{1}{2}, 1\right)$$

Now find the slopes between these midpoints:
$$m_1 = \frac{1-0}{\frac{9}{2}-2} = \frac{1}{\frac{5}{2}} = \frac{2}{5}$$
$$m_2 = \frac{2-1}{3-\frac{9}{2}} = \frac{1}{-\frac{3}{2}} = -\frac{2}{3}$$
$$m_3 = \frac{1-2}{\frac{1}{2}-3} = \frac{-1}{-\frac{5}{2}} = \frac{2}{5}$$
$$m_4 = \frac{1-0}{\frac{1}{2}-2} = \frac{1}{-\frac{3}{2}} = -\frac{2}{3}$$

Since $m_1 = m_3$ and $m_2 = m_4$, these midpoints form the vertices of a parallelogram.

45. In increasing order the slopes are $m_4 < m_2 < m_3 < m_1$. Note that steeper negative slopes (m_4) are less than moderately negative slopes (m_2), while moderately positive slopes (m_3) are less than steeply positive slopes (m_1).

47. **a.** This is true since MP is a transversal intersecting two parallel lines.

b. This is true since MP is a transversal intersecting L_1 and L_2.

c. This is true since MP is a transversal intersecting ST and PQ.

d. This is a restatement of part **c**, since $\triangle STU$ and $\triangle PQR$ will have congruent angles if and only if L_1 is parallel to L_2.

e. This is a restatement of part **d**, since similar triangles have corresponding sides which are proportional.

f. Since $m_1 = \frac{|\overline{UT}|}{|\overline{ST}|}$ and $m_2 = \frac{|\overline{RQ}|}{|\overline{PQ}|}$, this statement results from part **e**.

49.   Finding slopes between successive points:
$$m_1 = \frac{-2+3}{2+2} = \frac{1}{4}$$
$$m_2 = \frac{-1+2}{6-2} = \frac{1}{4}$$
$$m_3 = \frac{-1+3}{6+2} = \frac{2}{8} = \frac{1}{4}$$
Since these slopes are equal, the three points are collinear.

51.   If the two lines intersect at (a,b), the corresponding points from the text proof are $O(a,b)$, $P(a+1, b+m_1)$, and $Q(a+1, b+m_2)$. But note that the distances OP, OQ and PQ are all identical to the text proof, thus the rest of the proof is identical.

2.3 Equations of a Line

1.   Using the point-slope formula:
$$y - 2 = 3(x+1)$$
$$y - 2 = 3x + 3$$
$$y = 3x + 5$$
Sketching the graph:

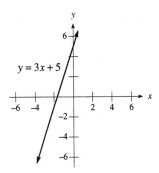

3.   Using the point-slope formula:
$$y + 3 = \tfrac{2}{5}(x+2)$$
$$y + 3 = \tfrac{2}{5}x + \tfrac{4}{5}$$
$$y = \tfrac{2}{5}x - \tfrac{11}{5}$$
Sketching the graph:

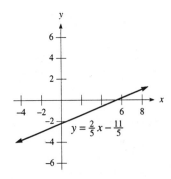

5.   Using the point-slope formula:
$$y - 4 = 4(x + 3)$$
$$y - 4 = 4x + 12$$
$$y = 4x + 16$$

7.   First find the slope: $m = \dfrac{3-6}{8-2} = \dfrac{-3}{6} = -\frac{1}{2}$
   Using the point-slope formula:
$$y - 6 = -\frac{1}{2}(x - 2)$$
$$y - 6 = -\frac{1}{2}x + 1$$
$$y = -\frac{1}{2}x + 7$$

9.   Using the point-slope formula:
$$y - 0 = 4(x - 0)$$
$$y = 4x$$

11.   The slope-intercept form is $y = -\frac{2}{3}x + 4$.

13.   First find the slope: $m = \dfrac{-1-0}{0-5} = \dfrac{-1}{-5} = \frac{1}{5}$
   The slope-intercept form is $y = \frac{1}{5}x - 1$.

15.   The equation is $x = -3$.

17.   The slope-intercept form is $y = -\frac{1}{3}x + 5$.

19.   Since line F has no x-intercept, it is a horizontal line. Thus its equation is $y = 5$.

21.   First find the slope: $m = \dfrac{-5+5}{-3-2} = \dfrac{0}{-5} = 0$
   Thus the equation is $y = -5$.

23.   The slope is -2.

25.   The slope is 2/3.

27.   Solving for y:
$$5y - 1 = x$$
$$5y = x + 1$$
$$y = \frac{1}{5}x + \frac{1}{5}$$
   The slope is $\frac{1}{5}$.

29.   Solving for y:
$$\frac{4x + 3y}{3} = 2$$
$$\frac{4}{3}x + y = 2$$
$$y = -\frac{4}{3}x + 2$$
   The slope is $-\frac{4}{3}$.

31.   Using the point-slope formula with $m = 4$:
$$y + 5 = 4(x - 1)$$
$$y + 5 = 4x - 4$$
$$y = 4x - 9$$

33. Using the point-slope formula with $m = \frac{4}{3}$:

$$y - 7 = \frac{4}{3}(x - 4)$$
$$y - 7 = \frac{4}{3}x - \frac{16}{3}$$
$$y = \frac{4}{3}x + \frac{5}{3}$$

35. First find the slope by solving for y:

$$5y - 4x = 9$$
$$5y = 4x + 9$$
$$y = \frac{4}{5}x + \frac{9}{5}$$

Using the point-slope formula with $m = \frac{4}{5}$:

$$y - 0 = \frac{4}{5}(x - 5)$$
$$y = \frac{4}{5}x - 4$$

37. First find the slope by solving for y:

$$6x - 7y = 6$$
$$-7y = -6x + 6$$
$$y = \frac{6}{7}x - \frac{6}{7}$$

Using the point-slope formula with $m = -\frac{7}{6}$:

$$y - 4 = -\frac{7}{6}(x - 7)$$
$$y - 4 = -\frac{7}{6}x + \frac{49}{6}$$
$$y = -\frac{7}{6}x + \frac{73}{6}$$

39. The slope is -1, so the equation is $y = -x$.

41. The line must be horizontal, so its equation is $y = 5$.

43. First find the slope and y-intercept by solving for y:

$$2x + 7y = 14$$
$$7y = -2x + 14$$
$$y = -\frac{2}{7}x + 2$$

The slope is $\frac{7}{2}$, so the equation is $y = \frac{7}{2}x + 2$.

45. Both y-intercepts are 0. The slope of $y = x$ is $m = 1$ and the slope of $y = 2x$ is $m = 2$:

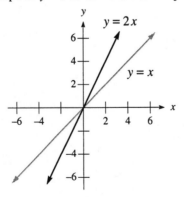

47. Both y-intercepts are -3. The slope of $y = x - 3$ is $m = 1$ and the slope of $y = 2x - 3$ is $m = 2$:

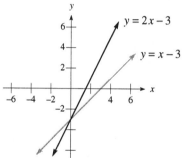

49. Both y-intercepts are 0. The slope of $y = x$ is $m = 1$ and the slope of $y = 2x$ is $m = 2$:

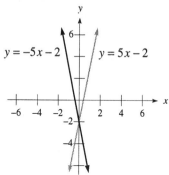

51. Both y-intercepts are -2. The slope of $y = 5x - 2$ is $m = 5$ and the slope of $y = -5x - 2$ is $m = -5$:

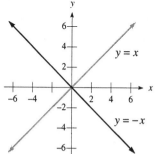

53. All three slopes are $m = 1$. The y-intercept of $y = x$ is 0, the y-intercept of $y = x - 3$ is -3, and the y-intercept of $y = x + 3$ is 3:

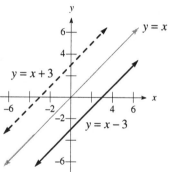

55.   All three slopes are $m = 5$. The y-intercept of $y = 5x$ is 0, the y-intercept of $y = 5x - 1$ is -1, and the y-intercept of $y = 5x + 2$ is 2:

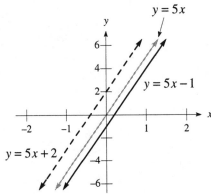

57.   First find the midpoint: $M = \left(\dfrac{-4+1}{2}, \dfrac{3+0}{2}\right) = \left(-\dfrac{3}{2}, \dfrac{3}{2}\right)$

Now find the slope: $m = \dfrac{0-3}{1+4} = -\dfrac{3}{5}$

Using the point-slope formula with $m = \dfrac{5}{3}$:

$$y - \tfrac{3}{2} = \tfrac{5}{3}\left(x + \tfrac{3}{2}\right)$$
$$y - \tfrac{3}{2} = \tfrac{5}{3}x + \tfrac{5}{2}$$
$$y = \tfrac{5}{3}x + 4$$

59.   The x-intercept is $\left(-\tfrac{2}{3}, 0\right)$ and the y-intercept is $\left(0, \tfrac{2}{5}\right)$.

Finding the midpoint: $M = \left(\dfrac{-\tfrac{2}{3}+0}{2}, \dfrac{0+\tfrac{2}{5}}{2}\right) = \left(-\tfrac{1}{3}, \tfrac{1}{5}\right)$

Now find the slope: $m = \dfrac{\tfrac{2}{5}-0}{0+\tfrac{2}{3}} = \dfrac{\tfrac{2}{5}}{\tfrac{2}{3}} = \dfrac{3}{5}$

Using the point-slope formula with $m = -\dfrac{5}{3}$:

$$y - \tfrac{1}{5} = -\tfrac{5}{3}\left(x + \tfrac{1}{3}\right)$$
$$y - \tfrac{1}{5} = -\tfrac{5}{3}x - \tfrac{5}{9}$$
$$y = -\tfrac{5}{3}x - \tfrac{16}{45}$$

61.   The equation is $C = 29 + 0.30m$, or $C = 0.30m + 29$.

63.   **a.**   Finding the slope: $m = \dfrac{750 - 600}{65 - 50} = \dfrac{150}{15} = 10$

Using the point-slope formula:
$$P - 600 = 10(x - 50)$$
$$P - 600 = 10x - 500$$
$$P = 10x + 100$$

b.   Substituting $x = 90$: $P = 10(90) + 100 = \$1000$

65. **a.** Finding the slope: $m = \dfrac{85-80}{18-15} = \dfrac{5}{3}$

Using the point-slope formula:
$$N - 80 = \tfrac{5}{3}(s-15)$$
$$N - 80 = \tfrac{5}{3}s - 25$$
$$N = \tfrac{5}{3}s + 55$$

b. Substituting $s = 25$: $N = \tfrac{5}{3}(25) + 55 \approx 97$

The jogger's heart rate would be approximately 97 beats per minute.

67. Finding the slope: $m = \dfrac{0-8500}{12-0} = -\dfrac{2125}{3}$

The equation is $V = -\dfrac{2125}{3}t + 8500$.

69. **a.** The equation is $C = 12{,}375 + 8.25d$, or $C = 8.25d + 12{,}375$.

b. The income is given by $I = 49.95d$.

c. Setting the income and costs equal:
$$49.95d = 8.25d + 12{,}375$$
$$41.7d = 12{,}375$$
$$d \approx 297$$

The company must rent the car for approximately 297 days.

d. The break-even point is where the graphs intersect:

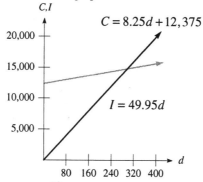

71. **a.** The slope is given by: $m = \dfrac{-10}{2.50} = -4$

Using the point-slope formula:
$$N - 275 = -4(P - 62.50)$$
$$N - 275 = -4P + 250$$
$$N = -4P + 525$$

b. Substituting $P = 75$: $N = -4(75) + 525 = 225$ players

c. Substituting $N = 240$:
$$-4P + 525 = 240$$
$$-4P = -285$$
$$P = 71.25$$

The price should be \$71.25 per player.

73.   Finding the slope: $m = \dfrac{100-0}{212-32} = \dfrac{100}{180} = \dfrac{5}{9}$

Using the point-slope formula:
$$C - 0 = \tfrac{5}{9}(F - 32)$$
$$C = \tfrac{5}{9}F - \tfrac{160}{9}$$

Substituting $F = 98.6°$: $C = \tfrac{5}{9}(98.6°) - \tfrac{160}{9} = 37°$

75.   **a.**   Finding the slope: $m = \dfrac{146.3 - 177.8}{39.1 - 47.5} = \dfrac{-31.5}{-8.4} = 3.75$

Using the point-slope formula:
$$h - 146.3 = 3.75(f - 39.1)$$
$$h - 146.3 = 3.75f - 146.625$$
$$h = 3.75f - 0.325$$

b.   Substituting $f = 52$: $h = 3.75(52) - 0.325 \approx 194.7$ cm

77.   The slope of the radius line is $\dfrac{-12-0}{5-0} = -\tfrac{12}{5}$, so the tangent line slope is $\tfrac{5}{12}$. Using the point-slope

formula:
$$y + 12 = \tfrac{5}{12}(x - 5)$$
$$y + 12 = \tfrac{5}{12}x - \tfrac{25}{12}$$
$$y = \tfrac{5}{12}x - \tfrac{169}{12}$$

Sketching the circle and the tangent line:

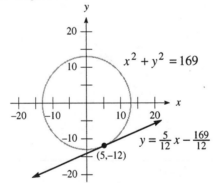

79.   Sketching the graph:

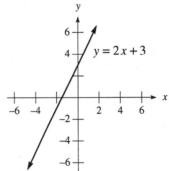

The answers to all three questions are $x < -\tfrac{3}{2}$.

81. You could find two points, such as the intercepts $\left(0, \frac{11}{5}\right)$ and $\left(-\frac{11}{7}, 0\right)$, and directly compute the slope. Alternatively, you could solve the equation for y, and the resulting x-coefficient is the slope. This latter method is probably the easiest.

83. Finding the slope: $m = \dfrac{b-0}{0-a} = -\dfrac{b}{a}$

The slope-intercept form is:

$$y = -\frac{b}{a}x + b$$
$$ay = -bx + ab$$
$$bx + ay = ab$$
$$\frac{bx}{ab} + \frac{ay}{ab} = \frac{ab}{ab}$$
$$\frac{x}{a} + \frac{y}{b} = 1$$

2.4 Relations and Functions

1. The ordered pairs are $\{(A,10),(B,20),(C,30)\}$. The domain is $\{A,B,C\}$ and the range is $\{10,20,30\}$. This relation represents a function.

3. The ordered pairs are $\{(6,A),(6,B),(10,B),(15,C),(19,D)\}$. The domain is $\{6,10,15,19\}$ and the range is $\{A,B,C,D\}$. This relation does not represent a function.

5. The ordered pairs are $\{(A,7),(B,7),(C,7),(D,7)\}$. The domain is $\{A,B,C,D\}$ and the range is $\{7\}$. This relation represents a function.

7. The domain is $\{-4,2,4\}$ and the range is $\{-5,-3,0,6\}$. This relation does not represent a function.

9. The domain is $\left\{-\sqrt{3},-1,0,1,\sqrt{3}\right\}$ and the range is $\{0,1,3\}$. This relation represents a function.

11. The domain is $\{0,1,2,3,4,5\}$ and the range is $\{5\}$. This relation represents a function.

13. The domain is all real numbers, or $(-\infty,\infty)$. This relation represents a function.

15. The domain is $\{x \mid x \neq 2\}$. This relation represents a function.

17. The domain is all real numbers, or $(-\infty,\infty)$. This relation represents a function.

19. The domain is $x \geq -2$, or $[-2,\infty)$. This relation represents a function.

21. The domain is $\left\{x \mid x \neq -\frac{4}{3}\right\}$. This relation represents a function.

23. The domain is $\{x \mid x \neq 0\}$. This relation represents a function.

25. Since $x^2 - 9 = (x+3)(x-3)$, form a sign chart:

$$\begin{array}{ccccc} + & & - & & + \\ \hline \text{For} \quad x=-6 & -3 & x=0 & 3 & x=6 \end{array}$$

So $x^2 - 9 > 0$ on the intervals $(-\infty,-3) \cup (3,\infty)$, which is the domain. This relation represents a function.

27. Since $x^2 + 16 > 0$ for all values of x, the domain is all real numbers, or $(-\infty,\infty)$. This relation represents a function.

29. Since there are no restrictions on the fifth root, the domain is all real numbers, or $(-\infty,\infty)$. This relation represents a function.

31. Solving the inequality:
$$8 - 5x \geq 0$$
$$-5x \geq -8$$
$$x \leq \tfrac{8}{5}$$

The domain is $\left(-\infty, \tfrac{8}{5}\right]$. This relation represents a function.

33. Since $|x - 4| \geq 0$, the domain is all real numbers, or $(-\infty, \infty)$. This relation represents a function.

35. For $\sqrt{x+5}$ we must have $x \geq -5$. The denominator requires that $x \neq 3$, so the domain is $[-5, 3) \cup (3, \infty)$. This relation represents a function.

37. Since $3x - 2x^2 = x(3 - 2x)$, form a sign chart:

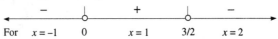

For $x = -1$ 0 $x = 1$ $3/2$ $x = 2$

So $3x - 2x^2 > 0$ on the interval $\left(0, \tfrac{3}{2}\right)$, which is the domain. This relation represents a function.

39. The domain is $\{x \mid x \neq 5\}$. This relation represents a function.

41. Since $x^2 - 16 = (x+4)(x-4)$, the domain is $\{x \mid x \neq -4, 4\}$. This relation represents a function.

43. This is the graph of a function.

45. This is not the graph of a function.

47. This is the graph of a function.

49. The domain is $(-\infty, \infty)$ and the range is $[-3, \infty)$.

51. The domain is $(-\infty, \infty)$ and the range is $(-\infty, 6]$.

53. The domain is $[-5, 5]$ and the range is $[-3, 3]$.

55. The cost is given by $C = 5(22) + 0.11m$, or $C = 0.11m + 110$.

57. Her total distance is given by: $d = 8t + 10(6 - t) = 8t + 60 - 10t = 60 - 2t$

59. **a.**   Graphing the equation:

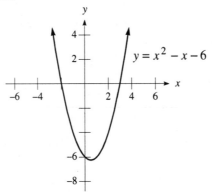

$y = x^2 - x - 6$

b.   Graphing the equation:

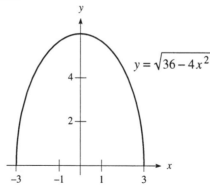

$y = \sqrt{36 - 4x^2}$

61.   A relation is a function when each domain value is paired with one, and only one, unique range value.

2.5 Function Notation

1.   Evaluating the function: $f(6) = 5(6) - 2 = 30 - 2 = 28$

3.   Evaluating the function: $h(10) = \sqrt{4(10) - 3} = \sqrt{40 - 3} = \sqrt{37}$

5.   Evaluating the function: $g(8) = 3(8)^2 - 4(8) + 1 = 192 - 32 + 1 = 161$

7.   Evaluating the function: $f\left(\frac{1}{3}\right) = 5\left(\frac{1}{3}\right) - 2 = \frac{5}{3} - 2 = -\frac{1}{3}$

9.   Evaluating the function: $f(x + 2) = 5(x + 2) - 2 = 5x + 10 - 2 = 5x + 8$

11.   Evaluating the function:
$$g(x - 1) = 3(x - 1)^2 - 4(x - 1) + 1 = 3x^2 - 6x + 3 - 4x + 4 + 1 = 3x^2 - 10x + 8$$

13.   Evaluating the function: $h\left(x^2\right) = \sqrt{4\left(x^2\right) - 3} = \sqrt{4x^2 - 3}$

15.   Evaluating the function: $g(3x) = 3(3x)^2 - 4(3x) + 1 = 27x^2 - 12x + 1$

17.   Evaluating the function: $f(4x + 7) = 5(4x + 7) - 2 = 20x + 35 - 7 = 20x + 33$

19.   Evaluating the function: $g(x + h) = 3(x + h)^2 - 4(x + h) + 1 = 3x^2 + 6xh + 3h^2 - 4x - 4h + 1$

21.   Evaluating the function: $F(9) = \dfrac{9}{9 + 1} = \dfrac{9}{10}$

23.   Evaluating the function: $G(15) = \dfrac{1}{\sqrt{15} - 1} = \dfrac{1}{\sqrt{14}} \cdot \dfrac{\sqrt{14}}{\sqrt{14}} = \dfrac{\sqrt{14}}{14}$

25.   Evaluating the function: $F\left(\frac{1}{2}\right) = \dfrac{\frac{1}{2}}{\frac{1}{2} + 1} = \dfrac{\frac{1}{2}}{\frac{3}{2}} = \dfrac{1}{3}$

27.   Evaluating the function: $G\left(t^2\right) = \dfrac{1}{\sqrt{t^2 - 1}}$

29.   Evaluating the function: $F(-1) = \dfrac{-1}{-1 + 1} = \dfrac{-1}{0} = \text{undefined}$

31.   Evaluating the function: $F(x + 1) = \dfrac{x + 1}{x + 1 + 1} = \dfrac{x + 1}{x + 2}$

33.   Evaluating the function: $5G(6) = 5 \cdot \dfrac{1}{\sqrt{6} - 1} = \dfrac{5}{\sqrt{5}} \cdot \dfrac{\sqrt{5}}{\sqrt{5}} = \dfrac{5\sqrt{5}}{5} = \sqrt{5}$

35. Evaluating the function: $F(40) = \dfrac{40}{40+1} = \dfrac{40}{41}$

37. Evaluating the function: $F\left(\dfrac{a}{3}\right) = \dfrac{\frac{a}{3}}{\frac{a}{3}+1} \cdot \dfrac{3}{3} = \dfrac{a}{a+3}$

39. Evaluating the function: $F(x-1) = \dfrac{x-1}{x-1+1} = \dfrac{x-1}{x}$

41. Evaluating the function: $f(5x+7) = 4(5x+7)-3 = 20x+28-3 = 20x+25$

43. Evaluating the function: $g(x-a) = \dfrac{1}{x-a}$

45. Evaluating the function: $g(x)-g(a) = \dfrac{1}{x}-\dfrac{1}{a}$

47. Evaluating the function: $g(x)-a = \dfrac{1}{x}-a$

49. Evaluating the function: $f(4)h(4) = (4 \cdot 4 - 3)\left(4^2 - 4\right) = 13 \cdot 12 = 156$

51. Evaluating the function: $f[h(4)] = f\left(4^2 - 4\right) = f(12) = 4(12)-3 = 45$

53. Evaluating the function: $g[h(x)] = g\left(x^2 - x\right) = \dfrac{1}{x^2 - x}$

55. Evaluating the function: $h(kx) = (kx)^2 - kx = k^2 x^2 - kx$

57. Evaluating the function: $f\big(g[h(3)]\big) = f\left(g\left(3^2 - 3\right)\right) = f(g(6)) = f\left(\tfrac{1}{6}\right) = 4\left(\tfrac{1}{6}\right)-3 = \tfrac{2}{3}-3 = -\tfrac{7}{3}$

59. Evaluating the function: $h\left(\dfrac{1}{x}\right) = \left(\dfrac{1}{x}\right)^2 - \dfrac{1}{x} = \dfrac{1}{x^2} - \dfrac{1}{x} = \dfrac{1-x}{x^2}$

61. Evaluating the function:

$$\dfrac{g(x)-g(5)}{x-5} = \dfrac{\left(x^2 - 3x + 2\right) - \left(5^2 - 3 \cdot 5 + 2\right)}{x-5}$$
$$= \dfrac{x^2 - 3x + 2 - 12}{x-5}$$
$$= \dfrac{x^2 - 3x - 10}{x-5}$$
$$= \dfrac{(x-5)(x+2)}{x-5}$$
$$= x+2$$

63. Evaluating the function:

$$\dfrac{f(x+3)-f(x)}{3} = \dfrac{[2 - 3(x+3)] - [2 - 3x]}{3} = \dfrac{2 - 3x - 9 - 2 + 3x}{3} = \dfrac{-9}{3} = -3$$

65. Evaluating the function:

$$\frac{g(x+h)-g(x)}{h} = \frac{\left[(x+h)^2 - 3(x+h)+2\right]-\left[x^2 - 3x + 2\right]}{h}$$

$$= \frac{\left(x^2 + 2xh + h^2 - 3x - 3h + 2\right)-\left(x^2 - 3x + 2\right)}{h}$$

$$= \frac{x^2 + 2xh + h^2 - 3x - 3h + 2 - x^2 + 3x - 2}{h}$$

$$= \frac{2xh + h^2 - 3h}{h}$$

$$= 2x + h - 3$$

67. **a.** Evaluating the function: $f(-3) = 3(-3) - 5 = -14$

 b. Evaluating the function: $f(1) = 1^2 - 1 = 0$

 c. Evaluating the function: $f(6) = 6^2 - 1 = 35$

69. **a.** Evaluating the function: $h(0) = |2 \bullet 0 - 1| = |-1| = 1$

 b. Evaluating the function: $h(3)$ is undefined

 c. Evaluating the function: $h(5) = \dfrac{5+1}{5-1} = \dfrac{6}{4} = \dfrac{3}{2}$

 d. Evaluating the function: $h(8)$ is undefined

71. **a.** Substituting $t = 1,2,3$:

$$h(1) = 50(1) - 16(1)^2 = 50 - 16 = 34 \text{ feet}$$

$$h(2) = 50(2) - 16(2)^2 = 100 - 64 = 36 \text{ feet}$$

$$h(3) = 50(3) - 16(3)^2 = 150 - 144 = 6 \text{ feet}$$

 b. Solving where $h(t) = 0$:

$$50t - 16t^2 = 0$$

$$2t(25 - 8t) = 0$$

$$t = 0, \tfrac{25}{8}$$

The object will hit the ground after $\frac{25}{8} = 3.125$ seconds.

73. **a.** $g(-5) = 1$

 b. $g(-2) = -1$

 c. $g(0) = -2$

 d. $g(1) = -\frac{1}{2}$

 e. $g(3) = 2$

 f. $g(x) = 0$ when $x = -6, -3, 2$

 g. There are 3 solutions to the equation $g(x) = 0$.

75. The function is:

$$A(h) = \begin{cases} 7 & \text{if } 0 \le h \le 35 \\ 7 + 0.12(h - 25) & \text{if } h > 35 \end{cases}$$

This simplifies to:

$$A(h) = \begin{cases} 7 & \text{if } 0 \le h \le 35 \\ 0.12h + 2.8 & \text{if } h > 35 \end{cases}$$

Graphing the function:

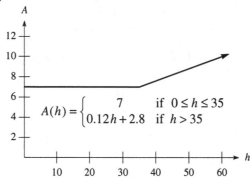

$$A(h) = \begin{cases} 7 & \text{if } 0 \le h \le 35 \\ 0.12h + 2.8 & \text{if } h > 35 \end{cases}$$

77. **a.**   The expression is $f(x) + 4$, not $f(x+4)$.
  b.   The expression is $f(x) \cdot (x+4)$, not $f(x+4)$.
  c.   The expression is $2f(x)$, not $f(2x)$.
  d.   The expression is $[f(x)]^2$, not $f(x^2)$.

2.6 Relating Functions to Their Graphs

1. **a.** $F(-7)$ is undefined
  b. $F(-5) = 4$
  c. $F(-2)$ is undefined
  d. $F(1) + F(4) = -3 + 0 = -3$
  e. $F(1+4) = F(5) = 1$
  f.   The domain is $(-7, -3] \cup [1, 6]$.
  g.   The range is $[-3, 2] \cup [3, 5]$.

3. **a.**   The zeros are approximately -1.2 and 3.2.
  b.   Solving where $x^2 - 2x - 4 = 0$:
$$x = \frac{-(-2) \pm \sqrt{(-2)^2 - 4(1)(-4)}}{2(1)} = \frac{2 \pm \sqrt{4+16}}{2} = \frac{2 \pm 2\sqrt{5}}{2} = 1 \pm \sqrt{5} \approx -1.2, 3.2$$

5. **a.**   The zeros are approximately -0.3 and 3.3.
  b.   Solving where $-x^2 + 3x + 1 = 0$:
$$x = \frac{-3 \pm \sqrt{3^2 - 4(-1)(1)}}{2(-1)} = \frac{-3 \pm \sqrt{9+4}}{-2} = \frac{3 \pm \sqrt{13}}{2} \approx -0.3, 3.3$$

7. **a.**   The zeros are approximately -0.6, 0, and 1.6.
  b.   Factoring yields $x(x^2 - x - 1) = 0$, so $x = 0$ is one zero. Using the quadratic formula:
$$x = \frac{-(-1) \pm \sqrt{(-1)^2 - 4(1)(-1)}}{2(1)} = \frac{1 \pm \sqrt{1+4}}{2} = \frac{1 \pm \sqrt{5}}{2} \approx -0.6, 1.6$$

9. **a.**   As $x \to \infty$, $F(x) \to 1$.
  b.   As $x \to -\infty$, $F(x) \to -4$.
  c.   As $x \to 3$, $F(x) \to \infty$.

11. **a.** f is increasing on the intervals $(-\infty, -6] \cup [-2, \infty)$.
  b. f is decreasing on the interval $[-6, -2]$.
  c. f is nonnegative on the intervals $[-7, -3] \cup [-1, \infty)$.

    d. f is positive on the intervals $(-7,-3)\cup(-1,\infty)$.

    e. f is negative on the intervals $(-\infty,-7)\cup(-3,-1)$.

    f.   The relative maximum is 3 when $x=-6$, and the relative minimum is -1 when $x=-2$.

13. **a.** f is increasing on the interval $[3,\infty)$.

    b. f is decreasing on the interval $(-\infty,3]$.

    c. f is nonnegative on the interval $(-\infty,\infty)$.

    d. f is positive on the intervals $(-\infty,3)\cup(3,\infty)$.

    e. f is never negative.

15. **a.** f is increasing on the intervals $(-\infty,0)\cup[0,2]$.

    b. f is decreasing on the interval $[5,\infty)$.

    c. f is constant on the interval $[2,5]$.

    d. f is positive on the intervals $(-\infty,0)\cup(0,7)$.

    e. f is negative on the interval $(7,\infty)$.

17. **a.** $f(-1)=3$

    b. $f(0)=5$

    c. $f(4)=0$

    d. $f(x)=-2$ for $x=-5,-3,5$

    e.   There are two values of x for which $f(x)=4$.

    f.   The relative minimum of f is -3 (at $x=-4$).

19. **a.** f has three zeros.

    b. $f(x)=2$ at three values of x.

    c.   There are no values of x for which $f(x)=6$.

    d.   The relative maximum of f is 5 (at $x=0$).

21. **a.** f is positive on the intervals $[-7,-6)\cup(-2,4)$.

    b. $f(x)<0$ on the intervals $(-6,-2)\cup(4,6]$.

    c. $f(x)\ge0$ on the intervals $[-7,-6]\cup[-2,4]$.

23. **a.** $f(3)-g(-7)=2-(-2)=4$

    b. $f(x)>g(x)$ on the intervals $[-7,-6)\cup(-2,4)$.

    c.   For $g(x)-f(x)$ to be nonnegative, $g(x)-f(x)\ge0$ and so $g(x)\ge f(x)$. This occurs on the intervals $[-6,-2]\cup[4,6]$.

25. **a.** $F(-2)=3$

    b. $F(0)=4$

    c. $F(2)=5$

    d. $F(x)=3$ for $x=-9,-2,4$

    e. $F(x)=0$ for $x=-7,-3,7$

    f. F has a relative minimum of -2 (at $x=-4$).

27. **a.** F has three zeros.

    b. $F(x)=2$ for three values of x.

    c. $F(x)=-6$ for one value of x.

    d. F has a relative maximum of 5 (at $x=2$).

29. **a.** F is positive on the intervals $(-\infty,-7)\cup(-3,7)$.

    b. $F(x)<0$ on the intervals $(-7,-3)\cup(7,\infty)$.

    c. $F(x)\ge0$ on the intervals $(-\infty,-7]\cup[-3,7]$.

31.   **a.** $F(-4) - G(-4) = -2 - 3 = -5$

      b. $F(x) > G(x)$ on the intervals $(-\infty, -7) \cup (-2, 7)$.

      c.   For $G(x) - F(x)$ to be nonnegative, $G(x) - F(x) \geq 0$ and so $G(x) \geq F(x)$. This occurs on the intervals $[-7, -2] \cup [7, \infty)$.

33.   **a.**   When grown alone, the P. Aurelia increases and then levels off near 100, while the P. Caudatum increases and then levels off near 65.

      b.   When grown together, the P. Aurelia increases and then levels off near 100, while the P. Caudatum increases up to 5 days, then decreases and eventually dies off.

      c.   The data supports Gause's principle, as the P. Aurelia survives while the P. Caudatum eventually dies out.

35.   **a.**   13 machines maximizes the profit.

      b.   The maximum profit is $700,000.

37.   Graphing the function:

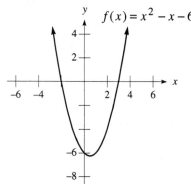

      a.   The zeros are at $x = -2, 3$.

      b. $f(x) > 0$ on the intervals $(-\infty, -2) \cup (3, \infty)$.

      c. $f(x) < 0$ on the interval $(-2, 3)$.

      d.   There is no relative maximum, and the relative minimum is -6.25.

39.   Graphing the function:

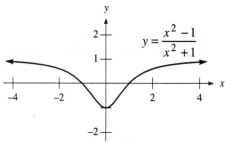

      a.   The zeros are at $x = -1, 1$.

      b. $y > 0$ on the intervals $(-\infty, -1) \cup (1, \infty)$.

      c. $y < 0$ on the interval $(-1, 1)$.

      d.   There is no relative maximum, and the relative minimum is -1.

41.   Graphing the function:

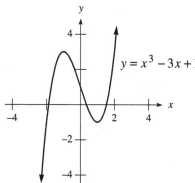

$$y = x^3 - 3x + 1$$

    a.   The zeros are at $x \approx -1.88, 0.35, 1.53$.

    b. $y > 0$ on the intervals $(-1.88, 0.35) \cup (1.53, \infty)$.

    c. $y < 0$ on the intervals $(-\infty, -1.88) \cup (0.35, 1.53)$.

    d.   The relative maximum is 3.00, and the relative minimum is –1.00.

43.   Graphing the function:

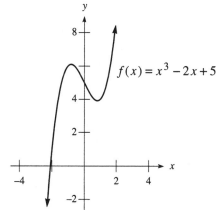

$$f(x) = x^3 - 2x + 5$$

Thus $f(x) = 0$ when $x \approx -2.09$.

45.   The equation is equivalent to $x^3 - 2x + 1 = 0$. Graphing the function:

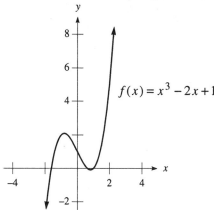

$$f(x) = x^3 - 2x + 1$$

Thus $f(x) = 0$ when $x \approx -1.62, 0.62, 1$.

47. The equation is equivalent to $x^4 + x^2 - 1 = 0$. Graphing the function:

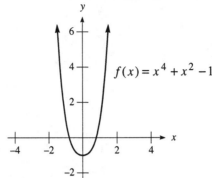

Thus $f(x) = 0$ when $x \approx -0.79, 0.79$.

49. The inequality is equivalent to $x^4 - 2x^2 - 3 \le 0$. Graphing the function:

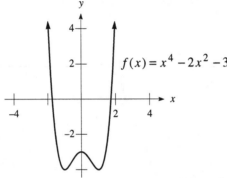

The x-intercepts are $x \approx -1.73, 1.73$, thus $f(x) \le 0$ on the interval $[-1.73, 1.73]$.

51. The inequality is equivalent to $x^3 - 5x^2 + x - 2 > 0$. Graphing the function:

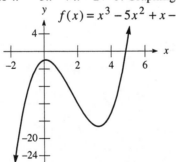

The x-intercept is $x \approx 4.88$, thus $f(x) > 0$ on the interval $(4.88, \infty)$.

2.7 Introduction to Graph Sketching: Symmetry

1. The graph has y-axis symmetry.
3. The graph has x-axis symmetry.
5. The graph has origin symmetry.
7. The graph does not exhibit any of these symmetries.

9. **a.** Completing the graph with y-axis symmetry:

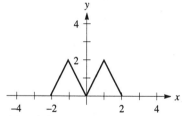

b. Completing the graph with origin symmetry:

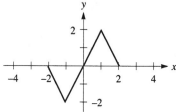

11. **a.** Completing the graph with y-axis symmetry:

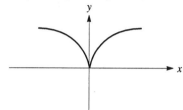

b. Completing the graph with origin symmetry:

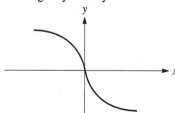

13. **a.** Sketching the reflection of $\overline{PQ}$ about the y-axis:

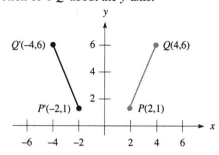

b.   Sketching the reflection of $\overline{PQ}$ about the x-axis:

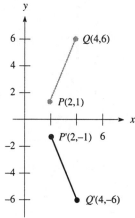

c.   Sketching the reflection of $\overline{PQ}$ about the origin:

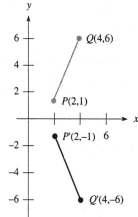

15.   **a.**   Sketching the reflection of $\overline{PQ}$ about the y-axis:

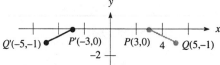

b.   Sketching the reflection of $\overline{PQ}$ about the x-axis:

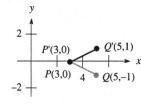

c. Sketching the reflection of $\overline{PQ}$ about the origin:

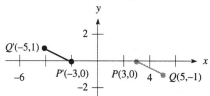

17. Since $f(-x) = 4(-x)^2 - 1 = 4x^2 - 1 = f(x)$, the graph has y-axis symmetry.

19. Since $h(-x) = (-x)^5 - 2(-x) = -x^5 + 2x = -h(x)$, the graph has origin symmetry.

21. Since $f(-x) = 2(-x)^2 - 3(-x) + 1 = 2x^2 + 3x + 1$, the graph has neither symmetry.

23. Since $F(-x) = \dfrac{5}{-x} = -\dfrac{5}{x} = -F(x)$, the graph has origin symmetry.

25. Since $F(-x) = -\dfrac{(-x)^4}{(-x)^2 - 1} = -\dfrac{x^4}{x^2 - 1} = F(x)$, the graph has y-axis symmetry.

27. Since $F(-x) = \dfrac{(-x)^3}{(-x)^2 - 1} = -\dfrac{x^3}{x^2 - 1} = -F(x)$, the graph has origin symmetry.

29. Since $H(-x) = \left| -x \right| = \left| x \right| = H(x)$, the graph has y-axis symmetry.

31. Since $f(-x) = \dfrac{-x}{4} - \dfrac{-x}{3} = -\dfrac{x}{4} + \dfrac{x}{3} = -f(x)$, the graph has y-axis symmetry.

33. Since $h(-x) = \dfrac{-x - 3}{4 - (-x)} = \dfrac{-x - 3}{4 + x}$, the graph has neither symmetry.

35. Graphing the curve:

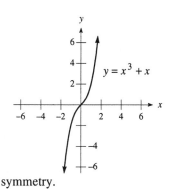

Note the curve exhibits origin symmetry.

37. Graphing the curve:

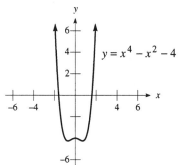

The curve exhibits y-axis symmetry.

39.   Graphing the curve:

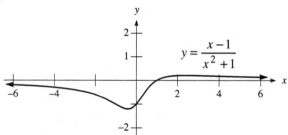

The curve exhibits neither type of symmetry.

41.   The graph must also pass the horizontal line test.

43.   **a.**   Since $f(-x)+g(-x)=f(x)+g(x)$, $f+g$ must be an even function.

   b.   Since $f(-x)+g(-x)=-f(x)-g(x)=-[f(x)+g(x)]$, $f+g$ must be an odd function.

   c.   Since $f(-x)+g(-x)=f(x)-g(x)$, $f+g$ is neither even nor odd.

Chapter 2 Review Exercises

1.   Graphing the equation:

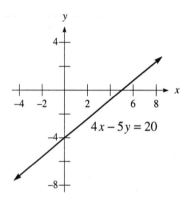

2.   Graphing the equation:

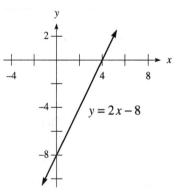

3. Graphing the equation:

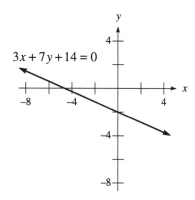

$$3x + 7y + 14 = 0$$

4. Graphing the equation:

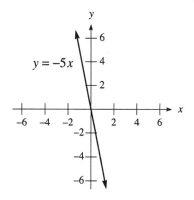

$$y = -5x$$

5. Graphing the equation:

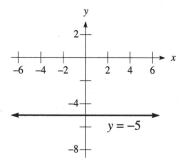

$$y = -5$$

6. Graphing the equation:

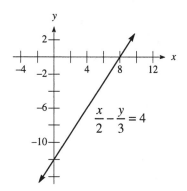

$$\frac{x}{2} - \frac{y}{3} = 4$$

7.   Graphing the equation:

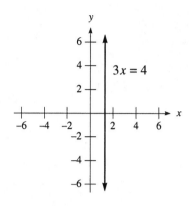

$3x = 4$

8.   Graphing the equation:

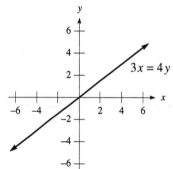

$3x = 4y$

9.   Graphing the equation:

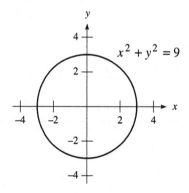

$x^2 + y^2 = 9$

10.   Graphing the equation:

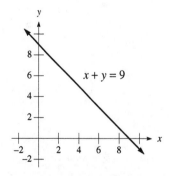

$x + y = 9$

11. Graphing the equation:

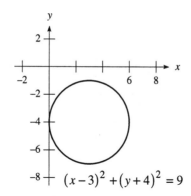

$$(x-3)^2 + (y+4)^2 = 9$$

12. First complete the square:

$$x^2 + y^2 - 6y = 1$$
$$x^2 + \left(y^2 - 6y + 9\right) = 1 + 9$$
$$x^2 + (y-3)^2 = 10$$

Graphing the equation:

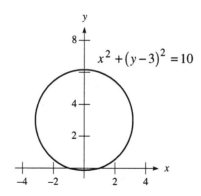

$$x^2 + (y-3)^2 = 10$$

13. Graphing the equation:

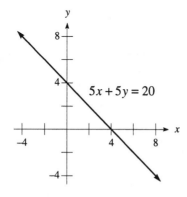

$$5x + 5y = 20$$

14.   Graphing the equation:

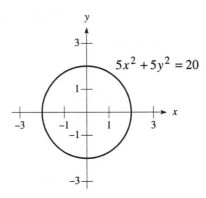

15.   First complete the square:
$$x^2 - 3x + y^2 + 2y = 2$$
$$\left(x^2 - 3x + \tfrac{9}{4}\right) + \left(y^2 + 2y + 1\right) = 2 + \tfrac{9}{4} + 1$$
$$\left(x - \tfrac{3}{2}\right)^2 + (y+1)^2 = \tfrac{21}{4}$$

Graphing the equation:

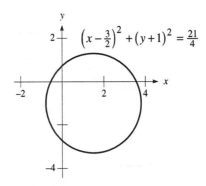

16.   First complete the square:
$$2x^2 + 2y^2 - 8x + 4y = 0$$
$$x^2 + y^2 - 4x + 2y = 0$$
$$\left(x^2 - 4x + 4\right) + \left(y^2 + 2y + 1\right) = 0 + 4 + 1$$
$$(x-2)^2 + (y+1)^2 = 5$$

Graphing the equation:

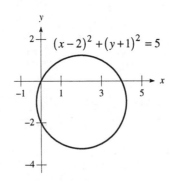

17. Graphing the equation:

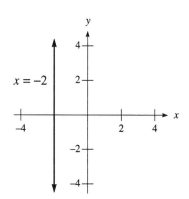

18. Graphing the equation:

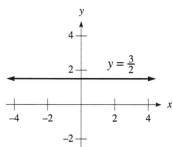

19. **a.** The graph is that of a function.
  b. The domain is all real numbers, or $(-\infty, \infty)$.
  c. The range is $[-5, \infty)$.

20. **a.** The graph is not that of a function.
  b. The domain is $(-\infty, 6]$.
  c. The range is all real numbers, or $(-\infty, \infty)$.

21. **a.** This graph is that of a function.
  b. The domain is $[-5, -2] \cup [3, 5]$.
  c. The range is $[-5, 4]$.

22. **a.** This graph is that of a function.
  b. The domain is all real numbers, or $(-\infty, \infty)$.
  c. The range is all real numbers, or $(-\infty, \infty)$.

23. When $x = 2$, $y = 2^4 = 16$. When $x = -2$, $y = (-2)^4 = 16$. This does define y as a function of x, since each value of x results in one unique value of y.

24. When $x = 16$, $y^4 = 16$, so $y = \pm 2$. This does not define y as a function of x, since $x = 16$ is assigned two different y-values ($y = 2$ and $y = -2$).

25. Finding the slope: $m = \dfrac{-5-4}{2+3} = -\dfrac{9}{5}$
  Using the point-slope formula:
$$y - 4 = -\tfrac{9}{5}(x+3)$$
$$y - 4 = -\tfrac{9}{5}x - \tfrac{27}{5}$$
$$y = -\tfrac{9}{5}x - \tfrac{7}{5}$$

26.   Solving for y:
$$4x - 7y = 2$$
$$-7y = -4x + 2$$
$$y = \tfrac{4}{7}x - \tfrac{2}{7}$$

So the perpendicular slope is $-\tfrac{7}{4}$. Using the point-slope formula:
$$y - 0 = -\tfrac{7}{4}(x - 2)$$
$$y = -\tfrac{7}{4}x + \tfrac{7}{2}$$

27.   Setting the slopes equal:
$$\frac{6+4}{t-1} = \frac{t+1}{5+3}$$
$$\frac{10}{t-1} = \frac{t+1}{8}$$
$$t^2 - 1 = 80$$
$$t^2 = 81$$
$$t = \pm 9$$

28.   Using the distance formula:
$$PQ = \sqrt{(3+1)^2 + (-2-3)^2} = \sqrt{16+25} = \sqrt{41}$$
$$QR = \sqrt{(13-3)^2 + (6+2)^2} = \sqrt{100+64} = \sqrt{164}$$
$$PR = \sqrt{(13+1)^2 + (6-3)^2} = \sqrt{196+9} = \sqrt{205}$$

Since $PQ^2 + QR^2 = 41 + 164 = 205 = PR^2$, by the Pythagorean theorem these three points are the vertices of a right triangle.

29.   The equation is $(x-(-4))^2 + (y-1)^2 = 7^2$, or $(x+4)^2 + (y-1)^2 = 49$.

30.   The center is $(0,-2)$ and the radius is $\sqrt{20} = 2\sqrt{5}$.

31.   Completing the square:
$$x^2 + 8x + y^2 - 2y = 8$$
$$\left(x^2 + 8x + 16\right) + \left(y^2 - 2y + 1\right) = 8 + 16 + 1$$
$$(x+4)^2 + (y-1)^2 = 25$$

The center is $(-4,1)$ and the radius is $\sqrt{25} = 5$.

32.   Completing the square:
$$x^2 - 4x + y^2 - 6y + 9 = 0$$
$$\left(x^2 - 4x + 4\right) + \left(y^2 - 6y + 9\right) = 0 + 4$$
$$(x-2)^2 + (y-3)^2 = 4$$

Graphing the circle with center $(2,3)$ and radius $\sqrt{4} = 2$:

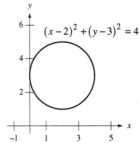

33. The equation is $(x-(-5))^2 +(y-4)^2 = \left(\tfrac{1}{2}\right)^2$, or $(x+5)^2 +(y-4)^2 = \tfrac{1}{4}$.

34. Completing the square:
$$x^2 +10x+y^2 = 33$$
$$\left(x^2 +10x+25\right)+y^2 = 33+25$$
$$(x+5)^2 +y^2 = 58$$
The center of the circle is $(-5,0)$, so the slope of the radius line drawn from $(2,-3)$ is given by:
$$m = \frac{-3-0}{2+5} = -\tfrac{3}{7}$$
Since the tangent line is perpendicular, the tangent slope is $\tfrac{7}{3}$. Using the point-slope formula:
$$y+3 = \tfrac{7}{3}(x-2)$$
$$y+3 = \tfrac{7}{3}x - \tfrac{14}{3}$$
$$y = \tfrac{7}{3}x - \tfrac{23}{3}$$

35. The center of the circle is the midpoint: $M = \left(\dfrac{1+0}{2}, \dfrac{-4+5}{2}\right) = \left(\tfrac{1}{2},\tfrac{1}{2}\right)$

Using the distance formula to find the radius: $r = \sqrt{\left(0-\tfrac{1}{2}\right)^2 +\left(5-\tfrac{1}{2}\right)^2} = \sqrt{\tfrac{1}{4}+\tfrac{81}{4}} = \sqrt{\tfrac{41}{2}}$

Thus the equation of the circle is: $\left(x-\tfrac{1}{2}\right)^2 +\left(y-\tfrac{1}{2}\right)^2 = \tfrac{41}{2}$

36. The line crosses the y-axis when $x = 0$:
$$3y -\tfrac{1}{2}(0) = 4$$
$$3y = 4$$
$$y = \tfrac{4}{3}$$
So the center is $\left(0,\tfrac{4}{3}\right)$, thus the equation is $(x-0)^2 +\left(y-\tfrac{4}{3}\right)^2 = 7^2$, or $x^2 +\left(y-\tfrac{4}{3}\right)^2 = 49$.

37. Solving the inequality:
$$5-3x \geq 0$$
$$-3x \geq -5$$
$$x \leq \tfrac{5}{3}$$
The domain is $\left(-\infty,\tfrac{5}{3}\right]$.

38. The domain is all real numbers, or $(-\infty,\infty)$.

39. Since $x^2 -3x-4 = (x-4)(x+1)$, the domain is $\{x \mid x \neq -1,4\}$.

40. Since $x^2 +9 \neq 0$ for all values of x, the domain is all real numbers, or $(-\infty,\infty)$.

41. Since $6x-x^2 = x(6-x)$, form the sign chart:

So $6x-x^2 \geq 0$ on the interval $[0,6]$, which is the domain.

42. For $x+6 \geq 0$, we must have $x \geq -6$. The denominator requires $x \neq 2$. Thus the domain is $[-6,2)\cup(2,\infty)$.

43. The domain is $(-4,\infty)$.

44. The domain is all real numbers, or $(-\infty,\infty)$.

45. Finding the function values:
$$f(-3) = -(-3)^2 + 4(-3) - 1 = -9 - 12 - 1 = -22$$
$$f(2x) = -(2x)^2 + 4(2x) - 1 = -4x^2 + 8x - 1$$
$$2f(x) = 2(-x^2 + 4x - 1) = -2x^2 + 8x - 2$$

46. Finding the function values:
$$g(-9) = \sqrt{7 - 2(-9)} = \sqrt{7 + 18} = \sqrt{25} = 5$$
$$g(0) = \sqrt{7 - 2(0)} = \sqrt{7 - 0} = \sqrt{7}$$
$$g\left(\tfrac{1}{2}\right) = \sqrt{7 - 2\left(\tfrac{1}{2}\right)} = \sqrt{7 - 1} = \sqrt{6}$$

47. Finding the function values:
$$h(2) = \sqrt{5}$$
$$h(10) = \sqrt{5}$$
$$h(-3) = \sqrt{5}$$

48. Finding the function values:
$$f(x+3) = 9 - 4(x+3)^2 = 9 - 4(x^2 + 6x + 9) = 9 - 4x^2 - 24x - 36 = -4x^2 - 24x - 27$$
$$f(x) + 3 = 9 - 4x^2 + 3 = -4x^2 + 12$$

49. Finding the function values:
$$g\left(\tfrac{1}{2}\right) = \frac{2\left(\tfrac{1}{2}\right) - 1}{\tfrac{1}{2} + 5} = \frac{1 - 1}{\tfrac{11}{2}} = 0$$
$$g(-5) = \frac{2(-5) - 1}{-5 + 5} = \frac{-11}{0} = \text{undefined}$$
$$g(t+1) = \frac{2(t+1) - 1}{t + 1 + 5} = \frac{2t + 2 - 1}{t + 6} = \frac{2t + 1}{t + 6}$$

50. Finding the function values:
$$H\left(-\tfrac{1}{2}\right) = \frac{-\tfrac{1}{2}}{4\left(-\tfrac{1}{2}\right)^2 + 1} = \frac{-\tfrac{1}{2}}{1 + 1} = \frac{-\tfrac{1}{2}}{2} = -\tfrac{1}{4}$$
$$H\left(\tfrac{1}{r}\right) = \frac{\tfrac{1}{r}}{4\left(\tfrac{1}{r}\right)^2 + 1} = \frac{\tfrac{1}{r}}{\tfrac{4}{r^2} + 1} \cdot \frac{r^2}{r^2} = \frac{r}{4 + r^2}$$
$$H(r+1) = \frac{r+1}{4(r+1)^2 + 1} = \frac{r+1}{4r^2 + 8r + 4 + 1} = \frac{r+1}{4r^2 + 8r + 5}$$

51. Finding the function values:
$$f(0) = 0^2 - 1 = 0 - 1 = -1$$
$$f(-1) = (-1)^2 - 1 = 1 - 1 = 0$$
$$f(-5) = -2(-5) + 7 = 10 + 7 = 17$$

52. Finding the function values:
$$f(3) = 3^2 = 9$$
$$f(4) = \sqrt{4} = 2$$
$$f(-2) = -(-2) = 2$$

53. Simplifying the expression:

$$\frac{f(x-4)-f(x)}{4} = \frac{\left[5(x-4)^2 - 6(x-4)+3\right]-\left[5x^2 - 6x+3\right]}{4}$$

$$= \frac{5x^2 - 40x + 80 - 6x + 24 + 3 - 5x^2 + 6x - 3}{4}$$

$$= \frac{-40x + 104}{4}$$

$$= -10x + 26$$

54. Simplifying the expression:

$$\frac{g(u+h)-g(u)}{h} = \frac{\left[9-7(u+h)\right]-\left[9-7u\right]}{h} = \frac{9-7u-7h-9+7u}{h} = \frac{-7h}{h} = -7$$

55. Simplifying the expression:

$$\frac{f(t+h)-f(t)}{h} = \frac{\dfrac{-3}{t+h}-\dfrac{-3}{t}}{h} \cdot \frac{t(t+h)}{t(t+h)} = \frac{-3t+3(t+h)}{ht(t+h)} = \frac{-3t+3t+3h}{ht(t+h)} = \frac{3h}{ht(t+h)} = \frac{3}{t(t+h)}$$

56. Simplifying the expression:

$$\frac{h(x-3)-h(x)}{3} = \frac{\dfrac{x-3+1}{x-3-1}-\dfrac{x+1}{x-1}}{3}$$

$$= \frac{\dfrac{x-2}{x-4}-\dfrac{x+1}{x-1}}{3} \cdot \frac{(x-4)(x-1)}{(x-4)(x-1)}$$

$$= \frac{(x-2)(x-1)-(x+1)(x-4)}{3(x-4)(x-1)}$$

$$= \frac{x^2 - 3x + 2 - x^2 + 3x + 4}{3(x-4)(x-1)}$$

$$= \frac{6}{3(x-4)(x-1)}$$

$$= \frac{2}{(x-4)(x-1)}$$

57. **a.** $f(-4)=-2, f(-2)=2, f(0)=3, f(1)=3$

 b. The zeros of f are $x = -5, -3, 4$.

 c. $f(x) > 0$ on the intervals $(-\infty, -5) \cup (-3, 4)$.

 d. f is negative on the intervals $(-5, -3) \cup (4, \infty)$.

 e. f is increasing on the interval $[-4, -1]$.

 f. f is decreasing on the interval $(-\infty, -4] \cup [2, \infty)$.

 g. f is constant on the interval $[-1, 2]$.

 h. The relative minimum value is -2 (at $x = -4$).

58. The graph does not exhibit any of these symmetries.

59. The graph is symmetric about the y-axis.

60. The graph is symmetric about the origin.

61. The graph is symmetric about the x-axis.

62. Since $f(-x) = \dfrac{-3(-x)}{(-x)^2+1} = \dfrac{3x}{x^2+1} = -f(x)$, the function exhibits origin symmetry.

63. Since $f(-x) = 2(-x)^2 - 5(-x) = 2x^2 + 5x$, the function exhibits neither type of symmetry.

64. Since $g(-x) = \dfrac{-x+4}{-x-4} = \dfrac{x-4}{x+4}$, the function exhibits neither type of symmetry.

65. Since $h(-x) = 7(-x)-(-x)^3 = -7x+x^3 = -h(x)$, the function exhibits origin symmetry.

66. Since $F(-x) = \sqrt{9-(-x)^2} = \sqrt{9-x^2} = F(x)$, the function exhibits y-axis symmetry.

67. Since $G(-x) = \dfrac{(-x)^2+4}{-x} = \dfrac{x^2+4}{-x} = -G(x)$, the function exhibits origin symmetry.

68. Expressing the cost as a function of g:
$$C(g) = \begin{cases} 1.32g & \text{if } g \le 150 \\ 1.32(150)+1.21(g-150) & \text{if } g > 150 \end{cases}$$
This simplifies to:
$$C(g) = \begin{cases} 1.32g & \text{if } g \le 150 \\ 1.21g+16.5 & \text{if } g > 150 \end{cases}$$

69. Finding the slope: $m = \dfrac{85-76}{5-3} = \dfrac{9}{2}$

Using the point-slope formula:
$$G-76 = \tfrac{9}{2}(t-3)$$
$$G-76 = \tfrac{9}{2}t-13.5$$
$$G = \tfrac{9}{2}t+62.5$$

Substituting $t = 6$: $G = \tfrac{9}{2}(6)+62.5 = 27+62.5 = 89.5$

70. The charge would be:
$$C(h) = \begin{cases} 75 & \text{if } 0 < h \le 1 \\ 75+95(h-1) & \text{if } h > 1 \end{cases}$$
This simplifies to:
$$C(h) = \begin{cases} 75 & \text{if } 0 < h \le 1 \\ 95h+20 & \text{if } h > 1 \end{cases}$$

71. The area is given by: $A(t) = (6+2.5t)^2 = 36+30t+6.25t^2$

72. Graphing the function:

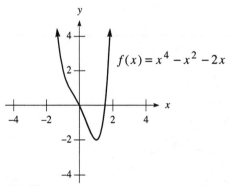

Thus $f(x) = 0$ when $x \approx 0, 1.52$.

73. The equation is equivalent to $x^3 - 6x - 3 = 0$. Graphing the function:

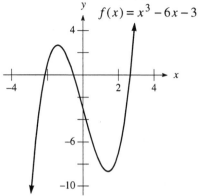

$$f(x) = x^3 - 6x - 3$$

Thus $f(x) = 0$ when $x \approx -2.15, -0.52, 2.67$.

74. Graphing the function:

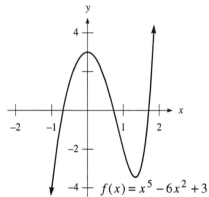

$$f(x) = x^5 - 6x^2 + 3$$

The x-intercepts are $x \approx -0.69, 0.73, 1.71$. Thus $f(x) > 0$ on the intervals $(-0.69, 0.73) \cup (1.71, \infty)$.

75. The inequality is equivalent to $x^4 - 2x^2 - 3x - 1 \le 0$. Graphing the function:

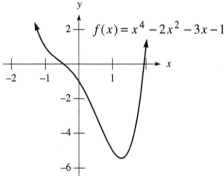

$$f(x) = x^4 - 2x^2 - 3x - 1$$

The x-intercepts are $x \approx -0.46, 1.95$. Thus $f(x) \le 0$ on the interval $[-0.46, 1.95]$.

Chapter 2 Practice Test

1.   **a.**   Evaluating the function: $f(-4) = -2(-4)^2 - 5(-4) + 3 = -32 + 20 + 3 = -9$

   b.   Evaluating the function:

$$f(x+4) = -2(x+4)^2 - 5(x+4) + 3$$
$$= -2x^2 - 16x - 32 - 5x - 20 + 3$$
$$= -2x^2 - 21x - 49$$

   c.   Evaluating the function: $f(4x) = -2(4x)^2 - 5(4x) + 3 = -32x^2 - 20x + 3$

   d.   Evaluating the function: $f\left(x^2\right) = -2\left(x^2\right)^2 - 5\left(x^2\right) + 3 = -2x^4 - 5x^2 + 3$

   e.   Evaluating the function:

$$\frac{f(x+h) - f(x)}{h} = \frac{\left[-2(x+h)^2 - 5(x+h) + 3\right] - \left[-2x^2 - 5x + 3\right]}{h}$$
$$= \frac{-2x^2 - 4xh - 2h^2 - 5x - 5h + 3 + 2x^2 + 5x - 3}{h}$$
$$= \frac{-4xh - 2h^2 - 5h}{h}$$
$$= -4x - 2h - 5$$

2.   **a.**   Finding the function value: $g\left(\frac{2}{3}\right) = \frac{2\left(\frac{2}{3}\right) + 1}{\frac{2}{3} - 2} = \frac{\frac{4}{3} + 1}{-\frac{4}{3}} = \frac{\frac{7}{3}}{-\frac{4}{3}} = -\frac{7}{4}$

   b.   Finding the function value: $g\left(\frac{1}{t}\right) = \frac{2\left(\frac{1}{t}\right) + 1}{\frac{1}{t} - 2} = \frac{\frac{2}{t} + 1}{\frac{1}{t} - 2} \cdot \frac{t}{t} = \frac{2 + t}{1 - 2t}$

3.   Finding the difference quotient:

$$\frac{F(x+5) - F(x)}{5} = \frac{\dfrac{x+5}{x+5-3} - \dfrac{x}{x-3}}{5}$$
$$= \frac{\dfrac{x+5}{x+2} - \dfrac{x}{x-3}}{5} \cdot \frac{(x+2)(x-3)}{(x+2)(x-3)}$$
$$= \frac{(x+5)(x-3) - x(x+2)}{5(x+2)(x-3)}$$
$$= \frac{x^2 + 2x - 15 - x^2 - 2x}{5(x+2)(x-3)}$$
$$= \frac{-15}{5(x+2)(x-3)}$$
$$= \frac{-3}{(x+2)(x-3)}$$

4. The x-intercept is 4 and the y-intercept is $-\frac{20}{3}$. Sketching the graph:

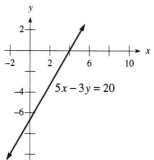

5. Using the points $(4,0)$ and $(0,-3)$, find the slope: $m = \dfrac{-3-0}{0-4} = \dfrac{3}{4}$

The slope-intercept form is $y = \frac{3}{4}x - 3$.

6. Solving for y:
$$3x - 7y + 10 = 0$$
$$-7y = -3x - 10$$
$$y = \tfrac{3}{7}x + \tfrac{10}{7}$$

The perpendicular slope is $-\frac{7}{3}$. Using the point-slope formula:
$$y + 5 = -\tfrac{7}{3}(x+2)$$
$$y + 5 = -\tfrac{7}{3}x - \tfrac{14}{3}$$
$$y = -\tfrac{7}{3}x - \tfrac{29}{3}$$

7. Completing the square:
$$x^2 - 2x + y^2 = 0$$
$$x^2 - 2x + 1 + y^2 = 0 + 1$$
$$(x-1)^2 + y^2 = 1$$

Sketching the graph:

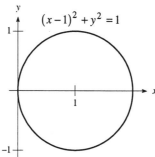

No, this is not the graph of a function (it fails the vertical line test).

8. The midpoint is the center of the circle: $M = \left(\dfrac{-3+1}{2}, \dfrac{4-3}{2}\right) = \left(-1, \tfrac{1}{2}\right)$

Using the distance formula to find the radius: $r = \sqrt{(-3+1)^2 + \left(4-\tfrac{1}{2}\right)^2} = \sqrt{4 + \tfrac{49}{4}} = \sqrt{\tfrac{65}{4}}$

Thus the equation of the circle is $(x - (-1))^2 + \left(y - \tfrac{1}{2}\right)^2 = \left(\sqrt{\tfrac{65}{4}}\right)^2$, or $(x+1)^2 + \left(y - \tfrac{1}{2}\right)^2 = \tfrac{65}{4}$.

9.   **a.**   The values are: $f(-5) = -2, f(-3) = 3, f(0) = -1, f(4) = 4$
        b.   The zeros of f occur at $x = -6, -4, -2, 2, 7$.
        c. f is positive on the intervals $(-\infty, -6) \cup (-4, -2) \cup (2, 7)$.
        d. $f(x) < 0$ on the intervals $(-6, -4) \cup (-2, 2) \cup (7, \infty)$.
        e. f is increasing on the intervals $[-5, -3] \cup [0, 3]$.
        f. f is decreasing on the intervals $(-\infty, -5] \cup [-3, 0] \cup [5, \infty)$.
        g. f is constant on the interval $[3, 5]$.
        h.   The relative maximum is 3 (occurring at $x = -3$).

10. **a.**   Completing the graph with y-axis symmetry:

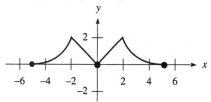

        b.   Completing the graph with origin symmetry:

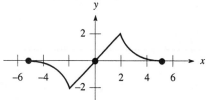

        c.   Completing the graph with x-axis symmetry:

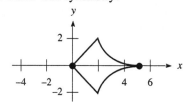

11. **a.**   Since $f(-x) = \dfrac{-x}{-x+5} = \dfrac{x}{x-5}$, the graph does not exhibit either type of symmetry.

        b.   Since $F(-x) = 3(-x)^2 - (-x)^4 = 3x^2 - x^4 = F(x)$, the graph exhibits y-axis symmetry.

        c.   Since $h(-x) = -x - \dfrac{1}{-x} = -x + \dfrac{1}{x} = -h(x)$, the graph exhibits origin symmetry.

12. Let the width be w and the length be $4w - 3$. The total cost of fencing is given by:
        $C(w) = 3.5(2w + 8w - 6) = 3.5(10w - 6) = 35w - 21$

13. **a.** Graphing the function:

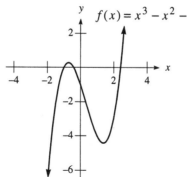

$$f(x) = x^3 - x^2 - 3x - 1$$

So $f(x) = 0$ for $x \approx -1, -0.41, 2.41$.

b. The inequality is equivalent to $x^4 - x^2 - 4 > 0$. Graphing the function:

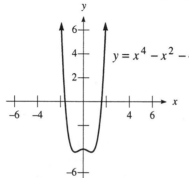

$$y = x^4 - x^2 - 4$$

The x-intercepts are $x \approx -1.60, 1.60$. So $f(x) > 0$ on the intervals $(-\infty, -1.60) \cup (1.60, \infty)$.

Chapter 3
Functions and Graphs: Part II

3.1 Basic Graphing Principles

1. The graph of $y = f(x) - 3$ will be displaced 3 units down, and the graph of $y = f(x - 3)$ will be displaced 3 units to the right. Sketching the graphs:

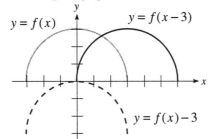

3. The graph of $y = h(x) - 1$ will be displaced 1 unit down, and the graph of $y = h(x) + 1$ will be displaced 1 unit up. Sketching the graphs:

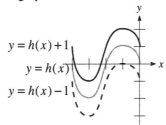

5. The graph will be $y = f(x)$ displaced 1 unit to the right and 2 units up:

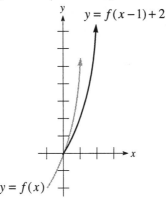

7.   The graph will be $y = h(x)$ displaced 2 units to the left and 3 units up:

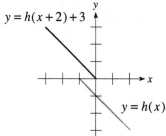

9.   The graph of $y = f(x)$ will be displaced 2 units down from $y = f(x) + 2$:

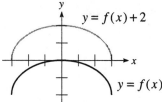

11.   The graph of $y = f(x)$ will be displaced 2 units to the left from $y = f(x-2)$:

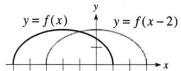

13.   The graph is $y = x^2$ displaced 2 units up:

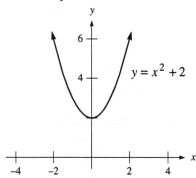

15.   The graph is $y = x^2$ displaced 1 unit to the right:

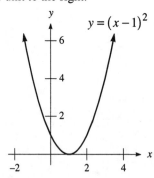

17. Graphing the function:

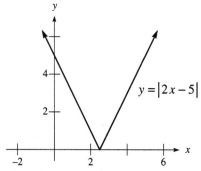

19. The graph is $y = |x|$ displaced 4 units to the left:

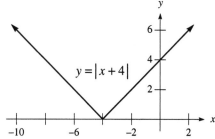

21. The graph is $y = x^2$ displaced 1 unit to the left and 4 units down:

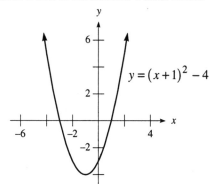

23. The graph is $y = x^2$ displaced 2 units to the right and 3 units down:

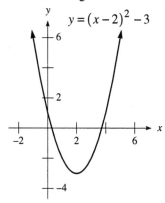

Note the x-intercepts occur when:

$$(x-2)^2 - 3 = 0$$
$$(x-2)^2 = 3$$
$$x - 2 = \pm\sqrt{3}$$
$$x = 2 \pm \sqrt{3}$$

25.   The graph is $y = |x|$ displaced 3 units down:

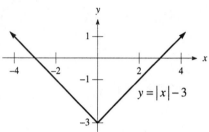

27.   Graphing the function:

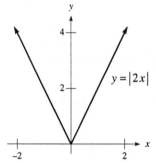

29.   The graph is $y = |x|$ displaced 1 unit to the right and 2 units up:

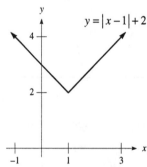

31.   The graph is the same as $y = |x - 2|$, which is $y = |x|$ displaced 2 units to the right:

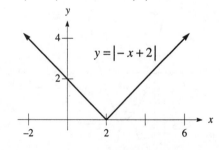

33. The graph is $y = |x|$ reflected across the x-axis, then displaced 5 units up:

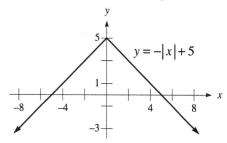

35. **a.** $f(-4) = 2$
b. $f(-1) = 2$
c. $f(0) = 1$
d. $f(2) = -1$
e. $f(5) = -1$
f. $f(x) = 2$ for all x in the interval $(-\infty, 0)$.
g. $f(x) = 1$ for $x = 0$
h. $f(x) = -1$ for all x in the interval $(0, \infty)$.
i. There are no values of x where $f(x) = 4$.

37. The domain is $(-\infty, \infty)$ and the range is $(-\infty, 0)$. Sketching the graph:

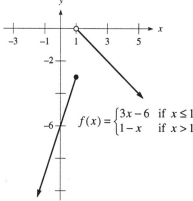

39. The domain is $[-2, 5)$ and the range is $[-6, -2) \cup \{4\}$. Sketching the graph:

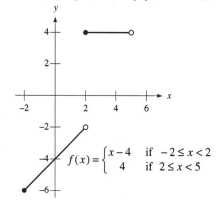

41. The domain is $(-\infty, 6) \cup (6, \infty)$ and the range is $(-\infty, 6)$. Sketching the graph:

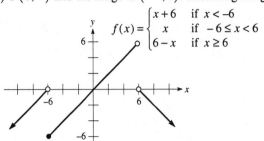

$$f(x) = \begin{cases} x+6 & \text{if } x < -6 \\ x & \text{if } -6 \le x < 6 \\ 6-x & \text{if } x \ge 6 \end{cases}$$

43. The domain is $(-\infty, \infty)$ and the range is $\{-4\} \cup [0, 4]$. Sketching the graph:

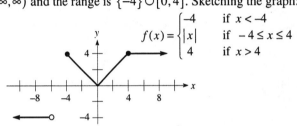

$$f(x) = \begin{cases} -4 & \text{if } x < -4 \\ |x| & \text{if } -4 \le x \le 4 \\ 4 & \text{if } x > 4 \end{cases}$$

45. The domain is $(-\infty, \infty)$ and the range is $\left\{0, \frac{7}{30}, \frac{9}{15}, 1\right\}$. Sketching the graph:

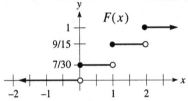

47. The domain is $(-\infty, \infty)$ and the range is $\left\{0, \frac{1}{5}\right\}$. Sketching the graph:

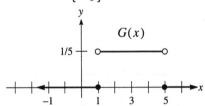

49. The graph of $y = -x^2$ will be $y = x^2$ reflected across the x-axis:

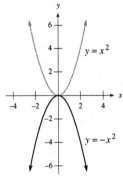

51. Yes. Reflect the graph of $y = f(x)$ across the x-axis.

3.2 More Graphing Principles: Types of Functions

1.   The graph of $y = -f(x)$ is a reflection of $y = f(x)$ across the x-axis, and the graph of $y = f(-x)$ is a reflection of $y = f(x)$ across the y-axis:

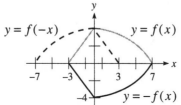

3.   The graph of $y = h(x) - 2$ is a displacement of $y = h(x)$ down 2 units, and the graph of $y = h(x-2)$ is a displacement of $y = h(x)$ to the right 2 units:

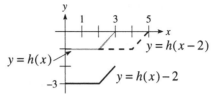

5.   The graph of $y = f(x-1)$ is a displacement of $y = f(x)$ to the right 1 unit, and the graph of $y = f(1-x)$ is a displacement of $y = f(x)$ to the left 1 unit, then a reflection across the y-axis:

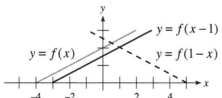

7.   The graph of $y = f(-x)$ is a reflection of $y = f(x)$ across the y-axis, and the graph of $y = -f(x)$ is a reflection of $y = f(x)$ across the x-axis:

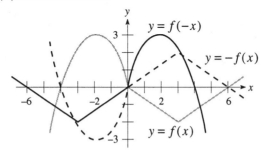

9.   The graph of $y = -f(x) + 2$ is a reflection of $y = f(x)$ across the x-axis, then a displacement up 2 units. The graph of $y = -f(x+2)$ is a displacement of $y = f(x)$ to the left 2 units, then a reflection across the x-axis:

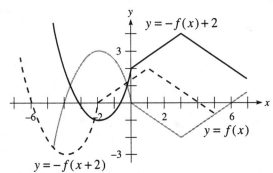

11.   The graph of $y = f(-x) - 3$ is a reflection of $y = f(x)$ across the y-axis, then a displacement down 3 units. The graph of $y = f(-x-3)$ is a displacement to the right 3 units, then a reflection across the y-axis:

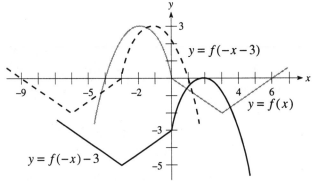

13.   This is the graph of $y = x^2$ displaced up 4 units:

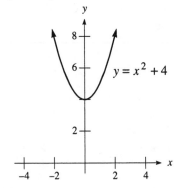

15. This is the graph of $y = x^2$ reflected across the *x*-axis, then displaced up 4 units:

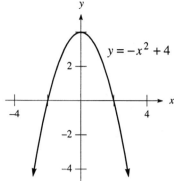

$$y = -x^2 + 4$$

17. This is the graph of $y = |x|$ displaced to the right 5 units:

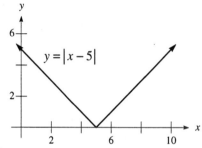

$$y = |x - 5|$$

19. This is the graph of $y = |x|$ reflected across the *x*-axis, then displaced up 3 units:

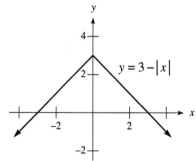

$$y = 3 - |x|$$

21. This is the graph of $y = x^2$ displaced to the left 1 unit, then displaced up 1 unit:

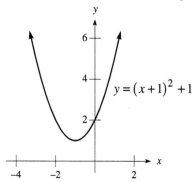

$$y = (x + 1)^2 + 1$$

23.   This is the graph of $y = |x|$ displaced to the right 9 units::

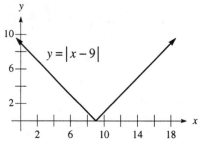

25.   This is the graph of $y = x^2$ displaced to the left 1 unit, then reflected across the x-axis:

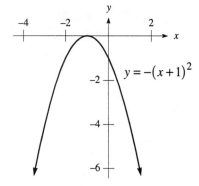

27.   Graphing the function:

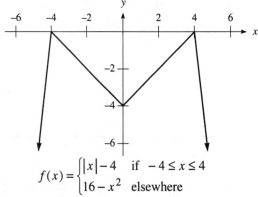

$$f(x) = \begin{cases} |x| - 4 & \text{if } -4 \le x \le 4 \\ 16 - x^2 & \text{elsewhere} \end{cases}$$

29.   The graph of $y = f(x)$ will be a reflection of $y = f(-x)$ across the y-axis:

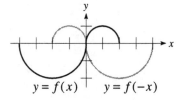

31.   The graph of $y = f(x)$ will be $y = -f(x) + 1$ displaced down 1 unit, then reflected across the x-axis:

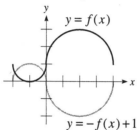

33.   This would be the graph of $y = f(x)$ displaced down 4 units.

35.   This would be the graph of $y = f(x)$ reflected across the y-axis.

37.   This would be the graph of $y = f(x)$ reflected across the x-axis, then displaced up 2 units.

39.   This would be the graph of $y = f(x)$ reflected across the x-axis, then displaced up 3 units.

41.   This would be the graph of $y = f(x)$ displaced to the left 3 units, then displaced up 4 units.

43.   This would be the graph of $y = f(x)$ displaced to the left 2 units, then reflected across the y-axis.

45.   The value is: $[6] = 6$

47.   The value is: $[-2.9] = -3$

49.   Graphing the function:

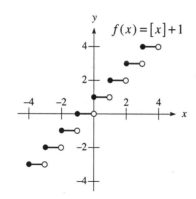

51.   Graphing the function:

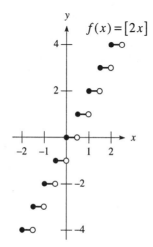

53.   Graphing the function:

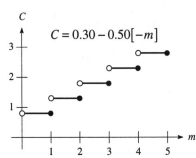

55.   Start with $y = x^2$, displace to the right h units, scaling by a factor of a (reflected across the x-axis if $a < 0$), then displace up k units.

3.3 Extracting Functions from Real-Life Situations

1.   **a.**   Using similar triangles:

$$\frac{y}{x+9} = \frac{6}{x}$$

$$y = \frac{6(x+9)}{x} = \frac{6x+54}{x}$$

The function is $y = \dfrac{6x+54}{x}$, for $x > 0$.

b.   Using the area formula: $A = \dfrac{1}{2}y(x+9) = \dfrac{1}{2} \cdot \dfrac{6(x+9)}{x}(x+9) = \dfrac{3(x+9)^2}{x}$

The function is $A = \dfrac{3(x+9)^2}{x}$, for $x > 0$.

3.   Using the Pythagorean theorem:

$$x^2 + 5^2 = (2r)^2$$
$$x^2 + 25 = 4r^2$$
$$x^2 = 4r^2 - 25$$
$$x = \sqrt{4r^2 - 25}$$

The area of the triangle is therefore: $A_\Delta = \frac{1}{2}(5)(x) = \frac{5}{2}\sqrt{4r^2 - 25}$

Since the shaded portion of the semicircle is the half-circle area minus this triangular area:

$$A = \tfrac{1}{2}\pi r^2 - \tfrac{5}{2}\sqrt{4r^2 - 25}$$

Since $4r^2 - 25 \geq 0$, the domain is $r \geq \frac{5}{2}$.

5.   Using the volume formula:

$$x^2 h = 80$$
$$h = \frac{80}{x^2}$$

The surface area is given by: $A = 2x^2 + 4xh = 2x^2 + 4x \cdot \dfrac{80}{x^2} = 2x^2 + \dfrac{320}{x} = \dfrac{2x^3 + 320}{x}$

The domain is $x > 0$.

7. The circumference is given by:
$$C = 2\pi r$$
$$r = \frac{C}{2\pi}$$

The area is given by: $A = \pi r^2 = \pi \left(\frac{C}{2\pi} \right)^2 = \frac{C^2}{4\pi}$

The domain is $C > 0$.

9. Using similar triangles:
$$\frac{s}{6} = \frac{s+x}{20}$$
$$20s = 6s + 6x$$
$$14s = 6x$$
$$s = \tfrac{3}{7}x$$

The domain is $x > 0$.

11. Using the slope formula:
$$\frac{b-1}{0-4} = \frac{0-1}{a-4}$$
$$(b-1)(a-4) = 4$$
$$b-1 = \frac{4}{a-4}$$
$$b = 1 + \frac{4}{a-4} = \frac{a-4}{a-4} + \frac{4}{a-4} = \frac{a}{a-4}$$

Using the area formula: $A = \frac{1}{2}ab = \frac{1}{2}a \cdot \frac{a}{a-4} = \frac{a^2}{2(a-4)}$

The domain is $a > 4$.

13. Let x represent the length of the rectangle, so:
$$2x + 2\pi r = 200$$
$$x + \pi r = 100$$
$$x = 100 - \pi r$$

Therefore the area is given by:
$$A = 2\left(\tfrac{1}{2}\pi r^2\right) + x \cdot 2r = \pi r^2 + 2r(100 - \pi r) = \pi r^2 + 200r - 2\pi r^2 = 200r - \pi r^2$$

Since $100 - \pi r > 0$, $r < \dfrac{100}{\pi}$, so the domain is $0 < r < \dfrac{100}{\pi}$.

15. The height of the balloon is $5t$ feet. Using the distance formula:
$$d^2 = (120)^2 + (5t)^2$$
$$d^2 = 14400 + 25t^2$$
$$d = \sqrt{14400 + 25t^2}$$
$$d = 5\sqrt{t^2 + 576}$$

The domain is $t \geq 0$.

17. a. Since s is the side of the square, $4s$ is the perimeter and thus $30 - 4s$ is the amount of the wire used for the circle. Since this represents the circumference of the circle:
$$2\pi r = 30 - 4s$$
$$r = \frac{30 - 4s}{2\pi}$$
$$r = \frac{15 - 2s}{\pi}$$

Thus the total combined area is given by:

$$A = s^2 + \pi r^2 = s^2 + \pi\left(\frac{15-2s}{\pi}\right)^2 = s^2 + \frac{(15-2s)^2}{\pi}$$

Since $r \geq 0$, $\frac{15-2s}{\pi} \geq 0$ and thus $s \leq \frac{15}{2}$. So the domain is $0 \leq s \leq \frac{15}{2}$.

b.   Since r is the radius of the circle, $2\pi r$ is the circumference and thus $30 - 2\pi r$ is the amount of wire used for the square. Since this represents the perimeter of the square:

$$4s = 30 - 2\pi r$$
$$s = \frac{30 - 2\pi r}{4}$$
$$s = \frac{15 - \pi r}{2}$$

Thus the total combined area is given by:

$$A = s^2 + \pi r^2 = \left(\frac{15-\pi r}{2}\right)^2 + \pi r^2 = \frac{(15-\pi r)^2}{4} + \pi r^2$$

Since $s \geq 0$, $\frac{15-\pi r}{2} \geq 0$ and thus $r \leq \frac{15}{\pi}$. So the domain is $0 \leq r \leq \frac{15}{\pi}$.

19.   The total revenue is given by: $R(n) = (3500 + 80n)(24 - 0.5n) = -40n^2 + 170n + 84000$
Since $24 - 0.5n \geq 0$, the domain is $0 \leq n \leq 48$.

21.   The total revenue is given by: $R(n) = (8500 - 225n)(9 + 0.5n) = -112.5n^2 + 2225n + 76,500$
Since $8500 - 225n \geq 0$, the domain is $0 \leq n \leq 37$.

23.   Let x represent each of the two equal sides, and $80 - 2x$ represent the third side. Using the area formula: $A(x) = x(80 - 2x) = 80x - 2x^2$
Since $80 - 2x > 0$, the domain is $0 < x < 40$.

25.   Let x represent the shorter sides (four sides) and y represent the longer sides (two sides). Since the total cost is $240:

$$5(2x + 2y) + 3(2x) = 240$$
$$10x + 10y + 6x = 240$$
$$16x + 10y = 240$$
$$5y + 8x = 120$$
$$5y = 120 - 8x$$
$$y = \frac{120 - 8x}{5}$$

The area is given by: $A(x) = xy = x\left(\frac{120 - 8x}{5}\right) = \frac{120x - 8x^2}{5} = -\frac{8}{5}x^2 + 24x$

Since $\frac{120 - 8x}{5} > 0$, $x < 15$, so the domain is $0 < x < 15$.

27. a.   The area is 400 sq. feet, therefore:

$$(2x)(2y) - xy = 400$$
$$3xy = 400$$
$$y = \frac{400}{3x}$$

The length of the partition is given by:

$$L = 2x + 2y + 2x + 2y = 4x + 4\left(\frac{400}{3x}\right) = 4x + \frac{1600}{3x} = \frac{12x^2 + 1600}{3x}$$

The domain is $x > 0$.

b.   The total cost is given by: $C = 18L = 18\left(\dfrac{12x^2 + 1600}{3x}\right) = \dfrac{72x^2 + 9600}{x}$

Again the domain is $x > 0$.

29.   Let l represent the length of AC. Using the Pythagorean theorem:

$$l^2 = 1^2 + x^2$$
$$l = \sqrt{x^2 + 1}$$

The total length of the cable is given by: $L(x) = l + (6 - x) = \sqrt{x^2 + 1} + 6 - x$
The domain is $0 \le x \le 6$.

31.   The interest can be written as:

$$I = \begin{cases} 0.0165b & \text{if } b \le 2000 \\ 0.0165(2000) + 0.0125(b - 2000) & \text{if } b > 2000 \end{cases}$$

This simplifies to:

$$I = \begin{cases} 0.0165b & \text{if } b \le 2000 \\ 0.0125b + 8 & \text{if } b > 2000 \end{cases}$$

33.   The volume of the rain gutter is given by: $V(x) = 18(x)(1 - 2x) = 18x - 36x^2$
Since $1 - 2x > 0$, $x < \frac{1}{2}$, so the domain is $0 < x < \frac{1}{2}$.

35.   The surface area is given by:

$$S = 4\pi r^2$$
$$r^2 = \frac{S}{4\pi}$$
$$r = \sqrt{\frac{S}{4\pi}}$$

The volume is given by: $V = \dfrac{4}{3}\pi r^3 = \dfrac{4}{3}\pi\left(\sqrt{\dfrac{S}{4\pi}}\right)^3 = \dfrac{4\pi S\sqrt{S}}{3 \cdot 8\pi\sqrt{\pi}} \cdot \dfrac{\sqrt{\pi}}{\sqrt{\pi}} = \dfrac{4\pi S\sqrt{\pi S}}{24\pi^2} = \dfrac{S\sqrt{\pi S}}{6\pi}$

The domain is $S \ge 0$.

37.   The volume is given by:

$$\pi r^2 \bullet 10 = V$$
$$r^2 = \frac{V}{10\pi}$$
$$r = \sqrt{\frac{V}{10\pi}}$$

The surface area is given by:

$$S = 2\pi rh + 2\pi r^2 = 20\pi\sqrt{\dfrac{V}{10\pi}} + 2\pi \bullet \dfrac{V}{10\pi} = \dfrac{20\pi\sqrt{10\pi V}}{10\pi} + \dfrac{V}{5} = 2\sqrt{10\pi V} + \dfrac{V}{5}$$

The domain is $V \ge 0$.

39.   The distance is given by:

$$D = \sqrt{(x + 1)^2 + (y - 2)^2}$$
$$= \sqrt{(x + 1)^2 + (9 - x^2 - 2)^2}$$
$$= \sqrt{(x + 1)^2 + (7 - x^2)^2}$$
$$= \sqrt{x^2 + 2x + 1 + 49 - 14x^2 + x^4}$$
$$= \sqrt{x^4 - 13x^2 + 2x + 50}$$

41. The distance is given by:

$$D = \sqrt{(x-5)^2 + (y-3)^2} = \sqrt{3-y+(y-3)^2} = \sqrt{3-y+y^2-6y+9} = \sqrt{y^2-7y+12}$$

The domain is $y \le 3$.

43. The distance is given by:

$$D = \sqrt{(x-0)^2 + (y-2)^2} = \sqrt{x^2 + \left(-\sqrt{x}\right)^2} = \sqrt{x^2+x}$$

The domain is $x \ge 0$.

3.4 Quadratic Functions

1. The axis is $x = 0$, the vertex is $(0,0)$, and the x- and y-intercepts are both 0. Sketching the graph:

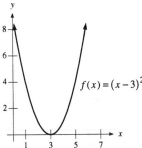

3. The axis is $x = 3$, the vertex is $(3,0)$, the x-intercept is 3, and the y-intercept is 9. Sketching the graph:

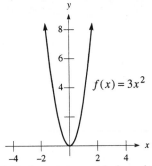

5. The axis is $x = 0$, the vertex is (0.8), there is no x-intercept, and the y-intercept is 8. Sketching the graph:

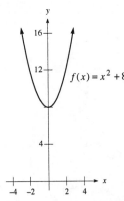

7.   The axis is $x = 3$, the vertex is (3,1), there is no *x*-intercept, and the *y*-intercept is 10. Sketching the graph:

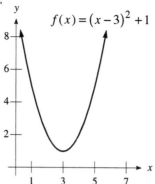

$$f(x) = (x-3)^2 + 1$$

9.   The axis is $x = 5$, the vertex is (5,4), the *x*-intercepts are $\dfrac{15 \pm 2\sqrt{3}}{3}$, and the *y*-intercept is –71. Sketching the graph:

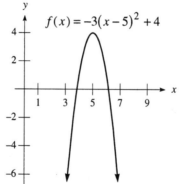

$$f(x) = -3(x-5)^2 + 4$$

11 .   The axis is $x = 4$, the vertex is (4,3), there is no *x*-intercept, and the *y*-intercept is 35. Sketching the graph:

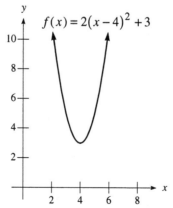

$$f(x) = 2(x-4)^2 + 3$$

13.   The axis is $x = 0$, the vertex is $(0,-4)$, the x-intercepts are ± 2, and the y-intercept is -4. Sketching the graph:

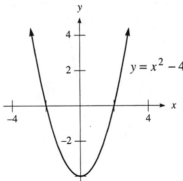

15.   The x-intercept is 4 and the y-intercept is -4. Sketching the graph:

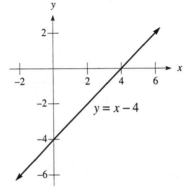

17.   The axis is $x = 0$, the vertex is $(0,10)$, the x-intercepts are $\pm\sqrt{5}$, and the y-intercept is 10. Sketching the graph:

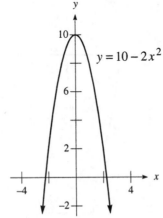

19.   Completing the square: $f(x) = 3x^2 - 9x = 3(x^2 - 3x) = 3\left(x^2 - 3x + \frac{9}{4}\right) - \frac{27}{4} = 3\left(x - \frac{3}{2}\right)^2 - \frac{27}{4}$

The axis is $x = \frac{3}{2}$, the vertex is $\left(\frac{3}{2}, -\frac{27}{4}\right)$, the x-intercepts are 0 and 3, and the y-intercept is 0.

Sketching the graph:

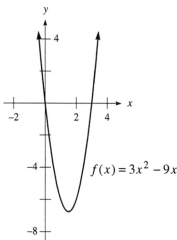

21.   Completing the square: $y = x^2 - 2x - 8 = 4(x^2 + 2x) = (x^2 - 2x + 1) - 1 - 8 = (x - 1)^2 - 9$

The axis is $x = 1$, the vertex is (1,–9), the x-intercepts are –2 and 4, and the y-intercept is –8.

Sketching the graph:

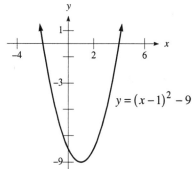

23.   Completing the square: $f(x) = x^2 - 10x + 21 = (x^2 - 10x + 25) - 25 + 21 = (x - 5)^2 - 4$

The axis is $x = 5$, the vertex is (5,–4), the x-intercepts are 3 and 7, and the y-intercept is 21.

Sketching the graph:

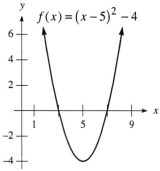

25. Completing the square: $y = 2x^2 + 12x + 16 = 2(x^2 + 6x + 9) - 18 + 16 = 2(x+3)^2 - 2$

The axis is $x = -3$, the vertex is $(-3,-2)$, the x-intercepts are -4 and -2, and the y-intercept is 16. Sketching the graph:

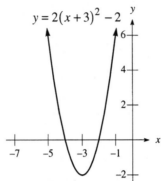

27. Completing the square: $f(x) = -2x^2 + 6x - 4 = -2(x^2 - 3x + \frac{9}{4}) + \frac{9}{2} - 4 = -2(x - \frac{3}{2})^2 + \frac{1}{2}$

The axis is $x = \frac{3}{2}$, the vertex is $(\frac{3}{2}, \frac{1}{2})$, the x-intercepts are 1 and 2, and the y-intercept is -4. Sketching the graph:

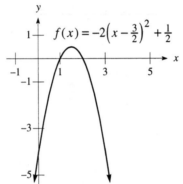

29. The axis is $x = 0$, the vertex is $(0,-1)$ the x-intercepts are $\pm\frac{1}{5}$, and the y-intercept is -1. Sketching the graph:

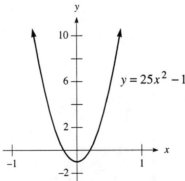

31. The minimum value is -9, which occurs when $x = 0$.
33. The maximum value is 9, which occurs when $x = 3$.

35. Using the vertex formula: $x = \dfrac{-4}{2(-1)} = 2$

The maximum value is: $f(2) = -4 + 8 + 32 = 36$

37. Using the vertex formula: $x = \dfrac{-6}{2(-2)} = \frac{3}{2}$

The maximum value is: $f\left(\frac{3}{2}\right) = -\frac{9}{2} + 9 - 18 = -\frac{27}{2}$

39. Using the vertex formula: $x = \dfrac{4}{2(1)} = 2$

The minimum value is: $f(2) = 4 - 8 - 32 = -36$

41. Using the vertex formula: $x = \dfrac{6}{2(2)} = \frac{3}{2}$

The minimum value is: $f\left(\frac{3}{2}\right) = \frac{9}{2} - 9 + 18 = \frac{27}{2}$

43. **a.** Sketching the graph:

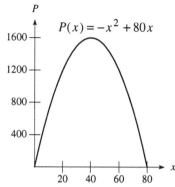

b. Using the vertex formula: $x = \dfrac{-80}{2(-1)} = 40$

Thus 40 items must be produced daily to maximize the profit.

c. The maximum profit is: $P(40) = -1600 + 3200 = \$1600$

45. Using the vertex formula: $x = \dfrac{0.48}{2(0.02)} = 12$

The minimum value is: $C(12) = 2.88 - 5.76 + 428.8 = 425.92$

The manufacturer should produce 12 tables to minimize the daily production cost at $425.92.

47. Using the vertex formula: $t = \dfrac{-864}{2(-16)} = 27$

The maximum value is: $s(27) = -11664 + 23328 = 11664$

It will take 27 seconds for the rocket to reach a maximum height of 11,664 feet.

49. If x is the dimension of one side and y is the dimension of the other side, then:

$$2x + 2y = 100$$
$$x + y = 50$$
$$y = 50 - x$$

The area is given by: $A = xy = x(50 - x) = -x^2 + 50x$

Using the vertex formula: $x = \dfrac{-50}{2(-1)} = 25$

Then $y = 50 - 25 = 25$, so the dimensions are 25 feet by 25 feet.

51.   Let x and $104 - x$ represent the two numbers. Their product is given by:
$$P = x(104 - x) = -x^2 + 104x$$
Using the vertex formula: $x = \dfrac{-104}{2(-1)} = 52$

The other number is $104 - 52 = 52$, so the two numbers are 52 and 52.

53.   Using the dimensions from the figure:
$$3x + 2y = 800$$
$$2y = 800 - 3x$$
$$y = 400 - \tfrac{3}{2}x$$

The area is given by: $A = xy = x\left(400 - \tfrac{3}{2}x\right) = -\tfrac{3}{2}x^2 + 400x$

Using the vertex formula: $x = \dfrac{-400}{2\left(-\frac{3}{2}\right)} = \dfrac{400}{3} = 133\tfrac{1}{3}$

Then $y = 400 - \tfrac{3}{2}\left(\tfrac{400}{3}\right) = 200$. The dimensions are $x = 133\tfrac{1}{3}$ feet and $y = 200$ feet.

55.   **a.**   Setting $p = 7$:
$$10 - \frac{x}{5} = 7$$
$$-\frac{x}{5} = -3$$
$$x = 15$$
So 15 items would be sold if the price were $7 per unit.

b.   Setting $x = 25$: $p = 10 - \dfrac{25}{5} = 10 - 5 = 5$

The price is $5 per unit if 25 units are to be sold.

c.   Sketching the graph:

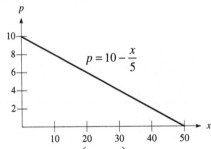

d.   The revenue function is: $R(x) = x\left(10 - \dfrac{x}{5}\right) = -\tfrac{1}{5}x^2 + 10x$

e.   Sketching the graph:

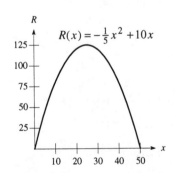

f.   Using the vertex formula: $x = \dfrac{-10}{2\left(-\frac{1}{5}\right)} = 25$

Then $p = 10 - \dfrac{25}{5} = 5$. So 25 items should be produced at a price of $5 each to maximize the revenue.

57.   The total perimeter is $2x + 2r + \frac{1}{2}(2\pi r) = 2x + 2r + \pi r$, so:

$$2x + 2r + \pi r = 12$$
$$2x + (2 + \pi)r = 12$$
$$2x = 12 - (2 + \pi)r$$
$$x = 6 - \frac{2 + \pi}{2} r$$

The total area is:

$$A = x(2r) + \frac{1}{2}\left(\pi r^2\right)$$
$$= 2rx + \frac{1}{2}\pi r^2$$
$$= 2r\left(6 - \frac{2 + \pi}{2} r\right) + \frac{1}{2}\pi r^2$$
$$= 12r - (2 + \pi)r^2 + \frac{1}{2}\pi r^2$$
$$= \left(-2 - \frac{\pi}{2}\right)r^2 + 12r$$

Using the vertex formula: $r = \dfrac{-12}{2\left(-2 - \dfrac{\pi}{2}\right)} = \dfrac{-12}{-4 - \pi} = \dfrac{12}{4 + \pi}$

Therefore: $x = 6 - \dfrac{2 + \pi}{2} \cdot \dfrac{12}{4 + \pi} = 6 - \dfrac{6(2 + \pi)}{4 + \pi} = \dfrac{24 + 6\pi - 12 - 6\pi}{4 + \pi} = \dfrac{12}{4 + \pi}$

The window will have a maximum area when $r = x = \dfrac{12}{4 + \pi}$. The maximum area is:

$$A = 2\left(\frac{12}{4 + \pi}\right)\left(\frac{12}{4 + \pi}\right) + \frac{1}{2}\pi\left(\frac{12}{4 + \pi}\right)^2 = \frac{288}{(4 + \pi)^2} + \frac{72\pi}{(4 + \pi)^2} = \frac{72(4 + \pi)}{(4 + \pi)^2} = \frac{72}{4 + \pi} \text{ ft}^2$$

59.   Let x represent the number of unoccupied units and $200 - x$ represent the number of occupied units. Let n represent the number of $30 rent increases, so $x = 5n$. The total rent collected is therefore:

$$R = (200 - x)(800 + 30n) = (200 - 5n)(800 + 30n) = -150n^2 + 2000n + 160000$$

The total cost is: $C = 320(200 - x) + 170(x) = 320(200 - 5n) + 170(5n) = 64000 - 750n$

The total profit is therefore:

$$P = R - C = \left(-150n^2 + 2000n + 160000\right) - (64000 - 750n) = -150n^2 + 2750n + 96000$$

Using the vertex formula: $n = \dfrac{-2750}{2(-150)} = \frac{55}{6} = 9\frac{1}{6}$

The company should have $9\frac{1}{6}$ rent increases to maximize profit, so the rent should be:

$$\$800 + \left(9\frac{1}{6}\right)(30) = \$800 + \$275 = \$1,075$$

61.   **a.**   Let n represent the number of $0.25 decreases in fare. The revenue is therefore:

$$R = (4 - 0.25n)(36000 + 3000n) = -750n^2 + 3000n + 144000$$

Using the vertex formula: $n = \dfrac{-3000}{2(-750)} = 2$

The transit authority should have 2 fare decreases, so the fare should be $4.00 - 2(0.25) = \$3.50$.

b.   Using a limit of 40,000 passengers:

$$36000 + 3000n = 40000$$
$$3000n = 4000$$
$$n = \tfrac{4}{3}$$

The transit authority should have $\tfrac{4}{3}$ fare decreases, so the fare should be:

$$\$4.00 - \tfrac{4}{3}(0.25) = \$3.67$$

63 .   Since $g(w) = -kw^2 + 21kw$, using the vertex formula: $w = \dfrac{-21k}{-2k} = 10.5$

The growth is a maximum when the infant's weight is 10.5 lb.

65 .   Substituting: $f\left(\dfrac{-B}{2A}\right) = A\left(\dfrac{-B}{2A}\right)^2 + B\left(\dfrac{-B}{2A}\right) + C = \dfrac{B^2}{4A} - \dfrac{B^2}{2A} + C = \dfrac{B^2 - 2B^2 + 4AC}{4A} = \dfrac{4AC - B^2}{4A}$

67 .   a.   Since $A > 0$, the parabola must face upward. Since $B^2 - 4AC < 0$, there are no real
x-intercepts. Such a parabola is:

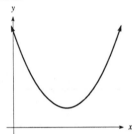

b.   Since $A < 0$, the parabola must face downward. Since $B^2 - 4AC < 0$, there are no real
x-intercepts. Such a parabola is:

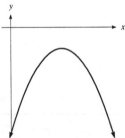

c.   No. Since $A > 0$, the parabola must open upward. If the vertex were below the x-axis, there
would be two x-intercepts and thus $B^2 - 4AC > 0$.

69 .   Sketching the graph:

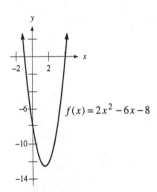

$$f(x) = 2x^2 - 6x - 8$$

a. Setting $f(x) = 0$:
$$2x^2 - 6x - 8 = 0$$
$$x^2 - 3x - 4 = 0$$
$$(x - 4)(x + 1) = 0$$
$$x = -1, 4$$

b. Looking at the graph, note that $f(x) > 0$ on the intervals $(-\infty, -1) \cup (4, \infty)$.

c. Looking at the graph, note that $f(x) < 0$ on the interval $(-1, 4)$.

d. The intervals are the same.

71. a. The maximum value is 4 (when $x = 0$).

b. The maximum value is $\sqrt{4} = 2$ (when $x = 0$).

73. a. Using the vertex formula: $x = \dfrac{-(-4)}{2(1)} = 2$

The minimum value is: $f(2) = 4 - 8 + 7 = 3$

b. The maximum value is $\frac{1}{3}$ (when $x = 2$).

75. Using the quadratic formula: $x = \dfrac{-B \pm \sqrt{B^2 - 4AC}}{2A}$

Now using the midpoint formula:

$$x = \frac{1}{2}\left(\frac{-B - \sqrt{B^2 - 4AC}}{2A} + \frac{-B + \sqrt{B^2 - 4AC}}{2A} \right) = \frac{1}{2}\left(\frac{-2B}{2A} \right) = -\frac{B}{2A}$$

77. Setting $y = 0$:
$$a(x - h)^2 + k = 0$$
$$a(x - h)^2 = -k$$
$$(x - h)^2 = -\frac{k}{a}$$
$$x - h = \pm\sqrt{-\frac{k}{a}}$$
$$x = h \pm \sqrt{-\frac{k}{a}}$$

The signs of a and k must be opposites for these x-intercepts to exist. That is, if the vertex of the parabola is above the x-axis ($k > 0$), the parabola must be pointed down ($a < 0$). If the vertex of the parabola is below the x-axis ($k < 0$), the parabola must be pointed up ($a > 0$).

3.5 Operations of Functions

1. Finding the value: $(g + r)(2) = g(2) + r(2) = [3(2) - 2] + [2^3 - 1] = 4 + 7 = 11$

3. Finding the value: $(h \cdot s)(-6) = h(-6) \cdot s(-6) = 5 \cdot \left(-\frac{1}{6}\right) = -\frac{5}{6}$

5. Finding the value: $\left(\dfrac{g}{f}\right)(4) = \dfrac{g(4)}{f(4)} = \dfrac{3(4) - 2}{2(4)^2 - 4 - 3} = \dfrac{10}{25} = \dfrac{2}{5}$

7. Finding the value: $\left(\dfrac{h}{r}\right)(-1) = \dfrac{h(-1)}{r(-1)} = \dfrac{5}{(-1)^3 - 1} = -\dfrac{5}{2}$

9. Finding the value: $(f \circ g)(4) = f[g(4)] = f(10) = 200 - 10 - 3 = 187$

11. Finding the value: $(h \circ s)(-6) = h[s(-6)] = h\left(-\frac{1}{6}\right) = 5$

13. Finding the value: $(f + g)(x) = f(x) + g(x) = 2x^2 - x - 3 + 3x - 2 = 2x^2 + 2x - 5$

15. Finding the function:

$$(f - g)(x) = f(x) - g(x) = \left(2x^2 - x - 3\right) - (3x - 2) = 2x^2 - x - 3 - 3x + 2 = 2x^2 - 4x - 1$$

The domain is all real numbers, or $(-\infty, \infty)$.

17. Finding the function: $\left(\dfrac{s}{h}\right)(x) = \dfrac{s(x)}{h(x)} = \dfrac{\frac{1}{x}}{5} = \dfrac{1}{5x}$

The domain is all real numbers except 0, or $(-\infty, 0) \cup (0, \infty)$.

19. Finding the function:

$$(r - f)(x) = r(x) - f(x) = \left(x^3 - 1\right) - \left(2x^2 - x - 3\right) = x^3 - 1 - 2x^2 + x + 3 = x^3 - 2x^2 + x + 2$$

The domain is all real numbers, or $(-\infty, \infty)$.

21. Finding the function: $(g \bullet t)(x) = g(x) \bullet t(x) = (3x - 2) \bullet \dfrac{4}{x + 2} = \dfrac{12x - 8}{x + 2}$

The domain is all real numbers except -2, or $(-\infty, -2) \cup (-2, \infty)$.

23. Finding the function:

$$\begin{aligned}(f \circ g)(x) &= f(g(x)) \\ &= f(3x - 2) \\ &= 2(3x - 2)^2 - (3x - 2) - 3 \\ &= 18x^2 - 24x + 8 - 3x + 2 - 3 \\ &= 18x^2 - 27x + 7\end{aligned}$$

The domain is all real numbers, or $(-\infty, \infty)$.

25. Finding the function: $(r \circ s)(x) = r(s(x)) = r\left(\dfrac{1}{x}\right) = \left(\dfrac{1}{x}\right)^3 - 1 = \dfrac{1}{x^3} - 1 = \dfrac{1 - x^3}{x^3}$

The domain is all real numbers except 0, or $(-\infty, 0) \cup (0, \infty)$.

27. Finding the function: $(h \circ t)(x) = h(t(x)) = h\left(\dfrac{4}{x + 2}\right) = 5$

The domain is all real numbers except -2, or $(-\infty, -2) \cup (-2, \infty)$.

29. Finding the function:

$$\begin{aligned}(r \circ g)(x) &= r(g(x)) \\ &= r(3x - 2) \\ &= (3x - 2)^3 - 1 \\ &= 27x^3 - 54x^2 + 36x - 8 - 1 \\ &= 27x^3 - 54x^2 + 36x - 9\end{aligned}$$

The domain is all real numbers, or $(-\infty, \infty)$.

31. Finding the function: $(s \circ t)(x) = s(t(x)) = s\left(\dfrac{4}{x + 2}\right) = \dfrac{1}{\frac{4}{x + 2}} = \dfrac{x + 2}{4}$

The domain is all real numbers except -2, or $(-\infty, -2) \cup (-2, \infty)$.

33. **a.** Finding the function: $(f \circ g)(x) = f(2x - 7) = \sqrt{2x - 7 + 1} = \sqrt{2x - 6}$

The domain is $x \geq 3$, or $[3, \infty)$.

   b. Finding the function: $(g \circ f)(x) = g\left(\sqrt{x + 1}\right) = 2\sqrt{x + 1} - 7$

The domain is $x \geq -1$, or $[-1, \infty)$.

35. **a.**   Finding the function:

$$(f \circ g)(t) = f\left(\frac{6}{t-3}\right) = \left(\frac{6}{t-3}\right)^2 + \frac{6}{t-3} = \frac{36}{(t-3)^2} + \frac{6(t-3)}{(t-3)^2} = \frac{6t+18}{(t-3)^2}$$

b.   Finding the function: $(g \circ f)(t) = g\left(t^2 + t\right) = \dfrac{6}{t^2 + t - 3}$

c.   Finding the function:

$$(f \circ f)(t) = f\left(t^2 + t\right) = \left(t^2 + t\right)^2 + \left(t^2 + t\right) = t^4 + 2t^3 + t^2 + t^2 + t = t^4 + 2t^3 + 2t^2 + t$$

d.   Finding the function:

$$(g \circ g)(t) = g\left(\frac{6}{t-3}\right) = \frac{6}{\dfrac{6}{t-3} - 3} = \frac{6(t-3)}{6 - 3(t-3)} = \frac{6(t-3)}{6 - 3t + 9} = \frac{6(t-3)}{3(5-t)} = \frac{2t-6}{5-t}$$

37. **a.**   Finding the function: $(h \circ g \circ f)(x) = h[g(2x+3)] = h\left(\sqrt{2x+3}\right) = \dfrac{1}{\sqrt{2x+3}}$

b.   Finding the function: $(f \circ h \circ g)(x) = f\left[h\left(\sqrt{x}\right)\right] = f\left(\dfrac{1}{\sqrt{x}}\right) = \dfrac{2}{\sqrt{x}} + 3 = \dfrac{2 + 3\sqrt{x}}{\sqrt{x}} = \dfrac{2\sqrt{x} + 3x}{x}$

c.   Finding the function:

$$(g \circ f \circ h)(x) = g\left[f\left(\frac{1}{x}\right)\right] = g\left(\frac{2}{x} + 3\right) = \sqrt{\frac{2}{x} + 3} = \frac{\sqrt{2+3x}}{\sqrt{x}} = \frac{\sqrt{3x^2 + 2x}}{x}$$

39.   Using $f(x) = x^3$ and $g(x) = x + 3$, then $h(x) = (f \circ g)(x)$.

41.   Using $f(x) = x^2$ and $g(x) = \dfrac{x+4}{x-1}$, then $h(x) = (f \circ g)(x)$.

43.   Using $f(x) = x + 6$ and $g(x) = \dfrac{1}{x}$, then $h(x) = (f \circ g)(x)$.

45.   Since $f(x) = 5x - 3$:

$$f(g(x)) = 2x + 7$$
$$5g - 3 = 2x + 7$$
$$5g = 2x + 10$$
$$g = \tfrac{2}{5}x + 2$$

So $g(x) = \tfrac{2}{5}x + 2$.

47.   Since $f(x) = 8 - 5x$:

$$f(g(x)) = x$$
$$8 - 5g = x$$
$$-5g = x - 8$$
$$g = -\tfrac{1}{5}x + \tfrac{8}{5}$$

So $g(x) = -\tfrac{1}{5}x + \tfrac{8}{5}$.

49. **a.**   Finding the composition:

$$(N \circ C)(h) = N(5h + 15)$$
$$= 3(5h + 15)^2 + 250(5h + 15) + 10,200$$
$$= 75h^2 + 450h + 675 + 1250h + 3750 + 10,200$$
$$= 75h^2 + 1700h + 14,625$$

b.   Substituting $h = 4$: $(N \circ C)(4) = 75(4)^2 + 1700(4) + 14,625 = 22,625$ bacteria

c.   Finding when $(N \circ C)(h) = 300,000$:

$$75h^2 + 1700h + 14,625 = 300,000$$
$$75h^2 + 1700h - 285,375 = 0$$
$$3h^2 + 68h - 11,415 = 0$$

Using the quadratic formula:

$$h = \frac{-68 \pm \sqrt{(68)^2 - 4(3)(-11415)}}{2(3)} = \frac{-68 \pm \sqrt{141604}}{6} \approx -74.1, 51.4$$

Neither of these values is within the domain $0 \le h \le 5$.

51. a.   The cost is $C(1) = 32(1) + 370 = 402$, so the price will be $P(402) = 1.3(402) = \$522.60$.

b.   The cost is $C(10) = 32(10) + 370 = 690$, so the price for 10 would be
$P(690) = 1.3(690) = \$897$. Since this is the price for 10 calculators, the price for each would
be $\dfrac{\$897}{10} = \$89.70$.

c.   The cost is $C(n) = 32n + 370$, so the price for n would be
$P(32n + 370) = 1.3(32n + 370) = 41.6n + 481$. Since this is the price for n calculators, the
price of each would be $P = \dfrac{41.6n + 481}{n} = 41.6 + \dfrac{481}{n}$.

53.   One example would be $f(x) = 2x$ and $g(x) = \frac{1}{2}x$. Then $(f \circ g)(x) = x$ and $(g \circ f)(x) = x$. In
general, these types of functions are called inverse functions, which will be explored in the next
section.

3.6 Inverse Functions

1.   The graph passes the horizontal line test, so it has an inverse.

3.   The graph fails the horizontal line test, so it does not have an inverse.

5.   The graph passes the horizontal line test, so it has an inverse.

7.   Finding the compositions:

$$(f \circ g)(x) = f(x+1) = x+1-1 = x$$
$$(g \circ f)(x) = g(x-1) = x-1+1 = x$$

Since $(f \circ g)(x) = (g \circ f)(x)$, f and g are inverse functions.

9.   Finding the compositions:

$$(f \circ g)(x) = f\left(\sqrt[3]{x}\right) = \left(\sqrt[3]{x}\right)^3 = x$$
$$(g \circ f)(x) = g\left(x^3\right) = \sqrt[3]{x^3} = x$$

Since $(f \circ g)(x) = (g \circ f)(x)$, f and g are inverse functions.

11.   Finding the compositions:

$$(f \circ g)(x) = f\left(3x - \tfrac{6}{5}\right) = \tfrac{1}{3}\left(3x - \tfrac{6}{5}\right) + \tfrac{2}{5} = x - \tfrac{2}{5} + \tfrac{2}{5} = x$$
$$(g \circ f)(x) = g\left(\tfrac{1}{3}x + \tfrac{2}{5}\right) = 3\left(\tfrac{1}{3}x + \tfrac{2}{5}\right) - \tfrac{6}{5} = x + \tfrac{6}{5} - \tfrac{6}{5} = x$$

Since $(f \circ g)(x) = (g \circ f)(x)$, f and g are inverse functions.

13.   Finding the compositions:

$$(f \circ g)(x) = f\left(\frac{4x}{4-x}\right) = \frac{\cdot 4\left(\dfrac{4x}{4-x}\right)}{\dfrac{4x}{4-x}+4} \cdot \frac{4-x}{4-x} = \frac{16x}{4x+16-4x} = \frac{16x}{16} = x$$

$$(g \circ f)(x) = g\left(\frac{4x}{x+4}\right) = \frac{4\left(\dfrac{4x}{x+4}\right)}{4-\dfrac{4x}{x+4}} \cdot \frac{x+4}{x+4} = \frac{16x}{4x+16-4x} = \frac{16x}{16} = x$$

Since $(f \circ g)(x) = (g \circ f)(x)$, f and g are inverse functions.

15.   Finding the composition: $(f \circ f)(x) = f\left(\dfrac{1}{x}\right) = \dfrac{1}{\dfrac{1}{x}} = x$

Since $(f \circ f)(x) = x$, f is its own inverse.

17.   Switching the roles of x and y:
$$5y - 9 = x$$
$$5y = x + 9$$
$$y = \tfrac{1}{5}x + \tfrac{9}{5}$$
So $f^{-1}(x) = \tfrac{1}{5}x + \tfrac{9}{5}$.

19.   Switching the roles of x and y:
$$2y^3 + 1 = x$$
$$2y^3 = x - 1$$
$$y^3 = \frac{x-1}{2}$$
$$y = \sqrt[3]{\frac{x-1}{2}} \cdot \frac{\sqrt[3]{4}}{\sqrt[3]{4}} = \frac{\sqrt[3]{4x-4}}{2}$$
So $F^{-1}(x) = \dfrac{\sqrt[3]{4x-4}}{2}$.

21.   Switching the roles of x and y:
$$\frac{1}{y+4} = x$$
$$y+4 = \frac{1}{x}$$
$$y = \frac{1}{x} - 4$$
$$y = \frac{1-4x}{x}$$
So $h^{-1}(x) = \dfrac{1-4x}{x}$.

23. Switching the roles of x and y:
$$2\sqrt{y} - 7 = x$$
$$2\sqrt{y} = x + 7$$
$$4y = (x+7)^2 \qquad \text{(for } x \geq -7\text{)}$$
$$y = \frac{(x+7)^2}{4}$$

So $f^{-1}(x) = \dfrac{(x+7)^2}{4}$ for $x \geq -7$.

25. Switching the roles of x and y:
$$y^2 - 1 = x$$
$$y^2 = x + 1$$
$$y = \sqrt{x+1}$$

So $h^{-1}(x) = \sqrt{x+1}$.

27. Switching the roles of x and y:
$$\frac{2y+5}{3y-2} = x$$
$$2y + 5 = 3xy - 2x$$
$$2x + 5 = 3xy - 2y$$
$$2x + 5 = y(3x - 2)$$
$$y = \frac{2x+5}{3x-2}$$

So $g^{-1}(x) = \dfrac{2x+5}{3x-2}$.

29. Switching the roles of x and y:
$$y^{3/5} = x$$
$$y = x^{5/3}$$

So $f^{-1}(x) = x^{5/3}$.

31. Switching the roles of x and y:
$$y^{-5/7} + 1 = x$$
$$y^{-5/7} = x - 1$$
$$y = (x-1)^{-7/5}$$

So $f^{-1}(x) = (x-1)^{-7/5}$.

33. Switching the roles of x and y:
$$7y - 6 = x$$
$$7y = x + 6$$
$$y = \tfrac{1}{7}x + \tfrac{6}{7}$$

So $f^{-1}(x) = \tfrac{1}{7}x + \tfrac{6}{7}$. Now finding the compositions:
$$f^{-1}[f(x)] = f^{-1}(7x-6) = \tfrac{1}{7}(7x-6) + \tfrac{6}{7} = x - \tfrac{6}{7} + \tfrac{6}{7} = x$$
$$f[f^{-1}(x)] = f\left(\tfrac{1}{7}x + \tfrac{6}{7}\right) = 7\left(\tfrac{1}{7}x + \tfrac{6}{7}\right) - 6 = x + 6 - 6 = x$$

Thus $f[f^{-1}(x)] = f^{-1}[f(x)] = x$.

35. Switching the roles of x and y:

$$1 - \frac{3}{y} = x$$

$$-\frac{3}{y} = x - 1$$

$$-\frac{y}{3} = \frac{1}{x-1}$$

$$y = \frac{-3}{x-1} = \frac{3}{1-x}$$

So $f^{-1}(x) = \frac{3}{1-x}$. Now finding the compositions:

$$f^{-1}[f(x)] = f^{-1}\left(1 - \frac{3}{x}\right) = \frac{3}{1 - \left(1 - \frac{3}{x}\right)} = \frac{3}{\frac{3}{x}} = x$$

$$f\left[f^{-1}(x)\right] = f\left(\frac{3}{1-x}\right) = 1 - \frac{3}{\frac{3}{1-x}} = 1 - (1-x) = 1 - 1 + x = x$$

Thus $f\left[f^{-1}(x)\right] = f^{-1}[f(x)] = x$.

37. Switching the roles of x and y:

$$\frac{5y+3}{1-2y} = x$$

$$5y + 3 = x - 2xy$$

$$5y + 2xy = x - 3$$

$$y(2x+5) = x - 3$$

$$y = \frac{x-3}{2x+5}$$

So $f^{-1}(x) = \frac{x-3}{2x+5}$. Now finding the compositions:

$$f^{-1}[f(x)] = f^{-1}\left(\frac{5x+3}{1-2x}\right) = \frac{\frac{5x+3}{1-2x} - 3}{2\left(\frac{5x+3}{1-2x}\right) + 5} = \frac{5x+3-3+6x}{10x+6+5-10x} = \frac{11x}{11} = x$$

$$f\left[f^{-1}(x)\right] = f\left(\frac{x-3}{2x+5}\right) = \frac{5\left(\frac{x-3}{2x+5}\right)+3}{1-\left(\frac{x-3}{2x+5}\right)} = \frac{5x-15+6x+15}{2x+5-2x+6} = \frac{11x}{11} = x$$

Thus $f\left[f^{-1}(x)\right] = f^{-1}[f(x)] = x$.

39. The inverse is $f^{-1}(x) = \sqrt{x}$. The restriction is needed so that the positive square root can be used. Note that domain of f = range of $f^{-1} = [0, \infty)$ and range of f = domain of $f^{-1} = [0, \infty)$.

41.   The function has an inverse. Sketching the graph:

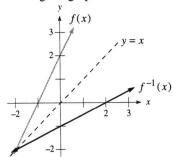

43.   The function has an inverse. Sketching the graph:

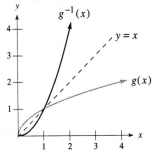

45.   The function has an inverse. Sketching the graph:

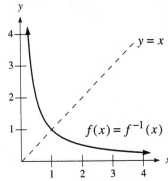

47.   The function has an inverse. Sketching the graph:

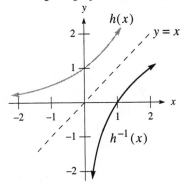

49. Finding $f(2)$: $f(2) = -2(2) + 7 = -4 + 7 = 3$
To find $f^{-1}(2)$, find where $f(x) = 2$:
$$-2x + 7 = 2$$
$$-2x = -5$$
$$x = \frac{5}{2}$$
So $f^{-1}(2) = \frac{5}{2}$.

51. Finding $f(2)$: $f(2) = 2^2 - 2(2) - 6 = 4 - 4 - 6 = -6$
To find $f^{-1}(2)$, find where $f(x) = 2$:
$$x^2 - 2x - 6 = 2$$
$$x^2 - 2x - 8 = 0$$
$$(x - 4)(x + 2) = 0$$
$$x = -2, 4$$
Since the domain specifies $x \geq 1$, $f^{-1}(2) = 4$.

53. From the graph, $y = 4$ when $x = 2$, so $f(2) = 4$. From the graph, $x = -3$ when $y = 2$, so $f^{-1}(2) = -3$.

55. Finding $f(2)$: $f(2) = 2^3 + 3(2) + 3 = 8 + 6 + 3 = 17$
To find $f^{-1}(2)$, find where $f(x) = 2$:
$$x^3 + 3x + 3 = 2$$
$$x^3 + 3x + 1 = 0$$
Using a graphing calculator yields $x \approx -0.32$, so $f^{-1}(2) \approx -0.32$.

57. Finding $f(2)$: $f(2) = 2(2)^4 + 2 - 3 = 32 + 2 - 3 = 31$
To find $f^{-1}(2)$, find where $f(x) = 2$:
$$2x^4 + x - 3 = 2$$
$$2x^4 + x - 5 = 0$$
Using a graphing calculator (where $x \geq 0$) yields $x \approx 1.18$, so $f^{-1}(2) \approx 1.18$.

59. Switching the roles of x and y:
$$3y - 6 = x$$
$$3y = x + 6$$
$$y = \frac{1}{3}x + 2$$
So the inverse is $y^{-1} = \frac{1}{3}x + 2$. Graphing both functions:

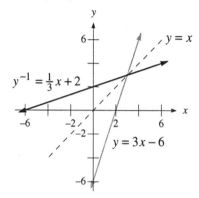

61. Switching the roles of x and y:

$$\sqrt[3]{y+4} = x$$
$$y+4 = x^3$$
$$y = x^3 - 4$$

So the inverse is $y^{-1} = x^3 - 4$. Graphing both functions:

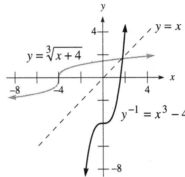

63. The implications are a restatement of the meaning of an inverse. Since $y = f(x)$, $f^{-1}(y) = x$ is just reformulating the idea that the inverse "undoes" what a function's action is. That is, x is transformed by f to $f(x) = y$. Then y is transformed by f^{-1} to $f^{-1}(y) = x$.

65. If a function is not one-to-one, then finding the inverse will not result in a unique solution. For example, if $f(x) = x^2$, then both $f^{-1}(x) = \sqrt{x}$ and $f^{-1}(x) = -\sqrt{x}$ are possible, which violates the rule that f^{-1} must be a function (with unique values). So, although $f(2) = 4$, $f^{-1}(4) = \pm 2$.

Chapter 3 Review Exercises

1. The graph will be $y = f(x)$ reflected across the y-axis:

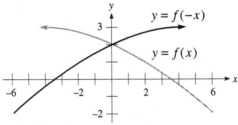

2. The graph will be $y = f(x)$ reflected across the x-axis:

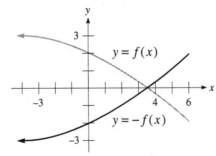

3. The graph will be $y = g(x)$ reflected across the x-axis:

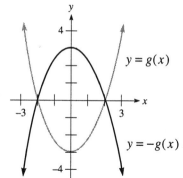

4. The graph will be $y = g(x)$ reflected across the y-axis:

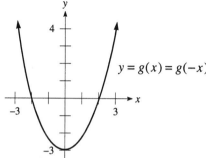

5. The graph will be $y = f(x)$ displaced 2 units to the left:

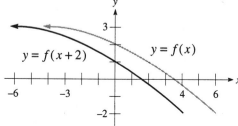

6. The graph will be $y = f(x)$ displaced 2 units up:

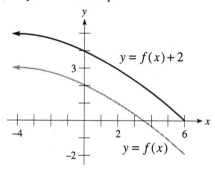

7.   The graph will be $y = f(x)$ reflected across the x-axis, then displaced 2 units up:

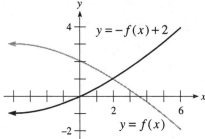

8.   The graph will be $y = g(x)$ displaced 3 units down:

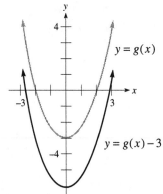

9.   The graph will be $y = g(x)$ displaced 3 units up:

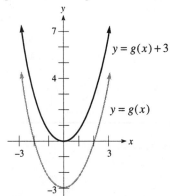

10.   The graph will be $y = f(x)$ displaced 2 units to the right, then reflected across the x-axis:

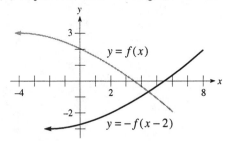

11.   The graph will be $y = g(x)$ displaced 3 units to the left, then reflected across the y-axis:

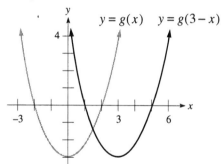

12.   The graph will be $y = f(x)$ displaced 2 units down:

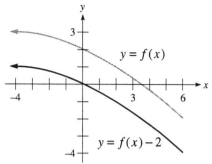

13.   The graph will be $y = f(x)$ reflected across the x-axis, then displaced 1 unit down:

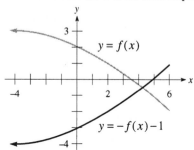

14.   The graph will be $y = g(x)$ reflected across the y-axis, then displaced 3 units up:

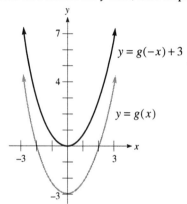

15. The graph will be $y = f(x)$ displaced 4 units to the right, then reflected across the x-axis:

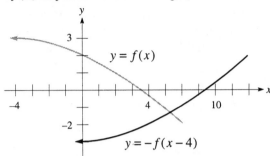

16. The graph will be $y = g(x)$ reflected across the x-axis, then displaced 1 unit down:

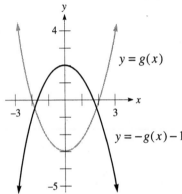

17. This is a reflection across the y-axis, so the equation is $y = f(-x)$.

18. This is a reflection across the x-axis, so the equation is $y = -f(x)$.

19. This is a displacement 2 units to the right, so the equation is $y = f(x-2)$.

20. This is a reflection across the x-axis, then a displacement 1 unit down, so the equation is
$y = -f(x) - 1$.

21. The vertex is $(4,6)$ and the axis of symmetry is $x = 4$.

22. Completing the square: $f(x) = x^2 - 6x + 5 = \left(x^2 - 6x + 9\right) - 9 + 5 = (x-3)^2 - 4$

The vertex is $(3,-4)$ and the axis of symmetry is $x = 3$.

23. Using the vertex formula: $x = \dfrac{9}{2(-2)} = -\dfrac{9}{4}$

Substituting for x: $y = -2\left(-\dfrac{9}{4}\right)^2 - 9\left(-\dfrac{9}{4}\right) = -\dfrac{81}{8} + \dfrac{81}{4} = \dfrac{81}{8}$

The maximum value is $\dfrac{81}{8}$.

24. Using the vertex formula: $x = \dfrac{4}{2(3)} = \dfrac{2}{3}$

Substituting for x: $y = 3\left(\dfrac{2}{3}\right)^2 - 4\left(\dfrac{2}{3}\right) + 2 = \dfrac{4}{3} - \dfrac{8}{3} + 2 = \dfrac{2}{3}$

The minimum value is $\dfrac{2}{3}$.

25. The *x*-intercepts are $\pm\sqrt{5}$ and the *y*-intercept is –5. Sketching the graph:

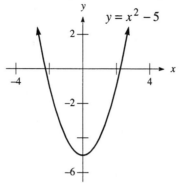

26. There are no *x*-intercepts, and the *y*-intercept is –11. Sketching the graph:

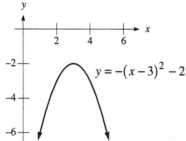

27. The *x*-intercept is 3 and the *y*-intercept is 6. Sketching the graph:

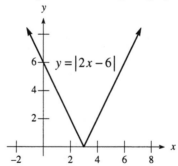

28. The *x*-intercepts are –4 and 2, and the *y*-intercept is –2. Sketching the graph:

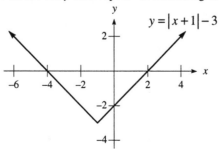

29. First complete the square:
$$f(x) = 2x^2 - 3x - 2 = 2\left(x^2 - \tfrac{3}{2}x\right) - 2 = 2\left(x^2 - \tfrac{3}{2}x + \tfrac{9}{16}\right) - \tfrac{9}{8} - 2 = 2\left(x - \tfrac{3}{4}\right)^2 - \tfrac{25}{8}$$
The x-intercepts are $-1/2$ and 2, and the y-intercept is -2. Sketching the graph:

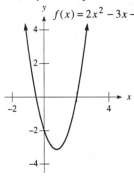

30. First complete the square:
$$y = -x^2 + 4x - 5 = -\left(x^2 - 4x\right) - 5 = -\left(x^2 - 4x + 4\right) + 4 - 5 = -(x - 2)^2 - 1$$
There are no x-intercepts, and the y-intercept is -5. Sketching the graph:

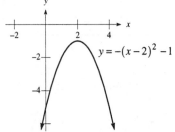

31. The x-intercepts are ± 2, and the y-intercept is 36. Sketching the graph:

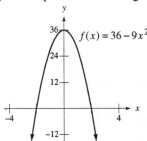

32. The x-intercepts are -1 and 3, and the y-intercept is -6. Sketching the graph:

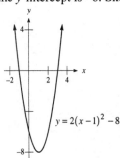

33. First complete the square:

$$y = 2x^2 - 10x = 2\left(x^2 - 5x\right) = 2\left(x^2 - 5x + \tfrac{25}{4}\right) - \tfrac{25}{2} = 2\left(x - \tfrac{5}{2}\right)^2 - \tfrac{25}{2}$$

The x-intercepts are 0 and 5, and the y-intercept is 0. Sketching the graph:

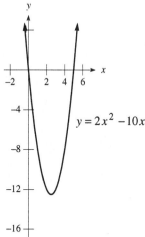

34. First complete the square: $y = x^2 - 2x - 3 = \left(x^2 - 2x + 1\right) - 1 - 3 = (x - 1)^2 - 4$

The x-intercepts are -1 and 3, and the y-intercept is -3. Sketching the graph:

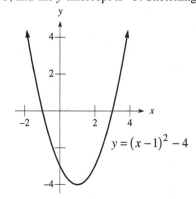

35. First complete the square:

$$y = 3x^2 - 6x + 2 = 3\left(x^2 - 2x\right) + 2 = 3\left(x^2 - 2x + 1\right) - 3 + 2 = 3(x - 1)^2 - 1$$

The x-intercepts are $\dfrac{3 \pm \sqrt{3}}{2}$, and the y-intercept is -4. Sketching the graph:

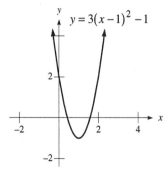

36. There are no x-intercepts, and the y-intercept is -4. Sketching the graph:

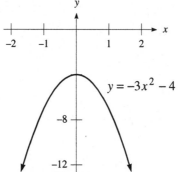

$y = -3x^2 - 4$

37. Finding the value: $(g+h)(2) = g(2) + h(2) = [5-2(2)] + [-3] = 1 + (-3) = -2$

38. Finding the value: $(r-t)(-3) = r(-3) - t(-3) = [(-3)^2 + 8] - \left[\dfrac{3}{-3-4}\right] = -19 + \dfrac{3}{7} = -\dfrac{130}{7}$

39. Finding the value: $(f \bullet s)(-6) = f(-6) \bullet s(-6) = [3(-6)^2 - 4(-6) + 1] \bullet \left[\dfrac{-2}{-6}\right] = 133 \bullet \dfrac{1}{3} = \dfrac{133}{6}$

40. Finding the value: $(t \bullet f)(3) = t(3) \bullet f(3) = \left[\dfrac{3}{3-4}\right] \bullet [3(3)^2 - 4(3) + 1] = -3 \bullet 16 = -48$

41. Finding the value: $\left(\dfrac{g}{f}\right)(4) = \dfrac{g(4)}{f(4)} = \dfrac{5-2(4)}{3(4)^2 - 4(4) + 1} = \dfrac{-3}{33} = -\dfrac{1}{11}$

42. Finding the value: $\left(\dfrac{S}{t}\right)(4) = \dfrac{s(4)}{t(4)} = \dfrac{-\dfrac{2}{4}}{\dfrac{3}{4-4}} = \dfrac{-\dfrac{1}{2}}{\dfrac{3}{0}} = $ undefined

43. Finding the value: $\left(\dfrac{r}{h}\right)(-1) = \dfrac{r(-1)}{h(-1)} = \dfrac{(-1)^3 + 8}{-3} = -\dfrac{7}{3}$

44. Finding the value: $(s-t)(8) = s(8) - t(8) = \dfrac{-2}{8} - \dfrac{3}{8-4} = -\dfrac{1}{4} - \dfrac{3}{4} = -1$

45. Finding the value: $(f \circ g)(4) = f[g(4)] = f(5 - 2 \bullet 4) = f(-3) = 3(-3)^2 - 4(-3) + 1 = 40$

46. Finding the value: $(g \circ f)(4) = g[f(4)] = g[3(4)^2 - 4(4) + 1] = g(33) = 5 - 2(33) = -61$

47. Finding the value: $(h \circ s)(-6) = h[s(-6)] = h\left(\dfrac{-2}{-6}\right) = h\left(\dfrac{1}{3}\right) = -3$

48. Finding the value: $(s \circ h)(-6) = s[h(8)] = s(-3) = \dfrac{-2}{-3} = \dfrac{2}{3}$

49. Finding the value: $(r \circ g)(a) = r[g(a)] = r(5 - 2a) = (5 - 2a)^3 + 8 = -8a^3 + 60a^2 - 150a + 133$

50. Finding the value: $(s \circ h)(c) = s[h(c)] = s(-3) = \dfrac{-2}{-3} = \dfrac{2}{3}$

51. Finding the function: $(f - g)(x) = f(x) - g(x) = 3x^2 - 4x + 1 - 5 + 2x = 3x^2 - 2x - 4$
The domain is all real numbers, or $(-\infty, \infty)$.

52. Finding the function: $(r \bullet s)(x) = r(x) \bullet s(x) = \left(x^3 + 8\right)\left(-\dfrac{2}{x}\right) = \dfrac{-2x^3 - 16}{x}$
The domain is all real numbers except 0, or $(-\infty, 0) \cup (0, \infty)$.

53. Finding the function: $\left(\dfrac{s}{h}\right)(x) = \dfrac{s(x)}{h(x)} = \dfrac{-\dfrac{2}{x}}{-3} = \dfrac{2}{3x}$

The domain is all real numbers except 0, or $(-\infty, 0) \cup (0, \infty)$.

54. Finding the function: $\left(\dfrac{h}{s}\right)(x) = \dfrac{h(x)}{s(x)} = \dfrac{-3}{-\dfrac{2}{x}} = \dfrac{3x}{2}$

The domain is all real numbers except 0, or $(-\infty, 0) \cup (0, \infty)$.

55. Finding the function: $(r - f)(x) = r(x) - f(x) = x^3 + 8 - 3x^2 + 4x - 1 = x^3 - 3x^2 + 4x + 7$

The domain is all real numbers, or $(-\infty, \infty)$.

56. Finding the function: $(s + t)(x) = s(x) + t(x) = -\dfrac{2}{x} + \dfrac{3}{x-4} = \dfrac{-2x + 8 + 3x}{x(x-4)} = \dfrac{x+8}{x(x-4)}$

The domain is all real numbers except 0 and 4, or $(-\infty, 0) \cup (0, 4) \cup (4, \infty)$.

57. Finding the function: $(g \cdot t)(x) = g(x) \cdot t(x) = (5 - 2x) \cdot \dfrac{3}{x-4} = \dfrac{15 - 6x}{x-4}$

The domain is all real numbers except 4, or $(-\infty, 4) \cup (4, \infty)$.

58. Finding the function: $\left(\dfrac{g}{t}\right)(x) = \dfrac{g(x)}{t(x)} = \dfrac{5 - 2x}{\dfrac{3}{x-4}} = \dfrac{(5 - 2x)(x - 4)}{3} = \dfrac{-2x^2 + 13x - 20}{3}$

The domain is all real numbers except 4, or $(-\infty, 4) \cup (4, \infty)$.

59. Finding the function:
$$(g \circ f)(x) = g[f(x)]$$
$$= g\left(3x^2 - 4x + 1\right)$$
$$= 5 - 2\left(3x^2 - 4x + 1\right)$$
$$= 5 - 6x^2 + 8x - 2$$
$$= -6x^2 + 8x + 3$$
The domain is all real numbers, or $(-\infty, \infty)$.

60. Finding the function:
$$(f \circ g)(x) = f[g(x)]$$
$$= f(5 - 2x)$$
$$= 3(5 - 2x)^2 - 4(5 - 2x) + 1$$
$$= 75 - 60x + 12x^2 - 20 + 8x + 1$$
$$= 12x^2 - 52x + 56$$
The domain is all real numbers, or $(-\infty, \infty)$.

61. Finding the function: $(s \circ r)(x) = s[r(x)] = s\left(x^3 + 8\right) = \dfrac{-2}{x^3 + 8}$

The domain is all real numbers except –2, or $(-\infty, -2) \cup (-2, \infty)$.

62. Finding the function: $(r \circ s)(x) = r[s(x)] = r\left(-\dfrac{2}{x}\right) = \left(-\dfrac{2}{x}\right)^3 + 8 = -\dfrac{8}{x^3} + 8 = \dfrac{8x^3 - 8}{x^3}$

The domain is all real numbers except 0, or $(-\infty, 0) \cup (0, \infty)$.

63. Finding the function: $(t \circ h)(x) = t[h(x)] = t(-3) = \dfrac{3}{-3 - 4} = -\dfrac{3}{7}$

The domain is all real numbers, or $(-\infty, \infty)$.

64. Finding the function: $(h \circ t)(x) = h[t(x)] = h\left(\dfrac{3}{x-4}\right) = -3$

The domain is all real numbers except 4, or $(-\infty, 4) \cup (4, \infty)$.

65. Finding the function: $(g \circ g)(x) = g[g(x)] = g(5 - 2x) = 5 - 2(5 - 2x) = 5 - 10 + 4x = 4x - 5$

The domain is all real numbers, or $(-\infty, \infty)$.

66. Finding the function: $(t \circ t)(x) = t[t(x)] = t\left(\dfrac{3}{x-4}\right) = \dfrac{3}{\dfrac{3}{x-4} - 4} = \dfrac{3x - 12}{3 - 4x + 16} = \dfrac{3x - 12}{19 - 4x}$

The domain is all real numbers except 4 and 19/4, or $(-\infty, 4) \cup \left(4, \frac{19}{4}\right) \cup \left(\frac{19}{4}, \infty\right)$.

67. Finding the function: $(t \circ s)(x) = t[s(x)] = t\left(-\dfrac{2}{x}\right) = \dfrac{3}{-\dfrac{2}{x} - 4} = \dfrac{3x}{-2 - 4x} = -\dfrac{3x}{4x + 2}$

The domain is all real numbers except $-1/2$ and 0, or $\left(-\infty, -\frac{1}{2}\right) \cup \left(-\frac{1}{2}, 0\right) \cup (0, \infty)$.

68. Finding the function: $(s \circ t)(x) = s[t(x)] = s\left(\dfrac{3}{x-4}\right) = \dfrac{-2}{\dfrac{3}{x-4}} = \dfrac{-2x + 8}{3}$

The domain is all real numbers except 4, or $(-\infty, 4) \cup (4, \infty)$.

69. Finding the function:
$$(s \circ g \circ t)(x) = (s \circ g)\left(\dfrac{3}{x-4}\right) = s\left(5 - \dfrac{6}{x-4}\right) = s\left(\dfrac{5x - 26}{x-4}\right) = \dfrac{-2}{\dfrac{5x-26}{x-4}} = \dfrac{-2x + 8}{5x - 26}$$

The domain is all real numbers except 4 and 26/5, or $(-\infty, 4) \cup \left(4, \frac{26}{5}\right) \cup \left(\frac{26}{5}, \infty\right)$.

70. Finding the function:
$$(t \circ s \circ g)(x) = (t \circ s)(5 - 2x) = t\left(\dfrac{-2}{5 - 2x}\right) = \dfrac{3}{\dfrac{-2}{5-2x} - 4} = \dfrac{15 - 6x}{-2 - 20 + 8x} = \dfrac{15 - 6x}{8x - 22}$$

The domain is all real numbers except 5/2 and 11/4, or $\left(-\infty, \frac{5}{2}\right) \cup \left(\frac{5}{2}, \frac{11}{4}\right) \cup \left(\frac{11}{4}, \infty\right)$.

71. Let $f(x) = \dfrac{1}{\sqrt{x}}$ and $g(x) = x + 2$, then $h(x) = (f \circ g)(x)$.

72. Let $f(x) = x^4$ and $g(x) = 3x + 2$, then $h(x) = (f \circ g)(x)$.

73. Let $f(x) = x^{-3}$ and $g(x) = 5x - 7$, then $h(x) = (f \circ g)(x)$.

74. Let $f(x) = \sqrt[3]{x}$ and $g(x) = \dfrac{x}{x+1}$, then $h(x) = (f \circ g)(x)$.

75. **a.**   Forming the composition:
$$(N \circ C)(h) = N(4h + 10)$$
$$= (4h + 10)^2 + 125(4h + 10) + 1000$$
$$= 16h^2 + 80h + 100 + 500h + 1250 + 1000$$
$$= 16h^2 + 580h + 2350$$

b.   Substituting $h = 4$: $(N \circ C)(4) = 16(4)^2 + 580(4) + 2350 = 4926$ bacteria

c. Setting $(N \circ C)(h) = 10,000$:

$$16h^2 + 580h + 2350 = 10,000$$
$$16h^2 + 580h - 7650 = 0$$
$$8h^2 + 290h - 3825 = 0$$

Using the quadratic formula:

$$h = \frac{-290 \pm \sqrt{290^2 - 4(8)(-3825)}}{2(8)} = \frac{-290 \pm \sqrt{206500}}{16} \approx -46.5, 10.3$$

Since the domain is $0 \le h \le 6$, neither of these values is in the domain.

76. a. Forming the composition: $(A \circ r)(t) = A\left(64 - t^2\right) = \pi\left(64 - t^2\right)^2$

b. Substituting $t = 4$: $(A \circ r)(4) = \pi\left(64 - 4^2\right)^2 = \pi(48)^2 = 2304\pi \approx 7,238$ square units

77. Using the slope formula on line segment $\overline{AB}$:

$$\frac{y - 0}{x - 0} = \frac{4 - 0}{3 - 0}$$
$$\frac{y}{x} = \frac{4}{3}$$
$$y = \frac{4}{3}x$$

The area of the triangle is given by: $A = \frac{1}{2}(12)(y) = 6\left(\frac{4}{3}x\right) = 8x$

Note the domain of this function is $0 \le x \le 12$.

78. Using the surface area formula:

$$2\pi r^2 + 2\pi rh = 20\pi$$
$$r^2 + rh = 10$$
$$rh = 10 - r^2$$
$$h = \frac{10 - r^2}{r}$$

The volume of the cylinder is given by: $V = \pi r^2 h = \pi r^2 \left(\dfrac{10 - r^2}{r}\right) = \pi r\left(10 - r^2\right) = 10\pi r - \pi r^3$

79. The width is x, the length is $2x$, and the height is h. Using the total surface area, we have:

$$2x(2x) + 2xh + 2(2x)(h) = 180$$
$$4x^2 + 2xh + 4xh = 180$$
$$4x^2 + 6xh = 180$$
$$6xh = 180 - 4x^2$$
$$h = \frac{90 - 2x^2}{3x}$$

The volume of the box is given by: $V = x(2x)(h) = 2x^2 \cdot \dfrac{90 - 2x^2}{3x} = \dfrac{180x - 4x^3}{3}$

80.   **a.**   Let x represent the shorter dimension (width) and y represent the longer dimension (length). Since the total cost is $240, we have:

$$6(2x+2y)+4(3x+y)=240$$
$$12x+12y+12x+4y=240$$
$$24x+16y=240$$
$$3x+2y+30$$
$$2y=30-3x$$
$$y=15-\tfrac{3}{2}x$$

Thus the area is given by: $A=xy=x\left(15-\tfrac{3}{2}x\right)=-\tfrac{3}{2}x^2+15x$

b.   Using the vertex formula: $x=\dfrac{-15}{2\left(-\tfrac{3}{2}\right)}=\dfrac{-15}{-3}=5$

So $y=15-\tfrac{3}{2}(5)=\tfrac{15}{2}$, thus the maximum area is: $A=xy=5\cdot\tfrac{15}{2}=\tfrac{75}{2}$ square feet

81.   Since the graph fails the horizontal line test, it does not have an inverse.

82.   Since the graph passes the horizontal line test, it does have an inverse.

83.   The domain of f = range of $f^{-1}=(-\infty,\infty)$, and the range of f = domain of $f^{-1}=(-\infty,\infty)$. Sketching the graphs:

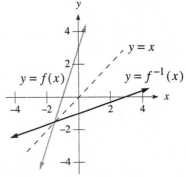

84.   The domain of f = range of $f^{-1}=[0,\infty)$, and the range of f = domain of $f^{-1}=(-\infty,0]$. Sketching the graphs:

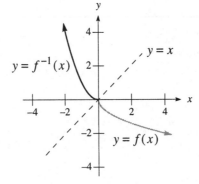

85. The domain of f = range of $f^{-1} = (-\infty, \infty)$, and the range of f = domain of $f^{-1} = (-\infty, \infty)$. Sketching the graphs:

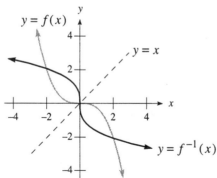

86. The domain of f = range of $f^{-1} = (0, \infty)$, and the range of f = domain of $f^{-1} = (-\infty, \infty)$. Sketching the graphs:

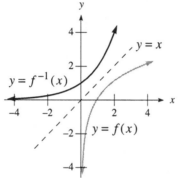

87. Switching the roles of x and y:

$$\tfrac{1}{4}y + 5 = x$$
$$\tfrac{1}{4}y = x - 5$$
$$y = 4x - 20$$

So $f^{-1}(x) = 4x - 20$. Verifying the compositions:

$$f^{-1}[f(x)] = f^{-1}\left(\tfrac{1}{4}x + 5\right) = 4\left(\tfrac{1}{4}x + 5\right) - 20 = x + 20 - 20 = x$$
$$f\left[f^{-1}(x)\right] = f(4x - 20) = \tfrac{1}{4}(4x - 20) + 5 = x - 5 + 5 = x$$

Thus $f\left[f^{-1}(x)\right] = f^{-1}[f(x)] = x$.

88. Switching the roles of x and y:

$$8 - 5y = x$$
$$-5y = x - 8$$
$$y = -\tfrac{1}{5}x + \tfrac{8}{5}$$

So $f^{-1}(x) = -\tfrac{1}{5}x + \tfrac{8}{5}$. Verifying the compositions:

$$f^{-1}[f(x)] = f^{-1}(8 - 5x) = -\tfrac{1}{5}(8 - 5x) + \tfrac{8}{5} = -\tfrac{8}{5} + x + \tfrac{8}{5} = x$$
$$f\left[f^{-1}(x)\right] = f\left(-\tfrac{1}{5}x + \tfrac{8}{5}\right) = 8 - 5\left(-\tfrac{1}{5}x + \tfrac{8}{5}\right) = 8 + x - 8 = x$$

Thus $f\left[f^{-1}(x)\right] = f^{-1}[f(x)] = x$.

89.   Switching the roles of x and y:

$$\frac{y+4}{y-3} = x$$
$$y+4 = xy-3x$$
$$3x+4 = xy-y$$
$$3x+4 = y(x-1)$$
$$y = \frac{3x+4}{x-1}$$

So $f^{-1}(x) = \frac{3x+4}{x-1}$. Verifying the compositions:

$$f^{-1}[f(x)] = f^{-1}\left(\frac{x+4}{x-3}\right) = \frac{3\left(\frac{x+4}{x-3}\right)+4}{\frac{x+4}{x-3}-1} = \frac{3x+12+4x-12}{x+4-x+3} = \frac{7x}{7} = x$$

$$f\left[f^{-1}(x)\right] = f\left(\frac{3x+4}{x-1}\right) = \frac{\frac{3x+4}{x-1}+4}{\frac{3x+4}{x-1}-3} = \frac{3x+4+4x-4}{3x+4-3x+3} = \frac{7x}{7} = x$$

Thus $f\left[f^{-1}(x)\right] = f^{-1}[f(x)] = x$.

90.   Switching the roles of x and y:

$$\sqrt{y-5} = x$$
$$y-5 = x^2 \qquad \text{if } x \geq 0$$
$$y = x^2 + 5$$

So $f^{-1}(x) = x^2 + 5$ if $x \geq 0$. Verifying the compositions:

$$f^{-1}[f(x)] = f^{-1}\left(\sqrt{x-5}\right) = \left(\sqrt{x-5}\right)^2 + 5 = x-5+5 = x$$
$$f\left[f^{-1}(x)\right] = f\left(x^2+5\right) = \sqrt{x^2+5-5} = \sqrt{x^2} = x \text{ (since } x \geq 0\text{)}$$

Thus $f\left[f^{-1}(x)\right] = f^{-1}[f(x)] = x$.

91.   Switching the roles of x and y:

$$\frac{3}{y+6} = x$$
$$\frac{y+6}{3} = \frac{1}{x}$$
$$y+6 = \frac{3}{x}$$
$$y = \frac{3}{x} - 6$$
$$y = \frac{3-6x}{x}$$

So $f^{-1}(x) = \dfrac{3-6x}{x}$. Verifying the compositions:

$$f^{-1}[f(x)] = f^{-1}\left(\frac{3}{x+6}\right) = \frac{3 - 6\left(\dfrac{3}{x+6}\right)}{\dfrac{3}{x+6}} = \frac{3x+18-18}{3} = \frac{3x}{3} = x$$

$$f[f^{-1}(x)] = f\left(\frac{3-6x}{x}\right) = \frac{3}{\dfrac{3-6x}{x}+6} = \frac{3x}{3-6x+6x} = \frac{3x}{3} = x$$

Thus $f[f^{-1}(x)] = f^{-1}[f(x)] = x$.

92. Switching the roles of x and y:

$$\frac{1}{y} - 2 = x$$

$$\frac{1}{y} = x + 2$$

$$y = \frac{1}{x+2}$$

So $f^{-1}(x) = \dfrac{1}{x+2}$. Verifying the compositions:

$$f^{-1}[f(x)] = f^{-1}\left(\frac{1}{x}-2\right) = \frac{1}{\dfrac{1}{x}-2+2} = \frac{1}{\dfrac{1}{x}} = x$$

$$f[f^{-1}(x)] = f\left(\frac{1}{x+2}\right) = \frac{1}{\dfrac{1}{x+2}} - 2 = x+2-2 = x$$

Thus $f[f^{-1}(x)] = f^{-1}[f(x)] = x$.

93. Finding $f(3)$: $f(3) = \dfrac{3+7}{2 \cdot 3} = \dfrac{10}{6} = \dfrac{5}{3}$

To find $f^{-1}(3)$, find where $f(x) = 3$:

$$\frac{x+7}{2x} = 3$$
$$x+7 = 6x$$
$$5x = 7$$
$$x = \tfrac{7}{5}$$

So $f^{-1}(3) = \tfrac{7}{5}$.

94. Finding $f(3)$: $f(3) = (3)^3 + 3 + 4 = 34$

To find $f^{-1}(3)$, find where $f(x) = 3$:

$$x^3 + x + 4 = 3$$
$$x^3 + x + 1 = 0$$

Using a graphing calculator, $x \approx -0.68$. So $f^{-1}(3) \approx -0.68$.

95. When $x = 3, y = 1$, so $f(3) = 1$. When $y = 3, x = -1$, so $f^{-1}(3) = -1$.

Chapter 3 Practice Test

1.   **a.**   This is the graph of $y = f(x)$ reflected across the y-axis:

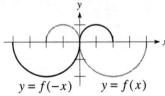

$$y = f(-x) \qquad y = f(x)$$

   b.   This is the graph of $y = f(x)$ reflected across the x-axis:

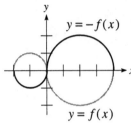

$$y = -f(x)$$

$$y = f(x)$$

   c.   This is the graph of $y = f(x)$ displaced 2 units down:

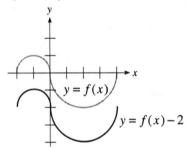

$$y = f(x)$$

$$y = f(x) - 2$$

   d.   This is the graph of $y = g(x)$ displaced 2 units up:

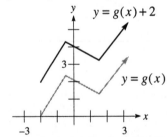

$$y = g(x) + 2$$

$$y = g(x)$$

   e.   This is the graph of $y = g(x)$ displaced 2 units to the left:

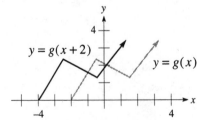

$$y = g(x + 2)$$

$$y = g(x)$$

f. This is the graph of $y = f(x)$ displaced 2 units to the right, then reflected across the x-axis:

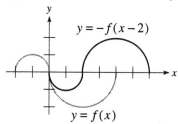

2. **a.** The vertex is $(-5,-3)$, and the axis of symmetry is $x = -5$.

b. Completing the square:

$$y = -2x^2 - 5x + 1 = -2\left(x^2 + \tfrac{5}{2}x\right) + 1 = -2\left(x^2 + \tfrac{5}{2}x + \tfrac{25}{16}\right) + \tfrac{25}{8} + 1 = -2\left(x + \tfrac{5}{4}\right)^2 + \tfrac{33}{8}$$

The vertex is $\left(-\tfrac{5}{4}, \tfrac{33}{8}\right)$, and the axis of symmetry is $x = -\tfrac{5}{4}$

3. **a.** The x-intercepts are ± 5 and the y-intercept is 25. Sketching the graph:

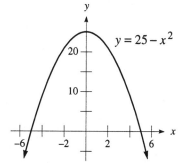

b. The x-intercept is -2 and the y-intercept is 8. Sketching the graph:

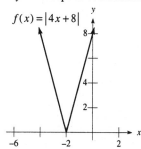

c. The x-intercepts are -5 and -1, and the y-intercept is -10. Sketching the graph:

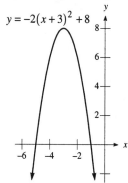

d.   First complete the square: $f(x) = x^2 + 4x - 5 = (x^2 + 4x + 4) - 4 - 5 = (x+2)^2 - 9$

The x-intercepts are –5 and 1, and the y-intercept is –5. Sketching the graph:

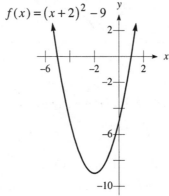

$$f(x) = (x+2)^2 - 9$$

e.   First complete the square:

$$f(x) = 3x^2 - 48x = 3(x^2 - 16x) = 3(x^2 - 16x + 64) - 192 = 3(x-8)^2 - 192$$

The x-intercepts are 0 and 16, and the y-intercept is –4. Sketching the graph:

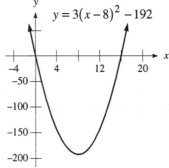

$$y = 3(x-8)^2 - 192$$

f.   First complete the square:

$$f(x) = -x^2 + 3x - 4 = -\left(x^2 - 3x\right) - 4 = -\left(x^2 - 3x + \tfrac{9}{4}\right) + \tfrac{9}{4} - 4 = -\left(x - \tfrac{3}{2}\right)^2 - \tfrac{7}{4}$$

There are no x-intercepts, and the y-intercept is 10. Sketching the graph:

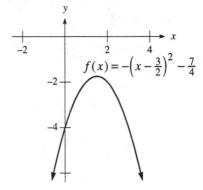

$$f(x) = -\left(x - \tfrac{3}{2}\right)^2 - \tfrac{7}{4}$$

g.   The x-intercepts are ± 10 and the y-intercept is 10. Sketching the graph:

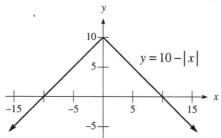

h.   First complete the square:

$$f(x) = 2x^2 - 5x - 1 = 2\left(x^2 - \tfrac{5}{2}x\right) - 1 = 2\left(x^2 - \tfrac{5}{2}x + \tfrac{25}{16}\right) - \tfrac{25}{8} - 1 = 2\left(x - \tfrac{5}{4}\right)^2 - \tfrac{33}{8}$$

The x-intercepts are $\dfrac{5 \pm \sqrt{33}}{4}$ and the y-intercept is -1. Sketching the graph:

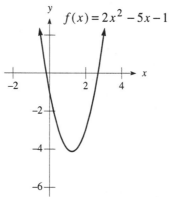

4.   **a.**   Finding the value: $(f \bullet h)(3) = f(3) \bullet h(3) = \left[7 - 3^2\right] \bullet \left[\sqrt{3 - 1}\right] = -2\sqrt{2}$

b.   Finding the value: $(g + h)(10) = g(10) + h(10) = \dfrac{3}{10 + 2} + \sqrt{10 - 1} = \tfrac{1}{4} + 3 = \tfrac{13}{4}$

c.   Finding the value: $\left(\dfrac{f}{g}\right)(-3) = \dfrac{f(-3)}{g(-3)} = \dfrac{7 - (-3)^2}{\dfrac{3}{-3 + 2}} = \dfrac{7 - 9}{-3} = \tfrac{2}{3}$

d.   Finding the function:

$$(f \circ g)(x) = f[g(x)] = f\left(\dfrac{3}{x + 2}\right) = 7 - \left(\dfrac{3}{x + 2}\right)^2 = \dfrac{7(x + 2)^2 - 9}{(x + 2)^2} = \dfrac{7x^2 + 28x + 19}{(x + 2)^2}$$

e.   Finding the function: $(g \circ f)(x) = g[f(x)] = g\left(7 - x^2\right) = \dfrac{3}{7 - x^2 + 2} = \dfrac{3}{9 - x^2}$

f.   Finding the function: $(g \circ g)(x) = g[g(x)] = g\left(\dfrac{3}{x + 2}\right) = \dfrac{3}{\dfrac{3}{x + 2} + 2} = \dfrac{3x + 6}{3 + 2x + 4} = \dfrac{3x + 6}{2x + 7}$

g.   Finding the function:

$$(g \circ f \circ h)(x) = (g \circ f)\left(\sqrt{x - 1}\right) = g\left(7 - \left(\sqrt{x - 1}\right)^2\right) = g(8 - x) = \dfrac{3}{8 - x + 2} = \dfrac{3}{10 - x}$$

Note the domain of this is $[1, \infty)$ in order for the radical to be defined.

5.   a.   Finding the composition: $(f \circ g)(x) = f\left(\dfrac{x}{x-1}\right) = \dfrac{\dfrac{2x}{x-1}}{\dfrac{x}{x-1}+1} = \dfrac{2x}{x+x-1} = \dfrac{2x}{2x-1}$

The domain is all real numbers except 1/2 and 1, or $\left(-\infty, \tfrac{1}{2}\right) \cup \left(\tfrac{1}{2}, 1\right) \cup (1, \infty)$.

   b.   Finding the composition: $(g \circ f)(x) = g\left(\dfrac{2x}{x+1}\right) = \dfrac{\dfrac{2x}{x+1}}{\dfrac{2x}{x+1}-1} = \dfrac{2x}{2x-x-1} = \dfrac{2x}{x-1}$

The domain is all real numbers except -1 and 1, or $(-\infty, -1) \cup (-1, 1) \cup (1, \infty)$.

6.   Since the diagonal of the rectangle is $2r$, using the Pythagorean theorem:

$$x^2 + 12^2 = (2r)^2$$
$$x^2 + 144 = 4r^2$$
$$x^2 = 4r^2 - 144$$
$$x = \sqrt{4r^2 - 144} = 2\sqrt{r^2 - 36}$$

The area is given by: $A = \pi r^2 - 12x = \pi r^2 - 12 \cdot 2\sqrt{r^2 - 36} = \pi r^2 - 24\sqrt{r^2 - 36}$

7.   The function passes the horizontal line test, so its inverse exists. Sketching the graph:

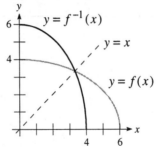

8.   a.   Switching the roles of x and y:

$$\frac{5y-4}{3} = x$$
$$5y - 4 = 3x$$
$$5y = 3x + 4$$
$$y = \tfrac{3}{5}x + \tfrac{4}{5}$$

So $f^{-1}(x) = \tfrac{3}{5}x + \tfrac{4}{5}$. Verifying the compositions:

$$f^{-1}[f(x)] = f^{-1}\left(\frac{5x-4}{3}\right) = \tfrac{3}{5}\left(\frac{5x-4}{3}\right) + \tfrac{4}{5} = x - \tfrac{4}{5} + \tfrac{4}{5} = x$$

$$f\left[f^{-1}(x)\right] = f\left(\tfrac{3}{5}x + \tfrac{4}{5}\right) = \frac{5\left(\tfrac{3}{5}x + \tfrac{4}{5}\right) - 4}{3} = \frac{3x + 4 - 4}{3} = \frac{3x}{3} = x$$

Thus $f\left[f^{-1}(x)\right] = f^{-1}[f(x)] = x$.

b.   Switching the roles of x and y:

$$\sqrt[3]{2y+9} = x$$
$$2y+9 = x^3$$
$$2y = x^3 - 9$$
$$y = \frac{x^3 - 9}{2}$$

So $f^{-1}(x) = \dfrac{x^3 - 9}{2}$. Verifying the compositions:

$$f^{-1}[f(x)] = f^{-1}\left(\sqrt[3]{2x+9}\right) = \frac{\left(\sqrt[3]{2x+9}\right)^3 - 9}{2} = \frac{2x+9-9}{2} = \frac{2x}{2} = x$$

$$f\left[f^{-1}(x)\right] = f\left(\frac{x^3-9}{2}\right) = \sqrt[3]{2\left(\frac{x^3-9}{2}\right)+9} = \sqrt[3]{x^3 - 9 + 9} = \sqrt[3]{x^3} = x$$

Thus $f\left[f^{-1}(x)\right] = f^{-1}[f(x)] = x$.

c.   Switching the roles of x and y:

$$\frac{2}{y} - 5 = x$$
$$\frac{2}{y} = x+5$$
$$\frac{y}{2} = \frac{1}{x+5}$$
$$y = \frac{2}{x+5}$$

So $f^{-1}(x) = \dfrac{2}{x+5}$. Verifying the compositions:

$$f^{-1}[f(x)] = f^{-1}\left(\frac{2}{x} - 5\right) = \frac{2}{\frac{2}{x} - 5 + 5} = \frac{2}{\frac{2}{x}} = \frac{2x}{2} = x$$

$$f\left[f^{-1}(x)\right] = f\left(\frac{2}{x+5}\right) = \frac{2}{\frac{2}{x+5}} - 5 = x+5-5 = x$$

Thus $f\left[f^{-1}(x)\right] = f^{-1}[f(x)] = x$.

9.   To find $f^{-1}(3)$, find where $f(x) = 3$:

$$\frac{5x}{x-1} = 3$$
$$5x = 3x - 3$$
$$2x = -3$$
$$x = -\frac{3}{2}$$

So $f^{-1}(3) = -\frac{3}{2}$.

Chapter 4
Polynomial, Rational, and Radical Functions

4.1 Polynomial Functions

1. Since there are 4 turning points, the degree is at least 5.
3. This cannot be the graph of a polynomial function, since it has a break in it.
5. Since there are no turning points, the degree is at least 1. Note this graph is a line, so the degree is exactly 1.
7. Since there are 3 turning points, the degree is at least 4.
9. The x-intercept is -1 and the y-intercept is 1. Sketching the graph:

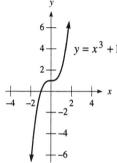

11. The x-intercept is 1 and the y-intercept is 1. Sketching the graph:

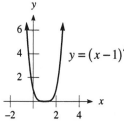

13. The x-intercept is 2 and the y-intercept is 2. Sketching the graph:

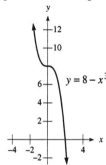

15 .   The x-intercepts are ± 2 and the y-intercept is -4. Sketching the graph:

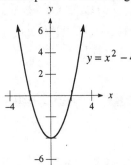

17 .   The x-intercept is -2 and the y-intercept is -32. Sketching the graph:

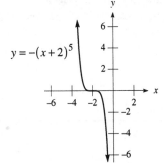

19 .   The x-intercept is 2 and the y-intercept is -16. Sketching the graph:

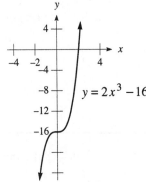

21 .   The x-intercept is 2 and the y-intercept is 2. Sketching the graph:

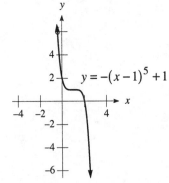

23 .   Sketching the graphs:

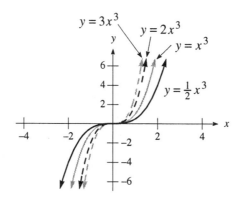

25 .   Sketching the graphs:

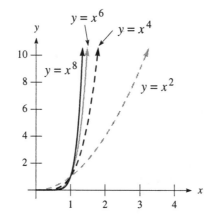

27 .   A possible graph is:

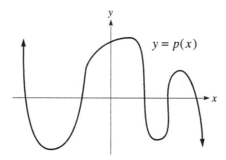

29 .   A possible graph is:

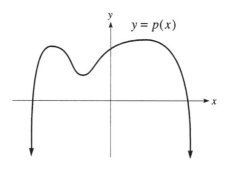

31.   Factoring: $x^3 - x^2 - 2x = x\left(x^2 - x - 2\right) = x(x+1)(x-2)$

The x-intercepts are $-1, 0, 2$. Sketching the graph:

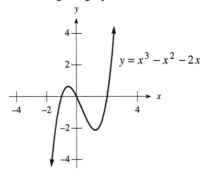

33.   Factoring: $x^2 - 6x + 8 = (x-4)(x-2)$

The x-intercepts are $2, 4$. Sketching the graph:

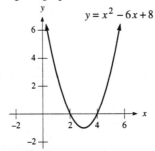

35.   Factoring: $x^3 - 9x^2 = x^2(x-9)$

The x-intercepts are $0, 9$. Sketching the graph:

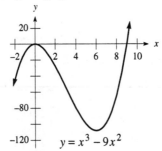

37.   Factoring: $x^4 + x^2 = x^2\left(x^2 + 1\right)$

The x-intercept is 0. Sketching the graph:

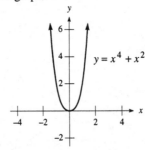

39. The *x*-intercepts are –2, –1, 0, 1. Sketching the graph:

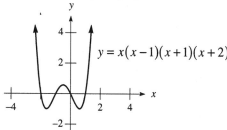

$$y = x(x-1)(x+1)(x+2)$$

41. The *x*-intercepts are –3, –1, 2, 4. Sketching the graph:

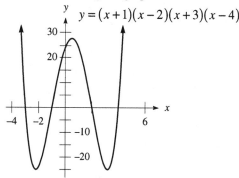

$$y = (x+1)(x-2)(x+3)(x-4)$$

43. Factoring: $x^3 + x^2 - 2x - 2 = x^2(x+1) - 2(x+1) = (x+1)(x^2 - 2)$

The *x*-intercepts are –1, $\pm\sqrt{2}$. Sketching the graph:

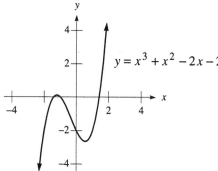

$$y = x^3 + x^2 - 2x - 2$$

45. Graphing the polynomial:

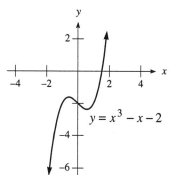

$$y = x^3 - x - 2$$

The graph has one real zero and two turning points. As $x \to -\infty$, $y \to -\infty$ and as $x \to \infty$, $y \to \infty$.

47. Graphing the polynomial:

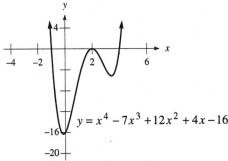

$$y = x^4 - 7x^3 + 12x^2 + 4x - 16$$

The graph has 3 real zeros and 3 turning points. As $x \to -\infty$, $y \to \infty$ and as $x \to \infty$, $y \to \infty$.

49. Graphing the polynomial:

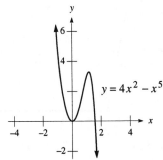

$$y = 4x^2 - x^5$$

The graph has 2 real zeros and 2 turning points. As $x \to -\infty$, $y \to \infty$ and as $x \to \infty$, $y \to -\infty$.

51. Graphing the polynomial:

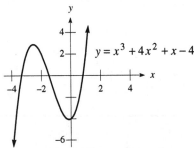

$$y = x^3 + 4x^2 + x - 4$$

The x-intercepts are -3.34, -1.47, and 0.81. The turning points are $(-2.54, 2.88)$ and $(-0.13, -4.06)$.

53. Graphing the polynomial:

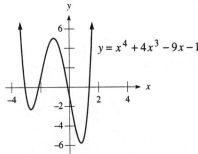

$$y = x^4 + 4x^3 - 9x - 1$$

The x-intercepts are -3.09, -2.14, -0.11, and 1.35. The turning points are $(-2.69, -2.29)$, $(-1.08, 5.04)$, and $(0.77, -5.75)$.

55. Graphing the polynomial:

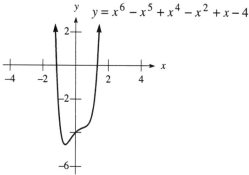

$$y = x^6 - x^5 + x^4 - x^2 + x - 4$$

The x-intercepts are -1.17 and 1.33. The turning point is $(-0.62, -4.71)$.

57. For $k = 1, 2, 3$, the graphs are:

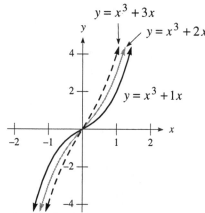

$$y = x^3 + 3x$$
$$y = x^3 + 2x$$
$$y = x^3 + 1x$$

All three graphs are similar to $y = x^3$. For $k = -1, -2, -3$, the graphs are:

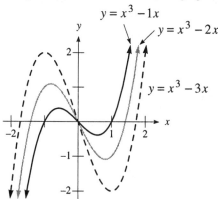

$$y = x^3 - 1x$$
$$y = x^3 - 2x$$
$$y = x^3 - 3x$$

As $|k|$ increases, the relative extrema become more pronounced. That is, the relative maximum value becomes larger and the relative minimum value becomes smaller.

59.   For $k = 1, 2, 3$, the graphs are:

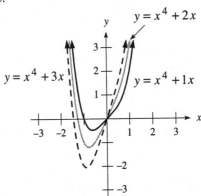

As k increases, the relative minimum value becomes smaller (more pronounced). For $k = -1, -2, -3$, the graphs are:

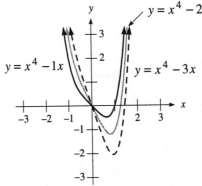

As $|k|$ increases, the relative minimum value becomes smaller (more pronounced).

61.   No. In order for the polynomial to have exactly two distinct real zeros, it must have a relative maximum or minimum value between those two zeros. The polynomial $y = 0$ has an infinite number of distinct zeros with no turning points, however.

63.   The values are:
$$p(10) = 870$$
$$p(100) = 1,899,600$$
$$p(1000) = 1,989,996,900$$
$$p(10,000) = 1,998,999,969,900$$

The values are becoming closer and closer to those of $2x^3$. Computing the values:
$$p(-100) = -2,099,800$$
$$p(-1000) = -2,009,997,100$$
$$p(-10,000) = -2,000,999,970,100$$
$$p(-100,000) = -2.00 \times 10^{15}$$

Again, the values are becoming closer and closer to those of $2x^3$. As $|x|$ gets very large, $p(x) \approx 2x^3$, thus $|p(x)|$ gets very large.

65.   If $a > 0$, $p(x) \to +\infty$ as $x \to +\infty$ and as $x \to -\infty$. If $a < 0$, $p(x) \to -\infty$ as $x \to +\infty$ and as $x \to -\infty$.

67. **a.** Graphing the curve:

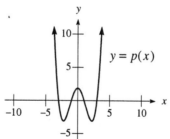

b. $p(x)$ appears to have 4 real zeros and 3 turning points.

c. It could be, but the $[-10, 10]$ by $[-10, 10]$ viewing rectangle may exclude (visually) some function behavior.

d. Graphing the curve:

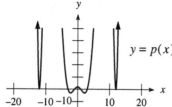

In this viewing window, $p(x)$ appears to have 6 zeros and 7 turning points.

4.2 More on Polynomial Functions and Mathematical Models

1. Sketching the graph:

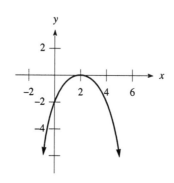

3. Sketching the graph:

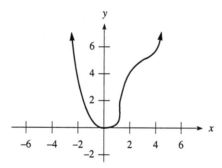

5.   Sketching the graph:

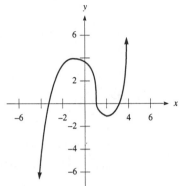

7.   Since $p(1) = -1$ and $p(2) = 6$, the zero is between 1 and 2.
Since $p(1.2) = -0.272$ and $p(1.3) = 0.197$, the zero is between 1.2 and 1.3.

9.   Since $p(-3) = 14$ and $p(-4) = -8$, the zero is between –3 and –4.
Since $p(-3.7) = 0.637$ and $p(-3.8) = -2.032$, the zero is between –3.8 and –3.7.

11.   Since $p(0) = -1$ and $p(1) = 2$, a zero is between 0 and 1.
Since $p(0.7) = -0.0739$ and $p(0.8) = 0.4336$, a zero is between 0.7 and 0.8.
Since $p(-2) = -1$ and $p(-3) = 26$, another zero is between –3 and –2.
Since $p(-2.1) = -0.0739$ and $p(-2.2) = 1.1296$, another zero is between –2.2 and –2.1.

13.   Finding where the two curves intersect:
$$x^3 = 9x$$
$$x^3 - 9x = 0$$
$$x\left(x^2 - 9\right) = 0$$
$$x(x+3)(x-3) = 0$$
$$x = -3, 0, 3$$
The points of intersection are (–3,–27), (0,0), and (3,27).

15.   Finding where the two curves intersect:
$$x^5 = 2x^4 + 3x^3$$
$$x^5 - 2x^4 - 3x^3 = 0$$
$$x^3\left(x^2 - 2x - 3\right) = 0$$
$$x^3(x-3)(x+1) = 0$$
$$x = -1, 0, 3$$
The points of intersection are (–1,–1), (0,0), and (3,243).

17.   Simplifying the difference quotient: $\dfrac{f(x) - f(2)}{x - 2} = \dfrac{x^3 - 8}{x - 2} = \dfrac{(x-2)\left(x^2 + 2x + 4\right)}{x - 2} = x^2 + 2x + 4$

19.   Simplifying the difference quotient:
$$\frac{f(x+h) - f(x)}{h} = \frac{(x+h)^4 - x^4}{h}$$
$$= \frac{x^4 + 4x^3h + 6x^2h^2 + 4xh^3 + h^4 - x^4}{h}$$
$$= \frac{4x^3h + 6x^2h^2 + 4xh^3 + h^4}{h}$$
$$= 4x^3 + 6x^2h + 4xh^2 + h^3$$

21.   We must find the smallest n such that:
$$\left(\tfrac{1}{2}\right)^n < 0.005$$
$$(0.5)^n < 0.005$$

The first n where this occurs is $n = 8$, since $(0.5)^8 \approx 0.0039 < 0.005$.

23.   Solving the inequality:
$$\left|\frac{x^3}{6}\right| < 0.001$$
$$\frac{|x|^3}{6} < 0.001$$
$$|x|^3 < 0.006$$
$$|x| < 0.182$$

25.   **a.**   The revenue is given by: $R(d) = d\left(430 - d^2\right) = 430d - d^3$

   b.   Graphing the revenue function:

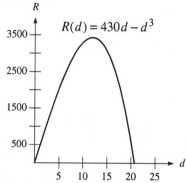

The revenue is maximized when $d \approx \$11.97$ per pound.

   c.   The maximum revenue is: $R(11.97) \approx \$3432$

27.   **a.**   The revenue is given by: $R(n) = n\left(2 + 0.45n - 0.001n^2\right) = 2n + 0.45n^2 - 0.001n^3$

   b.   Graphing the revenue function:

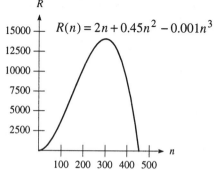

The revenue is maximized when $n \approx 302$, so the price is:
$$2 + 0.45(302) - 0.001(302)^2 \approx \$46.70 \text{ per item}$$

   c.   The maximum revenue is: $R(302) \approx \$14,102$

29. The surface area is given by:
$$S(w) = 2wl + 2wh + 2lh$$
$$= 2w(3w) + 2w(w - 6) + 2(3w)(w - 6)$$
$$= 6w^2 + 2w^2 - 12w + 6w^2 - 36w$$
$$= 14w^2 - 48w$$

Substituting $w = 20$ in., $w = 26$ in., and $w = 42$ in.:
$$s(20) = 14(20)^2 - 48(20) = 4,640 \text{ sq. in.}$$
$$s(26) = 14(26)^2 - 48(26) = 8,216 \text{ sq. in.}$$
$$s(42) = 14(42)^2 - 48(42) = 22,680 \text{ sq. in.}$$

31. **a.** Calculating $A(0)$, $A(5)$, $A(10)$, $A(15)$, $A(20)$, $A(25)$, and $A(30)$:
$$A(0) = 0.28(0) + 0.39(0)^2 - 0.01(0)^3 = 0 \text{ mg}$$
$$A(5) = 0.28(5) + 0.39(5)^2 - 0.01(5)^3 = 9.9 \text{ mg}$$
$$A(10) = 0.28(10) + 0.39(10)^2 - 0.01(10)^3 = 31.8 \text{ mg}$$
$$A(15) = 0.28(15) + 0.39(15)^2 - 0.01(15)^3 = 58.2 \text{ mg}$$
$$A(20) = 0.28(20) + 0.39(20)^2 - 0.01(20)^3 = 81.6 \text{ mg}$$
$$A(25) = 0.28(25) + 0.39(25)^2 - 0.01(25)^3 = 94.5 \text{ mg}$$
$$A(30) = 0.28(30) + 0.39(30)^2 - 0.01(30)^3 = 89.4 \text{ mg}$$

b. Calculating $A(23)$, $A(24)$, $A(25)$, $A(26)$, and $A(27)$:
$$A(23) = 0.28(23) + 0.39(23)^2 - 0.01(23)^3 = 91.1 \text{ mg}$$
$$A(24) = 0.28(24) + 0.39(24)^2 - 0.01(24)^3 = 93.1 \text{ mg}$$
$$A(25) = 0.28(25) + 0.39(25)^2 - 0.01(25)^3 = 94.5 \text{ mg}$$
$$A(26) = 0.28(26) + 0.39(26)^2 - 0.01(26)^3 = 95.2 \text{ mg}$$
$$A(27) = 0.28(27) + 0.39(27)^2 - 0.01(27)^3 = 95.0 \text{ mg}$$

The amount in the bloodstream is a maximum after 26 minutes.

33. **a.** The volume is given by: $V(x) = (20 - 2x)(20 - 2x)(x) = 4x^3 - 80x^2 + 400x$
The domain is $0 < x < 10$.

b. Graphing the volume function:

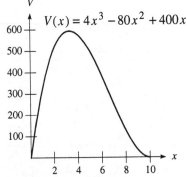

This has a maximum value when $x = \frac{10}{3} \approx 3.33$, so the dimensions of the box are 13.34 in. by 13.34 in. by 3.33 in.

35. The following table of values gives the function approximations:

Year	x	actual	(a).	(b)	(c)
1990	0	9.9	8.3	10.4	9.1
1991	1	10.3	9.37	10.55	10.5
1992	2	11.2	10.44	11	11.08
1993	3	11.9	11.51	11.75	11.52
1994	4	12.7	12.58	12.8	11.9
1995	5	14.3	13.65	14.15	12.23
1996	6	15.5	14.72	15.8	12.53
1997	7	17.9	15.79	17.75	12.80
1998	8	20.5	16.86	20	13.06

Function (b) seems to give the closest values to the actual values.

37. The zeroes are -3 and 2. Since $(x-2)^3$ has an exponent which is odd, the signs will change on either side of $x = 2$. However, since $(x+3)^2$ has an exponent which is even, the signs will stay the same on either side of $x = -3$.

4.3 Division of Polynomials and Synthetic Division

1. Using long division:

$$\begin{array}{r} x - 1 \\ x-4\overline{\smash)x^2 - 5x - 14} \\ \underline{x^2 - 4x} \\ -x - 14 \\ \underline{-x + 4} \\ -18 \end{array}$$

The quotient is $x - 1 - \dfrac{18}{x-4}$.

3. Using long division:

$$\begin{array}{r} x - 2 \\ x+9\overline{\smash)x^2 + 7x - 18} \\ \underline{x^2 + 9x} \\ -2x - 18 \\ \underline{-2x - 18} \\ 0 \end{array}$$

The quotient is $x - 2$.

5. Using long division:

$$\begin{array}{r} 2t - 5 \\ 3t-4\overline{\smash)6t^2 - 23t + 15} \\ \underline{6t^2 - 8t} \\ -15t + 15 \\ \underline{-15t + 20} \\ -5 \end{array}$$

The quotient is $2t - 5 - \dfrac{5}{3t-4}$.

7.   Using long division:

$$\begin{array}{r} 1 \\ y^2+1\overline{)y^2+2} \\ \underline{y^2+1} \\ 1 \end{array}$$

The quotient is $1+\dfrac{1}{y^2+1}$.

9.   Using long division:

$$\begin{array}{r} 2x^2-3x+1 \\ x+2\overline{)2x^3+x^2-5x+2} \\ \underline{2x^3+4x^2} \\ -3x^2-5x \\ \underline{-3x^2-6x} \\ x+2 \\ \underline{x+2} \\ 0 \end{array}$$

The quotient is $2x^2-3x+1$.

11.   Using long division:

$$\begin{array}{r} 3x^3+2x^2+2x+2 \\ x-1\overline{)3x^4-x^3+0x^2+0x+6} \\ \underline{3x^4-3x^3} \\ 2x^3+0x^2 \\ \underline{2x^3-2x^2} \\ 2x^2+0x \\ \underline{2x^2-2x} \\ 2x+6 \\ \underline{2x-2} \\ 8 \end{array}$$

The quotient is $3x^3+2x^2+2x+2+\dfrac{8}{x-1}$.

13.   Using long division:

$$\begin{array}{r} 4c^2-2c+1 \\ 2c+1\overline{)8c^3+0c^2+0c+1} \\ \underline{8c^3+4c^2} \\ -4c^2+0c \\ \underline{-4c^2-2c} \\ 2c+1 \\ \underline{2c+1} \\ 0 \end{array}$$

The quotient is $4c^2-2c+1$.

15. Using long division:

$$
\begin{array}{r}
2w^2 - w - 4 \\
w^2 + w + 1\,\overline{\smash{\big)}\,2w^4 + w^3 - 3w^2 - 5w + 4} \\
\underline{2w^4 + 2w^3 + 2w^2} \\
-w^3 - 5w^2 - 5w \\
\underline{-w^3 - w^2 - w} \\
-4w^2 - 4w + 4 \\
\underline{-4w^2 - 4w - 4} \\
8
\end{array}
$$

The quotient is $2w^2 - w - 4 + \dfrac{8}{w^2 + w + 1}$.

17. Using long division:

$$
\begin{array}{r}
x^3 + 2x^2 + 4x + 8 \\
x - 2\,\overline{\smash{\big)}\,x^4 + 0x^3 + 0x^2 + 0x - 16} \\
\underline{x^4 - 2x^3} \\
2x^3 + 0x^2 \\
\underline{2x^3 - 4x^2} \\
4x^2 + 0x \\
\underline{4x^2 - 8x} \\
8x - 16 \\
\underline{8x - 16} \\
0
\end{array}
$$

The quotient is $x^3 + 2x^2 + 4x + 8$.

19. Using long division:

$$
\begin{array}{r}
x^2 + 5x + 9 \\
x - 2\,\overline{\smash{\big)}\,x^3 + 3x^2 - x + 6} \\
\underline{x^3 - 2x^2} \\
5x^2 - x \\
\underline{5x^2 - 10x} \\
9x + 6 \\
\underline{9x - 18} \\
24
\end{array}
$$

The quotient is $x^2 + 5x + 9 + \dfrac{24}{x - 2}$.

21. Using synthetic division:

$$
\begin{array}{r|rrr}
2 & 1 & -8 & 7 \\
 & & 2 & -12 \\
\hline
 & 1 & -6 & -5
\end{array}
$$

The quotient is $x - 6$ and the remainder is -5.

23. Using synthetic division:

$$\begin{array}{r|rrrr} -2 & 2 & 1 & -3 & 15 \\ & & -4 & 6 & -6 \\ \hline & 2 & -3 & 3 & 9 \end{array}$$

The quotient is $2x^2 - 3x + 3$ and the remainder is 9.

25. Using synthetic division:

$$\begin{array}{r|rrrrr} 1 & 2 & 1 & -6 & -10 & 13 \\ & & 2 & 3 & -3 & -13 \\ \hline & 2 & 3 & -3 & -13 & 0 \end{array}$$

The quotient is $2x^3 + 3x^2 - 3x - 13$ and the remainder is 0.

27. Using synthetic division:

$$\begin{array}{r|rrr} -4 & 5 & 0 & -23 \\ & & -20 & 80 \\ \hline & 5 & -20 & 57 \end{array}$$

The quotient is $5x - 20$ and the remainder is 57.

29. Using synthetic division:

$$\begin{array}{r|rrrr} -3 & 2 & 4 & 0 & 6 \\ & & -6 & 6 & -18 \\ \hline & 2 & -2 & 6 & -12 \end{array}$$

The quotient is $2x^2 - 2x + 6$ and the remainder is -12.

31. Using synthetic division:

$$\begin{array}{r|rrrrr} 4 & 1 & 0 & 0 & 0 & -64 \\ & & 4 & 16 & 64 & 256 \\ \hline & 1 & 4 & 16 & 64 & 192 \end{array}$$

The quotient is $x^3 + 4x^2 + 16x + 64$ and the remainder is 192.

33. Using synthetic division:

$$\begin{array}{r|rrrr} 1/2 & 6 & -3 & -8 & 4 \\ & & 3 & 0 & -4 \\ \hline & 6 & 0 & -8 & 0 \end{array}$$

The quotient is $6x^2 - 8$ and the remainder is 0.

35. Using synthetic division:

$$\begin{array}{r|rrrr} 3/2 & 2 & -3 & 4 & -6 \\ & & 3 & 0 & 6 \\ \hline & 2 & 0 & 4 & 0 \end{array}$$

The quotient is $2x^2 + 4$ and the remainder is 0.

37. Using synthetic division:

$$\begin{array}{r|rrrr} -a & 1 & 0 & 0 & a^3 \\ & & -a & a^2 & -a^3 \\ \hline & 1 & -a & a^2 & 0 \end{array}$$

The quotient is $x^2 - ax + a^2$ and the remainder is 0.

39. Using synthetic division:

$$\begin{array}{r|rrrrr} a & 1 & 0 & 0 & 0 & -a^4 \\ & & a & a^2 & a^3 & a^4 \\ \hline & 1 & a & a^2 & a^3 & 0 \end{array}$$

The quotient is $x^3 + ax^2 + a^2x + a^3$ and the remainder is 0.

41. Using synthetic division:

$$\begin{array}{r|rrrr} 5 & 1 & 1 & 4 & -5 \\ & & 5 & 30 & 170 \\ \hline & 1 & 6 & 34 & 165 \end{array}$$

The quotient is $q(x) = x^2 + 6x + 34$ and the remainder is $R(x) = 165$.

43. Using synthetic division:

$$\begin{array}{r|rrrr} 2 & 1 & 0 & -k & 5 \\ & & 2 & 4 & 8-2k \\ \hline & 1 & 2 & 4-k & 13-2k \end{array}$$

Since the remainder is 1:
$$13 - 2k = 1$$
$$-2k = -12$$
$$k = 6$$

45. Using synthetic division:

$$\begin{array}{r|rrrr} -2 & 1 & 0 & k & 4 \\ & & -2 & 4 & -2k-8 \\ \hline & 1 & -2 & k+4 & -2k-4 \end{array}$$

Also:

$$\begin{array}{r|rrrr} 2 & 2 & k & 0 & -2 \\ & & 4 & 2k+8 & 4k+16 \\ \hline & 2 & k+4 & 2k+8 & 4k+14 \end{array}$$

Since these remainders are equal:
$$4k + 14 = -2k - 4$$
$$6k = -18$$
$$k = -3$$

47. Using synthetic division:

$$\begin{array}{r|rrrr} 2 & 2 & -5 & 6 & 4 \\ & & 4 & -2 & 8 \\ \hline & 2 & -1 & 4 & 12 \end{array}$$

The remainder is 12. Since $p(2) = 2(2)^3 - 5(2)^2 + 6(2) + 4 = 12$, note the remainder is equal to $p(2)$. Using synthetic division:

$$\begin{array}{r|rrrr} -2 & 2 & -5 & 6 & 4 \\ & & -4 & 18 & -48 \\ \hline & 2 & -9 & 24 & -44 \end{array}$$

The remainder is –44. Since $p(-2) = 2(-2)^3 - 5(-2)^2 + 6(-2) + 4 = -44$, note the remainder is equal to $p(-2)$.

4.4 Roots of Polynomial Equations: The Remainder, Factor, and Rational Root Theorems

1. Using synthetic division:

$$
\begin{array}{r|rrr}
-4 & 1 & -5 & 8 \\
   &   & -4 & 36 \\
\hline
   & 1 & -9 & 44
\end{array}
$$

The quotient is $x - 9$ and the remainder is $p(-4) = 44$.

3. Using synthetic division:

$$
\begin{array}{r|rrrr}
4 & 2 & -7 & 1 & 4 \\
  &   & 8 & 4 & 20 \\
\hline
  & 2 & 1 & 5 & 24
\end{array}
$$

The quotient is $2x^2 + x + 5$ and the remainder is $p(4) = 24$.

5. Using synthetic division:

$$
\begin{array}{r|rrrrr}
1 & -1 & 0 & 0 & 0 & 1 \\
  &   & -1 & -1 & -1 & -1 \\
\hline
  & -1 & -1 & -1 & -1 & 0
\end{array}
$$

The quotient is $-x^3 - x^2 - x - 1$ and the remainder is $p(1) = 0$.

7. Using synthetic division:

$$
\begin{array}{r|rrrrrr}
-1 & 1 & 0 & 0 & -2 & 0 & -3 \\
   &   & -1 & 1 & -1 & 3 & -3 \\
\hline
   & 1 & -1 & 1 & -3 & 3 & -6
\end{array}
$$

The quotient is $x^4 - x^3 + x^2 - 3x + 3$ and the remainder is $p(-1) = -6$.

9. Using synthetic division:

$$
\begin{array}{r|rrr}
4 & 1 & -5 & 3 \\
  &   & 4 & -4 \\
\hline
  & 1 & -1 & -1
\end{array}
$$

So $p(4) = -1$.

11. Using synthetic division:

$$
\begin{array}{r|rrrr}
-2 & 1 & -2 & 3 & -5 \\
   &   & -2 & 8 & -22 \\
\hline
   & 1 & -4 & 11 & -27
\end{array}
$$

So $p(-2) = -27$.

13. Using synthetic division:

$$
\begin{array}{r|rrrrr}
2 & 3 & 0 & -5 & -6 & -16 \\
  &   & 6 & 12 & 14 & 16 \\
\hline
  & 3 & 6 & 7 & 8 & 0
\end{array}
$$

So $p(2) = 0$.

15.   Using synthetic division:

-3	-1	0	6	0	0	0	23
		3	-9	9	-27	81	-243
	-1	3	-3	9	-27	81	-220

So $p(-3) = -220$.

17.   Using synthetic division:

$1/2$	8	4	0	-6	3
		4	4	2	-2
	8	8	4	-4	1

So $p\left(\frac{1}{2}\right) = 1$.

19.   Using synthetic division:

4	2	-11	10	8
		8	-12	-8
	2	-3	-2	0

So $p(x) = (x-4)\left(2x^2 - 3x - 2\right) = (x-4)(2x+1)(x-2)$.

21.   Using synthetic division:

$1/4$	4	-1	-36	9
		1	0	-9
	4	0	-36	0

So $r(x) = \left(x - \frac{1}{4}\right)\left(4x^2 - 36\right) = 4\left(x - \frac{1}{4}\right)(x+3)(x-3)$. The remaining zeros are -3 and 3.

23.   Using synthetic division:

3	1	-3	-19	87	-90
		3	0	-57	90
	1	0	-19	30	0

Using synthetic division again:

3	1	0	-19	30
		3	9	-30
	1	3	-10	0

So $p(x) = (x-3)^2\left(x^2 + 3x - 10\right) = (x-3)^2(x+5)(x-2)$. The roots of $p(x) = 0$ are therefore -5, 2, and 3.

25.   Using synthetic division:

3	1	-3	1	-3
		3	0	3
	1	0	1	0

So $p(x) = (x-3)\left(x^2 + 1\right)$. Since $x^2 + 1 = 0$ when $x = \pm i$, the other roots are $\pm i$.

27.   Factoring: $p(x) = x^3 - x^2 - 12x = x\left(x^2 - x - 12\right) = x(x-4)(x+3) = x(x-4)[x-(-3)]$
The zeros are -3, 0, and 4, all with multiplicity 1.

29 . **a.**   The polynomial is $p(x) = k(x-1)^3(x-2)^2(x-3)$. Its degree is 6.

b.   If $p(0) = 6$:

$$6 = k(0-1)^3(0-2)^2(0-3)$$
$$6 = k(-1)(4)(-3)$$
$$6 = 12k$$
$$k = \tfrac{1}{2}$$

So $p(x) = \tfrac{1}{2}(x-1)^3(x-2)^2(x-3)$.

31 .   The polynomial is $p(x) = kx^2(x+2)^2(x-1)$. Its degree is 5.

33 .   Since $2 - 3i$ is a root, $2 + 3i$ is also a root, and thus a factor is:

$$[x-(2-3i)][x-(2+3i)] = x^2 - 4x + 13$$

Using long division:

$$
\begin{array}{r}
2x+3 \\
x^2-4x+13\overline{\smash{\big)}\,2x^3-5x^2+14x+39} \\
\underline{2x^3-8x^2+26x} \\
3x^2-12x+39 \\
\underline{3x^2-12x+39} \\
0
\end{array}
$$

So $2x+3$ is another factor, thus $x = -\tfrac{3}{2}$ is another root. The remaining roots are $2+3i$ and $-\tfrac{3}{2}$.

35 .   Since $3 + 4i$ is a root, $3 - 4i$ is also a root, and thus a factor is:

$$[x-(3+4i)][x-(3-4i)] = x^2 - 6x + 25$$

Using long division:

$$
\begin{array}{r}
4x^2-4x+5 \\
x^2-6x+25\overline{\smash{\big)}\,4x^4-28x^3+129x^2-130x+125} \\
\underline{4x^4-24x^3+100x^2} \\
-4x^3+29x^2-130x \\
\underline{-4x^3+24x^2-100x} \\
5x^2-30x+125 \\
\underline{5x^2-30x+125} \\
0
\end{array}
$$

So $4x^2 - 4x + 5$ is another factor. Using the quadratic formula to find when $4x^2 - 4x + 5 = 0$:

$$x = \frac{4 \pm \sqrt{4^2 - 4(4)(5)}}{2(4)} = \frac{4 \pm \sqrt{-64}}{8} = \frac{4 \pm 8i}{8} = \tfrac{1}{2} \pm i$$

The remaining roots are $3 - 4i$, $\tfrac{1}{2} + i$, and $\tfrac{1}{2} - i$.

37 .   The possible rational roots are $\pm 1, \pm 2, \pm 5$, and ± 10. Using synthetic division:

$$
\begin{array}{r|rrrr}
1 & 1 & -4 & -7 & 10 \\
  &   & 1 & -3 & -10 \\
\hline
  & 1 & -3 & -10 & 0
\end{array}
$$

So the polynomial factors as $(x-1)\left(x^2 - 3x - 10\right) = (x-1)(x-5)(x+2)$. The rational roots are $-2, 1$, and 5.

39. Writing the equation as $2x^3 - 5x^2 - 28x + 15 = 0$, the possible rational roots are ± 1, ± 3, ± 5, ± 15, $\pm 1/2$, $\pm 3/2$, $\pm 5/2$, and $\pm 15/2$. Using synthetic division:

$$
\begin{array}{r|rrrr}
-3 & 2 & -5 & -28 & 15 \\
   &   & -6 & 33 & -15 \\
\hline
   & 2 & -11 & 5 & 0
\end{array}
$$

So the polynomial factors as $(x+3)\left(2x^2 - 11x + 5\right) = (x+3)(2x-1)(x-5)$. The rational roots are -3, $1/2$, and 5.

41. The possible rational roots are ± 1 and ± 2. Using synthetic division:

$$
\begin{array}{r|rrrr}
1 & 1 & -4 & 5 & -2 \\
  &   & 1 & -3 & 2 \\
\hline
  & 1 & -3 & 2 & 0
\end{array}
$$

So the polynomial factors as $(x-1)\left(x^2 - 3x + 2\right) = (x-1)^2(x-2)$. The rational roots are 1 and 2.

43. Writing the equation as $4x^3 - 3x - 1 = 0$, the possible rational roots are ± 1, $\pm 1/2$, and $\pm 1/4$. Using synthetic division:

$$
\begin{array}{r|rrrr}
1 & 4 & 0 & -3 & -1 \\
  &   & 4 & 4 & 1 \\
\hline
  & 4 & 4 & 1 & 0
\end{array}
$$

So the polynomial factors as $(x-1)\left(4x^2 + 4x + 1\right) = (x-1)(2x+1)^2$. The rational roots are $-1/2$ and 1.

45. The possible rational roots are ± 1, ± 2, and ± 4. Using synthetic division:

$$
\begin{array}{r|rrrrrr}
-1 & 1 & 0 & -4 & 1 & 0 & -4 \\
   &   & -1 & 1 & 3 & -4 & 4 \\
\hline
   & 1 & -1 & -3 & 4 & -4 & 0
\end{array}
$$

Using synthetic division again:

$$
\begin{array}{r|rrrrr}
2 & 1 & -1 & -3 & 4 & -4 \\
  &   & 2 & 2 & -2 & 4 \\
\hline
  & 1 & 1 & -1 & 2 & 0
\end{array}
$$

Using synthetic division again:

$$
\begin{array}{r|rrrr}
-2 & 1 & 1 & -1 & 2 \\
   &   & -2 & 2 & -2 \\
\hline
   & 1 & -1 & 1 & 0
\end{array}
$$

So the polynomial factors as $(x+1)(x-2)(x+2)\left(x^2 - x + 1\right)$. The rational roots are -2, -1, and 2.

47. The possible rational roots are ± 1, $\pm 1/2$, $\pm 1/3$, and $\pm 1/6$. Using synthetic division:

$$
\begin{array}{r|rrrr}
-1 & 6 & 1 & -4 & 1 \\
   &   & -6 & 5 & -1 \\
\hline
   & 6 & -5 & 1 & 0
\end{array}
$$

So the polynomial factors as $(x+1)\left(6x^2 - 5x + 1\right) = (x+1)(3x-1)(2x-1)$. The rational roots are -1, $1/3$, and $1/2$.

49. The possible rational roots are ± 1, ± 3, ± 5, and ± 15. None of these possible roots results in a 0 remainder, so there are no rational roots.

51. The possible rational roots are ±1, ±3, ±9, ±27, ±1/2, ±3/2, ±9/2, ±27/2, ±1/4, ±3/4, ±9/4, ±27/4, ±1/8, ±3/8, ±9/8, and ±27/8. Using synthetic division:

$$
\begin{array}{r|rrrr}
-3/4 & 8 & 18 & 45 & 27 \\
     &   & -6 & -9 & -27 \\
\hline
     & 8 & 12 & 36 & 0
\end{array}
$$

So the polynomial factors as $\left(x+\frac{3}{4}\right)\left(8x^2+12x+36\right)=(4x+3)\left(2x^2+3x+9\right)$. The only rational

root is $-3/4$.

53. The possible rational roots are ±1, ±2, ±4, ±1/3, ±2/3, ±4/3, ±1/9, ±2/9, and ±4/9. Using synthetic division:

$$
\begin{array}{r|rrrrrrr}
-1 & 9 & -18 & -28 & 38 & 39 & -4 & -4 \\
   &   & -9 & 27 & 1 & -39 & 0 & 4 \\
\hline
   & 9 & -27 & -1 & 39 & 0 & -4 & 0
\end{array}
$$

Using synthetic division again:

$$
\begin{array}{r|rrrrrr}
2 & 9 & -27 & -1 & 39 & 0 & -4 \\
  &   & 18 & -18 & -38 & 2 & 4 \\
\hline
  & 9 & -9 & -19 & 1   & 2 & 0
\end{array}
$$

Using synthetic division again:

$$
\begin{array}{r|rrrrr}
1/3 & 9 & -9 & -19 & 1 & 2 \\
    &   & 3 & -2 & -7 & -2 \\
\hline
    & 9 & -6 & -21 & -6 & 0
\end{array}
$$

Using synthetic division again:

$$
\begin{array}{r|rrrr}
-1/3 & 9 & -6 & -21 & -6 \\
     &   & -3 & 3   & 6 \\
\hline
     & 9 & -9 & -18 & 0
\end{array}
$$

So the polynomial factors as:
$$(x+1)(x-2)\left(x-\tfrac{1}{3}\right)\left(x+\tfrac{1}{3}\right)\left(9x^2-9x-18\right)=(x+1)(x-2)(3x-1)(3x+1)\left(x^2-x-2\right)$$
$$=(x+1)^2(x-2)^2(3x-1)(3x+1)$$

The rational roots are -1, $-1/3$, $1/3$, and 2.

55. Graphing the polynomial:

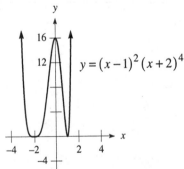

$$y=(x-1)^2(x+2)^4$$

The graph has a turning point at each zero with even multiplicity.

57. Graphing the polynomial:

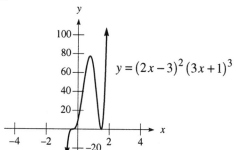

The graph has a turning point at the zero with even multiplicity, and an inflection point at the zero with odd multiplicity.

59. **a.** Graphing the polynomial:

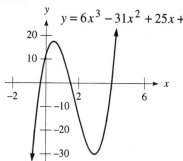

The zeros are –0.33, 1.5, and 4.

b. Using synthetic division:

$$
\begin{array}{r|rrrr}
4 & 6 & -31 & 25 & 12 \\
  &   & 24 & -28 & -12 \\
\hline
  & 6 & -7 & -3 & 0
\end{array}
$$

So $p(x) = (x-4)(6x^2 - 7x - 3) = (x-4)(3x+1)(2x-3)$. Thus the rational roots are $-1/3$, $3/2$, and 4. Though the graphing calculator approach was easier, the algebraic approach is certainly more accurate.

61. Graphing the polynomial:

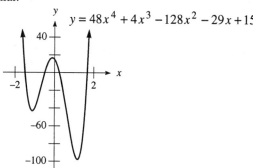

The zeros are approximately –1.5, –0.5, 0.2, and 1.7.

Using synthetic division:

$$
\begin{array}{r|rrrrr}
-3/2 & 48 & 4 & -128 & -29 & 15 \\
     &   & -72 & 102 & 39 & -15 \\
\hline
     & 48 & -68 & -26 & 10 & 0
\end{array}
$$

Using synthetic division again:

$$
\begin{array}{r|rrrr}
-1/2 & 48 & -68 & -26 & 10 \\
     &   & -24 & 46 & -10 \\
\hline
     & 48 & -92 & 20 & 0
\end{array}
$$

Therefore:

$$
\begin{aligned}
p(x) &= \left(x+\tfrac{3}{2}\right)\left(x+\tfrac{1}{2}\right)\left(48x^2 - 92x + 20\right) \\
&= (2x+3)(2x+1)\left(12x^2 - 23x + 5\right) \\
&= (2x+3)(2x+1)(4x-1)(3x-5)
\end{aligned}
$$

The zeros are $-3/2$, $-1/2$, $1/4$, and $5/3$.

63. Since $p(x) = (x-a)q(x) + p(a)$ by the division algorithm and the remainder theorem, and $p(a)$ is constant, then the leading term of $q(x)$ will have to be $a_n x^{n-1}$, thus its leading coefficient is a_n. Note that this implies the possible rational roots will still have factors of a_n as their denominators.

65. a.   If $f\left(\dfrac{p}{q}\right) = 0$, then:

$$
a_n\left(\frac{p}{q}\right)^n + a_{n-1}\left(\frac{p}{q}\right)^{n-1} + \ldots + a_1\left(\frac{p}{q}\right) + a_0 = 0
$$

$$
\frac{a_n p^n}{q^n} + \frac{a_{n-1}p^{n-1}}{q^{n-1}} + \ldots + \frac{a_1 p}{q} + a_0 = 0
$$

$$
a_n p^n + a_{n-1}p^{n-1}q + \ldots + a_1 pq^{n-1} + a_0 q^n = 0
$$

This last step resulted from multiplying the equation by q^n.

b.   Since p divides the first n terms, it must also divide the last term $a_0 q^n$. But $\dfrac{p}{q}$ was reduced to lowest terms, so p cannot divide q, thus p must divide a_0. Similarly, since q divides the last n terms, it must also divide the first term $a_n p^n$. Since q cannot divide p since $\dfrac{p}{q}$ is in lowest terms, q must divide a_n. This completes the proof.

67. A polynomial of even degree can have no x-intercepts. For example, $p(x) = x^4 + 6$ has no x-intercepts. Since complex roots come in conjugate pairs, however, a polynomial of odd degree can only have an even amount of complex roots. The remaining root must be real, therefore a polynomial of odd degree must have at least one x-intercept.

4.5 Rational Functions

1. Since $x^2 - 25 = (x+5)(x-5)$, the domain is all real numbers except -5 and 5, or $(-\infty, -5) \cup (-5, 5) \cup (5, \infty)$. There are no real zeros.

3. Since $2x^2 + 3x - 20 = (2x-5)(x+4)$, the domain is all real numbers except -4 and $5/2$, or $(-\infty, -4) \cup \left(-4, \tfrac{5}{2}\right) \cup \left(\tfrac{5}{2}, \infty\right)$. Since $x^2 - 4x = x(x-4)$, the zeros are $x = 0, 4$.

5. The domain is all real numbers except 0, or $(-\infty, 0) \cup (0, \infty)$. Since $x^2 - 2x + 5 \neq 0$ for any real number x, there are no real zeros.

7.   **a.**   Sketching the graph:

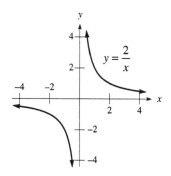

   b.   Sketching the graph:

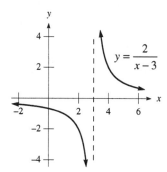

   c.   Sketching the graph:

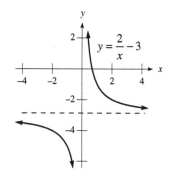

9.   **a.**   Sketching the graph:

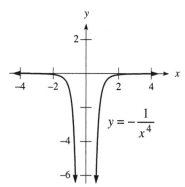

b.   Sketching the graph:

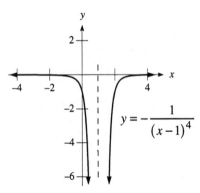

$$y = -\frac{1}{(x-1)^4}$$

c.   Sketching the graph:

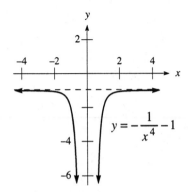

$$y = -\frac{1}{x^4} - 1$$

11 .   Sketching the graphs:

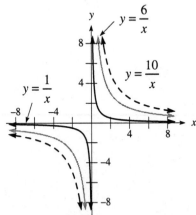

$$y = \frac{6}{x}$$

$$y = \frac{10}{x}$$

$$y = \frac{1}{x}$$

13 .   There are no intercepts. The horizontal asymptote is $y = 0$ and the vertical asymptote is $x = 0$:

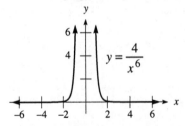

$$y = \frac{4}{x^6}$$

15. The *x*-intercept is 1/5 and there is no *y*-intercept. The horizontal asymptote is $y = -5$ and the vertical asymptote is $x = 0$:

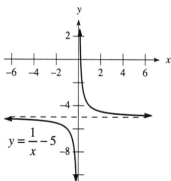

$$y = \frac{1}{x} - 5$$

17. There is no *x*-intercept and the *y*-intercept is 1/4. The horizontal asymptote is $y = 0$ and the vertical asymptote is $x = 2$:

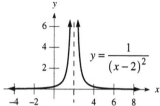

$$y = \frac{1}{(x-2)^2}$$

19. The *x*-intercept is 1/2 and there is no *y*-intercept. The horizontal asymptote is $y = -8$ and the vertical asymptote is $x = 0$:

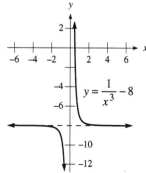

$$y = \frac{1}{x^3} - 8$$

21. The *x*-intercepts are $\pm 1/3$ and there is no *y*-intercept. The horizontal asymptote is $y = 9$ and the vertical asymptote is $x = 0$:

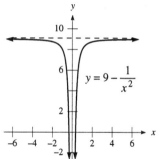

$$y = 9 - \frac{1}{x^2}$$

23. The *x*-intercept is 0 and the *y*-intercept is 0. The horizontal asymptote is $y = 1$ and the vertical asymptote is $x = -5$:

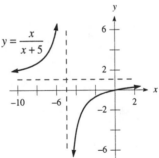

25. The *x*-intercept is 5/3 and there is no *y*-intercept. The horizontal asymptote is $y = 3$ and the vertical asymptote is $x = 0$:

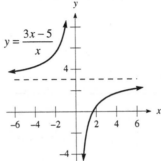

27. The *x*-intercept is –2 and the *y*-intercept is –1. The horizontal asymptote is $y = 1$ and the vertical asymptote is $x = 2$:

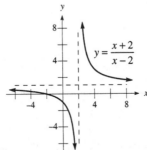

29. The *x*-intercept is 5 and the *y*-intercept is 5/4. The horizontal asymptote is $y = -1$ and the vertical asymptote is $x = -4$:

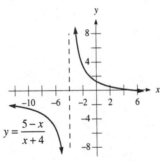

31. The x-intercept is 0 and the y-intercept is 0. The horizontal asymptote is $y = 0$ and the vertical asymptotes are $x = \pm 1$:

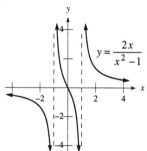

33. The x-intercept is 0 and the y-intercept is 0. The horizontal asymptote is $y = 1$ and the vertical asymptotes are $x = \pm 2$:

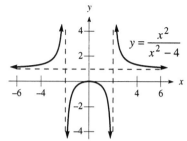

35. Rewrite the curve as $y = \dfrac{x-1}{(x-3)(x+1)}$. The x-intercept is 1 and the y-intercept is 1/3. The horizontal asymptote is $y = 0$ and the vertical asymptotes are $x = 3$ and $x = -1$:

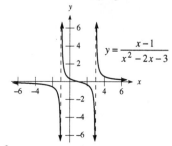

37. Rewrite the curve as $y = \dfrac{2-x}{(2x-3)(x+1)}$. The x-intercept is 2 and the y-intercept is –2/3. The horizontal asymptote is $y = 0$ and the vertical asymptotes are $x = -1$ and $x = \frac{3}{2}$:

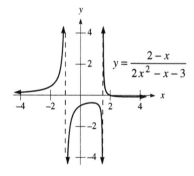

39.   Rewrite the curve as $y = \dfrac{(x-3)(x-1)}{x(x-2)}$. The x-intercepts are 1 and 3, and there is no y-intercept.

The horizontal asymptote is $y = 1$ and the vertical asymptotes are $x = 0$ and $x = 2$:

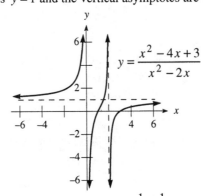

$$y = \frac{x^2 - 4x + 3}{x^2 - 2x}$$

41.   Simplifying the difference quotient: $\dfrac{f(x) - f(5)}{x - 5} = \dfrac{\dfrac{1}{x} - \dfrac{1}{5}}{x - 5} \cdot \dfrac{5x}{5x} = \dfrac{5 - x}{5x(x - 5)} = -\dfrac{1}{5x}$

43.   Simplifying the difference quotient:

$$\frac{f(x+h) - f(x)}{h} = \frac{\dfrac{1}{(x+h)^2} - \dfrac{1}{x^2}}{h} \cdot \frac{x^2(x+h)^2}{x^2(x+h)^2}$$

$$= \frac{x^2 - (x+h)^2}{hx^2(x+h)^2}$$

$$= \frac{x^2 - x^2 - 2xh - h^2}{hx^2(x+h)^2}$$

$$= \frac{-h(2x + h)}{hx^2(x+h)^2}$$

$$= \frac{-2x - h}{x^2(x+h)^2}$$

45.   First use synthetic division:

	1	0	−8	−3
3		3	9	3
	1	3	1	0

So $\dfrac{x^3 - 8x - 3}{x - 3} \approx x^2 + 3x + 1$. When x is near 3, the behavior is similar to $(3)^2 + 3(3) + 1 = 19$.

47.   Since the area is 60000, $xy = 60000$ and so $y = \dfrac{60000}{x}$. The length is given by:

$$L = 3x + 2y = 3x + 2\left(\frac{60000}{x}\right) = 3x + \frac{120000}{x} = \frac{3x^2 + 120000}{x}$$

49.   Simplifying: $\dfrac{(x^2 + 1)(2x) - (x^2 - 1)(2x)}{(x^2 + 1)^2} = \dfrac{2x(x^2 + 1 - x^2 + 1)}{(x^2 + 1)^2} = \dfrac{4x}{(x^2 + 1)^2}$

51. Simplifying:

$$\frac{x^3(2x-2)-\left(x^2-2x-3\right)3x^2}{x^6} = \frac{x^2\left[x(2x-2)-3\left(x^2-2x-3\right)\right]}{x^6}$$

$$= \frac{2x^2-2x-3x^2+6x+9}{x^4}$$

$$= \frac{-x^2+4x+9}{x^4}$$

53. Graphing the function:

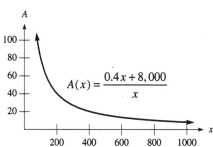

The horizontal asymptote is $A = 0.4$, which means that the average price per item approaches \$0.40 as the number of items increases.

55. Graphing the function:

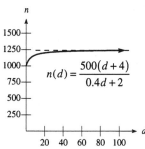

a. Substituting $d = 30$, $d = 60$, and $d = 90$:

$$n(30) = \frac{500(30+4)}{0.5(30)+2} \approx 1214 \text{ insects}$$

$$n(60) = \frac{500(60+4)}{0.5(60)+2} \approx 1231 \text{ insects}$$

$$n(90) = \frac{500(90+4)}{0.5(90)+2} \approx 1237 \text{ insects}$$

b. The horizontal asymptote is $n = 1250$. This means the number of insects approaches 1250 as the number of days increases.

57. **a.** Computing the values:

$$R(1) = \frac{6(1)}{6+1} \approx 0.86 \text{ ohms}$$

$$R(5) = \frac{6(5)}{6+5} \approx 2.73 \text{ ohms}$$

$$R(10) = \frac{6(10)}{6+10} = 3.75 \text{ ohms}$$

These values represent the combined resistance of the circuit when the second resistor is 1 ohm, 5 ohms, and 10 ohms.

b.   Graphing the function:

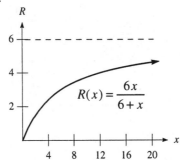

The horizontal asymptote is $R = 6$, which means the total resistance of the circuit approaches 6 ohms as the resistance of the second resistor increases.

59. **a.**   The domain is $g > 0$, or $(0, \infty)$.

   b.   Graphing the function:

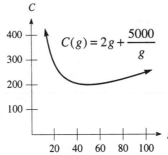

   c.   The cost initially decreases as g increases, then reaches a minimum value and increases steadily after that point.

   d.   The minimum cost is \$200 when 50 gallons of the chemical are stored.

61. **a.**   The domain is all real numbers except 3, or $(-\infty, 3) \cup (3, \infty)$.

   b.   Reducing the function: $f(x) = \dfrac{x^2 - x - 6}{x - 3} = \dfrac{(x-3)(x+2)}{x-3} = x + 2$, if $x \neq 3$

   The domain remains the same.

   c.   Graphing $y = \dfrac{x^2 - x - 6}{x - 3}$:

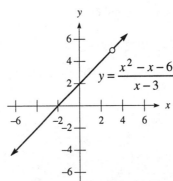

The graphs are not identical, as $y = x + 2$ does not have a "hole" at the point $(3, 5)$.

   d.   Since $f(x) = x + 2$ if $x \neq 3$, the graphs should be identical except for a "hole" at $(3, 5)$.

   e.   The "hole" generally will not appear on a graphing calculator.

63. **a.**   Dividing: $\dfrac{x^2+1}{x} = \dfrac{x^2}{x} + \dfrac{1}{x} = x + \dfrac{1}{x}$

Since $\dfrac{1}{x} \to 0$ as $x \to \pm\infty$, then $y \approx x$ as $x \to \pm\infty$.

b.   As $x \to 0$, $\dfrac{1}{x} \to \pm\infty$, so $y \approx \dfrac{1}{x}$ as $x \to 0$.

65. **a.**   Graphing the curve:

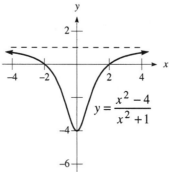

$$y = \dfrac{x^2-4}{x^2+1}$$

b.   Generally the curve appears accurate.

c.   The graph has a horizontal asymptote of $y = 1$, x-intercepts of ± 2, and a y-intercept of -4.

d.   Yes, the graphs generally agree.

67.   The horizontal asymptote will be $y = \dfrac{a}{b}$.

4.6 Radical Functions

1.   The x-intercept is -3 and the y-intercept is $\sqrt{3}$:

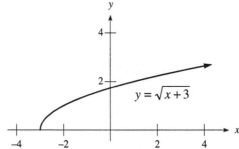

$$y = \sqrt{x+3}$$

3.   The x-intercept is 16 and the y-intercept is -4:

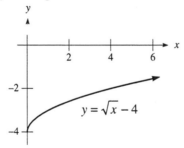

$$y = \sqrt{x} - 4$$

5.   The *x*-intercept is 3 and there is no *y*-intercept:

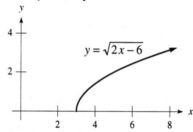

7.   The *x*-intercept is 9 and the *y*-intercept is −6:

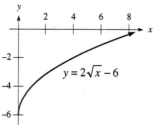

9.   The *x*-intercept is 0 and the *y*-intercept is 0:

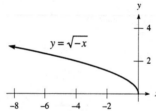

11.   The *x*-intercept is −2 and the *y*-intercept is $\sqrt[3]{2}$:

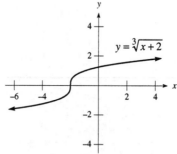

13.   The *x*-intercept is 27 and the *y*-intercept is −3:

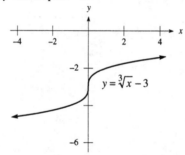

15 . The x-intercept is 1 and the y-intercept is 1:

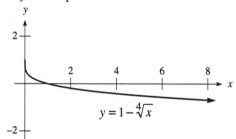

$$y = 1 - \sqrt[4]{x}$$

17 . The x-intercept is 1 and the y-intercept is –1:

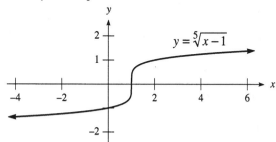

$$y = \sqrt[5]{x} - 1$$

19 . The x-intercept is 1 and the y-intercept is –1:

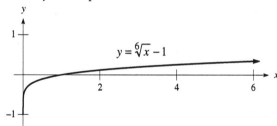

$$y = \sqrt[6]{x} - 1$$

21 . The x-intercepts are ±4 and the y-intercept is 4:

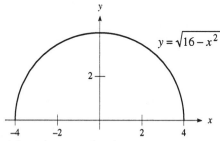

$$y = \sqrt{16 - x^2}$$

23 . The x-intercepts are ±4 and the y-intercept is –4:

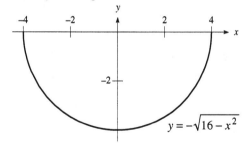

$$y = -\sqrt{16 - x^2}$$

25. Computing the difference quotient: $\dfrac{f(x)-f(4)}{x-4}=\dfrac{\sqrt{x}-2}{x-4}\cdot\dfrac{\sqrt{x}+2}{\sqrt{x}+2}=\dfrac{x-4}{(x-4)(\sqrt{x}+2)}=\dfrac{1}{\sqrt{x}+2}$

27. Computing the difference quotient:

$$\dfrac{f(t)-f(3)}{t-3}=\dfrac{\dfrac{1}{\sqrt{t}}-\dfrac{1}{\sqrt{3}}}{t-3}\cdot\dfrac{\sqrt{3t}}{\sqrt{3t}}$$

$$=\dfrac{\sqrt{3}-\sqrt{t}}{\sqrt{3t}\,(t-3)}\cdot\dfrac{\sqrt{3}+\sqrt{t}}{\sqrt{3}+\sqrt{t}}$$

$$=\dfrac{3-t}{\sqrt{3t}\,(t-3)\left(\sqrt{3}+\sqrt{t}\right)}$$

$$=\dfrac{-1}{\sqrt{3t}\left(\sqrt{3}+\sqrt{t}\right)}$$

29. Finding the intersection point:

$$\sqrt{x}=x-2$$
$$x=(x-2)^2$$
$$x=x^2-4x+4$$
$$0=x^2-5x+4$$
$$0=(x-4)(x-1)$$
$$x=4 \qquad\qquad (x=1 \text{ does not check})$$

The intersection point is (4,2).

31. Finding the intersection point:

$$\sqrt{x}-2=-x$$
$$\sqrt{x}=2-x$$
$$x=(2-x)^2$$
$$x=4-4x+x^2$$
$$0=x^2-5x+4$$
$$0=(x-4)(x-1)$$
$$x=1 \qquad\qquad (x=4 \text{ does not check})$$

The intersection point is (1,–1).

33. Finding the intersection point:

$$\sqrt{9-x}=x+3$$
$$9-x=(x+3)^2$$
$$9-x=x^2+6x+9$$
$$0=x^2+7x$$
$$0=x(x+7)$$
$$x=0 \qquad\qquad (x=-7 \text{ does not check})$$

The intersection point is (0,3).

35. Rationalizing: $\dfrac{\sqrt{x}-2}{x-4}\cdot\dfrac{\sqrt{x}+2}{\sqrt{x}+2}=\dfrac{x-4}{(x-4)(\sqrt{x}+2)}=\dfrac{1}{\sqrt{x}+2}$

Thus when x is near 4, the expression is near $\dfrac{1}{\sqrt{4}+2}=\tfrac{1}{4}$.

37. **a.**   Using the distance formula:
$$\sqrt{(x-2)^2+(y-3)^2}=\sqrt{(x-5)^2+(y-1)^2}$$

  b.   Simplifying by squaring each side:
$$(x-2)^2+(y-3)^2=(x-5)^2+(y-1)^2$$
$$x^2-4x+4+y^2-6y+9=x^2-10x+25+y^2-2y+1$$
$$-4x-6y+13=-10x-2y+26$$
$$6x-4y=13$$

To find the perpendicular bisector, first find the slope and midpoint for the two given points:
$$m=\frac{1-3}{5-2}=-\frac{2}{3}$$
$$M=\left(\frac{2+5}{2},\frac{3+1}{2}\right)=\left(\frac{7}{2},2\right)$$

So the perpendicular bisector has slope $=\frac{3}{2}$:
$$y-2=\frac{3}{2}\left(x-\frac{7}{2}\right)$$
$$y-2=\frac{3}{2}x-\frac{21}{4}$$
$$y=\frac{3}{2}x-\frac{13}{4}$$

This is the identical line to that found above.

39. **a.**   Draw the figure:

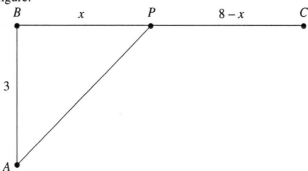

The total length of pipeline needed is: $L=\sqrt{9+x^2}+8-x$

  b.   The total cost is given by: $C=2D\sqrt{9+x^2}+D(8-x)$

  c.   Substituting $x=5$: $C=2D\sqrt{9+5^2}+D(8-5)=\left(2\sqrt{34}+3\right)D\approx14.7D$

41.   Since the height is 8:
$$r^2+8^2=s^2$$
$$r^2=s^2-64$$

The volume is given by: $V=\frac{1}{3}\pi r^2h=\frac{1}{3}\pi\left(s^2-64\right)8=\frac{8}{3}\pi\left(s^2-64\right)$

43.   Since the diagonal of the rectangle is $2r=36$ meters, by the Pythagorean theorem:
$$b^2+h^2=36^2$$
$$b^2=1296-h^2$$
$$b=\sqrt{1296-h^2}$$

The area is given by: $A=bh=h\sqrt{1296-h^2}$

45. **a.**   Substituting $L = 12$: $T = 2\pi\sqrt{\dfrac{12}{980}} \approx 0.7$ seconds

   b.   Substituting $T = 1$:

$$2\pi\sqrt{\frac{L}{980}} = 1$$

$$\sqrt{\frac{L}{980}} = \frac{1}{2\pi}$$

$$\frac{L}{980} = \frac{1}{4\pi^2}$$

$$L = \frac{980}{4\pi^2} \approx 24.8 \text{ cm}$$

47.   Simplifying: $x^{1/2} + (x+8)\frac{1}{2}x^{-1/2} = \sqrt{x} + \dfrac{x+8}{2\sqrt{x}} = \dfrac{2x+x+8}{2\sqrt{x}} = \dfrac{3x+8}{2\sqrt{x}}$

49.   Simplifying:

$$x\left(\tfrac{1}{2}\right)\left(x^2+4\right)^{-1/2}(2x) + \left(x^2+4\right)^{1/2} = \frac{x^2}{\sqrt{x^2+4}} + \frac{\sqrt{x^2+4}}{1} \cdot \frac{\sqrt{x^2+4}}{\sqrt{x^2+4}}$$

$$= \frac{x^2}{\sqrt{x^2+4}} + \frac{x^2+4}{\sqrt{x^2+4}}$$

$$= \frac{2x^2+4}{\sqrt{x^2+4}}$$

51.   Simplifying: $\sqrt[3]{x} + (x-3)x^{-2/3} = x^{1/3} + \dfrac{x-3}{x^{2/3}} = \dfrac{x+x-3}{x^{2/3}} = \dfrac{2x-3}{x^{2/3}} = \dfrac{2x-3}{\sqrt[3]{x^2}}$

4.7 Variation

1.   The variation equation is $y = kx$. Substituting to find k:

$$15 = k \cdot 8$$
$$k = \tfrac{15}{8}$$

So $y = \frac{15}{8}x$. Substituting $x = 25$: $y = \frac{15}{8} \cdot 25 = \frac{375}{8} = 46.875$

3.   The variation equation is $u = \dfrac{k}{t^2}$. Substituting to find k:

$$4 = \frac{k}{8^2}$$

$$4 = \frac{k}{64}$$

$$k = 256$$

So $u = \dfrac{256}{t^2}$. Substituting $t = 10$: $u = \dfrac{256}{10^2} = \dfrac{256}{100} = \dfrac{64}{25} = 2.56$

5.   The variation equation is $z = kmp$. Substituting to find k:

$$20 = k \cdot \tfrac{1}{2} \cdot 7$$

$$k = \tfrac{40}{7}$$

So $z = \frac{40}{7}mp$. Substituting $m = -3$ and $p = \frac{2}{9}$: $z = \frac{40}{7}(-3) \cdot \frac{2}{9} = -\frac{80}{21}$

7. The variation equation is $z = \dfrac{kx^2}{y^3}$. Substituting to find k:

$$1 = \frac{k \cdot 2^2}{3^3}$$
$$1 = \frac{4k}{27}$$
$$k = \frac{27}{4}$$

So $z = \dfrac{27x^2}{4y^3}$. Substituting $x = 3$ and $y = 2$: $z = \dfrac{27 \cdot 3^2}{4 \cdot 2^3} = \frac{243}{32}$

9. The variation equation is $y = \dfrac{k}{x}$. Then: $y = \dfrac{k}{4x} = \dfrac{1}{4} \cdot \dfrac{k}{x}$

So y is multiplied by a factor of $\frac{1}{4}$.

11. The variation equation is $y = k\sqrt{x}$. Then: $y = k\sqrt{9x} = 3 \cdot k\sqrt{x}$

So y is multiplied by a factor of 3.

13. The variation equation is $s = ktu$. Then: $s = k(2t)(3u) = 6ktu$

So s is multiplied by a factor of 6.

15. The variation equation is $F = kx$. Substituting to find k:

$$5 = k(0.4)$$
$$k = 12.5$$

So $F = 12.5x$. Substituting $x = 1.2$: $F = 12.5(1.2) = 15$

A 15 lb weight is necessary.

17. The variation equation is $F = kx$. Substituting to find k:

$$30 = k \cdot 6.5$$
$$k \approx 4.62$$

So $F = 4.62x$. Substituting $F = 42$:

$$42 = 4.62x$$
$$x = 9.1$$

It will stretch the spring 9.1 inches.

19. The variation equation is $I = \dfrac{k}{d^2}$. Substituting to find k:

$$50000 = \frac{k}{100^2}$$
$$k = 500000000$$

So $I = \dfrac{500000000}{d^2}$. Substituting $d = 200$: $I = \dfrac{500000000}{200^2} = 12,500$ lumens

21. The variation equation is $V = \dfrac{k}{P}$. Substituting to find k:

$$100 = \frac{k}{40}$$
$$k = 4000$$

So $V = \dfrac{4000}{P}$. Substituting $V = 30$:

$$30 = \frac{4000}{P}$$
$$30P = 4000$$
$$P \approx 133.33 \text{ lb / sq. in.}$$

23. The variation equation is $R = \dfrac{kl}{d^2}$. Substituting to find k:

$$80 = \frac{k \cdot 100}{(0.36)^2}$$
$$100k = 10.368$$
$$k = 0.10368$$

So $R = \dfrac{0.10368l}{d^2}$. Substituting $d = 0.36$ and $l = 40$: $R = \dfrac{0.10368(100)}{(0.36)^2} = 32$ ohms

25. The variation equation is $v = k\sqrt{l}$. Substituting to find k:

$$35 = k\sqrt{50}$$
$$k \approx 4.9497$$

So $v = 4.9497\sqrt{l}$. Substituting $l = 125$: $v = 4.9497\sqrt{125} \approx 55$ mph

27. The variation equation is $F = kwv^2$.
- **a.** Substituting $2v$: $F = kw(2v)^2 = 4kwv^2$
  So F is multiplied by a factor of 4.
- **b.** Substituting $2w$: $F = k(2w)v^2 = 2kwv^2$
  So F is doubled.
- **c.** Substituting $2v$ and $2w$: $F = k(2w)(2v)^2 = 8kwv^2$
  So F is multiplied by a factor of 8.

29. The variation equation is $L = \dfrac{kwd^2}{l}$. Substituting to find k:

$$750 = \frac{k(3)(5)^2}{10}$$
$$7500 = 75k$$
$$k = 100$$

So $L = \dfrac{100wd^2}{l}$. Substituting $w = 2.5$, $d = 4$, and $l = 15$: $L = \dfrac{100(2.5)(4)^2}{15} \approx 266.67$ lb

31. b. The force varies jointly with the two masses and inversely with the square of the distance.
- **b.** Substituting to find G:

$$6.67 \times 10^{-11} = G \cdot \frac{1 \cdot 1}{1^2}$$
$$G = 6.67 \times 10^{-11}$$

- **c.** Substituting (note that 6800 km = 6,800,000 m):

$$F = 6.67 \times 10^{-11} \cdot \frac{\left(5.98 \times 10^{24}\right)(1000)}{(6800000)^2} \approx 8,626 \text{ Newtons}$$

33. S varies jointly as r and h.

35. V varies directly as the cube of r.

37. z varies jointly as the square of x and the cube of y, and inversely as w.

39. d and r vary directly, d and t vary directly, while r and t vary inversely.

Chapter 4 Review Exercises

1.   The x-intercept is -2 and the y-intercept is 8:

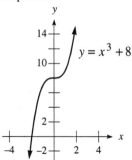

2.   The x-intercept is 2 and the y-intercept is 64:

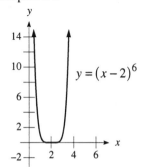

3.   The x-intercepts are -1 and 3, and the y-intercept is -15:

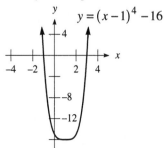

4.   The x-intercept is -1 and the y-intercept is 31:

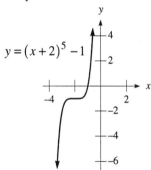

5.   Factoring: $y = x^3 - x^2 - 6x = x\left(x^2 - x - 6\right) = x(x-3)(x+2)$

The x-intercepts are –2, 0, and 3, and the y-intercept is 0:

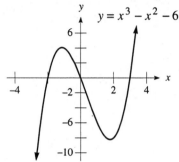

6.   Factoring: $y = x^4 - 4x^2 = x^2\left(x^2 - 4\right) = x^2(x+2)(x-2)$

The x-intercepts are –2, 0, and 2, and the y-intercept is 0:

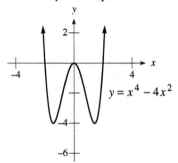

7.   Factoring: $y = x^2 - 6x + 9 = (x-3)^2$

The x-intercept is 3 and the y-intercept is 9:

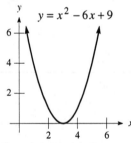

8.   The x-intercepts are –1 and 1, and the y-intercept is 1:

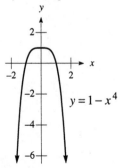

9. There is no x-intercept, the y-intercept is 1, the vertical asymptote is $x = 1$, and the horizontal asymptote is $y = 0$:

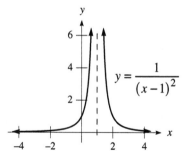

$$y = \frac{1}{(x-1)^2}$$

10. The x-intercept is 1/2, there is no y-intercept, the vertical asymptote is $x = 0$, and the horizontal asymptote is $y = 2$:

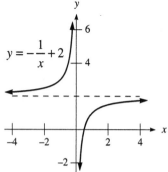

$$y = -\frac{1}{x} + 2$$

11. The x-intercept is $-5/3$, the y-intercept is $-5/2$, the vertical asymptote is $x = -2$, and the horizontal asymptote is $y = -3$:

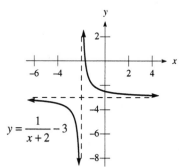

$$y = \frac{1}{x+2} - 3$$

12. There is no x-intercept, the y-intercept is 163/81, the vertical asymptote is $x = 3$, and the horizontal asymptote is $y = 2$:

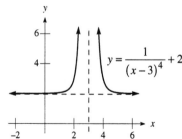

$$y = \frac{1}{(x-3)^4} + 2$$

13.   The x-intercept is -3, the y-intercept is $3/2$, the vertical asymptote is $x = -2$, and the horizontal asymptote is $y = 1$:

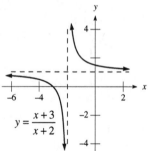

14.   The x-intercept is 0, the y-intercept is 0, the vertical asymptote is $x = -1$, and the horizontal asymptote is $y = 1$:

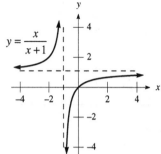

15.   The x-intercept is 0, the y-intercept is 0, the vertical asymptotes are $x = -1$ and $x = 1$, and the horizontal asymptote is $y = 0$:

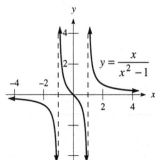

16.   There is no x-intercept, the y-intercept is $1/2$, the vertical asymptotes are $x = -2$ and $x = 2$, and the horizontal asymptote is $y = -1$:

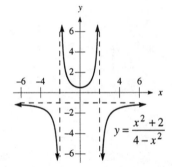

17. The *x*-intercept is 5/2 and there is no *y*-intercept:

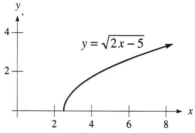

18. The *x*-intercept is 5 and the *y*-intercept is $1 + \sqrt[3]{2}$:

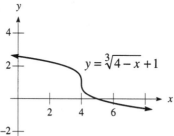

19. The *x*-intercept is 32 and the *y*-intercept is –2:

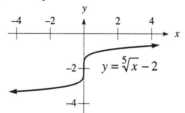

20. The *x*-intercept is –2 and the *y*-intercept is $\sqrt[4]{6}$:

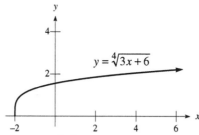

21. The *x*-intercept is 2 and the *y*-intercept is 8:

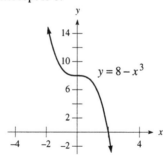

22. The x-intercept is 0 and the y-intercept is 0:

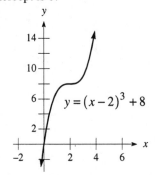

23. Using long division:

$$
\begin{array}{r}
x^2 + 2 \\
2x-3\overline{)2x^3 - 3x^2 + 4x - 7} \\
\underline{2x^3 - 3x^2} \\
4x - 7 \\
\underline{4x - 6} \\
-1
\end{array}
$$

The quotient is $x^2 + 2$ and the remainder is -1.

24. Using long division:

$$
\begin{array}{r}
x^2 - 1 \\
x^2+1\overline{)x^4 + 0x^2 - x + 1} \\
\underline{x^4 +\ x^2} \\
-x^2 - x + 1 \\
\underline{-x^2 \qquad -1} \\
-x + 2
\end{array}
$$

The quotient is $x^2 - 1$ and the remainder is $-x + 2$.

25. Using long division:

$$
\begin{array}{r}
x^3 + 1 \\
2x+1\overline{)2x^4 + x^3 + 2x + 1} \\
\underline{2x^4 + x^3} \\
2x + 1 \\
\underline{2x + 1} \\
0
\end{array}
$$

The quotient is $x^3 + 1$ and the remainder is 0.

26. Using long division:

$$
\begin{array}{r}
x^3 + 2x^2 - x - 1 \\
x^2 + x - 1{\overline{\smash{\big)}\,x^5 + 3x^4 + 0x^3 - 4x^2 + 0x + 2}} \\
\underline{x^5 + x^4 - x^3} \\
2x^4 + x^3 - 4x^2 \\
\underline{2x^4 + 2x^3 - 2x^2} \\
-x^3 - 2x^2 + 0x \\
\underline{-x^3 - x^2 + x} \\
-x^2 - x + 2 \\
\underline{-x^2 - x + 1} \\
1
\end{array}
$$

The quotient is $x^3 + 2x^2 - x - 1$ and the remainder is 1.

27. Using synthetic division:

$$
\begin{array}{r|rrrr}
3 & 2 & -3 & -4 & -15 \\
  &   & 6 & 9 & 15 \\
\hline
  & 2 & 3 & 5 & 0
\end{array}
$$

The quotient is $2x^2 + 3x + 5$ and the remainder is 0.

28. Using synthetic division:

$$
\begin{array}{r|rrrrr}
-2 & 1 & 0 & 0 & -5 & 3 \\
   &   & -2 & 4 & -8 & 26 \\
\hline
   & 1 & -2 & 4 & -13 & 29
\end{array}
$$

The quotient is $x^3 - 2x^2 + 4x - 13$ and the remainder is 29.

29. Using synthetic division:

$$
\begin{array}{r|rrrrrr}
2 & 1 & -2 & 1 & -3 & 5 & -4 \\
  &   & 2 & 0 & 2 & -2 & 6 \\
\hline
  & 1 & 0 & 1 & -1 & 3 & 2
\end{array}
$$

The quotient is $x^4 + x^2 - x + 3$ and the remainder is 2.

30. Using synthetic division:

$$
\begin{array}{r|rrrrrrr}
-1 & 1 & 0 & 0 & 0 & 0 & -1 & -3 \\
   &   & -1 & 1 & -1 & 1 & -1 & 2 \\
\hline
   & 1 & -1 & 1 & -1 & 1 & -2 & -1
\end{array}
$$

The quotient is $x^5 - x^4 + x^3 - x^2 + x - 2$ and the remainder is -1.

31. The possible rational roots are $\pm 1, \pm 2, \pm 3, \pm 4, \pm 6,$ and ± 12. Using synthetic division:

$$
\begin{array}{r|rrrr}
-1 & 1 & 0 & -13 & -12 \\
   &   & -1 & 1   & 12 \\
\hline
   & 1 & -1 & -12 & 0
\end{array}
$$

So $p(x) = (x+1)(x^2 - x - 12) = (x+1)(x-4)(x+3)$. The zeros are $-3, -1,$ and 4.

32. The possible rational roots are $\pm 1, \pm 2, \pm 3,$ and ± 6. Using synthetic division:

$$
\begin{array}{r|rrrr}
-2 & 1 & 2 & -3 & -6 \\
   &   & -2 & 0 & 6 \\
\hline
   & 1 & 0 & -3 & 0
\end{array}
$$

So $p(x) = (x+2)(x^2 - 3)$. The zeros are -2 and $\pm\sqrt{3}$.

33. The possible rational roots are ±1, ±2, ±5, and ±10. Using synthetic division:

$$
\begin{array}{r|rrrr}
5 & 1 & -6 & 7 & -10 \\
  &   & 5 & -5 & 10 \\
\hline
  & 1 & -1 & 2 & 0
\end{array}
$$

So $p(x) = (x-5)(x^2 - x - 2)$. Using the quadratic formula: $x = \dfrac{1 \pm \sqrt{1-8}}{2} = \dfrac{1 \pm \sqrt{-7}}{2} = \dfrac{1 \pm i\sqrt{7}}{2}$

The zeros are 5 and $\dfrac{1 \pm i\sqrt{7}}{2}$.

34. The possible rational roots are ±1, ±2, ±3, ±6, ±9, ±18, ±1/3, and ±2/3. Using synthetic division:

$$
\begin{array}{r|rrrr}
3 & 3 & -2 & -27 & 18 \\
  &   & 9 & 21 & -18 \\
\hline
  & 3 & 7 & -6 & 0
\end{array}
$$

So $p(x) = (x-3)(3x^2 + 7x - 6) = (x-3)(3x-2)(x+3)$. The zeros are –3, 2/3, and 3.

35. The possible rational roots are ±1, ±2, ±4, and ±1/2. Using synthetic division:

$$
\begin{array}{r|rrrrr}
-4 & 2 & 7 & -2 & 7 & -4 \\
   &   & -8 & 4 & -8 & 4 \\
\hline
   & 2 & -1 & 2 & -1 & 0
\end{array}
$$

Using synthetic division again:

$$
\begin{array}{r|rrrr}
1/2 & 2 & -1 & 2 & -1 \\
    &   & 1 & 0 & 1 \\
\hline
    & 2 & 0 & 2 & 0
\end{array}
$$

So $p(x) = (x+4)\left(x-\tfrac{1}{2}\right)(2x^2 + 2)$. Since $2x^2 + 2 = 0$ when $x = \pm i$, the zeros are –4, 1/2, and $\pm i$.

36. The possible rational roots are ±1, ±2, and ±4. Using synthetic division:

$$
\begin{array}{r|rrrrr}
-1 & 1 & -2 & -3 & 4 & 4 \\
   &   & -1 & 3 & 0 & -4 \\
\hline
   & 1 & -3 & 0 & 4 & 0
\end{array}
$$

Using synthetic division again:

$$
\begin{array}{r|rrrr}
2 & 1 & -3 & 0 & 4 \\
  &   & 2 & -2 & -4 \\
\hline
  & 1 & -1 & -2 & 0
\end{array}
$$

So $p(x) = (x+1)(x-2)(x^2 - x - 2) = (x+1)^2 (x-2)^2$. The zeros are –1 and 2, both with multiplicity 2.

37. $p(x)$ has one more zero of $5 + 2i$. One factor must be: $[x - (5-2i)][x - (5+2i)] = x^2 - 10x + 29$

Thus: $p(x) = (x+2)(x^2 - 10x + 29) = x^3 - 8x^2 + 9x + 58$

38. $p(x)$ has 2 more zeros: $2 - 3i$ and $1 + i$. One factor must be:

$[x - (2+3i)][x - (2-3i)] = x^2 - 4x + 13$

Another factor must be: $[x - (1-i)][x - (1+i)] = x^2 - 2x + 2$

Thus: $p(x) = (x^2 - 4x + 13)(x^2 - 2x + 2) = x^4 - 6x^3 + 23x^2 - 34x + 26$

39. Since the perimeter of the square is 20, $AD = 5$. Call the base of the triangle b, now using the Pythagorean theorem:

$$x^2 + b^2 = 5^2$$
$$b^2 = 25 - x^2$$
$$b = \sqrt{25 - x^2}$$

Thus the area of the triangle is given by: $A = \frac{1}{2}bx = \frac{1}{2}x\sqrt{25 - x^2}$

40. Let h represent the height. Since the volume is 100:

$$x^2 h = 100$$
$$h = \frac{100}{x^2}$$

Thus the surface area is given by: $S = 2x^2 + 4xh = 2x^2 + 4x \cdot \dfrac{100}{x^2} = 2x^2 + \dfrac{400}{x} = \dfrac{2x^3 + 400}{x}$

41. Finding the points of intersection:

$$\sqrt{4 - x} = x + 2$$
$$4 - x = (x + 2)^2$$
$$4 - x = x^2 + 4x + 4$$
$$0 = x^2 + 5x$$
$$0 = x(x + 5)$$
$$x = 0 \qquad (x = -5 \text{ does not check})$$

The intersection point is (0,2).

42. Finding the points of intersection:

$$\sqrt{x} - 3 = x - 9$$
$$\sqrt{x} = x - 6$$
$$x = (x - 6)^2$$
$$x = x^2 - 12x + 36$$
$$0 = x^2 - 13x + 36$$
$$0 = (x - 9)(x - 4)$$
$$x = 9 \qquad (x = 4 \text{ does not check})$$

The intersection point is (9,0).

43. Finding the points of intersection:

$$\sqrt{2x + 5} - 2 = x - 7$$
$$\sqrt{2x + 5} = x - 5$$
$$2x + 5 = (x - 5)^2$$
$$2x + 5 = x^2 - 10x + 25$$
$$0 = x^2 - 12x + 20$$
$$0 = (x - 10)(x - 2)$$
$$x = 10 \qquad (x = 2 \text{ does not check})$$

The intersection point is (10,3).

44. Finding the points of intersection:

$$4 - \sqrt{x} = 2x + 1$$
$$-\sqrt{x} = 2x - 3$$
$$x = (2x - 3)^2$$
$$x = 4x^2 - 12x + 9$$
$$0 = 4x^2 - 13x + 9$$
$$0 = (4x - 9)(x - 1)$$
$$x = 1 \qquad\qquad (x = \tfrac{9}{4} \text{ does not check})$$

The intersection point is (1,3).

45. The variation equation is $x = \dfrac{k}{\sqrt[3]{y}}$. Substituting to find k:

$$27 = \frac{k}{\sqrt[3]{27}}$$
$$27 = \frac{k}{3}$$
$$k = 81$$

So $x = \dfrac{81}{\sqrt[3]{y}}$. Substituting $y = 8$: $x = \dfrac{81}{\sqrt[3]{8}} = \tfrac{81}{2}$

46. The variation equation is $z = kxy^2$. Substituting to find k:

$$30 = k(5)2^2$$
$$30 = 20k$$
$$k = \tfrac{3}{2}$$

So $z = \tfrac{3}{2}xy^2$. Substituting $z = 10$ and $y = 3$:

$$10 = \tfrac{3}{2}x \cdot 3^2$$
$$20 = 27x$$
$$x = \tfrac{20}{27}$$

47. The variation equation is $d = kT^{3/2}$. Substituting to find k:

$$350 = k \cdot 4^{3/2}$$
$$350 = 8k$$
$$k = \tfrac{175}{4}$$

So $d = \tfrac{175}{4}T^{3/2}$. Substituting $d = 500$:

$$500 = \tfrac{175}{4}T^{3/2}$$
$$2000 = 175T^{3/2}$$
$$T^{3/2} \approx 11.43$$
$$T \approx 5.1$$

It will take approximately 5.1 seconds for the object to fall 500 feet.

48. The variation equation is $V = khr^2$. Substituting to find k:

$$4\pi = k(3)(2)^2$$
$$4\pi = 12k$$
$$k = \frac{\pi}{3}$$

So $V = \dfrac{\pi}{3}hr^2$. Substituting $h = 2$ and $r = 3$: $V = \dfrac{\pi}{3}(2)(3)^2 = 6\pi$ cubic inches

Chapter 4 Practice Test

1. **a.** The x-intercept is 5/3, the y-intercept is 5/2, the vertical asymptote is $x = 2$, and the horizontal asymptote is $y = 3$:

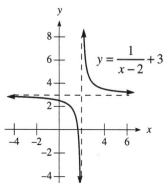

$$y = \frac{1}{x-2} + 3$$

b. There is no x-intercept, the y-intercept is -3, the vertical asymptote is $x = -1$, and the horizontal asymptote is $y = 0$:

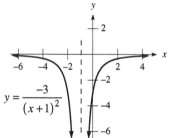

$$y = \frac{-3}{(x+1)^2}$$

c. The x-intercept is 4 and the y-intercept is 0:

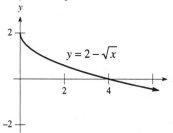

$$y = 2 - \sqrt{x}$$

d. The x-intercept is -3, the y-intercept is $-3/2$, the vertical asymptote is $x = 2$, and the horizontal asymptote is $y = 1$:

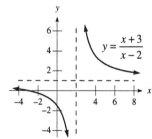

$$y = \frac{x+3}{x-2}$$

e.   Factoring: $y = x^3 - x^2 - 12x = x\left(x^2 - x - 12\right) = x(x - 4)(x + 3)$

The x-intercepts are $-3, 0,$ and 4, and the y-intercept is 0:

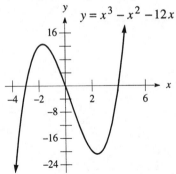

f.   The x-intercept is 0, the y-intercept is 0, the vertical asymptotes are $x = -3$ and $x = 3$, and the horizontal asymptote is $y = 0$:

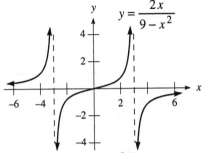

g.   The x-intercept is 1 and the y-intercept is $1 - \sqrt[3]{2}$:

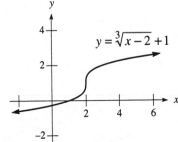

h.   The x-intercepts are -1 and 2, the y-intercept is 4:

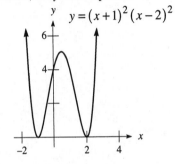

2.   Using long division:

$$
\begin{array}{r}
x^3 - 2x^2 - 5x - 1 \\
3x - 4 \overline{\smash{\big)}\ 3x^4 - 10x^3 - 7x^2 + 17x + 3} \\
\underline{3x^4 - 4x^3} \\
-6x^3 - 7x^2 \\
\underline{-6x^3 + 8x^2} \\
-15x^2 + 17x \\
\underline{-15x^2 + 20x} \\
-3x + 3 \\
\underline{-3x + 4} \\
-1
\end{array}
$$

The quotient is $x^3 - 2x^2 - 5x - 1 - \dfrac{1}{3x - 4}$.

3.   **a.**   Using synthetic division:

$$
\begin{array}{r|rrrrr}
-3 & 2 & -1 & -11 & 4 & 12 \\
   &   & -6 & 21 & -30 & 78 \\
\hline
   & 2 & -7 & 10 & -26 & 90
\end{array}
$$

Since the remainder is nonzero, $x + 3$ is not a factor of $p(x)$.

   b.   Using synthetic division:

$$
\begin{array}{r|rrrrr}
2 & 2 & -1 & -11 & 4 & 12 \\
  &   & 4 & 6 & -10 & -12 \\
\hline
  & 2 & 3 & -5 & -6 & 0
\end{array}
$$

Since the remainder is zero, $x - 2$ is a factor of $p(x)$.

   c.   Using synthetic division:

$$
\begin{array}{r|rrrrr}
1 & 2 & -1 & -11 & 4 & 12 \\
  &   & 2 & 1 & -10 & -6 \\
\hline
  & 2 & 1 & -10 & -6 & 6
\end{array}
$$

Since the remainder is nonzero, $x - 1$ is not a factor of $p(x)$.

   d.   Using synthetic division:

$$
\begin{array}{r|rrrrr}
-2 & 2 & -1 & -11 & 4 & 12 \\
   &   & -4 & 10 & 2 & -12 \\
\hline
   & 2 & -5 & -1 & 6 & 0
\end{array}
$$

Since the remainder is zero, $x + 2$ is a factor of $p(x)$.

4.   **a.**   The possible rational roots are ±1, ±2, ±3, ±4, ±6, ±12, ±1/3, ±2/3 and ±4/3. Using synthetic division:

$$
\begin{array}{r|rrrr}
3 & 3 & -20 & 29 & 12 \\
  &   & 9 & -33 & -12 \\
\hline
  & 3 & -11 & -4 & 0
\end{array}
$$

So $p(x) = (x - 3)(3x^2 - 11x + 4) = (x - 3)(3x + 1)(x - 4)$. The zeros are –1/3, 3, and 4.

b.   First factor $p(x) = x\left(2x^3 - 9x^2 + 19x - 15\right)$. The possible rational roots are $\pm 1, \pm 3, \pm 5, \pm 15,$
$\pm 1/2, \pm 3/2, \pm 5/2,$ and $\pm 15/2.$ Using synthetic division:

$$
\begin{array}{r|rrrr}
3/2 & 2 & -9 & 19 & -15 \\
    &   & 3 & -9 & 15 \\
\hline
    & 2 & -6 & 10 & 0
\end{array}
$$

So $p(x) = x\left(x - \frac{3}{2}\right)\left(2x^2 - 6x + 10\right) = x(2x - 3)\left(x^2 - 3x + 5\right)$. Using the quadratic formula:

$$x = \frac{3 \pm \sqrt{9 - 20}}{2} = \frac{3 \pm \sqrt{-11}}{2} = \frac{3 \pm i\sqrt{11}}{2}$$

The zeros are 0, 3/2, and $\dfrac{3 \pm i\sqrt{11}}{2}$.

5.   The variation equation is $P = \dfrac{kl}{r^4}$. Comparing $r = 2$ to $r = 3$:

$$\frac{P(2)}{P(3)} = \frac{\dfrac{kl}{2^4}}{\dfrac{kl}{3^4}} = \frac{3^4}{2^4} = \frac{81}{16} = 5.0625$$

The blood pressure would increase by a factor of 5.0625.

Chapter 5
Exponential and Logarithmic Functions

5.1 Exponential Functions

1.   Solving the equation:
$$2^{x-1} = 8$$
$$2^{x-1} = 2^3$$
$$x - 1 = 3$$
$$x = 4$$

3.   Solving the equation:
$$3^{2x} = 243$$
$$3^{2x} = 3^5$$
$$2x = 5$$
$$x = \tfrac{5}{2}$$

5.   Solving the equation:
$$9^t = 27$$
$$3^{2t} = 3^3$$
$$2t = 3$$
$$t = \tfrac{3}{2}$$

7.   Solving the equation:
$$\left(\tfrac{1}{2}\right)^{x+2} = 16$$
$$2^{-x-2} = 2^4$$
$$-x - 2 = 4$$
$$-x = 6$$
$$x = -6$$

9.   Solving the equation:
$$8^x = \sqrt{2}$$
$$2^{3x} = 2^{1/2}$$
$$3x = \tfrac{1}{2}$$
$$x = \tfrac{1}{6}$$

11. The domain is all real numbers, or $(-\infty, \infty)$.

13. For the radical to be defined:
$$2^x - 8 \geq 0$$
$$2^x \geq 8$$
$$2^x \geq 2^3$$
$$x \geq 3$$
The domain is $[3, \infty)$.

15. For the denominator to be nonzero:
$$2^{3x} - 2 \neq 0$$
$$2^{3x} \neq 2$$
$$3x \neq 1$$
$$x \neq \tfrac{1}{3}$$
The domain is all real numbers except $\tfrac{1}{3}$, or $\left(-\infty, \tfrac{1}{3}\right) \cup \left(\tfrac{1}{3}, \infty\right)$.

17. The domain is all real numbers, the range is $(0, \infty)$, there is no x-intercept, and the y-intercept is 1:

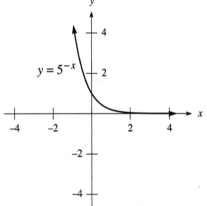

19. The domain is all real numbers, the range is $(-1, \infty)$, the x-intercept is 0, and the y-intercept is 0:

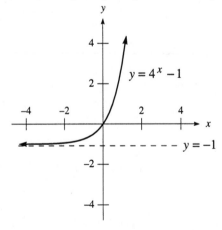

21. The domain is all real numbers, the range is $(-\infty, 9)$, the x-intercept is 2, and the y-intercept is 8:

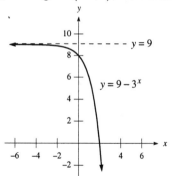

23. The domain is all real numbers, the range is $(-8, \infty)$, the x-intercept is 6, and the y-intercept is $-63/8$:

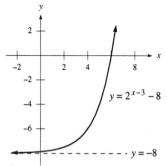

25. The domain is all real numbers, the range is $(-\infty, 1)$, the x-intercept is 0, and the y-intercept is 0:

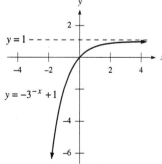

27. The domain is all real numbers, the range is $(0, \infty)$, there is no x-intercept, and the y-intercept is $e^{-2} \approx 0.14$:

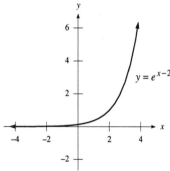

29.   The domain is all real numbers, the range is $(-\infty, 1)$, the x-intercept is 0, and the y-intercept is 0:

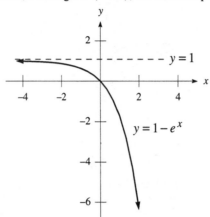

31.   Finding the composition: $(f \circ g)(x) = f\left(x^2 - 6x + 2\right) = 5^{x^2 - 6x + 2}$

33.   Finding the composition: $(f \circ t)(x) = f\left(\sqrt{x}\right) = 5^{\sqrt{x}}$

35.   Finding the composition: $(h \circ f)(x) = h\left(5^x\right) = \dfrac{1}{5^x + 1}$

37.   Finding the real zeros:
$$x \cdot 4^x = 0$$
$$x = 0$$

39.   Finding the real zeros:
$$9\left(2^x\right) - x^2\left(2^x\right) = 0$$
$$2^x\left(9 - x^2\right) = 0$$
$$2^x(3 + x)(3 - x) = 0$$
$$x = -3, 3$$

41.   Finding the real zeros:
$$\frac{1 - 3^x}{x^2} = 0$$
$$1 = 3^x$$
$$x = 0$$
But $x = 0$ results in a 0 denominator, so there are no real zeros.

43.   Using the hint, let $u = 3^x$:
$$u + u^{-1} = 2$$
$$u^2 + 1 = 2u$$
$$u^2 - 2u + 1 = 0$$
$$(u - 1)^2 = 0$$
$$u = 1$$
$$3^x = 1$$
$$x = 0$$

45.   Simplifying: $f(x + 3) = 2^{x+3} = 2^x \cdot 2^3 = 8f(x)$

47.   Simplifying: $\dfrac{f(x + 2) - f(x)}{2} = \dfrac{3^{x+2} - 3^x}{2} = \dfrac{9 \cdot 3^x - 3^x}{2} = \dfrac{8 \cdot 3^x}{2} = 4 \cdot 3^x$

49. Graphing the curves:

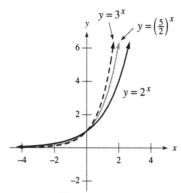

51. The population is given by $P(t) = 4 \cdot 2^{t/2}$. After 6 months, the population will be $P(6) = 4 \cdot 2^{6/2} = 4 \cdot 2^3 = 32$. After 2 years (24 months), the population will be $P(24) = 4 \cdot 2^{24/2} = 4 \cdot 2^{12} = 16{,}384$.

53. a. Graphing $B(t) = 20000\left(\frac{600}{20000}\right)^{t/20} = 20000\left(\frac{3}{100}\right)^{t/20}$:

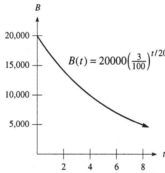

b. After 3 years, the value will be $B(3) = 20000\left(\frac{3}{100}\right)^{3/20} \approx \$11{,}819$.

55. a. Substituting $x = 500$: $p = 600 - 0.4e^{0.005 \cdot 500} \approx \$595.13$

b. Substituting $x = 900$: $p = 600 - 0.4e^{0.005 \cdot 900} \approx \$563.99$

57. a. Substituting $x = 0$: $P(0) = (14.69)3^0 = 14.69$ lb / in^2

b. Substituting $x = 29028$: $P(29028) = (14.69)3^{-0.000035 \cdot 29028} \approx 4.81$ lb / in^2

c. Substituting $x = -1290$: $P(-1290) = (14.69)3^{-0.000035 \cdot (-1290)} \approx 15.44$ lb / in^2

59. a. In 4 years: $P(4) = 12000 \cdot 2^{0.05 \cdot 4} \approx 13{,}784$
In 5 years: $P(5) = 12000 \cdot 2^{0.05 \cdot 5} \approx 14{,}270$

b. In 19 years: $P(19) = 12000 \cdot 2^{0.05 \cdot 19} \approx 23{,}182$
In 20 years: $P(20) = 12000 \cdot 2^{0.05 \cdot 20} \approx 24{,}000$

c. The ratios are:
$$\frac{P(5)}{P(4)} = \frac{14270}{13784} \approx 1.04 \qquad \frac{P(20)}{P(19)} = \frac{24000}{23182} \approx 1.04$$

d. The ratio is: $\dfrac{P(t+1)}{P(t)} = \dfrac{12000 \cdot 2^{0.05(t+1)}}{12000 \cdot 2^{0.05t}} = 2^{0.05t+0.05-0.05t} = 2^{0.05}$
The population after $t+1$ years is $2^{0.05}$ times the population after t years.

61. a.   The population is given by $N(t) = N_0 \cdot 2^{t/2.5}$.

  b.   Using $N_0 = 20$ and $t = 7$: $N(7) = 20 \cdot 2^{7/2.5} \approx 139$

63. a.   The amount of isotope still radioactive is given by $A(t) = A_0 \cdot 2^{-t/4}$.

  b.   Let $A_0 = 10$ mg. Substituting $t = 2$ and $t = 24$:
$$A(2) = 10 \cdot 2^{-2/4} \approx 7.07 \text{ mg}$$
$$A(24) = 10 \cdot 2^{-24/4} \approx 0.16 \text{ mg}$$

65.   Substituting $T = 40°$:
$$40 = 15 + 85e^{-0.4m}$$
$$25 = 85e^{-0.4m}$$
$$e^{-0.4m} = \tfrac{5}{17}$$
$$e^{0.4m} = \tfrac{17}{5}$$
$$m \approx 3.1 \text{ minutes}$$
The last value was found by using a graphing calculator.

67. a.   Graphing the curve:

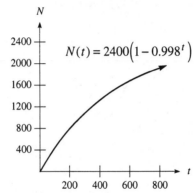

$$N(t) = 2400\left(1 - 0.998^t\right)$$

  b.   Substituting values for t:
$$N(400) = 2400\left(1 - 0.998^{400}\right) \approx 1322$$
$$N(600) = 2400\left(1 - 0.998^{600}\right) \approx 1678$$
$$N(800) = 2400\left(1 - 0.998^{800}\right) \approx 1916$$

  c.   In terms of percentages, these are:
$$t = 400: \quad \tfrac{1322}{2400} \approx 55\%$$
$$t = 600: \quad \tfrac{1678}{2400} \approx 70\%$$
$$t = 800: \quad \tfrac{1916}{2400} \approx 80\%$$

  d.   After 800 hours, 80% of the bulbs will have burned out. It might make more sense (since this is the average life of the bulb) to replace all of the bulbs at that time.

69. a.   When $t = 0$ the concentration is I, so I represents the initial concentration.

  b.   As t increases C becomes closer to F, so F represents the final concentration that C is approaching as time increases.

71.   Substituting $x = 5$: $I(5) = 12e^{-0.8(5)} \approx 0.22$ lumens

73. **a.** Substituting $d = 14$: $p(14) = 1 - e^{-0.0235 \cdot 14} \approx 28\%$

    b. Finding where $p(d) = 0.50$:

$$1 - e^{-0.0235d} = 0.50$$
$$e^{-0.0235d} = 0.50$$
$$d \approx 29 \quad \text{(obtained from the graph)}$$

It will take approximately 29 days.

75. **a.** Solving for x:

$$2^x = 16$$
$$2^x = 2^4$$
$$x = 4$$

The idea of one-to-oneness is used in concluding $x = 4$ from $2^x = 2^4$.

    b. Yes, but it is unclear what the solution is, since there is no rational number x such that $2^x = 15$. The next section will define a solution to this.

77. **a.** If $b = 1$, $b^x = 1$ for all x.

    b. If $b < 0$, the domain of $y = b^x$ becomes difficult, since $x = \frac{1}{3}$ would be a valid value while $x = \frac{1}{4}$ (which means the fourth root) would not. The domain of such a function would be totally disconnected (no intervals in the domain, only points), which is why we require $b > 0$ in our definition.

5.2 Logarithmic Functions

1. Written in exponential form: $7^2 = 49$

3. Written in logarithmic form: $\log_3 \frac{1}{81} = -4$

5. Written in exponential form: $\left(\frac{1}{4}\right)^{-3} = 64$

7. Written in logarithmic form: $\log_{27} \frac{1}{3} = -\frac{1}{3}$

9. Written in exponential form: $8^{-1/3} = \frac{1}{2}$

11. Written in logarithmic form: $\log_5 \sqrt[4]{5} = \frac{1}{4}$

13. Written in exponential form: $2^{-1} = \frac{1}{2}$

15. Written in exponential form: $\left(\frac{2}{3}\right)^{-3} = \frac{27}{8}$

17. Written in logarithmic form: $\log_4 1024 = 5$

19. Written in logarithmic form: $\log_{1/9} 3 = -\frac{1}{2}$

21. Let $x = \log_2 32$. Then:

$$2^x = 32$$
$$2^x = 2^5$$
$$x = 5$$

So $\log_2 32 = 5$.

23.   Let $x = \log_6 \frac{1}{6}$. Then:
$$6^x = \frac{1}{6}$$
$$6^x = 6^{-1}$$
$$x = -1$$
So $\log_6 \frac{1}{6} = -1$.

25.   Let $x = \log_8 16$. Then:
$$8^x = 16$$
$$2^{3x} = 2^4$$
$$3x = 4$$
$$x = \frac{4}{3}$$
So $\log_8 16 = \frac{4}{3}$.

27.   Let $x = \log_{16} \frac{1}{8}$. Then:
$$16^x = \frac{1}{8}$$
$$2^{4x} = 2^{-3}$$
$$4x = -3$$
$$x = -\frac{3}{4}$$
So $\log_{16} \frac{1}{8} = -\frac{3}{4}$.

29.   Let $x = \log_3 (-9)$. Then: $3^x = -9$

Since this is impossible, $\log_3 (-9)$ is undefined.

31.   Let $x = \log_4 \frac{1}{8}$. Then:
$$4^x = \frac{1}{8}$$
$$2^{2x} = 2^{-3}$$
$$2x = -3$$
$$x = -\frac{3}{2}$$
So $\log_4 \frac{1}{8} = -\frac{3}{2}$.

33.   Let $x = \log_{10} 0.0001$. Then:
$$10^x = 0.0001$$
$$10^x = 10^{-4}$$
$$x = -4$$
So $\log_{10} 0.0001 = -4$.

35.   Let $x = \log_5 \sqrt[3]{25}$. Then:
$$5^x = \sqrt[3]{25}$$
$$5^x = 5^{2/3}$$
$$x = \frac{2}{3}$$
So $\log_5 \sqrt[3]{25} = \frac{2}{3}$.

37.   Finding the logarithm: $\log_5 \left(\log_3 243 \right) = \log_5 \left(\log_3 3^5 \right) = \log_5 5 = 1$

39.   Finding the logarithm: $\log_2 \left(\log_9 3 \right) = \log_2 \left(\log_9 9^{1/2} \right) = \log_2 \frac{1}{2} = \log_2 \left(2^{-1} \right) = -1$

41.   Finding the logarithm: $4^{\log_4 7} = 7$

43.   Finding the logarithm: $\log_6 \dfrac{1}{\sqrt{6}} = \log_6 6^{-1/2} = -\dfrac{1}{2}$

45.   Finding the logarithm: $\log_b b^6 = 6$

47.   Solving for t:
$$\log_3 t = 4$$
$$t = 3^4$$
$$t = 81$$

49.   Solving for t:
$$\log_t \tfrac{1}{8} = -3$$
$$\tfrac{1}{8} = t^{-3}$$
$$2^{-3} = t^{-3}$$
$$t = 2$$

51.   Solving for t:
$$\log_{32} 16 = t$$
$$16 = 32^t$$
$$2^4 = 2^{5t}$$
$$5t = 4$$
$$t = \tfrac{4}{5}$$

53.   Finding the value: $F(6) = \log_2\left(6^2 - 4\right) = \log_2 32 = \log_2 2^5 = 5$

Since $x^2 - 4 = (x+2)(x-2) > 0$, the domain is $(-\infty, -2) \cup (2, \infty)$.

55.   The domain is $(3, \infty)$, the range is $(-\infty, \infty)$, the x-intercept is 4, there is no y-intercept, and the vertical asymptote is $x = 3$:

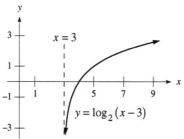

57.   The domain is $(-4, \infty)$, the range is $(-\infty, \infty)$, the x-intercept is -3, the y-intercept is $\log_5 4$, and the vertical asymptote is $x = -4$:

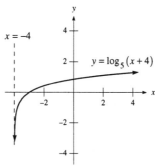

59. The domain is $(-\infty, 0)$, the range is $(-\infty, \infty)$, the x-intercept is -1, there is no y-intercept, and the vertical asymptote is $x = 0$:

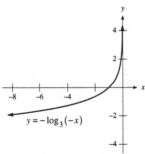

$$y = -\log_3(-x)$$

61. The domain is $(-\infty, 0) \cup (0, \infty)$, the range is $(-\infty, \infty)$, the x-intercepts are -1 and 1, there is no y-intercept, and the vertical asymptote is $x = 0$:

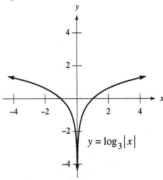

$$y = \log_3|x|$$

63. **a.** $\log_2 41$ is larger, since 2 must be raised to a larger exponent to yield 41 than to yield 35.

b. $\log_2 35$ is larger, since 2 must be raised to a larger exponent to yield 35 than 3 will be raised.

c.   Since $4^3 = 64$, $\log_4 60 < 3$. Since $3^3 = 27$, $\log_3 40 > 3$. Thus $\log_3 40$ is the larger value.

65. Writing each side in exponential form:

$$\log_{1/6} x = y \Rightarrow x = \left(\tfrac{1}{6}\right)^y$$

$$-\log_6 x = y \Rightarrow x = 6^{-y} = \left(\tfrac{1}{6}\right)^y$$

Note that both values are the same.

67. If $a < b$, then $\log_n a < \log_n b$.

If $a < b$, then $\log_a x > \log_b x$.

5.3 Properties of Logarithms: Logarithmic Equations

1.   Expanding the logarithm: $\log_4\left(x^2 yz^3\right) = \log_4 x^2 + \log_4 y + \log_4 z^3 = 2\log_4 x + \log_4 y + 3\log_4 z$

3.   Expanding the logarithm:

$$\log_b\left(\frac{x^3}{yz^4}\right) = \log_b x^3 - \log_b\left(yz^4\right)$$

$$= \log_b x^3 - \log_b y - \log_b z^4$$

$$= 3\log_b x - \log_b y - 4\log_b z$$

5. Expanding the logarithm:
$$\log_5 \sqrt{5x^3} = \log_5 \left(5x^3\right)^{1/2} = \tfrac{1}{2}\left[\log_5 5 + \log_5 x^3\right] = \tfrac{1}{2}\left[1 + 3\log_5 x\right] = \tfrac{1}{2} + \tfrac{3}{2}\log_5 x$$

7. $\log_b\left(x^3 + y^2 - z^5\right)$ cannot be expanded.

9. Expanding the logarithm: $\log_b\left(x^2 - 4\right) = \log_b\left[(x+2)(x-2)\right] = \log_b(x+2) + \log_b(x-2)$

11. Expanding the logarithm:
$$\log_4 \sqrt[3]{\frac{x-3}{2x^2}} = \log_4\left(\frac{x-3}{2x^2}\right)^{1/3}$$
$$= \tfrac{1}{3}\left[\log_4(x-3) - \log_4\left(2x^2\right)\right]$$
$$= \tfrac{1}{3}\left[\log_4(x-3) - \log_4 2 - \log_4 x^2\right]$$
$$= \tfrac{1}{3}\left[\log_4(x-3) - \tfrac{1}{2} - 2\log_4 x\right]$$
$$= \tfrac{1}{3}\log_4(x-3) - \tfrac{1}{6} - \tfrac{2}{3}\log_4 x$$

13. Expanding the logarithm:
$$\log_b \sqrt{\frac{x^2 - 16}{x^2 - 2x - 8}} = \log_b \sqrt{\frac{(x+4)(x-4)}{(x+2)(x-4)}}$$
$$= \log_b\left(\frac{x+4}{x+2}\right)^{1/2}$$
$$= \tfrac{1}{2}\log_b\left(\frac{x+4}{x+2}\right)$$
$$= \tfrac{1}{2}\log_b(x+4) - \tfrac{1}{2}\log_b(x+2)$$
Note that $x \neq 4$ in this expansion.

15. This statement is false. $\log_b\left(x^3 - y^4\right)$ cannot be expanded.

17. This statement is true.

19. Combining the logarithms: $4\log_b 2 + \log_b 3 = \log_b 2^4 + \log_b 3 = \log_b\left(2^4 \cdot 3\right) = \log_b 48$

21. Combining the logarithms: $\tfrac{1}{2}\log_b s - \tfrac{3}{2}\log_b t = \log_b s^{1/2} - \log_b t^{3/2} = \log_b \dfrac{s^{1/2}}{t^{3/2}} = \log_b \sqrt{\dfrac{s}{t^3}}$

23. Combining the logarithms: $\log_{10} 50 - \log_{10} 5 = \log_{10} \dfrac{50}{5} = \log_{10} 10 = 1$

25. Combining the logarithms: $\log_6 9 + \log_6 24 = \log_6(9 \cdot 24) = \log_6 216 = 3$

27. Combining the logarithms:
$$3\log_b x - 4\log_b y - 2\log_b z = \log_b x^3 - \log_b y^4 - \log_b z^2 = \log_b\left(\frac{x^3}{y^4 z^2}\right)$$

29. Combining the logarithms:
$$\tfrac{1}{4}\log_b x + \tfrac{1}{3}\log_b y - \tfrac{1}{2}\log_b z = \log_b x^{1/4} + \log_b y^{1/3} - \log_b z^{1/2} = \log_b\left(\frac{x^{1/4} y^{1/3}}{z^{1/2}}\right)$$

31. Simplifying the logarithm: $\log_{10} 16 = \log_{10} 2^4 = 4\log_{10} 2 = 4A$

33. Simplifying the logarithm:
$$\log_{10} \sqrt{6} = \log_{10}(2 \cdot 3)^{1/2} = \tfrac{1}{2}\log_{10} 2 + \tfrac{1}{2}\log_{10} 3 = \tfrac{1}{2}A + \tfrac{1}{2}B = \tfrac{1}{2}(A + B)$$

35. Simplifying the logarithm: $\log_{10} 300 = \log_{10}(3 \cdot 100) = \log_{10} 3 + \log_{10} 100 = B + 2$

37. Evaluating the logarithm: $\log_b 12 = \log_b \left(2^2 \cdot 3\right) = 2\log_b 2 + \log_b 3 = 2(0.69) + 1.09 = 2.47$

39. Evaluating the logarithm:

$$\log_b \tfrac{1}{18} = -\log_b 18 = -\log_b \left(2 \cdot 3^2\right) = -\log_b 2 - 2\log_b 3 = -0.69 - 2(1.09) = -2.87$$

41. Solving the equation:

$$\log_3 4 + \log_3 x = 2$$
$$\log_3 4x = 2$$
$$4x = 3^2$$
$$4x = 9$$
$$x = \tfrac{9}{4}$$

43. Solving the equation:

$$2\log_5 x = \log_5 49$$
$$\log_5 x^2 = \log_5 49$$
$$x^2 = 49$$
$$x = 7 \quad (x = -7 \text{ does not check})$$

45. Solving the equation:

$$\log_2 t - \log_2 (t - 2) = 3$$
$$\log_2 \frac{t}{t-2} = 3$$
$$\frac{t}{t-2} = 2^3$$
$$\frac{t}{t-2} = 8$$
$$t = 8t - 16$$
$$-7t = -16$$
$$t = \tfrac{16}{7}$$

47. Solving the equation:

$$\log_{10} 8 + \log_{10} x + \log_{10} x^2 = 3$$
$$\log_{10} \left(8x^3\right) = 3$$
$$8x^3 = 10^3$$
$$8x^3 = 1000$$
$$x^3 = 125$$
$$x = 5$$

49. Solving the equation:

$$\log_9 (2x + 7) - \log_9 (x - 1) = \log_9 (x - 7)$$
$$\log_9 \left(\frac{2x+7}{x-1}\right) = \log_9 (x - 7)$$
$$\frac{2x+7}{x-1} = x - 7$$
$$2x + 7 = x^2 - 8x + 7$$
$$0 = x^2 - 10x$$
$$0 = x(x - 10)$$
$$x = 10 \quad (x = 0 \text{ does not check})$$

51. Solving the equation:
$$\log_4 x - \log_4(x-4) = \log_4(x-6)$$
$$\log_4\left(\frac{x}{x-4}\right) = \log_4(x-6)$$
$$\frac{x}{x-4} = x-6$$
$$x = x^2 - 10x + 24$$
$$0 = x^2 - 11x + 24$$
$$0 = (x-8)(x-3)$$
$$x = 8 \quad (x = 3 \text{ does not check})$$

53. Solving the equation:
$$\tfrac{1}{2}\log_3 x = \log_3(x-6)$$
$$\log_3 \sqrt{x} = \log_3(x-6)$$
$$\sqrt{x} = x-6$$
$$x = x^2 - 12x + 36$$
$$0 = x^2 - 13x + 36$$
$$0 = (x-4)(x-9)$$
$$x = 9 \quad (x = 4 \text{ does not check})$$

55. Evaluating: $3^{2\log_3 5} = 3^{\log_3 5^2} = 5^2 = 25$

57. Sketching $f(x) = \log_b x^2$:

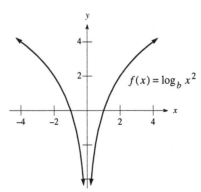

Sketching $g(x) = 2\log_b x$:

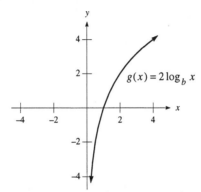

Note that these functions are not the same (not the same domains).

5.4 Common and Natural Logarithms; Exponential Equations and Change of Base

1. Simplifying: $\ln e^5 = 5\ln e = 5$

3. Simplifying: $\log\sqrt{10} = \log 10^{1/2} = \frac{1}{2}\log 10 = \frac{1}{2}$

5. Simplifying: $e^{\ln(x+1)} = x+1$

7. The domain is $(-\infty,\infty)$, the range is $(0,\infty)$, there is no x-intercept, the y-intercept is 10, and the asymptote is $y=0$:

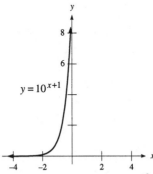

9. The domain is $(0,\infty)$, the range is $(-\infty,\infty)$, the x-intercept is e^2, there is no y-intercept, and the asymptote is $x=0$:

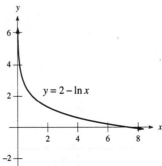

11. The domain is $(3,\infty)$, the range is $(-\infty,\infty)$, the x-intercept is 4, there is no y-intercept, and the asymptote is $x=3$:

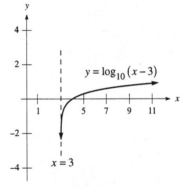

13.   The domain is $(-\infty,0)\cup(0,\infty)$, the range is $(-\infty,\infty)$, the x-intercepts are ± 1, there is no y-intercept, and the asymptote is $x = 0$:

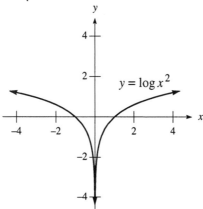

15.   Switching the roles of x and y:

$$e^{3y-1} = x$$
$$3y - 1 = \ln x$$
$$3y = \ln x + 1$$
$$y = \frac{\ln x + 1}{3}$$

So $f^{-1}(x) = \dfrac{\ln x + 1}{3}$.

17.   This is equivalent to finding the inverse. Switching the roles of x and y:

$$e^{\sqrt{y}} = x$$
$$\sqrt{y} = \ln x$$
$$y = (\ln x)^2$$

So $g(x) = (\ln x)^2$.

19.   Evaluating the logarithm: $\log_7 4 = \dfrac{\log 4}{\log 7} = \dfrac{\ln 4}{\ln 7} \approx 0.71$

21.   Evaluating the logarithm: $\log_{1/2} 5 = \dfrac{\log 5}{\log \frac{1}{2}} = \dfrac{\ln 5}{\ln \frac{1}{2}} \approx -2.32$

23.   Evaluating the logarithm: $\log_9 \frac{2}{3} = \dfrac{\log \frac{2}{3}}{\log 9} = \dfrac{\ln \frac{2}{3}}{\ln 9} \approx -0.18$

25.   Converting the logarithm: $\log_3 x = \dfrac{\log_9 x}{\log_9 3} = \dfrac{\log_9 x}{\log_9 9^{1/2}} = \dfrac{\log_9 x}{1/2} = 2\log_9 x$

27.   Converting the logarithm: $\log_{10} x = \dfrac{\ln x}{\ln 10} \approx 0.434\ln x$

29.   Solving the equation:

$$e^{1-2x} = 5$$
$$1 - 2x = \ln 5$$
$$1 - \ln 5 = 2x$$
$$x = \frac{1 - \ln 5}{2}$$

31. Solving the equation:
$$\frac{e^{3x+2}}{4} = 5$$
$$e^{3x+2} = 20$$
$$3x + 2 = \ln 20$$
$$3x = \ln 20 - 2$$
$$x = \frac{\ln 20 - 2}{3}$$

33. Solving the equation:
$$3^{4-x} = 7$$
$$\ln 3^{4-x} = \ln 7$$
$$(4 - x)\ln 3 = \ln 7$$
$$4\ln 3 - x\ln 3 = \ln 7$$
$$4\ln 3 - \ln 7 = x\ln 3$$
$$x = \frac{4\ln 3 - \ln 7}{\ln 3}$$

35. Solving the equation:
$$2^{2x-2} = 6^{-x}$$
$$\ln 2^{2x-2} = \ln 6^{-x}$$
$$(2x - 2)\ln 2 = -x\ln 6$$
$$(2\ln 2)x - 2\ln 2 = -x\ln 6$$
$$(2\ln 2 + \ln 6)x = 2\ln 2$$
$$x = \frac{2\ln 2}{2\ln 2 + \ln 6}$$

37. Graphing the function:

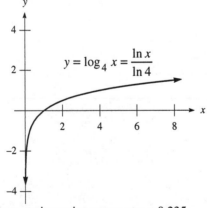

39. Graphing both curves, the intersection point occurs at $x \approx 0.235$.

41. Graphing both curves, the intersection points occur at $x \approx 1.764, 5$.

43. Graphing both curves, the intersection point occurs at $x \approx -2.032$.

45. Graphing both curves, the intersection points occur at $x \approx -1, 1.501$.

47. **a.** It appears there are two points of intersection at $x \approx -0.767$ and 2.

　　b. Note there are actually three zeros at $x \approx -0.767$, 2, and 4.

49.   Calculating the values:

$n = 10$: $\left(1 + \frac{1}{10}\right)^{10} \approx 2.59374$

$n = 20$: $\left(1 + \frac{1}{20}\right)^{20} \approx 2.65330$

$n = 50$: $\left(1 + \frac{1}{50}\right)^{50} \approx 2.69159$

$n = 100$: $\left(1 + \frac{1}{100}\right)^{100} \approx 2.70481$

$n = 1000$: $\left(1 + \frac{1}{1000}\right)^{1000} \approx 2.71692$

$n = 10,000$: $\left(1 + \frac{1}{10000}\right)^{10000} \approx 2.71815$

Note that $e \approx 2.71828$, which these values are approaching.

51.   Converting the logarithm: $\log_4 x = \dfrac{\log_{16} x}{\log_{16} 4} = \dfrac{\log_{16} x}{\log_{16} 16^{1/2}} = \dfrac{\log_{16} x}{1/2} = 2\log_{16} x$

The power to which 4 must be raised to yield x should be double the power to which 16 must be raised to yield x.

5.5 Applications

1.   No. If the original price is P, the price after raising its prices is $1.2P$. After reducing the prices, the new price is $0.8(1.2P) = 0.96P$. Thus the final price is 4% lower then the original price.

3.   Using the half-life:

$$\tfrac{1}{2} A_0 = A_0 e^{k \cdot 25000}$$
$$\tfrac{1}{2} = e^{25000k}$$
$$25000k = \ln \tfrac{1}{2}$$
$$k = \frac{\ln \tfrac{1}{2}}{25000} \approx -0.0002773$$

Thus the exponential decay model is $A = A_0 e^{-0.0002773t}$.

5.   Using the half-life:

$$\tfrac{1}{2} A_0 = A_0 e^{k \cdot 5730}$$
$$\tfrac{1}{2} = e^{5730k}$$
$$5730k = \ln \tfrac{1}{2}$$
$$k = \frac{\ln \tfrac{1}{2}}{5730} \approx -0.0001210$$

So $A = 100 e^{-0.0001210t}$. Substituting $t = 1000$: $A = 100 e^{-0.0001210 \cdot 1000} \approx 88.61$ grams

7.   Using the half-life:

$$\tfrac{1}{2} A_0 = A_0 e^{k(2.24)}$$
$$\tfrac{1}{2} = e^{2.24k}$$
$$2.24k = \ln \tfrac{1}{2}$$
$$k = \frac{\ln \tfrac{1}{2}}{2.24} \approx -0.30944$$

So $A = 40 e^{-0.30944t}$. Substituting $t = 30$: $A = 40 e^{-0.30944(30)} \approx 0.00372$ grams

9. Using the decay law:
$$60 = 80e^{k \cdot 100}$$
$$0.75 = e^{100k}$$
$$100k = \ln 0.75$$
$$k = \frac{\ln 0.75}{100} \approx -0.00288$$

So $A = A_0 e^{-0.00288t}$. Finding when $A = \frac{1}{5}A_0$:
$$\frac{1}{5}A_0 = A_0 e^{-0.00288t}$$
$$\frac{1}{5} = e^{-0.00288t}$$
$$-0.00288t = \ln \frac{1}{5}$$
$$t = \frac{\ln \frac{1}{5}}{-0.00288} \approx 559 \text{ years}$$

11. Substituting $A_0 = 100$ and $A = 200$:
$$200 = 100 \cdot 3^{t/20}$$
$$2 = 3^{t/20}$$
$$\ln 2 = \frac{t}{20} \ln 3$$
$$t = \frac{20 \ln 2}{\ln 3} \approx 12.6 \text{ hours}$$

Since the model indicates a tripling of the population every 20 hours, it will take 20 hours for the population to grow from 100 to 300.

13. Finding the annual growth rate:
$$248.7 = 226.5 e^{k \cdot 10}$$
$$\frac{248.7}{226.5} = e^{10k}$$
$$10k = \ln\left(\frac{248.7}{226.5}\right)$$
$$k = \frac{1}{10} \ln\left(\frac{248.7}{226.5}\right) \approx 0.0094$$

The annual growth rate is 0.94%.

15. Using the growth model $A = 5.8e^{0.018t}$, substitute $t = 29$: $A = 5.8e^{0.018(29)} \approx 9.78$ billion

17. Finding the growth rate:
$$1032 = 686.4 e^{k \cdot 24}$$
$$\frac{1032}{686.4} = e^{24k}$$
$$24k = \ln\left(\frac{1032}{686.4}\right)$$
$$k = \frac{1}{24} \ln\left(\frac{1032}{686.4}\right) \approx 0.0170$$

The growth rate is 1.70% per year. The growth rate is slightly lower than the world's growth rate of 1.8%.

19. Finding the time:

$$15,000 = 10,000\left(1 + \frac{0.074}{4}\right)^{4t}$$

$$1.5 = \left(1 + \frac{0.074}{4}\right)^{4t}$$

$$\ln 1.5 = 4t \ln\left(1 + \frac{0.074}{4}\right)$$

$$t = \frac{\ln 1.5}{4 \ln\left(1 + \frac{0.074}{4}\right)} \approx 5.39 \text{ years}$$

21. Computing the amount: $A = 3000\left(1 + \frac{0.067}{12}\right)^{12\cdot1} \approx 3207.29$

Thus the increase is $\frac{207.29}{3000} \approx 6.91\%$. This does not depend on the initial \$3000 invested.

23. Find where \$100 grows to \$106.7 for 1 year:

$$106.7 = 100e^{k\cdot1}$$

$$1.067 = e^k$$

$$k = \ln 1.067 \approx 0.0649$$

A 6.49% continuous rate will produce a 6.7% effective yield.

25. Substituting the values:

$$6000 = A_0 e^{0.058(5)}$$

$$6000 = A_0 e^{0.29}$$

$$A_0 = \frac{6000}{e^{0.29}} \approx 4489.58$$

The amount is \$4,489.58.

27. Finding the doubling time:

$$2000 = 1000e^{0.06t}$$

$$2 = e^{0.06t}$$

$$\ln 2 = 0.06t$$

$$t = \frac{\ln 2}{0.06} \approx 11.552 \text{ years}$$

The doubling time from example 2 was 11.553 years. Note that increasing the compounding from daily to continuously does not significantly affect the doubling time.

29. a. Substituting the values: $M = \dfrac{8000(0.09)}{12\left[1 - \left(1 + \dfrac{0.09}{12}\right)^{-12(4)}\right]} \approx \$199.08$

b. Over the four years, the amount paid back is: $48(199.08) = \$9,555.84$

31. a. Substituting the values: $M = \dfrac{80000(0.082)}{12\left[1 - \left(1 + \dfrac{0.082}{12}\right)^{-12(30)}\right]} \approx \$598.20$

b. Over the 30 years, the amount paid back is: $360(598.20) \approx \$215,352$

33. **a.** Substituting $I = 1000I_0$: $r = \log\dfrac{1000I_0}{I_0} = \log 1000 = 3$

b. Computing the intensity:

$$8.3 = \log\dfrac{I}{I_0}$$

$$10^{8.3} = \dfrac{I}{I_0}$$

$$I = 10^{8.3}I_0$$

The intensity is $10^{8.3}$ times the intensity of I_0.

35. The intensities compare as:

$$\dfrac{E_3}{E_2} = \dfrac{10^{10}I_0}{10^6 I_0} = 10^4 = 10,000$$

$$\dfrac{E_2}{E_1} = \dfrac{10^6 I_0}{10^2 I_0} = 10^4 = 10,000$$

E_3 is 10,000 times the intensity of E_2, and E_2 is 10,000 times the intensity of E_1.

37. Substituting: $pH = -\log\left(6.82 \times 10^{-5}\right) \approx 4.2$

39. Substituting: $pH = -\log\left(2.5 \times 10^{-6}\right) \approx 5.6$

41. **a.** Computing the intensity of a star with magnitude 2:

$$2 = 6 - 2.5\log\left(\dfrac{I}{I_0}\right)$$

$$-4 = -2.5\log\left(\dfrac{I}{I_0}\right)$$

$$1.6 = \log\left(\dfrac{I}{I_0}\right)$$

$$\dfrac{I}{I_0} = 10^{1.6}$$

$$I = 10^{1.6}I_0$$

Computing the intensity of a star with magnitude 1:

$$1 = 6 - 2.5\log\left(\dfrac{I}{I_0}\right)$$

$$-5 = -2.5\log\left(\dfrac{I}{I_0}\right)$$

$$2 = \log\left(\dfrac{I}{I_0}\right)$$

$$\dfrac{I}{I_0} = 10^2$$

$$I = 10^2 I_0$$

Calculating the ratio: $\dfrac{10^2 I_0}{10^{1.6}I_0} = 10^{0.4} \approx 2.51$

b.   Computing the intensity of a star with magnitude 5:

$$5 = 6 - 2.5\log\left(\frac{I}{I_0}\right)$$

$$-1 = -2.5\log\left(\frac{I}{I_0}\right)$$

$$0.4 = \log\left(\frac{I}{I_0}\right)$$

$$\frac{I}{I_0} = 10^{0.4}$$

$$I = 10^{0.4}\,I_0$$

Computing the intensity of a star with magnitude 4:

$$4 = 6 - 2.5\log\left(\frac{I}{I_0}\right)$$

$$-2 = -2.5\log\left(\frac{I}{I_0}\right)$$

$$0.8 = \log\left(\frac{I}{I_0}\right)$$

$$\frac{I}{I_0} = 10^{0.8}$$

$$I = 10^{0.8}\,I_0$$

Calculating the ratio: $\dfrac{10^{0.8}\,I_0}{10^{0.4}\,I_0} = 10^{0.4} \approx 2.51$

c.   Approximately 2.5 times greater.

d.   Substituting $40I_0$ for I: $M = 6 - 2.5\log\left(\dfrac{40I_0}{I_0}\right) = 6 - 2.5\log 40$

Comparing the magnitudes: $6 - 2.5\log 40 - 6 = -4$

Thus the magnitude decreases by 4 units.

43.   Substituting $I = 10^{-16}$: $N = 160 + 10\log\left(10^{-16}\right) = 0$ decibels

45.   **a.**   Finding the exponential rate:

$$3.6 = 1.5e^{k(20)}$$

$$\frac{3.6}{1.5} = e^{20k}$$

$$20k = \ln\left(\frac{3.6}{1.5}\right)$$

$$k = \tfrac{1}{20}\ln\left(\frac{3.6}{1.5}\right) \approx 0.0438$$

The growth rate is 4.38% per year.

b.   Finding the exponential rate:

$$20.4 = 12.1e^{k(10)}$$

$$\frac{20.4}{12.1} = e^{10k}$$

$$10k = \ln\left(\frac{20.4}{12.1}\right)$$

$$k = \frac{1}{10}\ln\left(\frac{20.4}{12.1}\right) \approx 0.0522$$

The growth rate is 5.22% per year.

47. **a.**   Using the half-life:

$$\tfrac{1}{2}A_0 = A_0 e^{k \cdot 5730}$$

$$\tfrac{1}{2} = e^{5730k}$$

$$5730k = \ln\tfrac{1}{2}$$

$$k = \frac{\ln\tfrac{1}{2}}{5730} \approx -0.000121$$

The decay model is $A = A_0 e^{-0.000121t}$.

b.   This model is valid since the half-life is 5730 years.

c.   Substituting $t = 1000$ and $A_0 = 60$:

$$A = 60e^{-0.000121(1000)} \approx 53.16 \text{ grams}$$

$$A = 60\left(\tfrac{1}{2}\right)^{1000/5730} \approx 53.16 \text{ grams}$$

d.   Substituting $A = 0.65A_0$:

$$0.65A_0 = A_0 e^{-0.000121t}$$

$$0.65 = e^{-0.000121t}$$

$$-0.000121t = \ln 0.65$$

$$t = \frac{\ln 0.65}{-0.000121} \approx 3561 \text{ years}$$

49.   Substituting $A = 0.76A_0$:

$$0.76A_0 = A_0 e^{-0.000121t}$$

$$0.76 = e^{-0.000121t}$$

$$-0.000121t = \ln 0.76$$

$$t = \frac{\ln 0.76}{-0.000121} \approx 2269 \text{ years}$$

51. **a.**   Substituting $t = 0$: $c(0) = 100 - 30e^0 = 70$ grams

b.   Substituting $t = 5$: $c(5) = 100 - 30e^{-5/10} \approx 81.8$ grams

c.   Substituting $t = 10$: $c(10) = 100 - 30e^{-10/10} \approx 89.0$ grams

d.   Substituting $t = 100$: $c(100) = 100 - 30e^{-100/10} \approx 99.999$ grams

53. **a.** Substituting values: $A = 100(1 - 0.048)^5 \approx \$78.20$

b. Substituting values:
$$50 = 100(1 - 0.048)^t$$
$$\tfrac{1}{2} = (0.952)^t$$
$$\ln \tfrac{1}{2} = t \ln 0.952$$
$$t = \frac{\ln \tfrac{1}{2}}{\ln 0.952} \approx 14.09 \text{ years}$$

55. Using Stirling's formula:
$$20! = \left(\frac{20}{e}\right)^{20} \sqrt{2\pi \cdot 20} = 2.422786847 \times 10^{18}$$
$$20! = 20 \cdot 19 \cdot 18 \cdot \ldots \cdot 3 \cdot 2 \cdot 1 = 2.432902008 \times 10^{18}$$

57. Computing the sound intensity of each machine:

$$90 = 160 + 10 \log I_1 \qquad\qquad 80 = 160 + 10 \log I_2$$
$$-70 = 10 \log I_1 \qquad\qquad -80 = 10 \log I_2$$
$$-7 = \log I_1 \qquad\qquad -8 = \log I_2$$
$$I_1 = 10^{-7} \qquad\qquad I_2 = 10^{-8}$$

So the combined intensity is $I_1 + I_2 = 10^{-7} + 10^{-8}$. Computing the combined intensity level:
$$N = 160 + 10 \log\left(10^{-7} + 10^{-8}\right) \approx 100.4 \text{ dB}$$

Yes, the level is slightly higher than the union regulation of 100 dB.

59. **a.** Since the half-life is 8 minutes, the decay model is $A = A_0 \left(\tfrac{1}{2}\right)^{t/8}$.

b. Substituting $A_0 = 80$ and $A = 7$:
$$7 = 80\left(\tfrac{1}{2}\right)^{t/8}$$
$$\tfrac{7}{80} = \left(\tfrac{1}{2}\right)^{t/8}$$
$$\ln \tfrac{7}{80} = \frac{t}{8} \ln \tfrac{1}{2}$$
$$t = \frac{8 \ln \tfrac{7}{80}}{\ln \tfrac{1}{2}} \approx 28.12 \text{ minutes}$$

Chapter 5 Review Exercises

1. The asymptote is $y = 0$, there is no x-intercept, and the y-intercept is $1/2$:

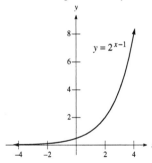

2.   The asymptote is $y = -1$, the x-intercept is 0, and the y-intercept is 0:

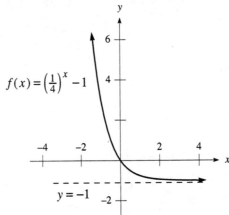

$$f(x) = \left(\tfrac{1}{4}\right)^x - 1$$

$$y = -1$$

3.   The asymptote is $x = -3$, the x-intercept is -2, and the y-intercept is $\log_{2/3} 3$:

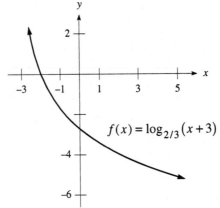

$$f(x) = \log_{2/3}(x + 3)$$

4.   The asymptote is $x = 0$, the x-intercept is 5^{-5}, and there is no y-intercept:

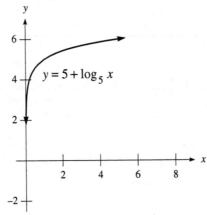

$$y = 5 + \log_5 x$$

5. The asymptote is $y = -1$, the x-intercept is -2, and the y-intercept is $e^2 - 1$:

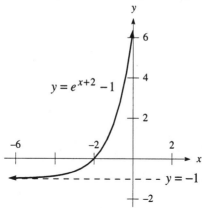

$y = e^{x+2} - 1$

$y = -1$

6. The asymptote is $x = 0$, the x-intercepts are $\pm\dfrac{1}{e} \approx \pm 0.368$, and there is no y-intercept:

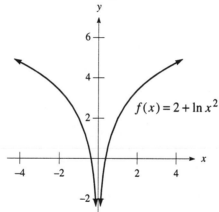

$f(x) = 2 + \ln x^2$

7. Writing in exponential form: $6^{-1} = \frac{1}{6}$

8. Writing in logarithmic form: $\log_9 3 = \frac{1}{2}$

9. Writing in logarithmic form: $\log_8 \frac{1}{4} = -\frac{2}{3}$

10. Writing in exponential form: $3^4 = 81$

11. Writing in exponential form: $b^6 = b^6$

12. Writing in logarithmic form: $\log_b t = \log_b t$

13. Evaluating the logarithm: $\log_{10} 10{,}000 = \log_{10} 10^4 = 4$

14. Evaluating the logarithm: $\log_{10} 0.000001 = \log_{10} 10^{-6} = -6$

15. Evaluating the logarithm: $\log_3 \frac{1}{9} = \log_3 3^{-2} = -2$

16. Evaluating the logarithm: $\log_{1/2} 8 = \log_{1/2} \left(\frac{1}{2}\right)^{-3} = -3$

17. Evaluating the logarithm: $\log_{32} 16 = \log_{32} 32^{4/5} = \frac{4}{5}$

18. Evaluating the logarithm: $\log_b \sqrt{b^3} = \log_b b^{3/2} = \frac{3}{2}$

19. Evaluating the logarithm: $\log_b 1 = 0$

20. Evaluating the logarithm: $\log_{1/5} 25 = \log_{1/5}\left(\frac{1}{5}\right)^{-2} = -2$

21. Expanding the logarithm:
$$\log_b\left(x^3 y^4 z^2\right) = \log_b x^3 + \log_b y^4 + \log_b z^2 = 3\log_b x + 4\log_b y + 2\log_b z$$

22. Expanding the logarithm:
$$\log_b\left(\frac{b^3}{\sqrt{xy}}\right) = \log_b b^3 - \log_b (xy)^{1/2}$$
$$= 3\log_b b - \frac{1}{2}\left(\log_b x + \log_b y\right)$$
$$= 3 - \frac{1}{2}\log_b x - \frac{1}{2}\log_b y$$

23. Expanding the logarithm:
$$\log_b \sqrt[3]{\frac{6x}{by^4}} = \log_b\left(\frac{6x}{by^4}\right)^{1/3}$$
$$= \frac{1}{3}\left[\log_b 6 + \log_b x - \log_b b - \log_b y^4\right]$$
$$= \frac{1}{3}\left[\log_b 6 + \log_b x - 1 - 4\log_b y\right]$$
$$= \frac{1}{3}\log_b 6 + \frac{1}{3}\log_b x - \frac{1}{3} - \frac{4}{3}\log_b y$$

24. $\log_b\left(x^2 + y^5\right)$ cannot be expanded.

25. $\dfrac{\sqrt{\log_b x}}{\sqrt[3]{\log_b y}}$ cannot be expanded.

26. Expanding the logarithm: $\log_b \dfrac{\sqrt{x}}{\sqrt[3]{y}} = \log_b \dfrac{x^{1/2}}{y^{1/3}} = \log_b x^{1/2} - \log_b y^{1/3} = \frac{1}{2}\log_b x - \frac{1}{3}\log_b y$

27. Solving the equation:
$$8^x = \frac{1}{64}$$
$$8^x = 8^{-2}$$
$$x = -2$$

28. Solving the equation:
$$\left(\frac{1}{9}\right)^x = 29$$
$$9^{-x} = 29$$
$$-x\ln 9 = \ln 29$$
$$x = -\frac{\ln 29}{\ln 9} \approx -1.533$$

29. Solving the equation:
$$\log_b (3x) + \log_b (x+2) = \log_b 9$$
$$\log_b\left(3x^2 + 6x\right) = \log_b 9$$
$$3x^2 + 6x = 9$$
$$x^2 + 2x - 3 = 0$$
$$(x+3)(x-1) = 0$$
$$x = 1 \qquad (x = -3 \text{ does not check})$$

30. Solving the equation:
$$\log_2 x + \log_2 (x+1) = 1$$
$$\log_2 \left(x^2 + x\right) = 1$$
$$x^2 + x = 2^1$$
$$x^2 + x - 2 = 0$$
$$(x+2)(x-1) = 0$$
$$x = 1 \qquad (x = -2 \text{ does not check})$$

31. Solving the equation:
$$7^{x-1} = 3$$
$$(x-1)\ln 7 = \ln 3$$
$$x\ln 7 - \ln 7 = \ln 3$$
$$x\ln 7 = \ln 3 + \ln 7$$
$$x = \frac{\ln 3 + \ln 7}{\ln 7} \approx 1.565$$

32. Solving the equation:
$$\log_2 (t+1) + \log_2 (t-1) = 3$$
$$\log_2 \left(t^2 - 1\right) = 3$$
$$t^2 - 1 = 2^3$$
$$t^2 = 9$$
$$t = 3 \qquad (t = -3 \text{ does not check})$$

33. Solving the equation:
$$\log_5 (6x) - \log_5 (x+2) = 1$$
$$\log_5 \left(\frac{6x}{x+2}\right) = 1$$
$$\frac{6x}{x+2} = 5^1$$
$$6x = 5x + 10$$
$$x = 10$$

34. Solving the equation:
$$3^x = 5^{x+2}$$
$$\ln 3^x = \ln 5^{x+2}$$
$$x\ln 3 = (x+2)\ln 5$$
$$x\ln 3 = x\ln 5 + 2\ln 5$$
$$x(\ln 3 - \ln 5) = 2\ln 5$$
$$x = \frac{2\ln 5}{\ln 3 - \ln 5} \approx -6.301$$

35. Solving the equation:
$$3^x = 5\left(2^x\right)$$
$$\ln 3^x = \ln\left(5 \cdot 2^x\right)$$
$$x\ln 3 = \ln 5 + x\ln 2$$
$$x(\ln 3 - \ln 2) = \ln 5$$
$$x = \frac{\ln 5}{\ln 3 - \ln 2} \approx 3.969$$

36. Solving the equation:
$$\log_4 x - \log_4 (x-4) = \log_4 (x-6)$$
$$\log_4 \left(\frac{x}{x-4}\right) = \log_4 (x-6)$$
$$\frac{x}{x-4} = x-6$$
$$x = x^2 - 10x + 24$$
$$0 = x^2 - 11x + 24$$
$$0 = (x-8)(x-3)$$
$$x = 8 \qquad (x = 3 \text{ does not check})$$

37. Solving the equation:
$$\tfrac{1}{2}\log_3 x = \log_3 (x-6)$$
$$\log_3 x = 2\log_3 (x-6)$$
$$\log_3 x = \log_3 (x-6)^2$$
$$x = x^2 - 12x + 36$$
$$0 = x^2 - 13x + 36$$
$$0 = (x-4)(x-9)$$
$$x = 9 \qquad (x = 4 \text{ does not check})$$

38. Solving the equation:
$$2\log_b x = \log_b (6x-5)$$
$$\log_b x^2 = \log_b (6x-5)$$
$$x^2 = 6x-5$$
$$x^2 - 6x + 5 = 0$$
$$(x-5)(x-1) = 0$$
$$x = 1, 5$$

39. Solving the equation:
$$8^{3x-2} = 9^{x+2}$$
$$(3x-2)\ln 8 = (x+2)\ln 9$$
$$(3\ln 8)x - 2\ln 8 = (\ln 9)x + 2\ln 9$$
$$(3\ln 8 - \ln 9)x = 2\ln 9 + 2\ln 8$$
$$x = \frac{2\ln 9 + 2\ln 8}{3\ln 8 - \ln 9} \approx 2.117$$

40. Solving the equation:
$$\log x = 2 + \log(x-1)$$
$$\log x - \log(x-1) = 2$$
$$\log\left(\frac{x}{x-1}\right) = 2$$
$$\frac{x}{x-1} = 10^2$$
$$\frac{x}{x-1} = 100$$
$$x = 100x - 100$$
$$-99x = -100$$
$$x = \frac{100}{99}$$

41. Solving the equation:
$$\log_b 125 = 3$$
$$125 = b^3$$
$$b = 5$$

42. Solving the equation:
$$\log_8 128 = x$$
$$128 = 8^x$$
$$2^7 = 2^{3x}$$
$$3x = 7$$
$$x = \tfrac{7}{3}$$

43. Finding each composition:
$$(f \circ g)(x) = f\left(\frac{3 + \ln x}{2}\right) = e^{2 \cdot \frac{3 + \ln x}{2} - 3} = e^{3 + \ln x - 3} = e^{\ln x} = x$$
$$(g \circ f)(x) = g\left(e^{2x-3}\right) = \frac{3 + \ln e^{2x-3}}{2} = \frac{3 + 2x - 3}{2} = \frac{2x}{2} = x$$
Thus f and g are inverse functions.

44. Switching the roles of x and y:
$$2^{y+3} = x$$
$$y + 3 = \log_2 x$$
$$y = -3 + \log_2 x$$
So $f^{-1}(x) = -3 + \log_2 x$.

45. Switching the roles of x and y:
$$5 + \log_3(y-1) = x$$
$$\log_3(y-1) = x - 5$$
$$y - 1 = 3^{x-5}$$
$$y = 1 + 3^{x-5}$$
So $f^{-1}(x) = 1 + 3^{x-5}$.

46. Converting logarithms: $\log_6 x = \dfrac{\log_3 x}{\log_3 6}$

47. Converting logarithms: $\log_7 11 = \dfrac{\ln 11}{\ln 7} \approx 1.23$

48. Computing the amount: $A = 2000\left(1 + \dfrac{0.065}{365}\right)^{365 \cdot 6} \approx \$2953.86$

49. Finding the time:
$$10,000 = A_0 e^{0.072(8)}$$
$$A_0 = \frac{10,000}{e^{0.576}} \approx \$5621.42$$

50. Finding the rate:
$$2000 = 1000 e^{k \cdot 6}$$
$$2 = e^{6k}$$
$$6k = \ln 2$$
$$k = \frac{\ln 2}{6} \approx 0.1155$$
The rate is 11.55%.

51. **a.**   Finding the growth rate:
$$2000 = 800e^{k\cdot 16}$$
$$2.5 = e^{16k}$$
$$16k = \ln 2.5$$
$$k = \frac{\ln 2.5}{16} \approx 0.0573$$
So $A = 800e^{0.0573t}$. Substituting $t = 10$: $A = 800e^{0.0573(10)} \approx 1418$ bacteria

b.   Finding the time:
$$3000 = 800e^{0.0573t}$$
$$3.75 = e^{0.0573t}$$
$$0.0573t = \ln 3.75$$
$$t = \frac{\ln 3.75}{0.0573} \approx 23 \text{ hours}$$

52.   Finding the doubling time:
$$2A_0 = A_0 \cdot 2^{0.06t/4}$$
$$2 = 2^{0.06t/4}$$
$$1 = \frac{0.06t}{4}$$
$$t = \frac{4}{0.06} \approx 66.67 \text{ hours}$$

53.   Finding the decay rate:
$$\tfrac{1}{2}A_0 = A_0 \cdot e^{k\cdot 53}$$
$$\tfrac{1}{2} = e^{53k}$$
$$53k = \ln \tfrac{1}{2}$$
$$k = \frac{\ln \tfrac{1}{2}}{53} \approx -0.0131$$
The decay equation is $A = A_0 e^{-0.0131t}$.

54.   Finding the half-life:
$$\tfrac{1}{2}A_0 = A_0 e^{k\cdot 5}$$
$$\tfrac{1}{2} = e^{5k}$$
$$5k = \ln \tfrac{1}{2}$$
$$k = \frac{\ln \tfrac{1}{2}}{5} \approx -0.1386$$
The decay equation is $A = A_0 e^{-0.1386t}$.

55.   If $r = 0.1$:
$$\tfrac{1}{2}A_0 = A_0 e^{-0.1t}$$
$$\tfrac{1}{2} = e^{-0.1t}$$
$$-0.1t = \ln \tfrac{1}{2}$$
$$t = \frac{\ln \tfrac{1}{2}}{-0.1} \approx 6.93 \text{ years}$$

If $r = 0.2$:

$$\tfrac{1}{2} A_0 = A_0 e^{-0.2t}$$

$$\tfrac{1}{2} = e^{-0.2t}$$

$$-0.2t = \ln \tfrac{1}{2}$$

$$t = \frac{\ln \tfrac{1}{2}}{-0.2} \approx 3.47 \text{ years}$$

The half-life is cut in half.

56. First find the intensity:

$$8.3 = \log \frac{I}{I_0}$$

$$10^{8.3} = \frac{I}{I_0}$$

$$I = 10^{8.3} I_0$$

For the new intensity which is $5 \cdot 10^{8.3} I_0$: $R = \log \dfrac{5 \cdot 10^{8.3} I_0}{I_0} = \log\left(5 \cdot 10^{8.3}\right) \approx 9.0$

57. Substituting $N = 100$:

$$100 = 160 + 10 \log I$$

$$-60 = 10 \log I$$

$$-6 = \log I$$

$$I = 10^{-6} \text{ watts / cm}^2$$

Chapter 5 Practice Test

1.   a.   The x-intercept is -3 and the y-intercept is $-7/2$:

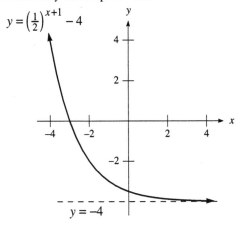

$$y = \left(\tfrac{1}{2}\right)^{x+1} - 4$$

$$y = -4$$

b.   The x-intercept is $1/3$ and there is no y-intercept:

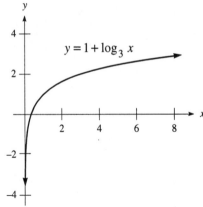

2.   Switching the roles of x and y:
$$2^{y+1} = x$$
$$y + 1 = \log_2 x$$
$$y = -1 + \log_2 x$$

So $f^{-1}(x) = -1 + \log_2 x$. Graphing both functions:

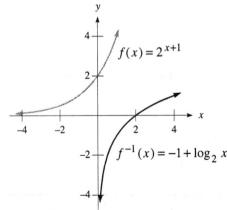

3.   **a.**   Evaluating the logarithm: $\log_9 \frac{1}{3} = \log_9 9^{-1/2} = -\frac{1}{2}$

 b.   Evaluating the logarithm: $\log_7 4 = \dfrac{\ln 4}{\ln 7} \approx 0.71$

 c.   Evaluating the logarithm: $\log_8 16 = \log_8 8^{4/3} = \frac{4}{3}$

 d.   Evaluating the logarithm: $\log_{1/2} \frac{1}{4} = \log_{1/2} \left(\frac{1}{2}\right)^2 = 2$

4.   **a.**   Solving the equation:
$$3^{2x-1} = 9$$
$$3^{2x-1} = 3^2$$
$$2x - 1 = 2$$
$$2x = 3$$
$$x = \frac{3}{2}$$

b. Solving the equation:

$$\log_3 (x-9) + \log_3 (x+1) = 2$$

$$\log_3 \left(x^2 - 8x - 9\right) = 2$$

$$x^2 - 8x - 9 = 3^2$$

$$x^2 - 8x - 9 = 9$$

$$x^2 - 8x - 18 = 0$$

Using the quadratic formula: $x = \dfrac{8 \pm \sqrt{64 - 4(-18)}}{2(1)} = \dfrac{8 \pm \sqrt{136}}{2} \approx -1.83, 9.83$

Since $x \approx -1.83$ does not check, the solution is $x \approx 9.83$.

c. Solving the equation:

$$\tfrac{1}{2}\log_2 x - \log_2 (x-3) = 1$$

$$\log_2 \sqrt{x} - \log_2 (x-3) = 1$$

$$\log_2 \dfrac{\sqrt{x}}{x-3} = 1$$

$$\dfrac{\sqrt{x}}{x-3} = 2$$

$$\sqrt{x} = 2x - 6$$

$$x = 4x^2 - 24x + 36$$

$$0 = 4x^2 - 25x + 36$$

$$0 = (4x-9)(x-4)$$

$$x = 4 \qquad \left(x = \tfrac{9}{4} \text{ does not check}\right)$$

d. Solving the equation:

$$5^{x+1} = 10$$

$$\ln 5^{x+1} = \ln 10$$

$$(x+1)\ln 5 = \ln 10$$

$$x \ln 5 + \ln 5 = \ln 10$$

$$x \ln 5 = \ln 10 - \ln 5$$

$$x = \dfrac{\ln 10 - \ln 5}{\ln 5} \approx 0.43$$

5. **a.** Finding the growth rate:

$$1500 = 600 e^{k \cdot 20}$$

$$2.5 = e^{20k}$$

$$20k = \ln 2.5$$

$$k = \dfrac{\ln 2.5}{20} \approx 0.0458$$

So $A = 600 e^{0.0458t}$. Substituting $t = 10$: $A = 600 e^{0.0458(10)} \approx 949$ bacteria

b. Finding the time:

$$6000 = 600 e^{0.0458t}$$

$$10 = e^{0.0458t}$$

$$\ln 10 = 0.0458t$$

$$t = \dfrac{\ln 10}{0.0458} \approx 50.26 \text{ hours}$$

6. Using the half-life:

$$\tfrac{1}{2}A_0 = A_0 e^{k \cdot 750}$$

$$\tfrac{1}{2} = e^{750k}$$

$$750k = \ln \tfrac{1}{2}$$

$$k = \frac{\ln \tfrac{1}{2}}{750} \approx -0.000924$$

So $A = A_0 e^{-0.000924t}$. Finding the amount remaining: $A = 100 e^{-0.000924(200)} \approx 83.12$ grams

Chapter 6
Trigonometry

6.1 Angle Measurement and Two Special Triangles

1. Converting to degrees: $\dfrac{\pi}{6} \cdot \dfrac{180°}{\pi} = 30°$

3. Converting to degrees: $\dfrac{\pi}{3} \cdot \dfrac{180°}{\pi} = 60°$

5. Converting to degrees: $-\dfrac{5\pi}{2} \cdot \dfrac{180°}{\pi} = -450°$

7. Converting to degrees: $\dfrac{11\pi}{6} \cdot \dfrac{180°}{\pi} = 330°$

9. Converting to degrees: $\dfrac{3\pi}{2} \cdot \dfrac{180°}{\pi} = 270°$

11. Converting to degrees: $3 \cdot \dfrac{180°}{\pi} \approx 171.9°$

13. Converting to degrees: $-2 \cdot \dfrac{180°}{\pi} \approx -114.6°$

15. Converting to degrees: $-\dfrac{\pi}{12} \cdot \dfrac{180°}{\pi} = -15°$

17. Converting to radians: $150° \cdot \dfrac{\pi}{180°} = \dfrac{5\pi}{6}$

19. Converting to radians: $-120° \cdot \dfrac{\pi}{180°} = -\dfrac{2\pi}{3}$

21. Converting to radians: $18° \cdot \dfrac{\pi}{180°} = \dfrac{\pi}{10}$

23. Converting to radians: $-40° \cdot \dfrac{\pi}{180°} = -\dfrac{2\pi}{9}$

25. Converting to radians: $210° \cdot \dfrac{\pi}{180°} = \dfrac{7\pi}{6}$

27. The arc length and area are given by:
$$s = 12 \cdot \dfrac{\pi}{3} = 4\pi \text{ cm}$$
$$A = \tfrac{1}{2}(12)^2 \cdot \dfrac{\pi}{3} = 24\pi \text{ cm}^2$$

29. Using the arc length formula:

$$\pi = r \cdot \frac{\pi}{6}$$

$$r = 6 \text{ cm}$$

31. Using the area formula:

$$32\pi = \frac{1}{2}(8)^2 \cdot \theta$$

$$32\pi = 32\theta$$

$$\theta = \pi \text{ radians}$$

33. Since $120° = \frac{2\pi}{3}$ radians, using the arc length formula:

$$24\pi = r \cdot \frac{2\pi}{3}$$

$$r = 36 \text{ meters}$$

Using the area formula: $A = \frac{1}{2}(36)^2 \cdot \frac{2\pi}{3} = 432\pi \text{ m}^2$

35. Using the arc length formula:

$$9.5 = 4.3\theta$$

$$\theta \approx 2.2 \text{ radians}$$

37. In 1 minute, the radians are: $700 \cdot 2\pi = 1400\pi$ radians
In 1 minute, the degrees are: $700 \cdot 360° = 252,000°$

39. **a.** Converting to rpm: $\dfrac{180°}{1 \text{ sec}} \cdot \dfrac{60 \text{ sec}}{1 \text{ min}} \cdot \dfrac{\pi \text{ radians}}{180°} \cdot \dfrac{1 \text{ rev}}{2\pi \text{ radians}} = 30$ rpm

   b. Converting to rpm: $\dfrac{180°}{1 \text{ day}} \cdot \dfrac{1 \text{ day}}{24 \text{ hours}} \cdot \dfrac{1 \text{ hour}}{60 \text{ min}} \cdot \dfrac{1 \text{ rev}}{360°} = \frac{1}{2880}$ rpm ≈ 0.00035 rpm

41. Since $150 \text{ rpm} = \dfrac{150 \text{ rev}}{1 \text{ min}} \cdot \dfrac{2\pi \text{ radians}}{1 \text{ rev}} \cdot \dfrac{60 \text{ min}}{1 \text{ hour}} = 18000\pi \dfrac{\text{radians}}{\text{hour}}$, the bicycle turns a central angle
of 18000π radians in 1 hour. Finding the arc length: $s = 1 \cdot 18000\pi = 18000\pi$ feet

43. Finding the central angle:

$$2 = 4\theta$$

$$\theta = \frac{1}{2} \text{ radian}$$

Converting to degrees: $\theta = \dfrac{1}{2} \cdot \dfrac{180°}{\pi} = \dfrac{90°}{\pi} \approx 28.6°$

45. Finding the arc length: $s = 5 \cdot \dfrac{5\pi}{2} = \dfrac{25\pi}{2}$ inches ≈ 39.3 inches

47. Note that $B = 90° - 30° = 60°$. Since BC is the shortest side, $|BC| = 8$ and $|AC| = 8\sqrt{3}$.

49. Since $|AC| = |BC|$, $A = B = 45°$ and $|AB| = 10\sqrt{2}$.

51. Note that $B = 90° - 60° = 30°$. Since $|BC| = 4\sqrt{3}$, $|AC| = 4$ and $|AB| = 8$.

53. Note that $C = \dfrac{\pi}{4}$. Also $|AB| = |BC| = \dfrac{10}{\sqrt{2}} \cdot \dfrac{\sqrt{2}}{\sqrt{2}} = 5\sqrt{2}$.

55. The radian measure of an angle is the arc length measured on the circle divided by the radius of the circle. Alternatively, if a circle of radius 1 were used, the radian measure of an angle is the arc length of the central angle.

57. Converting the formula: $s = r \cdot \theta_{\text{radians}} = r \cdot \theta_{\text{degrees}} \cdot \dfrac{\pi \text{ radians}}{180°} = \dfrac{\pi}{180} r\theta_{\text{degrees}}$

59. An angle of 5 radians is one whose arc length is 5 times the radius of the circle.

6.2 The Trigonometric Functions

1. The reference angle is $\dfrac{\pi}{3}$:

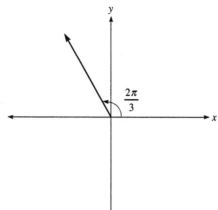

3. The reference angle is $\dfrac{\pi}{6}$:

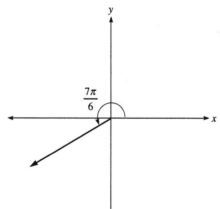

5. The reference angle is 45°:

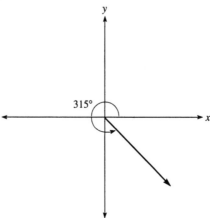

7.   The reference angle is 60°:

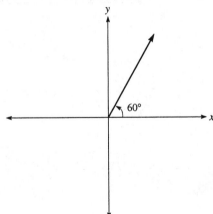

9.   The terminal side falls on the positive *y*-axis:

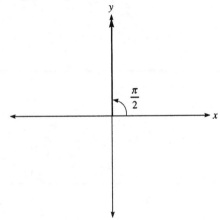

11.   The reference angle is $\frac{\pi}{6}$:

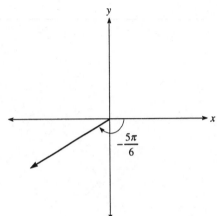

13. The terminal side falls on the positive *x*-axis:

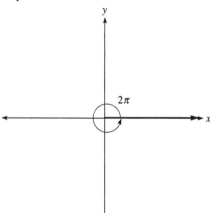

15. The reference angle is 45°:

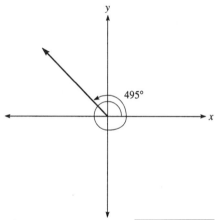

17. The terminal side passes through $\left(-\sqrt{3},1\right)$, so $r=\sqrt{\left(-\sqrt{3}\right)^2+(1)^2}=\sqrt{4}=2$. Therefore:

$$\sin\frac{5\pi}{6}=\frac{y}{r}=\frac{1}{2}$$

19. The terminal side passes through $\left(-1,-\sqrt{3}\right)$, so $r=\sqrt{(-1)^2+\left(-\sqrt{3}\right)^2}=\sqrt{4}=2$. Therefore:

$$\tan\frac{4\pi}{3}=\frac{y}{x}=\frac{-\sqrt{3}}{-1}=\sqrt{3}$$

21. The terminal side passes through $(0,-1)$, so $r=1$. Therefore: $\sin\frac{3\pi}{2}=\frac{y}{r}=\frac{-1}{1}=-1$

23. The terminal side passes through $(0,1)$, so $r=1$. Therefore: $\tan\frac{\pi}{2}=\frac{y}{x}=\frac{1}{0}=$ undefined

25. The terminal side passes through $(-1,-1)$, so $r=\sqrt{(-1)^2+(-1)^2}=\sqrt{2}$. Therefore:

$$\sec 225°=\frac{r}{x}=\frac{\sqrt{2}}{-1}=-\sqrt{2}$$

27. The terminal side passes through $(0,-1)$, so $r=1$. Therefore: $\cos 270°=\frac{x}{r}=\frac{0}{1}=0$

29. The terminal side passes through $(-1,1)$, so $r = \sqrt{(-1)^2 + (1)^2} = \sqrt{2}$. Therefore:

$$\cot \frac{3\pi}{4} = \frac{x}{y} = \frac{-1}{1} = -1$$

31. The terminal side passes through $\left(\sqrt{3},1\right)$, so $r = \sqrt{\left(\sqrt{3}\right)^2 + (1)^2} = \sqrt{4} = 2$. Therefore:

$$\sin 30° = \frac{y}{r} = \frac{1}{2}$$

33. The terminal side passes through $(-1,1)$, so $r = \sqrt{(-1)^2 + (1)^2} = \sqrt{2}$. Therefore:

$$\sec(-225°) = \frac{r}{x} = \frac{\sqrt{2}}{-1} = -\sqrt{2}$$

35. The terminal side passes through $(0,1)$, so $r = 1$. Therefore: $\sin\left(-\frac{3\pi}{2}\right) = \frac{y}{r} = \frac{1}{1} = 1$

37. Finding the value: $\sec\theta = \dfrac{1}{\cos\theta} = \dfrac{1}{2/3} = \dfrac{3}{2}$

39. Finding the value: $\sin\theta = \dfrac{1}{\csc\theta} = -\dfrac{1}{3}$

41. Given $y = 3$ and $r = 5$, $x = -\sqrt{25-9} = -\sqrt{16} = -4$. Therefore: $\cos\theta = \dfrac{x}{r} = -\dfrac{4}{5}$

43. Given $x = 3$ and $y = -1$, $r = \sqrt{9+1} = \sqrt{10}$. Therefore: $\cos\theta = \dfrac{x}{r} = \dfrac{3}{\sqrt{10}} = \dfrac{3\sqrt{10}}{10}$

45. Given $y = \sqrt{3}$ and $r = 2$, $x = -\sqrt{4-3} = -\sqrt{1} = -1$. Therefore: $\tan\theta = \dfrac{y}{x} = \dfrac{\sqrt{3}}{-1} = -\sqrt{3}$

47. Given $x = -5$ and $r = 13$, $y = \sqrt{169-25} = \sqrt{144} = 12$. Therefore: $\tan\theta = \dfrac{y}{x} = -\dfrac{12}{5}$

49. Given $x = -4$ and $y = -3$, $r = \sqrt{16+9} = \sqrt{25} = 5$. Therefore: $\sin\theta = \dfrac{y}{r} = -\dfrac{3}{5}$

51. Given $x = -4$ and $r = 5$, $y = \sqrt{25-16} = \sqrt{9} = 3$. Therefore: $\tan\theta = \dfrac{y}{x} = -\dfrac{3}{4}$

53. Given $y = -3$ and $x = -1$, $r = \sqrt{9+1} = \sqrt{10}$. Therefore: $\cos\theta = \dfrac{x}{r} = -\dfrac{1}{\sqrt{10}} = -\dfrac{\sqrt{10}}{10}$

55. $\sin\theta$ is defined for all real numbers.

57. $\tan\theta$ is defined for $\theta \ne k \cdot \dfrac{\pi}{2}$, where k is any odd integer.

59. Two coterminal angles are $\dfrac{8\pi}{3}$ and $-\dfrac{4\pi}{3}$.

61. Two coterminal angles are $470°$ and $-250°$.

63. Two coterminal angles are $\dfrac{7\pi}{2}$ and $-\dfrac{\pi}{2}$.

65. Three such angles are: $\theta = \dfrac{\pi}{6}, \dfrac{5\pi}{6}, \dfrac{13\pi}{6}$

67. Three such angles are: $\theta = \dfrac{2\pi}{3}, \dfrac{5\pi}{3}, \dfrac{8\pi}{3}$

69. Finding the value: $\sin^2\dfrac{\pi}{3} + \cos^2\dfrac{\pi}{3} = \left(\dfrac{\sqrt{3}}{2}\right)^2 + \left(\dfrac{1}{2}\right)^2 = \dfrac{3}{4} + \dfrac{1}{4} = 1$

71. Finding the value: $\sin^2 \dfrac{\pi}{2} + \cos^2 \dfrac{\pi}{2} = (1)^2 + (0)^2 = 1 + 0 = 1$

73. Completing the table:

θ	$\sin\theta$	$\cos\theta$	$\tan\theta$
0	0	1	0
$\dfrac{\pi}{6}$	$\dfrac{1}{2}$	$\dfrac{\sqrt{3}}{2}$	$\dfrac{1}{\sqrt{3}} = \dfrac{\sqrt{3}}{3}$
$\dfrac{\pi}{4}$	$\dfrac{1}{\sqrt{2}} = \dfrac{\sqrt{2}}{2}$	$\dfrac{1}{\sqrt{2}} = \dfrac{\sqrt{2}}{2}$	1
$\dfrac{\pi}{3}$	$\dfrac{\sqrt{3}}{2}$	$\dfrac{1}{2}$	$\sqrt{3}$
$\dfrac{\pi}{2}$	1	0	undefined
$\dfrac{2\pi}{3}$	$\dfrac{\sqrt{3}}{2}$	$-\dfrac{1}{2}$	$-\sqrt{3}$
$\dfrac{3\pi}{4}$	$\dfrac{1}{\sqrt{2}} = \dfrac{\sqrt{2}}{2}$	$-\dfrac{1}{\sqrt{2}} = -\dfrac{\sqrt{2}}{2}$	-1
$\dfrac{5\pi}{6}$	$\dfrac{1}{2}$	$-\dfrac{\sqrt{3}}{2}$	$-\dfrac{1}{\sqrt{3}} = -\dfrac{\sqrt{3}}{3}$
π	0	-1	0

75. No. Since $\sin\theta = \dfrac{y}{r}$ and $y \le r$, $-1 \le \sin\theta \le 1$. Thus $\sin\theta \ne 2$.

77. Since $\csc\theta = \dfrac{1}{\sin\theta}$, $\csc\theta \ge 1$ or $\csc\theta \le -1$. Similarly, since $\sec\theta = \dfrac{1}{\cos\theta}$, $\sec\theta \ge 1$ or $\sec\theta \le -1$.

79. Computing the values:

$\sin\dfrac{\pi}{4} = \dfrac{\sqrt{2}}{2}$ $\qquad\qquad\qquad$ $\sin\left(-\dfrac{\pi}{4}\right) = -\dfrac{\sqrt{2}}{2}$

$\sin\dfrac{2\pi}{3} = \dfrac{\sqrt{3}}{2}$ $\qquad\qquad\qquad$ $\sin\left(-\dfrac{2\pi}{3}\right) = -\dfrac{\sqrt{3}}{2}$

$\sin\dfrac{11\pi}{6} = \dfrac{1}{2}$ $\qquad\qquad\qquad$ $\sin\left(-\dfrac{11\pi}{6}\right) = -\dfrac{1}{2}$

Yes, it appears that $\sin(-\theta) = -\sin\theta$ for all angles θ.

81. Since $\tan(-\theta) = \dfrac{\sin(-\theta)}{\cos(-\theta)} = \dfrac{-\sin\theta}{\cos\theta} = -\tan\theta$, it is true that $\tan(-\theta) = -\tan\theta$ for all angles θ.

6.3 Right Triangle Trigonometry and Applications

1. Finding the six trigonometric functions:

$\sin\theta = \dfrac{6}{10} = \dfrac{3}{5}$ $\qquad\qquad\qquad$ $\csc\theta = \dfrac{1}{\sin\theta} = \dfrac{5}{3}$

$\cos\theta = \dfrac{8}{10} = \dfrac{4}{5}$ $\qquad\qquad\qquad$ $\sec\theta = \dfrac{1}{\cos\theta} = \dfrac{5}{4}$

$\tan\theta = \dfrac{6}{8} = \dfrac{3}{4}$ $\qquad\qquad\qquad$ $\cot\theta = \dfrac{1}{\tan\theta} = \dfrac{4}{3}$

3.   The remaining side is $\sqrt{13^2 - 12^2} = \sqrt{169 - 144} = \sqrt{25} = 5$. Finding the six trigonometric functions:

$$\sin\theta = \frac{12}{13} \qquad\qquad \csc\theta = \frac{1}{\sin\theta} = \frac{13}{12}$$

$$\cos\theta = \frac{5}{13} \qquad\qquad \sec\theta = \frac{1}{\cos\theta} = \frac{13}{5}$$

$$\tan\theta = \frac{12}{5} \qquad\qquad \cot\theta = \frac{1}{\tan\theta} = \frac{5}{12}$$

5.   The remaining side is $\sqrt{4^2 + 5^2} = \sqrt{16 + 25} = \sqrt{41}$. Finding the six trigonometric functions:

$$\sin\theta = \frac{5}{\sqrt{41}} = \frac{5\sqrt{41}}{41} \qquad\qquad \csc\theta = \frac{1}{\sin\theta} = \frac{\sqrt{41}}{5}$$

$$\cos\theta = \frac{4}{\sqrt{41}} = \frac{4\sqrt{41}}{41} \qquad\qquad \sec\theta = \frac{1}{\cos\theta} = \frac{\sqrt{41}}{4}$$

$$\tan\theta = \frac{5}{4} \qquad\qquad \cot\theta = \frac{1}{\tan\theta} = \frac{4}{5}$$

7.   The remaining side is $\sqrt{7^2 + 24^2} = \sqrt{49 + 576} = \sqrt{625} = 25$. Finding the six trigonometric functions:

$$\sin\theta = \frac{24}{25} \qquad\qquad \csc\theta = \frac{1}{\sin\theta} = \frac{25}{24}$$

$$\cos\theta = \frac{7}{25} \qquad\qquad \sec\theta = \frac{1}{\cos\theta} = \frac{25}{7}$$

$$\tan\theta = \frac{24}{7} \qquad\qquad \cot\theta = \frac{1}{\tan\theta} = \frac{7}{24}$$

9.   The remaining side is $\sqrt{6^2 - 3^2} = \sqrt{36 - 9} = \sqrt{27} = 3\sqrt{3}$. Finding the six trigonometric functions:

$$\sin\theta = \frac{3\sqrt{3}}{6} = \frac{\sqrt{3}}{2} \qquad\qquad \csc\theta = \frac{1}{\sin\theta} = \frac{2}{\sqrt{3}} = \frac{2\sqrt{3}}{3}$$

$$\cos\theta = \frac{3}{6} = \frac{1}{2} \qquad\qquad \sec\theta = \frac{1}{\cos\theta} = 2$$

$$\tan\theta = \frac{3\sqrt{3}}{3} = \sqrt{3} \qquad\qquad \cot\theta = \frac{1}{\tan\theta} = \frac{1}{\sqrt{3}} = \frac{\sqrt{3}}{3}$$

11.   Finding the values: $\sin A = \frac{4}{5}$ and $\cos B = \frac{4}{5}$

13.   The remaining side is $\sqrt{\left(\sqrt{34}\right)^2 - 5^2} = \sqrt{34 - 25} = \sqrt{9} = 3$. Finding the values:

$$\tan\alpha = \frac{3}{5} \text{ and } \cot\beta = \frac{3}{5}$$

15.   The remaining side is $\sqrt{6^2 - 4^2} = \sqrt{36 - 16} = \sqrt{20} = 2\sqrt{5}$. Finding the values:

$$\sec P = \frac{6}{2\sqrt{5}} = \frac{3}{\sqrt{5}} = \frac{3\sqrt{5}}{5} \text{ and } \csc Q = \frac{6}{2\sqrt{5}} = \frac{3}{\sqrt{5}} = \frac{3\sqrt{5}}{5}$$

17.   Solving for x:

$$\tan 20° = \frac{x}{40}$$
$$x = 40\tan 20° \approx 14.6$$

19. Solving for x:

$$\sin 18° = \frac{50}{x}$$
$$x \sin 18° = 50$$
$$x = \frac{50}{\sin 18°} \approx 161.8$$

21. Solving for x:

$$\cos 72° = \frac{x}{8}$$
$$x = 8 \cos 72° \approx 2.5$$

23. Solving for θ:

$$\sin \theta = \frac{6}{11}$$
$$\theta = \sin^{-1}\left(\frac{6}{11}\right) \approx 33°$$

25. Solving for θ:

$$\cos \theta = \frac{8}{10}$$
$$\theta = \cos^{-1}\left(\frac{8}{10}\right) \approx 37°$$

27. Finding the value: $\sin 123° \approx 0.8387$

29. Finding the value: $\tan 351° \approx -0.1584$

31. Finding the value: $\sec 225° = \dfrac{1}{\cos 225°} = -\dfrac{1}{\cos 45°} = -\dfrac{1}{1/\sqrt{2}} = -\sqrt{2}$

33. Finding the value: $\cos 158° \approx -0.9272$

35. Finding the value: $\tan 111° \approx -2.6051$

37. Finding the value: $\sec 150° = \dfrac{1}{\cos 150°} = -\dfrac{1}{\cos 30°} = -\dfrac{1}{\sqrt{3}/2} = -\dfrac{2}{\sqrt{3}} = -\dfrac{2\sqrt{3}}{3}$

39. Solving for θ: $\theta = \sin^{-1}(0.4384) \approx 26°$

41. Solving for x: $x = \tan^{-1}(1.428) \approx 55°$

43. Solving for θ: $\theta = \sec^{-1}(3.420) = \cos^{-1}\left(\dfrac{1}{3.420}\right) \approx 73°$

45. Solving for x: $x = \sin^{-1}(0.9630) \approx 74°$

47. Solving for θ: $\theta = \tan^{-1}(0.5275) \approx 28°$

49. Using the cofunction relationship: $\sin 76° = \cos(90° - 76°) = \cos 14°$

51. Using the cofunction relationship: $\csc 88° = \sec(90° - 88°) = \sec 2°$

53. Using the cofunction relationship: $\cos \dfrac{3\pi}{7} = \sin\left(\dfrac{\pi}{2} - \dfrac{3\pi}{7}\right) = \sin \dfrac{\pi}{14}$

55. Evaluating the expression: $\sin^2 30° + \cos^2 30° = \left(\dfrac{1}{2}\right)^2 + \left(\dfrac{\sqrt{3}}{2}\right)^2 = \dfrac{1}{4} + \dfrac{3}{4} = 1$

57. Evaluating the expression: $\csc^2 45° - \cot^2 45° = \left(\sqrt{2}\right)^2 - (1)^2 = 2 - 1 = 1$

59. Let h represent the height the ladder reaches. Then:

$$\sin 37° = \frac{h}{25}$$
$$h = 25 \sin 37° \approx 15.0$$

The ladder reaches a height of 15.0 feet.

61. Let h represent the height of the flagpole. Then:

$$\tan 48° = \frac{h}{50}$$
$$h = 50 \tan 48° \approx 55.5$$

The flagpole is 55.5 feet tall.

63. Let l represent the length of the wire. Then:

$$\sin 50° = \frac{30}{l}$$
$$l \sin 50° = 30$$
$$l = \frac{30}{\sin 50°} \approx 39.2$$

The wire is 39.2 feet long.

65. First draw the figure:

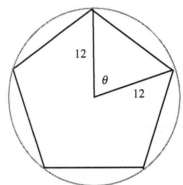

Note that $\theta = \frac{360°}{5} = 72°$. The area of the triangle is: $\frac{1}{2}(12)(12)\sin 72° = 72 \sin 72°$

Since the pentagon consists of 5 such triangles, its area is:
$$A = 5(72 \sin 72°) = 360 \sin 72° \approx 342.4 \text{ cm}^2$$

67. Let d represent the distance the balloon traveled in the 2 minutes. Then:

$$\tan 70° = \frac{800}{d}$$
$$d \tan 70° = 800$$
$$d = \frac{800}{\tan 70°} \approx 291.2$$

The speed is therefore: $\frac{291.2 \text{ m}}{2 \text{ min}} \cdot \frac{60 \text{ min}}{1 \text{ hr}} \cdot \frac{1 \text{ km}}{1000 \text{ m}} \approx 9 \text{ km / hour}$

69. Let θ represent the angle of elevation to the sun. Then:

$$\tan \theta = \frac{6}{9} = \frac{2}{3}$$
$$\theta = \tan^{-1}\left(\frac{2}{3}\right) \approx 34°$$

The angle of elevation is 34°.

71. First draw the figure:

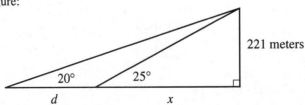

Then:
$$\tan 25° = \frac{221}{x}$$
$$x \tan 25° = 221$$
$$x = \frac{221}{\tan 25°}$$

Therefore:
$$\tan 20° = \frac{221}{d + x}$$
$$d + x = \frac{221}{\tan 20°}$$
$$d + \frac{221}{\tan 25°} = \frac{221}{\tan 20°}$$
$$d = \frac{221}{\tan 20°} - \frac{221}{\tan 25°} \approx 133.3$$

She has traveled 133.3 meters.

73. Let l represent the length of the antenna. Therefore:
$$\tan 24° = \frac{l + 30}{100}$$
$$l + 30 = 100 \tan 24°$$
$$l = 100 \tan 24° - 30 \approx 14.5$$

The antenna is approximately 14.5 feet long.

75. Drawing the figure:

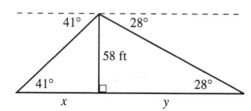

Using each triangle:

$$\tan 41° = \frac{58}{x} \qquad\qquad \tan 28° = \frac{58}{y}$$
$$x \tan 41° = 58 \qquad\qquad y \tan 28° = 58$$
$$x = \frac{58}{\tan 41°} \qquad\qquad y = \frac{58}{\tan 28°}$$

The distance between the ships is therefore: $x + y = \dfrac{58}{\tan 41°} + \dfrac{58}{\tan 28°} \approx 175.8$ feet

77. Let h represent the altitude. Then:
$$\sin 41.4° = \frac{4000}{4000 + h}$$
$$(4000 + h) \sin 41.4° = 4000$$
$$4000 \sin 41.4° + h \sin 41.4° = 4000$$
$$h \sin 41.4° = 4000 - 4000 \sin 41.4°$$
$$h = \frac{4000 - 4000 \sin 41.4°}{\sin 41.4°} \approx 2048.6$$

The altitude of the satellite is approximately 2048.6 miles.

79 .   First draw the figure:

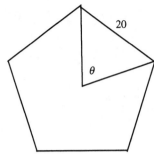

Note that $\theta = \dfrac{360°}{5} = 72°$. Now construct the triangle:

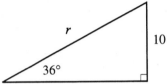

Then:

$$\sin 36° = \frac{10}{r}$$
$$r\sin 36° = 10$$
$$r = \frac{10}{\sin 36°}$$

From the original figure, the area of the triangle is: $\dfrac{1}{2}r^2 \sin 72° = \dfrac{1}{2}\left(\dfrac{10}{\sin 36°}\right)^2 \sin 72° = \dfrac{50\sin 72°}{\sin^2 36°}$

Since the pentagon consists of five such triangles, its area is:

$$A = 5\left(\frac{50\sin 72°}{\sin^2 36°}\right) = \frac{250\sin 72°}{\sin^2 36°} \approx 688.2 \text{ mm}^2$$

81 .   First draw the figure:

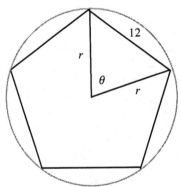

Note that $\theta = \dfrac{360°}{6} = 60°$. Now construct the triangle:

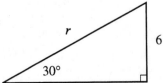

Then:

$$\sin 30° = \frac{6}{r}$$

$$r\sin 30° = 6$$

$$r = \frac{6}{\sin 30°} = \frac{6}{1/2} = 12$$

The area of the circle is therefore: $\pi r^2 = \pi(12)^2 = 144\pi \approx 452.4$ in.^{2}

83. First draw the figure:

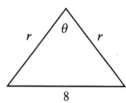

Note that $\theta = \dfrac{360°}{5} = 72°$. Now construct the triangle:

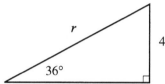

Then:

$$\sin 36° = \frac{4}{r}$$

$$r\sin 36° = 4$$

$$r = \frac{4}{\sin 36°}$$

The area of the larger triangle is therefore: $\frac{1}{2}r^2 \sin\theta = \frac{1}{2}\left(\dfrac{4}{\sin 36°}\right)^2 \sin 72° = \dfrac{8\sin 72°}{\sin^2 36°}$

Since the pentagon consists of 5 such triangles, its area is: $A = 5\left(\dfrac{8\sin 72°}{\sin^2 36°}\right) = \dfrac{40\sin 72°}{\sin^2 36°}$

The shaded area is therefore: $\pi\left(\dfrac{4}{\sin 36°}\right)^2 - \dfrac{40\sin 72°}{\sin^2 36°} = \dfrac{16\pi - 40\sin 72°}{\sin^2 36°} \approx 35.4$ cm^2

85. The hypotenuse of the triangle is $2r$ and let the base be x. Then:

$$8^2 + x^2 = (2r)^2$$

$$64 + x^2 = 4r^2$$

$$x^2 = 4r^2 - 64$$

$$x = \sqrt{4r^2 - 64} = 2\sqrt{r^2 - 16}$$

The area of the figure is therefore: $\frac{1}{2}\pi r^2 + \frac{1}{2}(8)(x) = \frac{1}{2}\pi r^2 + 4\cdot 2\sqrt{r^2 - 16} = \frac{1}{2}\pi r^2 + 8\sqrt{r^2 - 16}$

87. **a.** First find AD and CD using $\triangle ADC$:

$$\cos\theta = \frac{AD}{10} \qquad\qquad \sin\theta = \frac{CD}{10}$$

$$AD = 10\cos\theta \qquad\qquad CD = 10\sin\theta$$

Therefore the area is given by: $A = \frac{1}{2}(AD)(CD) = \frac{1}{2}(10\cos\theta)(10\sin\theta) = 50\sin\theta\cos\theta$

b. First find BC using $\triangle ACB$:

$$\tan\theta = \frac{BC}{10}$$
$$BC = 10\tan\theta$$

Therefore the area is given by: $A = \frac{1}{2}(AC)(BC) = \frac{1}{2}(10)(10\tan\theta) = 50\tan\theta$

c. Subtracting these two areas: $A = 50\tan\theta - 50\sin\theta\cos\theta = 50(\tan\theta - \sin\theta\cos\theta)$

89. Using the smaller triangle:

$$\sin\theta = \frac{10}{BC}$$
$$BC\sin\theta = 10$$
$$BC = \frac{10}{\sin\theta}$$

Using $\triangle ACB$:

$$\tan\theta = \frac{AC}{BC}$$
$$\tan\theta = \frac{AC}{10/\sin\theta}$$
$$AC = \frac{10\tan\theta}{\sin\theta} = \frac{10\sin\theta}{\sin\theta\cos\theta} = \frac{10}{\cos\theta} = 10\sec\theta$$

91. Let $PQ = 2 + x$. Using the upper triangle:

$$\tan\theta = \frac{x}{5}$$
$$x = 5\tan\theta$$

Thus $PQ = 2 + 5\tan\theta$.

93. Drawing the figure:

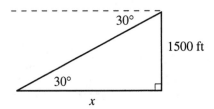

Then:

$$\tan 30° = \frac{1500}{x}$$
$$x\tan 30° = 1500$$
$$x = \frac{1500}{\tan 30°} = \frac{1500}{1/\sqrt{3}} = 1500\sqrt{3} \approx 2598.1$$

The plant is 2598.1 feet from the factory.

95. Using the two triangles:

$$\tan\alpha = \frac{1.5}{PC} \qquad\qquad \tan\alpha = \frac{2}{PD}$$
$$PC\tan\alpha = 1.5 \qquad\qquad PD\tan\alpha = 2$$
$$PC = \frac{1.5}{\tan\alpha} \qquad\qquad PD = \frac{2}{\tan\alpha}$$

97. Drawing the triangle:

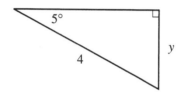

Thus:

$$\sin 5° = \frac{y}{4}$$
$$y = 4\sin 5° \approx 0.35 \text{ miles} = 1841 \text{ feet}$$

The change in elevation is 1841 feet.

99. Let θ be the angle. Then:

$$\sin\theta = \frac{5}{30} = \frac{1}{6}$$
$$\theta = \sin^{-1}\left(\frac{1}{6}\right) \approx 9.6°$$

101. Let d represent the distance to travel. Then:

$$\sin 15° = \frac{5000}{d}$$
$$d\sin 15° = 5000$$
$$d = \frac{5000}{\sin 15°}$$

Since the plane is traveling at $170\dfrac{\text{feet}}{\sec}$, let t represent the time:

$$170t = \frac{5000}{\sin 15°}$$
$$t = \frac{5000}{170\sin 15°} = \frac{500}{17\sin 15°} \approx 113.6$$

The plane must travel for 113.6 seconds, or 1 minute 53.6 seconds.

6.4 The Trigonometric Functions as Functions of Real Numbers

1. Performing the operations: $2\cos\theta(\cos\theta - 1) = 2\cos^2\theta - 2\cos\theta$

3. Performing the operations: $\sin 5\theta(\sin 5\theta + 4) = \sin^2 5\theta + 4\sin 5\theta$

5. Performing the operations: $(\csc\theta - 2)^2 = (\csc\theta - 2)(\csc\theta - 2) = \csc^2\theta - 4\csc\theta + 4$

7. Performing the operations: $(\cos 4\theta + 1)^2 = (\cos 4\theta + 1)(\cos 4\theta + 1) = \cos^2 4\theta + 2\cos 4\theta + 1$

9. Performing the operations: $(5\sin\theta - 2)(\sin\theta + 3) = 5\sin^2\theta + 13\sin\theta - 6$

11. Performing the operations: $(\cos 2\theta + 3)(\cos 2\theta - 3) = \cos^2 2\theta - 9$

13. Performing the operations: $(\sin 3\theta + 2)(\sin 5\theta - 4) = \sin 3\theta\sin 5\theta + 2\sin 5\theta - 4\sin 3\theta - 8$

15. Factoring the expression: $\sin^2\theta - \sin\theta - 2 = (\sin\theta - 2)(\sin\theta + 1)$

17. Factoring the expression: $\tan^2\theta + 2\tan\theta - 8 = (\tan\theta + 4)(\tan\theta - 2)$

19. Factoring the expression: $2\csc^2\theta - 5\csc\theta - 3 = (2\csc\theta + 1)(\csc\theta - 3)$

21. Combining the fractions: $\dfrac{2}{\sin\theta} + \dfrac{3}{\cos\theta} = \dfrac{2\cos\theta}{\sin\theta\cos\theta} + \dfrac{3\sin\theta}{\cos\theta\sin\theta} = \dfrac{2\cos\theta + 3\sin\theta}{\cos\theta\sin\theta}$

23. Combining the fractions: $\dfrac{\sin\theta}{\cos^2\theta} - \dfrac{1}{\sin\theta} = \dfrac{\sin^2\theta}{\sin\theta\cos^2\theta} - \dfrac{\cos^2\theta}{\sin\theta\cos^2\theta} = \dfrac{\sin^2\theta - \cos^2\theta}{\sin\theta\cos^2\theta}$

25. Combining the fractions: $\dfrac{1}{\cos^2\theta}+1=\dfrac{1}{\cos^2\theta}+\dfrac{\cos^2\theta}{\cos^2\theta}=\dfrac{1+\cos^2\theta}{\cos^2\theta}$

27. Combining the fractions: $\dfrac{1}{\tan\theta}+\sec\theta=\dfrac{1}{\tan\theta}+\dfrac{\sec\theta\tan\theta}{\tan\theta}=\dfrac{1+\sec\theta\tan\theta}{\tan\theta}$

29. Combining the fractions: $1-\dfrac{1}{\cot^2\theta}=\dfrac{\cot^2\theta}{\cot^2\theta}-\dfrac{1}{\cot^2\theta}=\dfrac{\cot^2\theta-1}{\cot^2\theta}$

Alternatively: $1-\dfrac{1}{\cot^2\theta}=1-\tan^2\theta$

31. Combining the fractions: $\dfrac{3}{\sin 2\theta}+\dfrac{2}{\sin 3\theta}=\dfrac{3\sin 3\theta}{\sin 2\theta\sin 3\theta}+\dfrac{2\sin 2\theta}{\sin 2\theta\sin 3\theta}=\dfrac{3\sin 3\theta+2\sin 2\theta}{\sin 2\theta\sin 3\theta}$

33. Combining the fractions: $\dfrac{1}{\cos^2\theta}+\dfrac{\tan^2\theta}{2\cos\theta}=\dfrac{2}{2\cos^2\theta}+\dfrac{\tan^2\theta\cos\theta}{2\cos^2\theta}=\dfrac{2+\tan^2\theta\cos\theta}{2\cos^2\theta}$

35. Combining the fractions: $\sin\theta+\dfrac{1}{\sin\theta}=\dfrac{\sin^2\theta}{\sin\theta}+\dfrac{1}{\sin\theta}=\dfrac{\sin^2\theta+1}{\sin\theta}$

37. Simplifying the fraction: $\dfrac{5\cos\theta}{3\cos^2\theta+2\cos\theta}=\dfrac{5\cos\theta}{\cos\theta(3\cos\theta+2)}=\dfrac{5}{3\cos\theta+2}$

39. Simplifying the fraction: $\dfrac{1-\sin^2\theta}{(1-\sin\theta)^2}=\dfrac{(1+\sin\theta)(1-\sin\theta)}{(1-\sin\theta)^2}=\dfrac{1+\sin\theta}{1-\sin\theta}$

41. Simplifying the fraction: $\dfrac{\sin^2 4\theta}{\sin^2 4\theta-\sin 4\theta}=\dfrac{\sin^2 4\theta}{\sin 4\theta(\sin 4\theta-1)}=\dfrac{\sin 4\theta}{\sin 4\theta-1}$

43. The other side is $\sqrt{3^2-x^2}=\sqrt{9-x^2}$. Thus: $\sin\theta=\dfrac{\sqrt{9-x^2}}{3}$

45. The other side is $\sqrt{y^2-5^2}=\sqrt{y^2-25}$. Thus: $\tan\theta=\dfrac{5}{\sqrt{y^2-25}}$

47. The other side is $\sqrt{a^2+10^2}=\sqrt{a^2+100}$. Thus: $\sec\theta=\dfrac{1}{\cos\theta}=\dfrac{\sqrt{a^2+100}}{a}$

49. Since $\sin\theta=\dfrac{x}{4}$, $x=4\sin\theta$.

51. Since $\tan\theta=\dfrac{y}{4}$, $y=4\tan\theta$.

53. Since $\sec\theta=\dfrac{5}{s}$, $s\sec\theta=5$ and thus $s=\dfrac{5}{\sec\theta}$.

55. Remember to set your calculator to radian mode to perform these calculations.
  a.   Substituting $t=0$: $d=6\cos 0=6$ inches
  b.   Substituting $t=2$: $d=6\cos 8\approx -0.87$ inches
  c.   Substituting $t=5$: $d=6\cos 20\approx 2.45$ inches

57. Using the approximation: $\sin 0.1=0.1-\dfrac{(0.1)^3}{6}\approx 0.0998$

The error is $\left|\dfrac{(0.1)^5}{120}\right|=8.3\times 10^{-8}$.

59.   The remaining side is $\sqrt{3^2 - x^2} = \sqrt{9 - x^2}$. Now note that: $\tan\theta = \dfrac{x}{\sqrt{9 - x^2}}$ and $\sin\theta = \dfrac{x}{3}$

Therefore: $\dfrac{x^2}{\sqrt{9 - x^2}} = \dfrac{x}{\sqrt{9 - x^2}} \cdot \dfrac{x}{3} \cdot 3 = \tan\theta \cdot \sin\theta \cdot 3 = 3\sin\theta\tan\theta$

61.   The remaining side is $\sqrt{x^2 - 3^2} = \sqrt{x^2 - 9}$. Now note that: $\cos\theta = \dfrac{3}{x}$ and $\tan\theta = \dfrac{\sqrt{x^2 - 9}}{3}$

Therefore: $\dfrac{x\sqrt{x^2 - 9}}{9} = \dfrac{\sqrt{x^2 - 9}}{3} \cdot \dfrac{x}{3} = \tan\theta \cdot \dfrac{1}{\cos\theta} = \tan\theta\sec\theta$

63.   Computing the values:

$$f\left(\frac{\pi}{3}\right) = \sin\left(\frac{2\pi}{3}\right) = \frac{\sqrt{3}}{2}$$

$$g\left(\frac{\pi}{3}\right) = 2\sin\left(\frac{\pi}{3}\right) = 2 \cdot \frac{\sqrt{3}}{2} = \sqrt{3}$$

They are not equal, nor should they be, since multiplying the angle by 2 is not the same as multiplying the function value by 2.

65.   Simplifying: $\sin^2\theta + \cos^2\theta = y^2 + x^2 = x^2 + y^2 = 1$

This last equality is due to the fact that the unit circle has an equation of $x^2 + y^2 = 1$.

Chapter 6 Review Exercises

1.   Converting to degrees: $\dfrac{\pi}{12} \cdot \dfrac{180°}{\pi} = 15°$

2.   Converting to radians: $80° \cdot \dfrac{\pi}{180°} = \dfrac{4\pi}{9}$

3.   Converting to radians: $200° \cdot \dfrac{\pi}{180°} = \dfrac{10\pi}{9}$

4.   Converting to degrees: $\dfrac{\pi}{9} \cdot \dfrac{180°}{\pi} = 20°$

5.   Converting to degrees: $-\dfrac{3\pi}{5} \cdot \dfrac{180°}{\pi} = -108°$

6.   Converting to radians: $-400° \cdot \dfrac{\pi}{180°} = -\dfrac{20\pi}{9}$

7.   Converting to radians: $330° \cdot \dfrac{\pi}{180°} = \dfrac{11\pi}{6}$

8.   Converting to degrees: $\dfrac{3\pi}{2} \cdot \dfrac{180°}{\pi} = 270°$

9.   Converting to degrees: $\dfrac{7\pi}{4} \cdot \dfrac{180°}{\pi} = 315°$

10.   Converting to degrees: $-\dfrac{2\pi}{3} \cdot \dfrac{180°}{\pi} = -120°$

11.   Finding the value: $\tan\dfrac{5\pi}{6} = -\tan\dfrac{\pi}{6} = -\dfrac{1}{\sqrt{3}} = -\dfrac{\sqrt{3}}{3}$

12. Finding the value: $\sec \dfrac{7\pi}{4} = \sec \dfrac{\pi}{4} = \dfrac{1}{\cos \dfrac{\pi}{4}} = \dfrac{1}{1/\sqrt{2}} = \sqrt{2}$

13. Finding the value: $\sin 240° = -\sin 60° = -\dfrac{\sqrt{3}}{2}$

14. Finding the value: $\cos 180° = -1$

15. Finding the value: $\sin \dfrac{\pi}{2} = 1$

16. Finding the value: $\tan 0 = 0$

17. Finding the value: $\csc 125° = \dfrac{1}{\sin 125°} \approx 1.2208$

18. Finding the value: $\cos(-143°) \approx -0.7986$

19. Finding the value: $\cot(-300°) = \cot 60° = \dfrac{1}{\tan 60°} = \dfrac{1}{\sqrt{3}} = \dfrac{\sqrt{3}}{3}$

20. Finding the value: $\cos \dfrac{\pi}{3} = \dfrac{1}{2}$

21. Finding the value: $\sin \dfrac{\pi}{9} \approx 0.3420$

22. Finding the value: $\csc \dfrac{11\pi}{6} = \dfrac{1}{\sin \dfrac{11\pi}{6}} = -\dfrac{1}{\sin \dfrac{\pi}{6}} = -\dfrac{1}{1/2} = -2$

23. Finding the value: $\tan 3 \approx -0.1425$
24. Finding the value: $\sin 2 \approx 0.9093$

25. Finding the value: $\sec \dfrac{\pi}{2} = \dfrac{1}{\cos \dfrac{\pi}{2}} = \dfrac{1}{0} = $ undefined

26. Finding the value: $\tan \dfrac{3\pi}{2} = \dfrac{-1}{0} = $ undefined

27. Finding the value: $\cos\left(-\dfrac{\pi}{15}\right) \approx 0.9781$

28. Finding the value: $\sec 206° = \dfrac{1}{\cos 206°} \approx -1.1126$

29. Finding the value: $\sin \dfrac{5\pi}{4} = -\sin \dfrac{\pi}{4} = -\dfrac{1}{\sqrt{2}} = -\dfrac{\sqrt{2}}{2}$

30. Finding the value: $\cos \dfrac{7\pi}{6} = -\cos \dfrac{\pi}{6} = -\dfrac{\sqrt{3}}{2}$

31. Solving for x:
$$\cos 60° = \dfrac{x}{50}$$
$$\dfrac{1}{2} = \dfrac{x}{50}$$
$$x = 25$$

32. Solving for θ:
$$\cos \theta = \tfrac{4}{8} = \tfrac{1}{2}$$
$$\theta = \cos^{-1}\left(\tfrac{1}{2}\right) = \dfrac{\pi}{3}$$

33. Solving for θ:
$$\sin\theta = \frac{6}{12} = \frac{1}{2}$$
$$\theta = \sin^{-1}\left(\frac{1}{2}\right) = \frac{\pi}{6}$$

34. Solving for x:
$$\tan 28° = \frac{x}{24}$$
$$x = 24\tan 28° \approx 12.76$$

35. Solving for x:
$$\tan 72° = \frac{30}{x}$$
$$x\tan 72° = 30$$
$$x = \frac{30}{\tan 72°} \approx 9.75$$

36. Solving for x:
$$\sin\frac{\pi}{4} = \frac{x}{32}$$
$$\frac{\sqrt{2}}{2} = \frac{x}{32}$$
$$x = 16\sqrt{2} \approx 22.63$$

37. Solving for θ:
$$\cos\theta = \frac{6}{6\sqrt{2}} = \frac{1}{\sqrt{2}}$$
$$\theta = \cos^{-1}\left(\frac{1}{\sqrt{2}}\right) = \frac{\pi}{4}$$

38. Solving for θ:
$$\tan\theta = \frac{5}{2} = 2.5$$
$$\theta = \tan^{-1}(2.5) \approx 68.20°$$

39. Finding the area: $A = \frac{1}{2}(8)(12)\sin 40° = 48\sin 40° \approx 30.9$ sq. units

40. Using the area formula: $A = \frac{1}{2}(18)^2 \cdot \frac{\pi}{9} = 18\pi$ cm^2

41. Note that $100° = 100° \cdot \frac{\pi}{180°} = \frac{5\pi}{9}$. Using the arc length formula: $s = 9 \cdot \frac{5\pi}{9} = 5\pi$ inches

42. Using the arc length formula:
$$3 = 12 \cdot \theta$$
$$\theta = \frac{1}{4} \text{ radian}$$

43. Let h represent the height the ladder will reach. Then:
$$\sin 65° = \frac{h}{22}$$
$$h = 22\sin 65° \approx 19.9$$
The ladder reaches a height of 19.9 feet.

44. Drawing the figure:

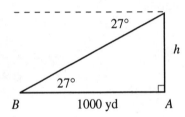

Therefore:

$$\tan 27° = \frac{h}{1000}$$
$$h = 1000 \tan 27° \approx 509.5 \text{ yd} = 1529 \text{ feet}$$

The altitude of the helicopter is 1529 feet.

45. Let d represent the distance the jet travels during the 20 seconds. Therefore:

$$\tan 15° = \frac{10000}{d}$$
$$d \tan 15° = 10000$$
$$d = \frac{10000}{\tan 15°} \approx 37320.5$$

The speed of the jet is therefore: $\dfrac{37320.5 \text{ feet}}{20 \text{ sec}} \cdot \dfrac{1 \text{ mile}}{5280 \text{ feet}} \cdot \dfrac{60 \text{ sec}}{1 \text{ min}} \cdot \dfrac{60 \text{ min}}{1 \text{ hr}} \approx 1272 \text{ mph}$

46. Drawing the figure:

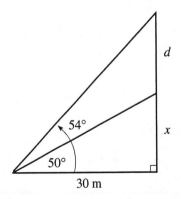

Therefore:

$$\tan 50° = \frac{x}{30}$$
$$x = 30 \tan 50°$$

Also:

$$\tan 54° = \frac{d + x}{30}$$
$$d + x = 30 \tan 54°$$
$$d + 30 \tan 50° = 30 \tan 54°$$
$$d = 30 \tan 54° - 30 \tan 50° \approx 5.5$$

The antenna is approximately 5.5 feet long.

47.   Finding the base of each triangle:

$$\tan 36° = \frac{16}{x_1}$$

$$x_1 \tan 36° = 16$$

$$x_1 = \frac{16}{\tan 36°}$$

$$\tan 55° = \frac{22}{x_2}$$

$$x_2 \tan 55° = 22$$

$$x_2 = \frac{22}{\tan 55°}$$

Therefore: $x = x_1 + x_2 = \dfrac{16}{\tan 36°} + \dfrac{22}{\tan 55°} \approx 37.4$

The two poles are approximately 37.4 feet apart.

48.   The central angle of the triangle is $\dfrac{360°}{3} = 120°$. Drawing the triangle:

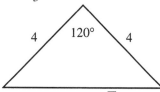

The area of this triangle is: $\frac{1}{2}(4)(4)\sin 120° = 8 \cdot \dfrac{\sqrt{3}}{2} = 4\sqrt{3}$

Since $\triangle ABC$ consists of 3 such triangles, its area is $3(4\sqrt{3}) = 12\sqrt{3}$. The shaded area is therefore:

$$\pi(4)^2 - 12\sqrt{3} = 16\pi - 12\sqrt{3} \approx 29.5 \text{ square units}$$

49.   Performing the operations: $\dfrac{1}{\sin\theta} + \dfrac{4}{\cos 2\theta} = \dfrac{\cos 2\theta}{\sin\theta\cos 2\theta} + \dfrac{4\sin\theta}{\sin\theta\cos 2\theta} = \dfrac{\cos 2\theta + 4\sin\theta}{\sin\theta\cos 2\theta}$

50.   Performing the operations: $(2\tan\theta - 3)(3\tan\theta + 4) = 6\tan^2\theta - \tan\theta - 12$

51.   Performing the operations: $\dfrac{2\sec^2\theta - 9\sec\theta + 9}{\sec^2\theta - 9} = \dfrac{(2\sec\theta - 3)(\sec\theta - 3)}{(\sec\theta + 3)(\sec\theta - 3)} = \dfrac{2\sec\theta - 3}{\sec\theta + 3}$

52.   Performing the operations:

$$\frac{\cot\theta}{\cot\theta + 1} - \frac{1}{\cot^2\theta + \cot\theta} = \frac{\cot\theta}{\cot\theta + 1} - \frac{1}{\cot\theta(\cot\theta + 1)}$$

$$= \frac{\cot^2\theta}{\cot\theta(\cot\theta + 1)} - \frac{1}{\cot\theta(\cot\theta + 1)}$$

$$= \frac{\cot^2\theta - 1}{\cot\theta(\cot\theta + 1)}$$

$$= \frac{(\cot\theta + 1)(\cot\theta - 1)}{\cot\theta(\cot\theta + 1)}$$

$$= \frac{\cot\theta - 1}{\cot\theta}$$

Chapter 6 Practice Test

1.   Converting to degree measure: $\dfrac{3\pi}{5} \cdot \dfrac{180°}{\pi} = 108°$

2.   Converting to radian measure: $80° \cdot \dfrac{\pi}{180°} = \dfrac{4\pi}{9}$

3.   Finding AB:

$$\cos 30° = \frac{12}{AB}$$

$$\frac{\sqrt{3}}{2} = \frac{12}{AB}$$

$$AB = \frac{24}{\sqrt{3}} = 8\sqrt{3}$$

4.   **a.**   Finding the value: $\sin\dfrac{5\pi}{4} = -\sin\dfrac{\pi}{4} = -\dfrac{1}{\sqrt{2}} = -\dfrac{\sqrt{2}}{2}$

   b.   Finding the value: $\cos\dfrac{2\pi}{3} = -\cos\dfrac{\pi}{3} = -\dfrac{1}{2}$

   c.   Finding the value: $\tan\dfrac{7\pi}{6} = \tan\dfrac{\pi}{6} = \dfrac{1}{\sqrt{3}} = \dfrac{\sqrt{3}}{3}$

   d.   Finding the value: $\csc 180° = \dfrac{1}{\sin 180°} = \dfrac{1}{0} = $ undefined

   e.   Finding the value: $\tan 270° = \dfrac{-1}{0} = $ undefined

   f.   Finding the value: $\sec 39° = \dfrac{1}{\cos 39°} \approx 1.2868$

   g.   Finding the value: $\sin 84° \approx 0.9945$

   h.   Finding the value: $\cot 134° = \dfrac{1}{\tan 134°} \approx -0.9657$

5.   Using the arc length formula: $s = 10 \cdot \dfrac{\pi}{5} = 2\pi$ cm

6.   Using the area formula: $A = \frac{1}{2}(10)(6)\sin 20° = 30\sin 20° \approx 10.3$ cm^2

7.   Solving for x:

$$\tan 52° = \frac{100}{x}$$

$$x \tan 52° = 100$$

$$x = \frac{100}{\tan 52°} \approx 78.1$$

8.   Solving for θ:

$$\sin\theta = \tfrac{3}{8}$$

$$\theta = \sin^{-1}\left(\tfrac{3}{8}\right) \approx 22°$$

9.   Drawing the figure:

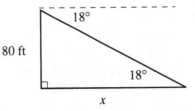

Thus:

$$\tan 18° = \frac{80}{x}$$
$$x \tan 18° = 80$$
$$x = \frac{80}{\tan 18°} \approx 246$$

The boat is approximately 246 feet from the lighthouse.

10. Drawing the figure:

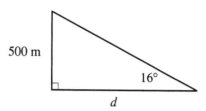

Thus:

$$\tan 16° = \frac{500}{d}$$
$$d \tan 16° = 500$$
$$d = \frac{500}{\tan 16°} \approx 1743 \text{ meters}$$

So the speed of the person was: $\dfrac{1743 \text{ meters}}{6 \text{ minutes}} \approx 291$ meters / minute

11. Simplifying: $\dfrac{1}{\sin 2\theta} + \dfrac{1}{2\sin\theta} = \dfrac{2\sin\theta}{2\sin\theta\sin 2\theta} + \dfrac{\sin 2\theta}{2\sin\theta\sin 2\theta} = \dfrac{2\sin\theta + \sin 2\theta}{2\sin\theta\sin 2\theta}$

Chapter 7
The Trigonometric Functions

7.1 The Sine and Cosine Functions and Their Graphs

1. The amplitude is 3, the period is 2π, and the frequency is 1.

3. The amplitude is 1, the period is $\dfrac{2\pi}{5}$, and the frequency is 5.

5. The amplitude is 1/4, the period is $\dfrac{2\pi}{2} = \pi$, and the frequency is 2.

7. The amplitude is 3, the period is $\dfrac{2\pi}{1/7} = 14\pi$, and the frequency is 1/7.

9. The amplitude is 3, the period is 2π, the frequency is 1, and the phase shift is $\dfrac{\pi}{3}$.

11. The amplitude is 4, the period is $\dfrac{2\pi}{3}$, the frequency is 3, and the phase shift is $\dfrac{-\pi/4}{3} = -\dfrac{\pi}{12}$.

13. Sketching the graph:

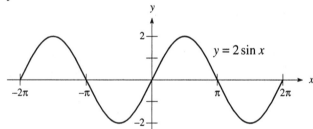

15. Sketching the graph:

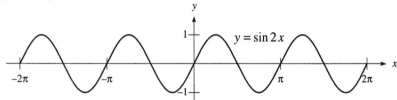

17.   Sketching the graph:

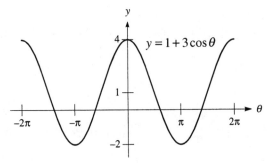

19.   Sketching the graph:

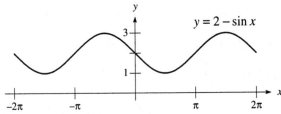

21.   Sketching the graph:

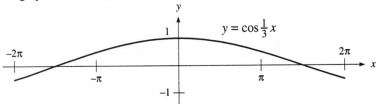

23.   Sketching the graph:

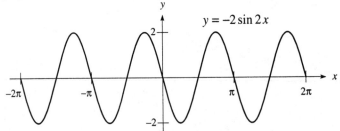

25.   Sketching the graph:

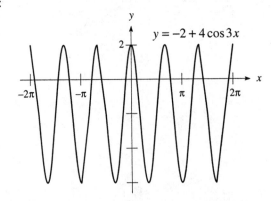

27.   Sketching the graph:

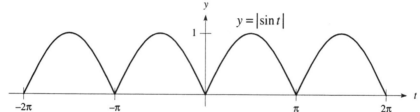

29.   Sketching the graph:

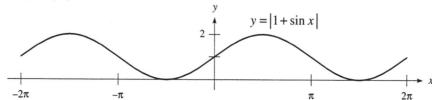

31.   Sketching the graph:

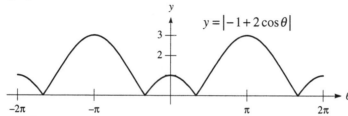

33.   Sketching the graph:

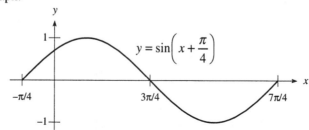

35.   Sketching the graph:

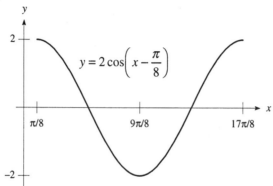

37. Sketching the graph:

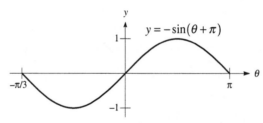

39. Sketching the graph:

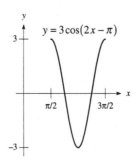

41. Sketching the graph:

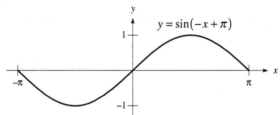

43. Sketching the graph:

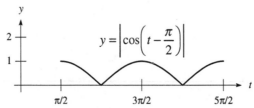

45. The amplitude is 3, so $A = 3$. The period is π, so $B = 2$. Since this is a sine curve, the equation is $y = 3\sin 2x$.

47. The amplitude is 1 (but reflected), so $A = -1$. The period is 2π, so $B = 1$. Since this is a cosine curve, the equation is $y = -\cos x$.

49. Graphing the function:

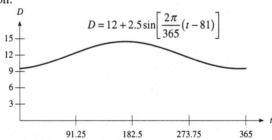

The amplitude is 2.5, the period is 365, the phase shift is 81, and the vertical shift is 12.

51. Graphing the function:

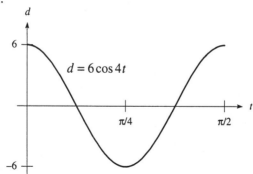

The amplitude is 6, the period is $\dfrac{2\pi}{4} = \dfrac{\pi}{2}$, the phase shift is 0, and the vertical shift is 0.

53. The amplitude is $\frac{1}{2}(96° - 40°) = 28°$ and the vertical shift is $\frac{1}{2}(96° + 40°) = 68°$. The phase shift is 222, corresponding to August 10, and the period is $2(222 - 41) = 362$ days. Thus $F(t) = 68 + 28\cos\left[\dfrac{2\pi}{362}(t - 222)\right]$. Sketching the graph:

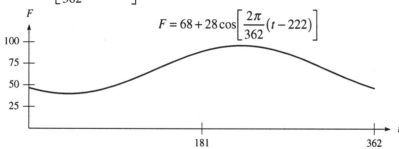

55. The maximum is $210 + 34.8(1) = 244.8$ thousand gallons (244,800 gallons), and the minimum is $210 + 34.8(-1) = 175.2$ thousand gallons (175,200 gallons).

57. Below are two lines with slope = 2/5. Note the difference in their appearance:

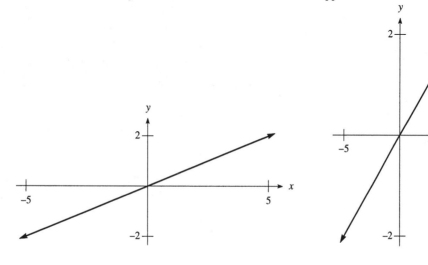

59.   Drawing the graph:

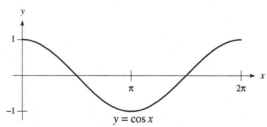

Note that $\cos x > 0$ when $0 < x < \dfrac{\pi}{2}$ (first quadrant) or $\dfrac{3\pi}{2} < x < 2\pi$ (fourth quadrant), while

$\cos x < 0$ when $\dfrac{\pi}{2} < x < \pi$ (second quadrant) or $\pi < x < \dfrac{3\pi}{2}$ (third quadrant).

61.   Graphing the curve:

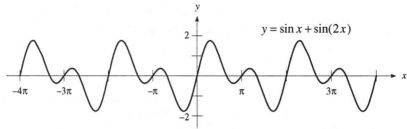

The period appears to be 2π.

63.   Graphing the curves:

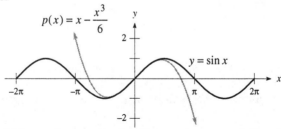

Note that $p(x)$ is a fairly good approximation for $y = \sin x$, especially when x-values are close to 0.

7.2 The Tangent, Secant, Cosecant, and Cotangent Functions and Their Graphs

1.   Sketching the graph:

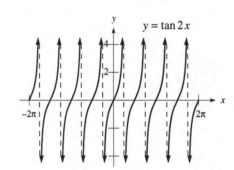

3. Sketching the graph:

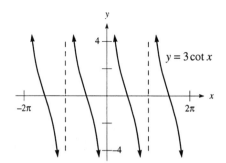

5. Sketching the graph:

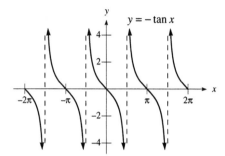

7. Sketching the graph:

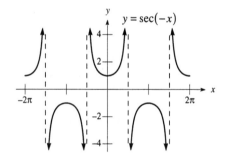

9. Sketching the graph:

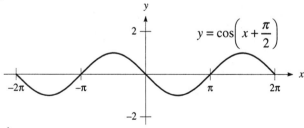

11. Sketching the graph:

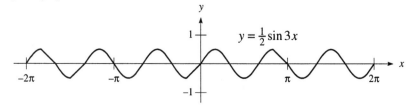

13.   Sketching the graph:

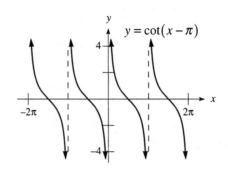

15.   Sketching the graph:

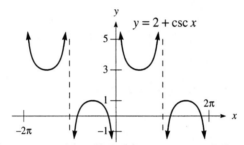

17.   Sketching the graph:

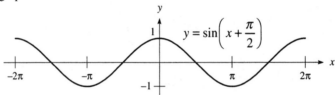

19.   Sketching the graph:

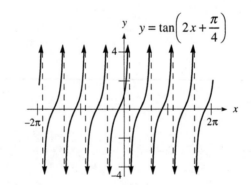

21.   Sketching the graph:

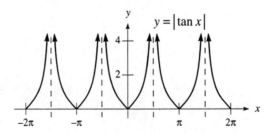

23. Sketching the graphs:

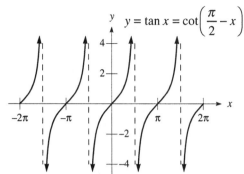

$$y = \tan x = \cot\left(\frac{\pi}{2} - x\right)$$

Note that the two graphs are identical.

25. Sketching the graph:

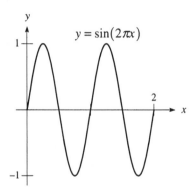

$$y = \sin(2\pi x)$$

The period is $\dfrac{2\pi}{2\pi} = 1$.

27. No. None of the graphs of the trigonometric functions pass the horizontal line test, so they are not one-to-one. As you will see later, with domain restrictions these functions will have inverse functions.

7.3 Basic Identities

1. Simplifying the expression: $\sin\theta\cot\theta = \sin\theta \bullet \dfrac{\cos\theta}{\sin\theta} = \cos\theta$

3. Simplifying the expression: $\tan\theta\csc\theta = \dfrac{\sin\theta}{\cos\theta} \bullet \dfrac{1}{\sin\theta} = \dfrac{1}{\cos\theta}$

5. Simplifying the expression: $\cos^2\theta\left(\tan^2\theta + 1\right) = \cos^2\theta \bullet \sec^2\theta = \cos^2\theta \bullet \dfrac{1}{\cos^2\theta} = 1$

7. Simplifying the expression: $\dfrac{\tan\theta}{\cot\theta} = \dfrac{\dfrac{\sin\theta}{\cos\theta}}{\dfrac{\cos\theta}{\sin\theta}} \bullet \dfrac{\sin\theta\cos\theta}{\sin\theta\cos\theta} = \dfrac{\sin^2\theta}{\cos^2\theta}$

9. Simplifying the expression: $\dfrac{1 - \tan\theta}{\cos\theta} = \dfrac{1 - \dfrac{\sin\theta}{\cos\theta}}{\cos\theta} \bullet \dfrac{\cos\theta}{\cos\theta} = \dfrac{\cos\theta - \sin\theta}{\cos^2\theta}$

11. Rewriting in terms of $\sin\theta$: $\dfrac{\cos^2\theta}{\sin\theta} = \dfrac{1-\sin^2\theta}{\sin\theta}$

13. Rewriting in terms of $\sin\theta$: $\sin\theta - \csc\theta = \sin\theta - \dfrac{1}{\sin\theta} = \dfrac{\sin^2\theta - 1}{\sin\theta}$

15. Rewriting in terms of $\cos\theta$: $\cos\theta - \sec\theta = \cos\theta - \dfrac{1}{\cos\theta} = \dfrac{\cos^2\theta - 1}{\cos\theta}$

17. Rewriting in terms of $\cos\theta$:

$$\cos^2\theta - \sin^2\theta = \cos^2\theta - \left(1 - \cos^2\theta\right) = \cos^2\theta - 1 + \cos^2\theta = 2\cos^2\theta - 1$$

19. Verifying the identity: $\dfrac{\sin\theta}{\tan\theta} = \dfrac{\sin\theta}{\dfrac{\sin\theta}{\cos\theta}} \cdot \dfrac{\cos\theta}{\cos\theta} = \dfrac{\sin\theta\cos\theta}{\sin\theta} = \cos\theta$

21. Verifying the identity: $\sin\theta\cot\theta\sec\theta = \sin\theta \cdot \dfrac{\cos\theta}{\sin\theta} \cdot \dfrac{1}{\cos\theta} = \dfrac{\sin\theta\cos\theta}{\sin\theta\cos\theta} = 1$

23. Verifying the identity:

$$\sin\alpha + \cos\alpha\cot\alpha = \sin\alpha + \cos\alpha \cdot \dfrac{\cos\alpha}{\sin\alpha} = \sin\alpha + \dfrac{\cos^2\alpha}{\sin\alpha} = \dfrac{\sin^2\alpha + \cos^2\alpha}{\sin\alpha} = \dfrac{1}{\sin\alpha} = \csc\alpha$$

25. Verifying the identity: $(1+\sin w)(1-\sin w) = 1 - \sin^2 w = \cos^2 w = \dfrac{1}{\sec^2 w}$

27. Verifying the identity:

$$\sin x(\csc x - \sin x) = \sin x\csc x - \sin^2 x = \sin x \cdot \dfrac{1}{\sin x} - \sin^2 x = 1 - \sin^2 x = \cos^2 x$$

29. Verifying the identity:

$$\sec\beta - \cos\beta = \dfrac{1}{\cos\beta} - \cos\beta = \dfrac{1-\cos^2\beta}{\cos\beta} = \dfrac{\sin^2\beta}{\cos\beta} = \dfrac{\sin\beta}{\cos\beta} \cdot \sin\beta = \tan\beta\sin\beta$$

31. Verifying the identity:

$$\sin^4\theta - \cos^4\theta = \left(\sin^2\theta + \cos^2\theta\right)\left(\sin^2\theta - \cos^2\theta\right) = 1 \cdot \left(\sin^2\theta - \cos^2\theta\right) = \sin^2\theta - \cos^2\theta$$

33. Verifying the identity:

$$\cot^2\theta - \cos^4\theta\csc^2\theta = \dfrac{\cos^2\theta}{\sin^2\theta} - \cos^4\theta \cdot \dfrac{1}{\sin^2\theta}$$
$$= \dfrac{\cos^2\theta - \cos^4\theta}{\sin^2\theta}$$
$$= \dfrac{\cos^2\theta\left(1 - \cos^2\theta\right)}{\sin^2\theta}$$
$$= \dfrac{\cos^2\theta \cdot \sin^2\theta}{\sin^2\theta}$$
$$= \cos^2\theta$$

35. Verifying the identity: $1 - \dfrac{1}{\csc^2\theta} = 1 - \dfrac{1}{\dfrac{1}{\sin^2\theta}} = 1 - \sin^2\theta = \cos^2\theta = \dfrac{1}{\sec^2\theta}$

37. Verifying the identity: $\dfrac{\sin x + \cos x}{\cos x} = \dfrac{\sin x}{\cos x} + \dfrac{\cos x}{\cos x} = \tan x + 1 = 1 + \tan x$

39. Verifying the identity: $\dfrac{\sin t + \tan t}{\sin t} = \dfrac{\sin t + \dfrac{\sin t}{\cos t}}{\sin t} = 1 + \dfrac{1}{\cos t} = 1 + \sec t$

41. Verifying the identity:

$$\frac{\cos x}{1+\sin x}+\frac{\cos x}{1-\sin x}=\frac{\cos x(1-\sin x)+\cos x(1+\sin x)}{(1+\sin x)(1-\sin x)}$$

$$=\frac{\cos x-\cos x\sin x+\cos x+\cos x\sin x}{1-\sin^2 x}$$

$$=\frac{2\cos x}{\cos^2 x}$$

$$=\frac{2}{\cos x}$$

$$=2\sec x$$

43. Verifying the identity: $\tan\theta(\tan\theta+\cot\theta)=\tan^2\theta+\tan\theta\bullet\dfrac{1}{\tan\theta}=\tan^2\theta+1=\sec^2\theta$

45. Verifying the identity:

$$(\tan u+\cot u)(\cos u+\sin u)=\tan u\cos u+\tan u\sin u+\cot u\cos u+\cot u\sin u$$

$$=\frac{\sin u}{\cos u}\bullet\cos u+\frac{\sin u}{\cos u}\bullet\sin u+\frac{\cos u}{\sin u}\bullet\cos u+\frac{\cos u}{\sin u}\bullet\sin u$$

$$=\sin u+\frac{\sin^2 u}{\cos u}+\frac{\cos^2 u}{\sin u}+\cos u$$

$$=\left(\sin u+\frac{\cos^2 u}{\sin u}\right)+\left(\cos u+\frac{\sin^2 u}{\cos u}\right)$$

$$=\frac{\sin^2 u+\cos^2 u}{\sin u}+\frac{\cos^2 u+\sin^2 u}{\cos u}$$

$$=\frac{1}{\cos u}+\frac{1}{\sin u}$$

$$=\sec u+\csc u$$

47. Verifying the identity: $\dfrac{\cot x+\tan x}{\sec x}=\dfrac{\dfrac{\cos x}{\sin x}+\dfrac{\sin x}{\cos x}}{\dfrac{1}{\cos x}}\bullet\dfrac{\sin x\cos x}{\sin x\cos x}=\dfrac{\cos^2 x+\sin^2 x}{\sin x}=\dfrac{1}{\sin x}=\csc x$

49. Verifying the identity:

$$\frac{1+\cos u}{\sin u}+\frac{\sin u}{1+\cos u}=\frac{(1+\cos u)^2+\sin^2 u}{\sin u(1+\cos u)}$$

$$=\frac{1+2\cos u+\cos^2 u+\sin^2 u}{\sin u(1+\cos u)}$$

$$=\frac{2+2\cos u}{\sin u(1+\cos u)}$$

$$=\frac{2(1+\cos u)}{\sin u(1+\cos u)}$$

$$=\frac{2}{\sin u}$$

$$=2\csc u$$

51. Verifying the identity: $\dfrac{\cot A\cos A}{\csc^2 A-1}=\dfrac{\cot A\cos A}{\cot^2 A}=\dfrac{\cos A}{\cot A}=\dfrac{\cos A}{\dfrac{\cos A}{\sin A}}\bullet\dfrac{\sin A}{\sin A}=\dfrac{\cos A\sin A}{\cos A}=\sin A$

53. Verifying the identity: $\left(\sin^2\theta+\cos^2\theta\right)^6=(1)^6=1$

55. Verifying the identity: $\dfrac{1}{\tan\alpha+\cot\alpha}=\dfrac{1}{\dfrac{\sin\alpha}{\cos\alpha}+\dfrac{\cos\alpha}{\sin\alpha}}\cdot\dfrac{\sin\alpha\cos\alpha}{\sin\alpha\cos\alpha}=\dfrac{\sin\alpha\cos\alpha}{\sin^2\alpha+\cos^2\alpha}=\sin\alpha\cos\alpha$

57. Verifying the identity:

$$\sec\gamma+\tan\gamma=\dfrac{1}{\cos\gamma}+\dfrac{\sin\gamma}{\cos\gamma}$$
$$=\dfrac{1+\sin\gamma}{\cos\gamma}\cdot\dfrac{1-\sin\gamma}{1-\sin\gamma}$$
$$=\dfrac{1-\sin^2\gamma}{\cos\gamma(1-\sin\gamma)}$$
$$=\dfrac{\cos^2\gamma}{\cos\gamma(1-\sin\gamma)}$$
$$=\dfrac{\cos\gamma}{1-\sin\gamma}$$

59. Verifying the identity:

$$\dfrac{\sin A+\tan A}{1+\sec A}=\dfrac{\sin A+\dfrac{\sin A}{\cos A}}{1+\dfrac{1}{\cos A}}\cdot\dfrac{\cos A}{\cos A}=\dfrac{\sin A\cos A+\sin A}{\cos A+1}=\dfrac{\sin A(\cos A+1)}{\cos A+1}=\sin A$$

61. Verifying the identity:

$$\dfrac{\tan^2 B}{\sec B+1}=\dfrac{\dfrac{\sin^2 B}{\cos^2 B}}{\dfrac{1}{\cos B}+1}\cdot\dfrac{\cos^2 B}{\cos^2 B}$$
$$=\dfrac{\sin^2 B}{\cos B+\cos^2 B}$$
$$=\dfrac{1-\cos^2 B}{\cos B(1+\cos B)}$$
$$=\dfrac{(1+\cos B)(1-\cos B)}{\cos B(1+\cos B)}$$
$$=\dfrac{1-\cos B}{\cos B}$$

63. Verifying the identity: $\dfrac{\sec A+\csc A}{\sec A-\csc A}=\dfrac{\dfrac{1}{\cos A}+\dfrac{1}{\sin A}}{\dfrac{1}{\cos A}-\dfrac{1}{\sin A}}\cdot\dfrac{\sin A\cos A}{\sin A\cos A}=\dfrac{\sin A+\cos A}{\sin A-\cos A}$

65. Verifying the identity: $\ln|\sec x|=\ln\left|\dfrac{1}{\cos x}\right|=\ln|\cos x|^{-1}=-\ln|\cos x|$

67. Since θ must lie in the second quadrant where $\cos\theta<0$:

$$\cos\theta=-\sqrt{1-\sin^2\theta}=-\sqrt{1-\left(\tfrac{4}{5}\right)^2}=-\sqrt{\tfrac{9}{25}}=-\tfrac{3}{5}$$

69. Since B must lie in the second quadrant where $\sec B<0$:

$$\sec B=-\sqrt{1+\tan^2 B}=-\sqrt{1+\left(-\tfrac{2}{3}\right)^2}=-\sqrt{\tfrac{13}{9}}=-\dfrac{\sqrt{13}}{3}$$

71. Since $\csc\theta = -\frac{7}{3}$, $\sin\theta = -\frac{3}{7}$. Since θ must lie in the third quadrant where $\cos\theta < 0$:

$$\cos\theta = -\sqrt{1-\sin^2\theta} = -\sqrt{1-\left(-\frac{3}{7}\right)^2} = -\sqrt{\frac{40}{49}} = -\frac{2\sqrt{10}}{7}$$

Therefore: $\tan\theta = \dfrac{\sin\theta}{\cos\theta} = \dfrac{-\frac{3}{7}}{-2\sqrt{10}\Big/7} = \dfrac{3}{2\sqrt{10}} = \dfrac{3\sqrt{10}}{20}$

73. If θ is a quadrantal angle, either the points $(\pm 1,0)$ or $(0,\pm 1)$ lie on the terminal side of θ. In all four cases $\sin^2\theta + \cos^2\theta = 1$.

75. No. Since $\sin\theta \le 1$ and $\cos\theta \le 1$, it is impossible for $\sin\theta = 4$ and $\cos\theta = 5$. The correct values are

$$\sin\theta = \frac{4}{\sqrt{41}} \text{ and } \cos\theta = \frac{5}{\sqrt{41}}, \text{ in which case } \tan\theta = \frac{4\Big/\sqrt{41}}{5\Big/\sqrt{41}} = \frac{4}{5}.$$

7.4 Trigonometric Equations

1. Solving the equation:
$$\sin x = \tfrac{1}{2}$$
$$x = \frac{\pi}{6}, \frac{5\pi}{6}$$

3. Solving the equation:
$$\cos x = -\frac{\sqrt{3}}{2}$$
$$x = \frac{5\pi}{6}, \frac{7\pi}{6}$$

5. Solving the equation:
$$\sin x = 1$$
$$x = \frac{\pi}{2}$$

7. Solving the equation:
$$\tan\theta + 1 = 0$$
$$\tan\theta = -1$$
$$x = \frac{3\pi}{4}, \frac{7\pi}{4}$$

9. Solving the equation:
$$\csc x = 1$$
$$\sin x = 1$$
$$x = \frac{\pi}{2}$$

11. Solving the equation:
$$\tan\theta = \sqrt{3}$$
$$x = \frac{2\pi}{3}, \frac{5\pi}{3}$$

13. Solving the equation:
$$3\cos x + 1 = 5$$
$$3\cos x = 4$$
$$\cos x = \tfrac{4}{3} > 1$$
There is no solution to the equation.

15. Solving the equation:
$$\sqrt{3}\tan t = 1$$
$$\tan t = \frac{1}{\sqrt{3}}$$
$$t = \frac{\pi}{6}, \frac{7\pi}{6}$$

17. Solving the equation:
$$\cos x = 0$$
$$x = \frac{\pi}{2}, \frac{3\pi}{2}$$

19. Solving the equation:
$$5\sec\theta = 2$$
$$\sec\theta = \tfrac{2}{5}$$
$$\cos\theta = \tfrac{5}{2} > 1$$
There is no solution to the equation.

21. Solving the equation:
$$\sin^2 x + 4 = 5$$
$$\sin^2 x = 1$$
$$\sin x = \pm 1$$
$$x = \frac{\pi}{2}, \frac{3\pi}{2}$$

23. Solving the equation:
$$4\cos^2 x - 3 = 0$$
$$4\cos^2 x = 3$$
$$\cos^2 x = \tfrac{3}{4}$$
$$\cos x = \pm\frac{\sqrt{3}}{2}$$
$$x = \frac{\pi}{6}, \frac{5\pi}{6}, \frac{7\pi}{6}, \frac{11\pi}{6}$$

25. Solving the equation:
$$6\cot^2 w + 1 = 3$$
$$6\cot^2 w = 2$$
$$\cot^2 w = \tfrac{1}{3}$$
$$\tan^2 w = 3$$
$$\tan w = \pm\sqrt{3}$$
$$w = \frac{\pi}{3}, \frac{2\pi}{3}, \frac{4\pi}{3}, \frac{5\pi}{3}$$

27. Solving the equation:
$$\cos^2 x + 2 = 3\cos x$$
$$\cos^2 x - 3\cos x + 2 = 0$$
$$(\cos x - 1)(\cos x - 2) = 0$$
$$\cos x = 1 \qquad (\cos x \neq 2)$$
$$x = 0$$

29. Solving the equation:
$$\tan^2 \theta = \tan \theta$$
$$\tan^2 \theta - \tan \theta = 0$$
$$\tan \theta (\tan \theta - 1) = 0$$
$$\tan \theta = 0, 1$$
$$\theta = 0, \frac{\pi}{4}, \pi, \frac{5\pi}{4}$$

31. Solving the equation:
$$2 \sin t = \sin t \cos t$$
$$2 \sin t - \sin t \cos t = 0$$
$$\sin t (2 - \cos t) = 0$$
$$\sin t = 0 \qquad (\cos t \neq 2)$$
$$t = 0, \pi$$

33. Solving the equation:
$$\sin x = \cos x$$
$$\tan x = 1$$
$$x = \frac{\pi}{4}, \frac{5\pi}{4}$$

35. Solving the equation:
$$\cos w - \sec w = 0$$
$$\cos w - \frac{1}{\cos w} = 0$$
$$\cos^2 w - 1 = 0$$
$$\cos^2 w = 1$$
$$\cos w = \pm 1$$
$$w = 0, \pi$$

37. Solving the equation:
$$\sin x + 3 = -2 \csc x$$
$$\sin x + 3 = \frac{-2}{\sin x}$$
$$\sin^2 x + 3 \sin x = -2$$
$$\sin^2 x + 3 \sin x + 2 = 0$$
$$(\sin x + 2)(\sin x + 1) = 0$$
$$\sin x = -1 \qquad (\sin x \neq -2)$$
$$x = \frac{3\pi}{2}$$

39.   Solving the equation:

$$2\cos^2 x - \sin x - 1 = 0$$
$$2\left(1 - \sin^2 x\right) - \sin x - 1 = 0$$
$$2 - 2\sin^2 x - \sin x - 1 = 0$$
$$-2\sin^2 x - \sin x + 1 = 0$$
$$2\sin^2 x + \sin x - 1 = 0$$
$$(2\sin x - 1)(\sin x + 1) = 0$$
$$\sin x = \tfrac{1}{2}, -1$$
$$x = \frac{\pi}{6}, \frac{5\pi}{6}, \frac{3\pi}{2}$$

41.   Solving the equation:

$$3\tan^2 x - \sec^2 x = 5$$
$$3\tan^2 x - \left(\tan^2 x + 1\right) = 5$$
$$2\tan^2 x - 1 = 5$$
$$2\tan^2 x = 6$$
$$\tan^2 x = 3$$
$$\tan x = \pm\sqrt{3}$$
$$x = \frac{\pi}{3}, \frac{2\pi}{3}, \frac{4\pi}{3}, \frac{5\pi}{3}$$

43.   Solving the equation:

$$\sin\theta + \cos\theta = 1$$
$$\cos\theta = 1 - \sin\theta$$
$$\cos^2\theta = (1 - \sin\theta)^2$$
$$\cos^2\theta = 1 - 2\sin\theta + \sin^2\theta$$
$$1 - \sin^2\theta = 1 - 2\sin\theta + \sin^2\theta$$
$$0 = 2\sin^2\theta - 2\sin\theta$$
$$0 = 2\sin\theta(\sin\theta - 1)$$
$$\sin\theta = 0, 1$$
$$\theta = 0, \frac{\pi}{2}, \pi$$

Note that $\theta = \pi$ does not check in the original equation, so the solutions are $\theta = 0, \dfrac{\pi}{2}$.

45.   Solving the equation:

$$5\sin x = 3$$
$$\sin x = \tfrac{3}{5}$$
$$x \approx 37°, 143°$$

47.   Solving the equation:

$$7 - 4\tan x = 14$$
$$-4\tan x = 7$$
$$\tan x = -\tfrac{7}{4}$$
$$x \approx 120°, 300°$$

49. Solving the equation:
$$9 \csc x = -20$$
$$\csc x = -\tfrac{20}{9}$$
$$\sin x = -\tfrac{9}{20}$$
$$x \approx 207°, 333°$$

51. Using the quadratic formula: $\cos x = \dfrac{1 \pm \sqrt{1 - 4(-1)}}{2} = \dfrac{1 \pm \sqrt{5}}{2} \approx -0.6180, 1.6180$

So $\cos x \approx -0.6180$, thus $x \approx 2.24, 4.05$.

53. Using the quadratic formula: $\tan t = \dfrac{3 \pm \sqrt{9 - 4(2)(-4)}}{2(2)} = \dfrac{3 \pm \sqrt{41}}{4} \approx -0.8508, 2.3508$

If $\tan t \approx -0.8508$, then $t \approx 139.61°, 319.61°$. If $\tan t \approx 2.3508$, then $t \approx 66.96°, 246.96°$.
Thus the solutions are $t \approx 66.96°, 139.61°, 246.96°, 319.61°$.

55. Solving the equation:
$$2 \sin x = 1$$
$$\sin x = \tfrac{1}{2}$$
$$x = \frac{\pi}{6} + 2n\pi, \frac{5\pi}{6} + 2n\pi$$

57. Solving the equation:
$$\tan^2 x = 3$$
$$\tan x = \pm\sqrt{3}$$
$$x = \frac{\pi}{3} + n\pi, \frac{2\pi}{3} + n\pi$$

59. Solving the equation:
$$3 \csc^2 x = 4$$
$$\csc^2 x = \tfrac{4}{3}$$
$$\sin^2 x = \tfrac{3}{4}$$
$$\sin x = \pm\frac{\sqrt{3}}{2}$$
$$x = \frac{\pi}{3} + n\pi, \frac{2\pi}{3} + n\pi$$

61. Solving the equation:
$$5 \cos x = 0$$
$$\cos x = 0$$
$$x = \frac{\pi}{2} + n\pi$$

63. Solving the equation:
$$5 \sin x = -5$$
$$\sin x = -1$$
$$x = \frac{3\pi}{2} + 2n\pi$$

65. Solving the equation:
$$\sin 3x = 1$$
$$3x = \frac{\pi}{2} + 2n\pi$$
$$x = \frac{\pi}{6} + \frac{2n\pi}{3}$$

67. Solving the equation:
$$2\cos 4x = -\sqrt{2}$$
$$\cos 4x = -\frac{\sqrt{2}}{2}$$
$$4x = \frac{3\pi}{4} + 2n\pi, \frac{5\pi}{4} + 2n\pi$$
$$x = \frac{3\pi}{16} + \frac{n\pi}{2}, \frac{5\pi}{16} + \frac{n\pi}{2}$$

69. For the denominator to be nonzero, we must have $\sin\theta \neq 1$, so $\theta \neq \frac{\pi}{2} + 2n\pi$.

71. The domain is all real numbers.

73. For the denominator to be nonzero, we must have $\cos^2 x \neq \frac{3}{4}$, so $\cos x \neq \pm\frac{\sqrt{3}}{2}$. Thus the domain is $x \neq \frac{\pi}{6} + n\pi, \frac{5\pi}{6} + n\pi$.

75. For the denominator to be nonzero, we must have $\csc t \neq 2$, or $\sin t \neq \pm\frac{1}{2}$. Thus the domain is $t \neq \frac{\pi}{6} + n\pi, \frac{5\pi}{6} + n\pi$.

77. We must have $\sec\theta \geq 0$, so $\cos\theta > 0$. The domain is $\left[0, \frac{\pi}{2}\right) \cup \left(\frac{3\pi}{2}, 2\pi\right)$.

79. We must have $\csc\theta \neq 0$, which occurs everywhere (except where $\csc\theta$ is undefined). The domain is $\theta \neq 0, \pi$, or $(0, \pi) \cup (\pi, 2\pi)$.

81. We must have $\tan\theta \neq 1$, so $\theta \neq \frac{\pi}{4}, \frac{5\pi}{4}$. The domain is $\left[0, \frac{\pi}{4}\right) \cup \left(\frac{\pi}{4}, \frac{5\pi}{4}\right) \cup \left(\frac{5\pi}{4}, 2\pi\right)$.

83. We must have $\cos^2\theta \neq \frac{1}{2}$, so $\cos\theta \neq \pm\frac{1}{\sqrt{2}}$. The domain is $\theta \neq \frac{\pi}{4}, \frac{3\pi}{4}, \frac{5\pi}{4}, \frac{7\pi}{4}$.

85. Using a graphing calculator, the solutions are $x \approx -2.28, 0, 2.28$.
87. Using a graphing calculator, the solutions are $x \approx -3.80, -3.02, 5.31$.
89. Using a graphing calculator, the solutions are $x \approx 0.60, 1.27, 2.43$.
91. Using a graphing calculator, the solutions are $x \approx -1.39, 0, 1.39$.

7.5 The Inverse Trigonometric Functions

1. Let $\theta = \sin^{-1}\left(\frac{\sqrt{3}}{2}\right)$. Then $\sin\theta = \frac{\sqrt{3}}{2}$ and $\theta \in \left[-\frac{\pi}{2}, \frac{\pi}{2}\right]$, so $\theta = \frac{\pi}{3}$. Thus $\sin^{-1}\left(\frac{\sqrt{3}}{2}\right) = \frac{\pi}{3}$.

3. Let $\theta = \arccos\left(-\frac{1}{2}\right)$. Then $\cos\theta = -\frac{1}{2}$ and $\theta \in [0, \pi]$, so $\theta = \frac{2\pi}{3}$. Thus $\arccos\left(-\frac{1}{2}\right) = \frac{2\pi}{3}$.

5. Let $\theta = \tan^{-1}\left(-\sqrt{3}\right)$. Then $\tan\theta = -\sqrt{3}$ and $\theta \in \left(-\frac{\pi}{2}, \frac{\pi}{2}\right)$, so $\theta = -\frac{\pi}{3}$. Thus $\tan^{-1}\left(-\sqrt{3}\right) = -\frac{\pi}{3}$.

7.   Let $\theta = \sec^{-1}(2)$. Then $\sec\theta = 2$ and $\theta \in \left[0, \dfrac{\pi}{2}\right) \cup \left(\dfrac{\pi}{2}, \pi\right]$, so $\theta = \dfrac{\pi}{3}$. Thus $\sec^{-1}(2) = \dfrac{\pi}{3}$.

9.   Using a calculator, $\sin^{-1}\left(\dfrac{2}{5}\right) \approx 0.4115$.

11.   Let $\theta = \csc^{-1}(-1)$. Then $\csc\theta = -1$ and $\theta \in \left[-\dfrac{\pi}{2}, 0\right) \cup \left(0, \dfrac{\pi}{2}\right]$, so $\theta = -\dfrac{\pi}{2}$. Thus

$$\csc^{-1}(-1) = -\dfrac{\pi}{2}.$$

13.   Let $\theta = \cot^{-1}(0)$. Then $\cot\theta = 0$ and $\theta \in (0, \pi)$, so $\theta = \dfrac{\pi}{2}$. Thus $\cot^{-1}(0) = \dfrac{\pi}{2}$.

15.   Let $\theta = \arcsin(1)$. Then $\sin\theta = 1$ and $\theta \in \left[-\dfrac{\pi}{2}, \dfrac{\pi}{2}\right]$, so $\theta = \dfrac{\pi}{2}$. Thus $\arcsin(1) = \dfrac{\pi}{2}$.

17.   Finding the value: $\sin\left(\cos^{-1}\dfrac{\sqrt{3}}{2}\right) = \sin\dfrac{\pi}{6} = \dfrac{1}{2}$

19.   Finding the value: $\tan\left(\sin^{-1}\left(-\dfrac{1}{2}\right)\right) = \tan\left(-\dfrac{\pi}{6}\right) = -\dfrac{1}{\sqrt{3}} = -\dfrac{\sqrt{3}}{3}$

21.   Finding the value: $\cos\left(\cos^{-1} 1\right) = \cos(0) = 1$

23.   Let $\theta = \tan^{-1}\left(\dfrac{1}{2}\right)$, so $\tan\theta = \dfrac{1}{2}$. Drawing a triangle:

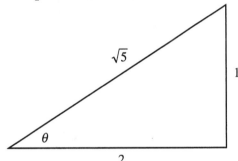

So $\cos\left(\tan^{-1}\left(\dfrac{1}{2}\right)\right) = \cos\theta = \dfrac{2}{\sqrt{5}} = \dfrac{2\sqrt{5}}{5}$.

25.   Let $\theta = \sin^{-1}\left(\dfrac{2}{5}\right)$, so $\sin\theta = \dfrac{2}{5}$. Drawing a triangle:

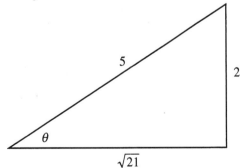

So $\tan\left(\sin^{-1}\left(\dfrac{2}{5}\right)\right) = \tan\theta = \dfrac{2}{\sqrt{21}} = \dfrac{2\sqrt{21}}{21}$.

27. Let $\theta = \csc^{-1}(1)$, so $\csc\theta = 1$. Then $\theta = \dfrac{\pi}{2}$, thus $\sec\!\left(\csc^{-1}(1)\right) = \sec\dfrac{\pi}{2}$, which is undefined.

29. Since $-\dfrac{\pi}{2} \le \dfrac{\pi}{3} \le \dfrac{\pi}{2}$, $\sin^{-1}\!\left(\sin\dfrac{\pi}{3}\right) = \dfrac{\pi}{3}$.

31. Finding the value: $\cos^{-1}\!\left(\cos\dfrac{5\pi}{3}\right) = \cos^{-1}\!\left(\dfrac{1}{2}\right) = \dfrac{\pi}{3}$

33. Finding the value: $\cos^{-1}\!\left(\cos\!\left(-\dfrac{\pi}{3}\right)\right) = \cos^{-1}\!\left(\dfrac{1}{2}\right) = \dfrac{\pi}{3}$

35. Finding the value: $\tan\!\left(\csc^{-1}1\right) = \tan\dfrac{\pi}{2}$, which is undefined

37. Finding the value: $\sin^{-1}(\cos 82°) = \sin^{-1}(\sin 8°) = 8°$

39. Since $\cos\theta = \dfrac{x}{6}$, $\theta = \cos^{-1}\!\left(\dfrac{x}{6}\right)$.

41. Since $\tan\theta = \dfrac{x}{5}$, $\theta = \tan^{-1}\!\left(\dfrac{x}{5}\right)$.

43. Since $\cos\theta = \dfrac{\sqrt{16-x^2}}{4}$, $\sqrt{16-x^2} = 4\cos\theta$.

45. Since $\cos\theta = \dfrac{4}{\sqrt{x^2+16}}$, $\sqrt{x^2+16} = \dfrac{4}{\cos\theta} = 4\sec\theta$.

47. Simplifying where $u = 3\sin\theta$:
$$u\sqrt{9-u^2} = 3\sin\theta\sqrt{9-9\sin^2\theta} = 3\sin\theta\sqrt{9\cos^2\theta} = 3\sin\theta \cdot 3\cos\theta = 9\sin\theta\cos\theta$$

49. Simplifying where $x = 2\sec\theta$:
$$x^2\sqrt{x^2-4} = 4\sec^2\theta\sqrt{4\sec^2\theta - 4} = 4\sec^2\theta\sqrt{4\tan^2\theta} = 4\sec^2\theta \cdot 2\tan\theta = 8\sec^2\theta\tan\theta$$

51. Simplifying where $u = 4\tan\theta$: $\dfrac{u^2}{16+u^2} = \dfrac{16\tan^2\theta}{16+16\tan^2\theta} = \dfrac{16\tan^2\theta}{16\sec^2\theta} = \dfrac{\dfrac{\sin^2\theta}{\cos^2\theta}}{\dfrac{1}{\cos^2\theta}} = \sin^2\theta$

53. Since $\theta = \sin^{-1}\!\left(\dfrac{x}{5}\right)$, $\sin\theta = \dfrac{x}{5}$. Drawing the triangle:

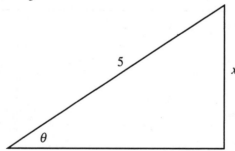

55. Since $\theta = \sec^{-1}\left(\dfrac{x}{a}\right)$, $\sec\theta = \dfrac{x}{a}$. Drawing the triangle:

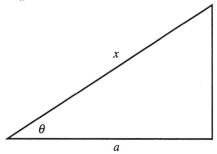

57. Drawing the triangle:

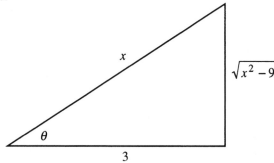

Therefore $\dfrac{\sqrt{x^2-9}}{x} = \sin\theta$.

59. Yes. The inverse of $y = \sin kx$ is $y = \dfrac{\sin^{-1}x}{k}$ on the interval $\left[-\dfrac{\pi}{2k}, \dfrac{\pi}{2k}\right]$.

Chapter 7 Review Exercises

1. Sketching the graph:

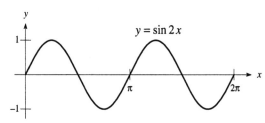

2. Sketching the graph:

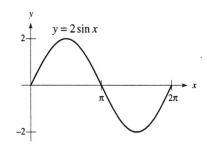

3.   Sketching the graph:

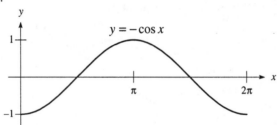

4.   Sketching the graph:

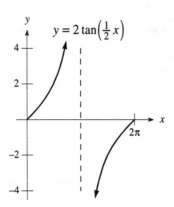

5.   Sketching the graph:

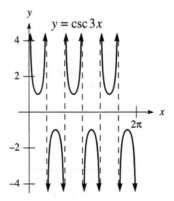

6.   Sketching the graph:

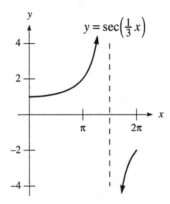

7. Sketching the graph:

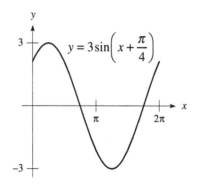

8. Sketching the graph:

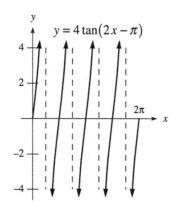

9. Sketching the graph:

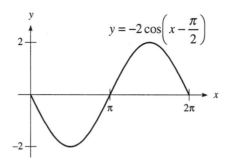

10. Sketching the graph:

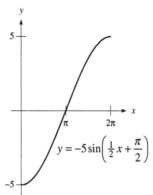

11.   Sketching the graph:

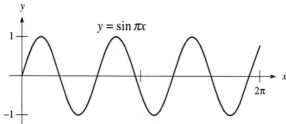

$$y = \sin \pi x$$

12.   Sketching the graph:

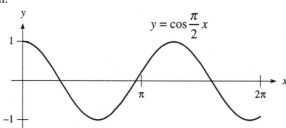

$$y = \cos \frac{\pi}{2} x$$

13.   Changing the angle: $\sin(-80°) = -\sin 80°$

14.   Changing the angle: $\cos\left(-\dfrac{5\pi}{7}\right) = \cos\dfrac{5\pi}{7} = -\cos\dfrac{2\pi}{7}$

15.   Changing the angle: $\tan\left(-\dfrac{9\pi}{8}\right) = \tan\dfrac{7\pi}{8} = -\tan\dfrac{\pi}{8}$

16.   Changing the angle: $\cot(-329°) = \cot 31°$

17.   Changing the angle: $\csc(-187°) = \csc 173° = \csc 7°$

18.   Changing the angle: $\sec(-256°) = \sec 104° = -\sec 76°$

19.   Verifying the identity: $\sin\theta + \sin\theta\cot^2\theta = \sin\theta\left(1+\cot^2\theta\right) = \sin\theta\csc^2\theta = \sin\theta \cdot \dfrac{1}{\sin^2\theta} = \dfrac{1}{\sin\theta}$

20.   Verifying the identity: $\dfrac{\sec\beta}{\csc\beta} + \dfrac{\sin\beta}{\cos\beta} = \dfrac{1/\cos\beta}{1/\sin\beta} + \dfrac{\sin\beta}{\cos\beta} = \dfrac{\sin\beta}{\cos\beta} + \dfrac{\sin\beta}{\cos\beta} = \tan\beta + \tan\beta = 2\tan\beta$

21.   Verifying the identity:

$$\dfrac{1+\sec x}{\sin x + \tan x} = \dfrac{1+\dfrac{1}{\cos x}}{\sin x + \dfrac{\sin x}{\cos x}} \cdot \dfrac{\cos x}{\cos x} = \dfrac{\cos x + 1}{\sin x \cos x + \sin x} = \dfrac{\cos x + 1}{\sin x(\cos x + 1)} = \dfrac{1}{\sin x} = \csc x$$

22.   Verifying the identity: $\dfrac{1}{\tan A + \cot A} = \dfrac{1}{\dfrac{\sin A}{\cos A} + \dfrac{\cos A}{\sin A}} \cdot \dfrac{\sin A \cos A}{\sin A \cos A} = \dfrac{\sin A \cos A}{\sin^2 A + \cos^2 A} = \sin A \cos A$

23. Verifying the identity:

$$\tan^2 \alpha - \sin^2 \alpha = \frac{\sin^2 \alpha}{\cos^2 \alpha} - \sin^2 \alpha$$

$$= \frac{\sin^2 \alpha - \sin^2 \alpha \cos^2 \alpha}{\cos^2 \alpha}$$

$$= \frac{\sin^2 \alpha \left(1 - \cos^2 \alpha\right)}{\cos^2 \alpha}$$

$$= \frac{\sin^2 \alpha}{\cos^2 \alpha} \cdot \sin^2 \alpha$$

$$= \tan^2 \alpha \sin^2 \alpha$$

24. Verifying the identity:

$$\sec \gamma - \frac{\cos \gamma}{1 + \sin \gamma} = \frac{1}{\cos \gamma} - \frac{\cos \gamma}{1 + \sin \gamma}$$

$$= \frac{1 + \sin \gamma - \cos^2 \gamma}{\cos \gamma (1 + \sin \gamma)}$$

$$= \frac{\sin \gamma + \sin^2 \gamma}{\cos \gamma (1 + \sin \gamma)}$$

$$= \frac{\sin \gamma (1 + \sin \gamma)}{\cos \gamma (1 + \sin \gamma)}$$

$$= \frac{\sin \gamma}{\cos \gamma}$$

$$= \tan \gamma$$

25. Solving the equation:

$$2 \cos x = -1$$

$$\cos x = -\tfrac{1}{2}$$

$$x = \frac{2\pi}{3} + 2n\pi, \frac{4\pi}{3} + 2n\pi$$

26. Solving the equation:

$$\tan^2 x = 3$$

$$\tan x = \pm\sqrt{3}$$

$$x = \frac{\pi}{3} + n\pi, \frac{2\pi}{3} + n\pi$$

27. Solving the equation:

$$3 \csc^2 x = 6$$

$$\csc^2 x = 2$$

$$\sin^2 x = \tfrac{1}{2}$$

$$\sin x = \pm\frac{1}{\sqrt{2}}$$

$$x = \frac{\pi}{4} + \frac{n\pi}{2}$$

28. Solving the equation:
$$\cot x = -1$$
$$x = \frac{3\pi}{4} + n\pi$$

29. Solving the equation:
$$4\sin x = 2\sqrt{3}$$
$$\sin x = \frac{\sqrt{3}}{2}$$
$$x = \frac{\pi}{3} + 2n\pi, \frac{2\pi}{3} + 2n\pi$$

30. Solving the equation:
$$5\sec^2 x = 5$$
$$\sec^2 x = 1$$
$$\cos^2 x = 1$$
$$\cos x = \pm 1$$
$$x = n\pi$$

31. Solving the equation:
$$\sqrt{3}\sec x - 1 = 1$$
$$\sqrt{3}\sec x = 2$$
$$\sec x = \frac{2}{\sqrt{3}}$$
$$\cos x = \frac{\sqrt{3}}{2}$$
$$x = \frac{\pi}{6} + 2n\pi, \frac{11\pi}{6} + 2n\pi$$

32. Solving the equation:
$$\sin^2 x + 1 = 2$$
$$\sin^2 x = 1$$
$$\sin x = \pm 1$$
$$x = \frac{\pi}{2} + n\pi$$

33. Solving the equation:
$$2\sin^2 x = 1$$
$$\sin^2 x = \tfrac{1}{2}$$
$$\sin x = \pm\frac{1}{\sqrt{2}}$$
$$x = \frac{\pi}{4}, \frac{3\pi}{4}, \frac{5\pi}{4}, \frac{7\pi}{4}$$

34. Solving the equation:
$$4\cos^2 x = 3$$
$$\cos^2 x = \tfrac{3}{4}$$
$$\cos x = \pm\frac{\sqrt{3}}{2}$$
$$x = \frac{\pi}{6}, \frac{5\pi}{6}, \frac{7\pi}{6}, \frac{11\pi}{6}$$

35. Solving the equation:
$$\cos^2\theta\sin\theta = \cos\theta$$
$$\cos^2\theta\sin\theta - \cos\theta = 0$$
$$\cos\theta(\cos\theta\sin\theta - 1) = 0$$

Now $\cos\theta = 0$ when $\theta = \dfrac{\pi}{2}, \dfrac{3\pi}{2}$. Now solve the other portion:
$$\cos\theta\sin\theta = 1$$
$$\cos^2\theta\sin^2\theta = 1$$
$$\left(1 - \sin^2\theta\right)\sin^2\theta = 1$$
$$\sin^2\theta - \sin^4\theta = 1$$
$$\sin^4\theta - \sin^2\theta + 1 = 0$$

The quadratic formula yields $\sin\theta = \dfrac{1 \pm \sqrt{-3}}{2}$, which has no solution. The only solutions are
$$\theta = \frac{\pi}{2}, \frac{3\pi}{2}.$$

36. Solving the equation:
$$\sin^2 x - \sin x - 2 = 0$$
$$(\sin x - 2)(\sin x + 1) = 0$$
$$\sin x = -1 \qquad (\sin x = 2 \text{ is impossible})$$

37. Solving the equation:
$$1 - \sin^2 x - \cos x = 6$$
$$\cos^2 x - \cos x = 6$$
$$\cos^2 x - \cos x - 6 = 0$$
$$(\cos x - 3)(\cos x + 2) = 0$$
$$\cos x \neq 3, -2$$

There is no solution to this equation.

38. Solving the equation:
$$\tan^2\theta\cot\theta = \tan\theta$$
$$\tan^2\theta \bullet \frac{1}{\tan\theta} = \tan\theta$$
$$\tan\theta = \tan\theta$$

This equation is true for all real numbers θ in which $\tan\theta$, $\cot\theta$ are defined (that is, no multiples of $\dfrac{\pi}{2}$).

39. Solving the equation:
$$2\sin^2\theta + 1 = 3\sin\theta$$
$$2\sin^2\theta - 3\sin\theta + 1 = 0$$
$$(2\sin\theta - 1)(\sin\theta - 1) = 0$$
$$\sin\theta = \tfrac{1}{2}, 1$$
$$\theta = \frac{\pi}{6}, \frac{\pi}{2}, \frac{5\pi}{6}$$

40. Solving the equation:
$$\sec^2\theta + 2 = 3\sec\theta$$
$$\sec^2\theta - 3\sec\theta + 2 = 0$$
$$(\sec\theta - 2)(\sec\theta - 1) = 0$$
$$\sec\theta = 2, 1$$
$$\cos\theta = \tfrac{1}{2}, 1$$
$$\theta = 0, \frac{\pi}{3}, \frac{5\pi}{3}$$

41. Solving the equation:
$$2\csc^2\theta - 3\csc\theta = 2$$
$$2\csc^2\theta - 3\csc\theta - 2 = 0$$
$$(2\csc\theta + 1)(\csc\theta - 2) = 0$$
$$\csc\theta = -\tfrac{1}{2}, 2$$
$$\sin\theta = \tfrac{1}{2} \qquad \left(\sin\theta = -2 \text{ is impossible}\right)$$
$$\theta = \frac{\pi}{6}, \frac{5\pi}{6}$$

42. Solving the equation:
$$12\sin^2\theta - 7\sin\theta + 1 = 0$$
$$(4\sin\theta - 1)(3\sin\theta - 1) = 0$$
$$\sin\theta = \tfrac{1}{4}, \tfrac{1}{3}$$
$$\theta \approx 0.2527, 0.3398, 2.8018, 2.8889$$

43. Solving the equation:
$$5\sin^2\theta = 3$$
$$\sin^2\theta = \tfrac{3}{5}$$
$$\sin\theta = \pm\sqrt{\tfrac{3}{5}}$$
$$\theta \approx 0.8861, 2.2555, 4.0277, 5.3971$$

44. Solving the equation:
$$2\tan^2\theta = 14$$
$$\tan^2\theta = 7$$
$$\tan\theta = \pm\sqrt{7}$$
$$\theta \approx 1.2094, 1.9322, 4.3510, 5.0738$$

45. Solving the equation:
$$4\cos x = 1$$
$$\cos x = \tfrac{1}{4}$$
$$x \approx 76°, 284°$$

46. Solving the equation:
$$\cot^2 x = 10$$
$$\cot x = \pm\sqrt{10}$$
$$\tan x = \pm\frac{1}{\sqrt{10}}$$
$$x \approx 18°, 162°, 198°, 342°$$

47. Solving the equation:
$$\sec\theta\csc\theta = \sec\theta$$
$$\sec\theta\csc\theta - \sec\theta = 0$$
$$\sec\theta(\csc\theta - 1) = 0$$
Since $\sec\theta \neq 0$, the only possible solutions occur when $\csc\theta = 1$, or $\sin\theta = 1$. But then $\cos\theta = 0$, so $\sec\theta$ is undefined. Thus the equation has no solution.

48. Solving the equation:
$$2\cos^2\theta + 1 = 3\cos\theta$$
$$2\cos^2\theta - 3\cos\theta + 1 = 0$$
$$(2\cos\theta - 1)(\cos\theta - 1) = 0$$
$$\cos\theta = \tfrac{1}{2}, 1$$
$$\theta = 0°, 60°, 300°$$

49. Solving the equation:
$$2\sin^2\theta + 1 = 3\sin\theta$$
$$2\sin^2\theta - 3\sin\theta + 1 = 0$$
$$(2\sin\theta - 1)(\sin\theta - 1) = 0$$
$$\sin\theta = \tfrac{1}{2}, 1$$
$$\theta = 30°, 90°, 150°$$

50. Solving the equation:
$$2\csc^2 x + 3\csc x - 9 = 0$$
$$(2\csc x - 3)(\csc x + 3) = 0$$
$$\csc x = \tfrac{3}{2}, -3$$
$$\sin x = \tfrac{2}{3}, -\tfrac{1}{3}$$
$$x \approx 42°, 138°, 199°, 341°$$

51. Let $\theta = \sin^{-1}\left(-\dfrac{\sqrt{3}}{2}\right)$. Then $\sin\theta = -\dfrac{\sqrt{3}}{2}$ and $\theta \in \left[-\dfrac{\pi}{2}, \dfrac{\pi}{2}\right]$, so $\theta = -\dfrac{\pi}{3}$. Thus
$$\sin^{-1}\left(-\dfrac{\sqrt{3}}{2}\right) = -\dfrac{\pi}{3}.$$

52. Let $\theta = \tan^{-1}\left(\dfrac{1}{\sqrt{3}}\right)$. Then $\tan\theta = \dfrac{1}{\sqrt{3}}$ and $\theta \in \left(-\dfrac{\pi}{2}, \dfrac{\pi}{2}\right)$, so $\theta = \dfrac{\pi}{6}$. Thus $\tan^{-1}\left(\dfrac{1}{\sqrt{3}}\right) = \dfrac{\pi}{6}$.

53. Let $\theta = \sec^{-1}\left(\sqrt{2}\right)$. Then $\sec\theta = \sqrt{2}$ and $\theta \in \left[0, \dfrac{\pi}{2}\right) \cup \left(\dfrac{\pi}{2}, \pi\right]$, so $\theta = \dfrac{\pi}{4}$. Thus $\sec^{-1}\left(\sqrt{2}\right) = \dfrac{\pi}{4}$.

54. Let $\theta = \csc^{-1}(-2)$. Then $\csc\theta = -2$ and $\theta \in \left[-\dfrac{\pi}{2}, 0\right) \cup \left(0, \dfrac{\pi}{2}\right]$, so $\theta = -\dfrac{\pi}{6}$. Thus
$$\csc^{-1}(-2) = -\dfrac{\pi}{6}.$$

55. Let $\theta = \cot^{-1}(0)$. Then $\cot\theta = 0$ and $\theta \in (0, \pi)$, so $\theta = \dfrac{\pi}{2}$. Thus $\cot^{-1}(0) = \dfrac{\pi}{2}$.

56. Let $\theta = \cos^{-1}(-1)$. Then $\cos\theta = -1$ and $\theta \in [0, \pi]$, so $\theta = \pi$. Thus $\cos^{-1}(-1) = \pi$.

57. Finding the value: $\sin^{-1}\left(\cos\dfrac{2\pi}{3}\right) = \sin^{-1}\left(-\tfrac{1}{2}\right) = -\dfrac{\pi}{6}$

58. Finding the value: $\tan\left(\sin^{-1} 1\right) = \tan\dfrac{\pi}{2}$, which is undefined

59. Finding the value: $\sec\left(\cot^{-1}\left(-\dfrac{\sqrt{3}}{3}\right)\right) = \sec\left(\dfrac{2\pi}{3}\right) = -2$

60. Finding the value: $\csc^{-1}(\sin 45°) = \csc^{-1}\left(\dfrac{\sqrt{2}}{2}\right)$, which is undefined

61. Finding the value: $\cos^{-1}\left(\cos\dfrac{5\pi}{6}\right) = \cos^{-1}\left(-\dfrac{\sqrt{3}}{2}\right) = \dfrac{5\pi}{6}$

62. Finding the value: $\tan^{-1}\left(\tan\left(-\dfrac{\pi}{3}\right)\right) = \tan^{-1}\left(-\sqrt{3}\right) = -\dfrac{\pi}{3}$

63. Finding the value: $\sin^{-1}\left(\sin\dfrac{5\pi}{3}\right) = \sin^{-1}\left(-\dfrac{\sqrt{3}}{2}\right) = -\dfrac{\pi}{3}$

64. Finding the value: $\sec^{-1}\left(\sec\left(-\dfrac{\pi}{6}\right)\right) = \sec^{-1}\left(\dfrac{2}{\sqrt{3}}\right) = \dfrac{\pi}{6}$

65. Since $\cot\theta = \dfrac{x}{7}$, $\theta = \cot^{-1}\left(\dfrac{x}{7}\right)$.

66. Simplifying where $x = 3\sin\theta$:
$$x\sqrt{9-x^2} = 3\sin\theta\sqrt{9-9\sin^2\theta} = 3\sin\theta\sqrt{9\cos^2\theta} = 3\sin\theta \cdot 3\cos\theta = 9\sin\theta\cos\theta$$

67. Since $\tan\theta = \dfrac{\sqrt{x^2-64}}{8}$, $\sqrt{x^2-64} = 8\tan\theta$.

68. Since $\theta = \sin^{-1}\left(\dfrac{x}{7}\right)$, $\sin\theta = \dfrac{x}{7}$. Drawing the triangle:

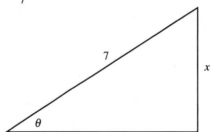

Chapter 7 Practice Test

1. For $\tan 3x$ to be defined, $3x \neq \dfrac{\pi}{2} + n\pi$, so $x \neq \dfrac{\pi}{6} + \dfrac{n\pi}{3}$, which is the domain. The range is all real numbers.

2. Sketching the graph:

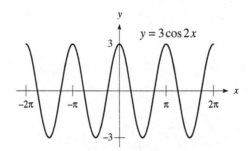

3.   Sketching the graph:

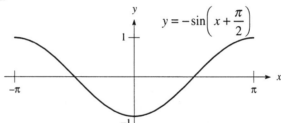

4.   Sketching the graph:

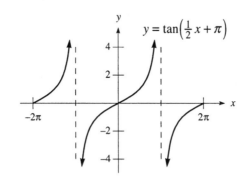

5.   Sketching the graph:

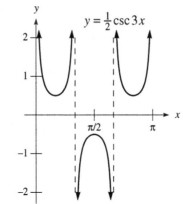

6.   **a.**   Verifying the identity:

$$\frac{\tan \theta}{\sec \theta - \cos \theta} = \frac{\dfrac{\sin \theta}{\cos \theta}}{\dfrac{1}{\cos \theta} - \cos \theta} \cdot \frac{\cos \theta}{\cos \theta} = \frac{\sin \theta}{1 - \cos^2 \theta} = \frac{\sin \theta}{\sin^2 \theta} = \frac{1}{\sin \theta}$$

$$\frac{\sec \theta}{\tan \theta} = \frac{\dfrac{1}{\cos \theta}}{\dfrac{\sin \theta}{\cos \theta}} \cdot \frac{\cos \theta}{\cos \theta} = \frac{1}{\sin \theta}$$

Since both sides simplify to the same quantity, they are equal and the identity is verified.

b.   Verifying the identity:

$$\frac{1+\sec A}{\tan A + \sin A} = \frac{1 + \dfrac{1}{\cos A}}{\dfrac{\sin A}{\cos A} + \sin A} \cdot \frac{\cos A}{\cos A}$$

$$= \frac{\cos A + 1}{\sin A + \sin A \cos A}$$

$$= \frac{\cos A + 1}{\sin A (1 + \cos A)}$$

$$= \frac{1}{\sin A}$$

$$= \csc A$$

7.   a.   Solving the equation:

$$2\cos x + 1 = 0$$

$$2\cos x = -1$$

$$\cos x = -\tfrac{1}{2}$$

$$x = \frac{2\pi}{3} + 2n\pi, \frac{4\pi}{3} + 2n\pi$$

b.   Solving the equation:

$$\tan^2 \theta = 3$$

$$\tan \theta = \pm\sqrt{3}$$

$$\theta = \frac{\pi}{3} + n\pi, \frac{2\pi}{3} + n\pi$$

8.   a.   Solving the equation:

$$\sin^2 x = \sqrt{3}\sin x \cos x$$

$$\sin^2 x - \sqrt{3}\sin x \cos x = 0$$

$$\sin x\left(\sin x - \sqrt{3}\cos x\right) = 0$$

If $\sin x = 0$, then $x = 0, \pi$. If $\sin x - \sqrt{3}\cos x = 0$, then $\sin x = \sqrt{3}\cos x$ and thus $\tan x = \sqrt{3}$, so $x = \dfrac{\pi}{3}, \dfrac{4\pi}{3}$. Thus the solutions are $x = 0, \dfrac{\pi}{3}, \pi, \dfrac{4\pi}{3}$.

b.   Solving the equation:

$$3\sin^2 x - \sin x = 4$$

$$3\sin^2 x - \sin x - 4 = 0$$

$$(3\sin x - 4)(\sin x + 1) = 0$$

$$\sin x = -1 \qquad \left(\sin x = \tfrac{4}{3} \text{ is impossible}\right)$$

$$x = \frac{3\pi}{2}$$

9.   For the denominator to be nonzero, we must have $\sec x + 1 \neq 0$, so $\sec x \neq -1$, thus $x \neq \pi + 2n\pi$.

10.   a.   Changing the angle: $\sin(-127°) = -\sin 127° = -\sin 53°$

b.   Changing the angle: $\cos\left(-\dfrac{7\pi}{8}\right) = \cos\dfrac{7\pi}{8} = -\cos\dfrac{\pi}{8}$

11.   **a.**   Let $\theta = \sin^{-1}\left(-\dfrac{\sqrt{3}}{2}\right)$. Then $\sin\theta = -\dfrac{\sqrt{3}}{2}$ and $\theta \in \left[-\dfrac{\pi}{2}, \dfrac{\pi}{2}\right]$, so $\theta = -\dfrac{\pi}{3}$. Thus

$\sin^{-1}\left(-\dfrac{\sqrt{3}}{2}\right) = -\dfrac{\pi}{3}$.

b.   Let $\theta = \arccos(-1)$. Then $\cos\theta = -1$ and $\theta \in [0, \pi]$, so $\theta = \pi$. Thus $\arccos(-1) = \pi$.

c.   Let $\theta = \tan^{-1}(0)$. Then $\tan\theta = 0$ and $\theta \in \left(-\dfrac{\pi}{2}, \dfrac{\pi}{2}\right)$, so $\theta = 0$. Thus $\tan^{-1}(0) = 0$.

d.   Finding the value: $\sin\left(\sec^{-1}\sqrt{2}\right) = \sin\left(\dfrac{\pi}{4}\right) = \dfrac{\sqrt{2}}{2}$

e.   Finding the value: $\csc^{-1}\left(\csc\left(-\dfrac{\pi}{3}\right)\right) = \csc^{-1}\left(-\dfrac{2}{\sqrt{3}}\right) = -\dfrac{\pi}{3}$

f.   Finding the value: $\cos^{-1}\left(\cos\dfrac{5\pi}{6}\right) = \cos^{-1}\left(-\dfrac{\sqrt{3}}{2}\right) = \dfrac{5\pi}{6}$

12.   Since $\cos\theta = \dfrac{x}{10}, \theta = \cos^{-1}\left(\dfrac{x}{10}\right)$.

13.   Simplifying when $x = 4\tan\theta$: $\dfrac{x^2}{x^2 + 16} = \dfrac{16\tan^2\theta}{16\tan^2\theta + 16} = \dfrac{16\tan^2\theta}{16\sec^2\theta} = \dfrac{\dfrac{\sin^2\theta}{\cos^2\theta}}{\dfrac{1}{\cos^2\theta}} = \sin^2\theta$

Chapter 8
More Trigonometry and Its Applications

8.1 The Addition Formulas

1. Using the addition formula:
$$\cos 105° = \cos(60° + 45°) = \cos 60° \cos 45° - \sin 60° \sin 45° = \frac{1}{2} \cdot \frac{\sqrt{2}}{2} - \frac{\sqrt{3}}{2} \cdot \frac{\sqrt{2}}{2} = \frac{\sqrt{2} - \sqrt{6}}{4}$$

3. Using the addition formula:
$$\tan \frac{5\pi}{12} = \tan\left(\frac{\pi}{4} + \frac{\pi}{6}\right)$$
$$= \frac{\tan \frac{\pi}{4} + \tan \frac{\pi}{6}}{1 - \tan \frac{\pi}{4} \tan \frac{\pi}{6}}$$
$$= \frac{1 + \frac{\sqrt{3}}{3}}{1 - 1 \cdot \frac{\sqrt{3}}{3}} \cdot \frac{3}{3}$$
$$= \frac{3 + \sqrt{3}}{3 - \sqrt{3}} \cdot \frac{3 + \sqrt{3}}{3 + \sqrt{3}}$$
$$= \frac{9 + 6\sqrt{3} + 3}{9 - 3}$$
$$= \frac{12 + 6\sqrt{3}}{6}$$
$$= 2 + \sqrt{3}$$

5. Using the addition formula:
$$\sin 165° = \sin(120° + 45°)$$
$$= \sin 120° \cos 45° + \cos 120° \sin 45°$$
$$= \frac{\sqrt{3}}{2} \cdot \frac{\sqrt{2}}{2} + \left(-\frac{1}{2}\right) \cdot \frac{\sqrt{2}}{2}$$
$$= \frac{\sqrt{6} - \sqrt{2}}{4}$$

7.   Using the addition formula:

$$\cos\frac{17\pi}{12} = \cos\left(\frac{3\pi}{4} + \frac{2\pi}{3}\right)$$

$$= \cos\frac{3\pi}{4}\cos\frac{2\pi}{3} - \sin\frac{3\pi}{4}\sin\frac{2\pi}{3}$$

$$= \left(-\frac{\sqrt{2}}{2}\right)\cdot\left(-\frac{1}{2}\right) - \frac{\sqrt{2}}{2}\cdot\frac{\sqrt{3}}{2}$$

$$= \frac{\sqrt{2}-\sqrt{6}}{4}$$

9.   Simplifying the expression: $\sin 23°\cos 37° + \cos 23°\sin 37° = \sin(23° + 37°) = \sin 60° = \dfrac{\sqrt{3}}{2}$

11.   Simplifying the expression: $\cos\dfrac{4\pi}{5}\cos\dfrac{3\pi}{10} + \sin\dfrac{4\pi}{5}\sin\dfrac{3\pi}{10} = \cos\left(\dfrac{4\pi}{5} - \dfrac{3\pi}{10}\right) = \cos\dfrac{\pi}{2} = 0$

13.   Simplifying the expression: $\dfrac{\tan 50° + \tan 10°}{1 - \tan 50°\tan 10°} = \tan(50° + 10°) = \tan 60° = \sqrt{3}$

15.   Verifying the identity: $\sin(\pi - \theta) = \sin\pi\cos\theta - \cos\pi\sin\theta = 0\cdot\cos\theta - (-1)\cdot\sin\theta = \sin\theta$

17.   Verifying the identity: $\cos(\pi + \theta) = \cos\pi\cos\theta - \sin\pi\sin\theta = -1\cdot\cos\theta - 0\cdot\sin\theta = -\cos\theta$

19.   Verifying the identity: $\tan\left(x - \dfrac{\pi}{4}\right) = \dfrac{\tan x - \tan\dfrac{\pi}{4}}{1 + \tan x\cdot\tan\dfrac{\pi}{4}} = \dfrac{\tan x - 1}{1 + \tan x\cdot 1} = \dfrac{\tan x - 1}{\tan x + 1}$

21.   Verifying the identity: $\cos\left(\alpha + \dfrac{\pi}{2}\right) = \cos\alpha\cos\dfrac{\pi}{2} - \sin\alpha\sin\dfrac{\pi}{2} = \cos\alpha\cdot 0 - \sin\alpha\cdot 1 = -\sin\alpha$

23.   Verifying the identity: $\sin(2\pi - \theta) = \sin 2\pi\cos\theta - \cos 2\pi\sin\theta = 0\cdot\cos\theta - 1\cdot\sin\theta = -\sin\theta$

25.   Verifying the identity:

$$\cos\left(u + \frac{\pi}{4}\right) = \cos u\cos\frac{\pi}{4} - \sin u\sin\frac{\pi}{4} = \cos u\cdot\frac{1}{\sqrt{2}} - \sin u\cdot\frac{1}{\sqrt{2}} = \frac{\cos u - \sin u}{\sqrt{2}}$$

27.   Verifying the identity:

$$\sin(A + B) + \sin(A - B) = (\sin A\cos B + \cos A\sin B) + (\sin A\cos B - \cos A\sin B)$$

$$= \sin A\cos B + \sin A\cos B$$

$$= 2\sin A\cos B$$

29.   Since $\sin A = -\frac{3}{5}$ and A is in quadrant III, $\cos A = -\frac{4}{5}$. Since $\cos B = \frac{5}{12}$ and B is in quadrant IV,

$\sin B = -\dfrac{\sqrt{119}}{12}$.

a.   Using the addition formula:

$$\sin(A + B) = \sin A\cos B + \cos A\sin B = \left(-\frac{3}{5}\right)\left(\frac{5}{12}\right) + \left(-\frac{4}{5}\right)\left(-\frac{\sqrt{119}}{12}\right) = \frac{-15 + 4\sqrt{119}}{60}$$

b.   Using the addition formula:

$$\cos(A + B) = \cos A\cos B - \sin A\sin B = \left(-\frac{4}{5}\right)\left(\frac{5}{12}\right) - \left(-\frac{3}{5}\right)\left(-\frac{\sqrt{119}}{12}\right) = \frac{-20 - 3\sqrt{119}}{60}$$

c.   Since $\sin(A + B) > 0$ and $\cos(A + B) < 0$, $A + B$ must lie in quadrant II.

31. Since $\sec t = -\frac{5}{4}$ and t is in quadrant II, $\cos t = -\frac{4}{5}$, $\sin t = \frac{3}{5}$, and $\tan t = -\frac{3}{4}$. Since $\cot u = \frac{2}{\sqrt{5}}$

and u is in quadrant III, $\tan u = \frac{\sqrt{5}}{2}$, $\sin u = -\frac{\sqrt{5}}{3}$, and $\cos u = -\frac{2}{3}$.

a. Using the addition formula:

$$\sin(u - t) = \sin u \cos t - \cos u \sin t = \left(-\frac{\sqrt{5}}{3}\right) \cdot \left(-\frac{4}{5}\right) - \left(-\frac{2}{3}\right) \cdot \frac{3}{5} = \frac{4\sqrt{5} + 6}{15}$$

b. Using the addition formula: $\tan(u - t) = \dfrac{\tan u - \tan t}{1 + \tan u \tan t} = \dfrac{\dfrac{\sqrt{5}}{2} - \left(-\dfrac{3}{4}\right)}{1 + \dfrac{\sqrt{5}}{2} \cdot \left(-\dfrac{3}{4}\right)} \cdot \dfrac{8}{8} = \dfrac{4\sqrt{5} + 6}{8 - 3\sqrt{5}}$

33. Since $\sec t = -\frac{5}{4}$ and $\sin t > 0$, $\cos t = -\frac{4}{5}$ and $\sin t = \frac{3}{5}$. Since $\cos r = \frac{4}{7}$ and $\tan r < 0$,

$\sin r = -\dfrac{\sqrt{33}}{7}$.

a. Using the addition formula:

$$\sin(r + t) = \sin r \cos t + \cos r \sin t = \left(-\frac{\sqrt{33}}{7}\right)\left(-\frac{4}{5}\right) + \frac{4}{7} \cdot \frac{3}{5} = \frac{4\sqrt{33} + 12}{35}$$

b. Using the addition formula:

$$\cos(t - r) = \cos t \cos r + \sin t \sin r = \left(-\frac{4}{5}\right)\left(\frac{4}{7}\right) + \frac{3}{5} \cdot \left(-\frac{\sqrt{33}}{7}\right) = \frac{-16 - 3\sqrt{33}}{35}$$

35. a. Since $A + B + C = 180°$, $A + B = 180° - C$, so:

$\sin(A + B) = \sin(180° - C)$
$= \sin 180° \cos C - \cos 180° \sin C$
$= 0 \cdot \cos C - (-1) \cdot \sin C$
$= \sin C$

b. Since $A + B + C = 180°$, $A + B = 180° - C$, so:

$\cos(A + B) = \cos(180° - C)$
$= \cos 180° \cos C - \sin 180° \sin C$
$= -1 \cdot \cos C - 0 \cdot \sin C$
$= -\cos C$

c. Since $A + B + C = 180°$, $A + B = 180° - C$, so:

$$\tan(A + B) = \tan(180° - C) = \frac{\tan 180° - \tan C}{1 + \tan 180° \tan C} = \frac{0 - \tan C}{1 + 0 \cdot \tan C} = -\tan C$$

Alternatively, using parts (a) and (b): $\tan(A + B) = \dfrac{\sin(A + B)}{\cos(A + B)} = \dfrac{\sin C}{-\cos C} = -\tan C$

37. Simplifying the difference quotient:

$$\frac{f(\theta + h) - f(\theta)}{h} = \frac{\sin(\theta + h) - \sin \theta}{h}$$

$$= \frac{\sin \theta \cos h + \cos \theta \sin h - \sin \theta}{h}$$

$$= \frac{\sin \theta (\cos h - 1) + \cos \theta \sin h}{h}$$

$$= \sin \theta \left(\frac{\cos h - 1}{h}\right) + \cos \theta \left(\frac{\sin h}{h}\right)$$

39. Verifying the identity:
$$\sin(A+B)+\sin(A-B)=(\sin A\cos B+\cos A\sin B)+(\sin A\cos B-\cos A\sin B)$$
$$=\sin A\cos B+\sin A\cos B$$
$$=2\sin A\cos B$$

41. Verifying the identity:
$$\sin(A+B)\sin(A-B)=(\sin A\cos B+\cos A\sin B)(\sin A\cos B-\cos A\sin B)$$
$$=\sin^2 A\cos^2 B-\cos^2 A\sin^2 B$$
$$=\sin^2 A\left(1-\sin^2 B\right)-\left(1-\sin^2 A\right)\sin^2 B$$
$$=\sin^2 A-\sin^2 A\sin^2 B-\sin^2 B+\sin^2 A\sin^2 B$$
$$=\sin^2 A-\sin^2 B$$

43. Using the addition formula:
$$\sin(A+B+C)=\sin(A+B)\cos C+\cos(A+B)\sin C$$
$$=(\sin A\cos B+\cos A\sin B)\cos C+(\cos A\cos B-\sin A\sin B)\sin C$$
$$=\sin A\cos B\cos C+\cos A\sin B\cos C+\cos A\cos B\sin C-\sin A\sin B\sin C$$

45. Substituting $-B$ for B and the fact that $\tan(-B)=-\tan B$:
$$\tan(A-B)=\frac{\tan A+\tan(-B)}{1-\tan A\tan(-B)}=\frac{\tan A-\tan B}{1+\tan A\tan B}$$

8.2 The Double-Angle and Half-Angle Formulas

1. Since $\sin\theta=-\frac{24}{25}$ and θ is in quadrant IV, $\cos\theta=\frac{7}{25}$. Using the double-angle formula:
$$\sin 2\theta=2\sin\theta\cos\theta=2\left(-\tfrac{24}{25}\right)\left(\tfrac{7}{25}\right)=-\tfrac{336}{625}$$

3. Using the double-angle formula: $\tan 2\theta=\dfrac{2\tan\theta}{1-\tan^2\theta}=\dfrac{2\left(\frac{3}{5}\right)}{1-\left(\frac{3}{5}\right)^2}=\dfrac{\frac{6}{5}}{1-\frac{9}{25}}\cdot\dfrac{25}{25}=\dfrac{30}{25-9}=\dfrac{30}{16}=\dfrac{15}{8}$

5. Since $\sin\theta=-\frac{3}{5}$ and $\pi<\theta<\dfrac{3\pi}{2}$, $\cos\theta=-\frac{4}{5}$. Also note that $\dfrac{\pi}{2}<\dfrac{\theta}{2}<\dfrac{3\pi}{4}$, so $\dfrac{\theta}{2}$ lies in quadrant II.

    a. Using the half-angle formula: $\sin\dfrac{\theta}{2}=\sqrt{\dfrac{1-\cos\theta}{2}}=\sqrt{\dfrac{1-\left(-\frac{4}{5}\right)}{2}}=\sqrt{\dfrac{\frac{9}{5}}{2}}=\sqrt{\dfrac{9}{10}}=\dfrac{3}{\sqrt{10}}$

    b. Using the half-angle formula: $\cos\dfrac{\theta}{2}=-\sqrt{\dfrac{1+\cos\theta}{2}}=-\sqrt{\dfrac{1+\left(-\frac{4}{5}\right)}{2}}=-\sqrt{\dfrac{\frac{1}{5}}{2}}=-\sqrt{\dfrac{1}{10}}=-\dfrac{1}{\sqrt{10}}$

    c. Dividing: $\tan\dfrac{\theta}{2}=\dfrac{\sin\frac{\theta}{2}}{\cos\frac{\theta}{2}}=\dfrac{3/\sqrt{10}}{-1/\sqrt{10}}=-3$

7. Verifying the identity: $(\sin x+\cos x)^2=\sin^2 x+2\sin x\cos x+\cos^2 x=1+2\sin x\cos x=1+\sin 2x$

9. Verifying the identity: $\dfrac{2}{\cot A-\tan A}=\dfrac{2}{\dfrac{1}{\tan A}-\tan A}\cdot\dfrac{\tan A}{\tan A}=\dfrac{2\tan A}{1-\tan^2 A}=\tan 2A$

11. Verifying the identity: $\dfrac{\sec^2\theta}{2-\sec^2\theta} = \dfrac{\dfrac{1}{\cos^2\theta}}{2-\dfrac{1}{\cos^2\theta}} \cdot \dfrac{\cos^2\theta}{\cos^2\theta} = \dfrac{1}{2\cos^2\theta-1} = \dfrac{1}{\cos 2\theta} = \sec 2\theta$

13. Verifying the identity: $\cot x\sin 2x = \dfrac{\cos x}{\sin x} \cdot 2\sin x\cos x = 2\cos^2 x$

15. Verifying the identity:
$$\begin{aligned}
\sin 3x &= \sin(x+2x)\\
&= \sin x\cos 2x + \cos x\sin 2x\\
&= \sin x\left(1-2\sin^2 x\right) + \cos x\left(2\sin x\cos x\right)\\
&= \sin x - 2\sin^3 x + 2\sin x\cos^2 x\\
&= \sin x - 2\sin^3 x + 2\sin x\left(1-\sin^2 x\right)\\
&= \sin x - 2\sin^3 x + 2\sin x - 2\sin^3 x\\
&= 3\sin x - 4\sin^3 x
\end{aligned}$$

17. Verifying the identity:
$$\begin{aligned}
\cos 4x &= \cos^2 2x - \sin^2 2x\\
&= \left(2\cos^2 x-1\right)^2 - 4\sin^2 x\cos^2 x\\
&= 4\cos^4 x - 4\cos^2 x + 1 - 4\left(1-\cos^2 x\right)\cos^2 x\\
&= 4\cos^4 x - 4\cos^2 x + 1 - 4\cos^2 x + 4\cos^4 x\\
&= 8\cos^4 x - 8\cos^2 x + 1
\end{aligned}$$

19. Verifying the identity: $\dfrac{\sec\theta-1}{2\sec\theta} = \dfrac{\dfrac{1}{\cos\theta}-1}{\dfrac{2}{\cos\theta}} \cdot \dfrac{\cos\theta}{\cos\theta} = \dfrac{1-\cos\theta}{2} = \sin^2\dfrac{\theta}{2}$

21. Verifying the identity:
$$\tan\theta + \cot\theta = \dfrac{\sin\theta}{\cos\theta} + \dfrac{\cos\theta}{\sin\theta} = \dfrac{\sin^2\theta + \cos^2\theta}{\sin\theta\cos\theta} = \dfrac{1}{\sin\theta\cos\theta} = \dfrac{2}{2\sin\theta\cos\theta} = \dfrac{2}{\sin 2\theta}$$

23. Solving the equation:
$$\begin{aligned}
\sin 2x &= \sin x\\
2\sin x\cos x &= \sin x\\
2\sin x\cos x - \sin x &= 0\\
\sin x(2\cos x - 1) &= 0\\
\sin x = 0,\ \cos x &= \tfrac{1}{2}\\
x = 0, \dfrac{\pi}{3}, \pi, \dfrac{5\pi}{3}
\end{aligned}$$

25. Solving the equation:
$$\tan 2x = \tan x$$
$$\frac{2\tan x}{1-\tan^2 x} = \tan x$$
$$2\tan x = \tan x - \tan^3 x$$
$$\tan^3 x + \tan x = 0$$
$$\tan x\left(\tan^2 x + 1\right) = 0$$
$$\tan x = 0 \qquad \left(\tan^2 x = -1 \text{ is impossible}\right)$$
$$x = 0, \pi$$

27. Solving the equation:
$$\sin^3 2x = \sin 2x$$
$$\sin^3 2x - \sin 2x = 0$$
$$\sin 2x\left(\sin^2 2x - 1\right) = 0$$
$$\sin 2x = 0, -1, 1$$
$$2x = 0, \frac{\pi}{2}, \pi, \frac{3\pi}{2}, 2\pi, \frac{5\pi}{2}, 3\pi, \frac{7\pi}{2}$$
$$x = 0, \frac{\pi}{4}, \frac{\pi}{2}, \frac{3\pi}{4}, \pi, \frac{5\pi}{4}, \frac{3\pi}{2}, \frac{7\pi}{4}$$

29. Solving the equation:
$$\cos 4x = \sin 2x$$
$$1 - 2\sin^2 2x = \sin 2x$$
$$2\sin^2 2x + \sin 2x - 1 = 0$$
$$(2\sin 2x - 1)(\sin 2x + 1) = 0$$
$$\sin 2x = \tfrac{1}{2}, -1$$
$$2x = \frac{\pi}{6}, \frac{5\pi}{6}, \frac{3\pi}{2}, \frac{13\pi}{6}, \frac{17\pi}{6}, \frac{7\pi}{2}$$
$$x = \frac{\pi}{12}, \frac{5\pi}{12}, \frac{3\pi}{4}, \frac{13\pi}{12}, \frac{17\pi}{12}, \frac{7\pi}{4}$$

31. Solving the equation:
$$\sin x + \cos x = 1$$
$$(\sin x + \cos x)^2 = 1$$
$$\sin^2 x + 2\sin x \cos x + \cos^2 x = 1$$
$$2\sin x \cos x + 1 = 1$$
$$2\sin x \cos x = 0$$
$$\sin x = 0, \cos x = 0$$
$$x = 0, \frac{\pi}{2}, \pi, \frac{3\pi}{2}$$

Since the equation was squared, these solutions must be checked. Note that $x = \pi, \frac{3\pi}{2}$ do not check,

so the solutions are $x = 0, \frac{\pi}{2}$.

33. Solving the equation:

$$2\sin\frac{\theta}{2} = 1$$

$$\sin\frac{\theta}{2} = \tfrac{1}{2}$$

$$\frac{\theta}{2} = \frac{\pi}{6}, \frac{5\pi}{6}$$

$$\theta = \frac{\pi}{3}, \frac{5\pi}{3}$$

35. Solving the equation:

$$\sin\frac{t}{2} = 1 - \cos t$$

$$\sin^2\frac{t}{2} = (1 - \cos t)^2$$

$$\frac{1 - \cos t}{2} = 1 - 2\cos t + \cos^2 t$$

$$1 - \cos t = 2 - 4\cos t + 2\cos^2 t$$

$$2\cos^2 t - 3\cos t + 1 = 0$$

$$(2\cos t - 1)(\cos t - 1) = 0$$

$$\cos t = \tfrac{1}{2}, 1$$

$$t = 0, \frac{\pi}{3}, \frac{5\pi}{3}$$

These solutions check in the original equation.

37. Since $\dfrac{\pi}{12}$ lies in quadrant I, $\cos\dfrac{\pi}{12} > 0$ and thus:

$$\cos\frac{\pi}{12} = \sqrt{\frac{1 + \cos\dfrac{\pi}{6}}{2}} = \sqrt{\frac{1 + \dfrac{\sqrt{3}}{2}}{2}} = \sqrt{\frac{2 + \sqrt{3}}{4}} = \frac{\sqrt{2 + \sqrt{3}}}{2}$$

39. Since $67.5°$ lies in quadrant I, $\tan 67.5° > 0$ and thus:

$$\tan 67.5° = \sqrt{\frac{1 - \cos 135°}{1 + \cos 135°}} = \sqrt{\frac{1 - \left(-\dfrac{\sqrt{2}}{2}\right)}{1 + \left(-\dfrac{\sqrt{2}}{2}\right)}} = \sqrt{\frac{2 + \sqrt{2}}{2 - \sqrt{2}}}$$

41. Using the alternative form of the double-angle identity:

$$\cos^4 t = \left(\cos^2 t\right)^2$$

$$= \left(\frac{1 + \cos 2t}{2}\right)^2$$

$$= \frac{1 + 2\cos 2t + \cos^2 2t}{4}$$

$$= \tfrac{1}{4} + \tfrac{1}{2}\cos 2t + \tfrac{1}{4}\left(\frac{1 + \cos 4t}{2}\right)$$

$$= \tfrac{1}{4} + \tfrac{1}{2}\cos 2t + \tfrac{1}{8} + \tfrac{1}{8}\cos 4t$$

$$= \tfrac{1}{8}\cos 4t + \tfrac{1}{2}\cos 2t + \tfrac{3}{8}$$

43. The area of the end is $\frac{1}{2}(2)(2)\sin\theta = 2\sin\theta$, so the volume is given by:

$$V = (2\sin\theta)(8) = 16\sin\theta \ \text{ft}^3$$

45. Using the double-angle formula: $\sin 2\theta = 2\sin\theta\cos\theta = 2\cdot\dfrac{x}{\sqrt{16+x^2}}\cdot\dfrac{4}{\sqrt{16+x^2}} = \dfrac{8x}{x^2+16}$

47. Note that $\tan\theta = \dfrac{x}{3}$ and $\tan 2\theta = \dfrac{6}{3} = 2$, so by the double-angle formula:

$$\tan 2\theta = \frac{2\tan\theta}{1-\tan^2\theta}$$

$$2 = \frac{2\cdot\dfrac{x}{3}}{1-\left(\dfrac{x}{3}\right)^2}$$

$$2 = \frac{\dfrac{2x}{3}}{1-\dfrac{x^2}{9}}$$

$$2 = \frac{6x}{9-x^2}$$

$$18 - 2x^2 = 6x$$

$$2x^2 + 6x - 18 = 0$$

$$x^2 + 3x - 9 = 0$$

Using the quadratic formula: $x = \dfrac{-3\pm\sqrt{9-4(-9)}}{2} = \dfrac{-3\pm\sqrt{45}}{2} = \dfrac{-3\pm 3\sqrt{5}}{2} \approx -4.85, 1.85$

Since $x > 0$, $x = \dfrac{-3+3\sqrt{5}}{2} \approx 1.85$.

49. Using the double-angle formula: $\cos 2\theta = 1 - 2\sin^2\theta = 1 - 2\left(\dfrac{a}{4}\right)^2 = 1 - \dfrac{a^2}{8} = \dfrac{8-a^2}{8}$

51. Deriving the formula:

$$\begin{aligned}
\sin 3\theta &= \sin(\theta + 2\theta)\\
&= \sin\theta\cos 2\theta + \cos\theta\sin 2\theta\\
&= \sin\theta\left(1 - 2\sin^2\theta\right) + \cos\theta(2\sin\theta\cos\theta)\\
&= \sin\theta - 2\sin^3\theta + 2\sin\theta\cos^2\theta\\
&= \sin\theta - 2\sin^3\theta + 2\sin\theta\left(1 - \sin^2\theta\right)\\
&= \sin\theta - 2\sin^3\theta + 2\sin\theta - 2\sin^3\theta\\
&= 3\sin\theta - 4\sin^3\theta
\end{aligned}$$

53. Starting with the double-angle formula, substitute $A = \dfrac{\theta}{2}$:

$$\cos 2A = 2\cos^2 A - 1$$
$$1 + \cos 2A = 2\cos^2 A$$
$$\cos^2 A = \frac{1 + \cos 2A}{2}$$
$$\cos^2 \frac{\theta}{2} = \frac{1 + \cos\theta}{2}$$
$$\cos\frac{\theta}{2} = \pm\sqrt{\frac{1 + \cos\theta}{2}}$$

55. Starting with the two formulas:
$$\cos(A - B) = \cos A \cos B + \sin A \sin B$$
$$\cos(A + B) = \cos A \cos B - \sin A \sin B$$
Subtracting yields:
$$\cos(A - B) - \cos(A + B) = 2\sin A \sin B$$
$$\tfrac{1}{2}\left[\cos(A - B) - \cos(A + B)\right] = \sin A \sin B$$

57. a. Using the formula for $\sin A \cos B$:
$$\sin 5x \cos 3x = \tfrac{1}{2}\left[\sin(5x + 3x) + \sin(5x - 3x)\right] = \tfrac{1}{2}(\sin 8x + \sin 2x)$$

b. Using the formula for $\sin A \sin B$:
$$\sin 5x \sin 3x = \tfrac{1}{2}\left[\cos(5x - 3x) - \cos(5x + 3x)\right] = \tfrac{1}{2}(\cos 2x - \cos 8x)$$

c. Using the formula for $\cos A \cos B$:
$$\cos x \cos\frac{x}{2} = \tfrac{1}{2}\left[\cos\left(x + \frac{x}{2}\right) + \cos\left(x - \frac{x}{2}\right)\right] = \tfrac{1}{2}\left(\cos\frac{3x}{2} + \cos\frac{x}{2}\right)$$

59. a. Using the formula for $\sin u + \sin v$:
$$\sin 5x + \sin 3x = 2\sin\frac{5x + 3x}{2}\cos\frac{5x - 3x}{2} = 2\sin\frac{8x}{2}\cos\frac{2x}{2} = 2\sin 4x \cos x$$

b. Using the formula for $\cos u - \cos v$:
$$\cos 7x - \cos 4x = -2\sin\frac{7x + 4x}{2}\sin\frac{7x - 4x}{2} = -2\sin\frac{11x}{2}\sin\frac{3x}{2}$$

c. Using the formula for $\sin u - \sin v$:
$$\sin 3x - \sin\tfrac{1}{2}x = 2\cos\frac{3x + \tfrac{1}{2}x}{2}\sin\frac{3x - \tfrac{1}{2}x}{2} = 2\cos\frac{7x}{4}\sin\frac{5x}{4}$$

61. Using the addition formula:
$$\sin 105° = \sin(60° + 45°) = \sin 60° \cos 45° + \cos 60° \sin 45° = \frac{\sqrt{3}}{2}\cdot\frac{\sqrt{2}}{2} + \frac{1}{2}\cdot\frac{\sqrt{2}}{2} = \frac{\sqrt{6} + \sqrt{2}}{4}$$

Using the half-angle formula (note that $\sin 105° > 0$):

$$\sin 105° = \sqrt{\frac{1 - \cos 210°}{2}} = \sqrt{\frac{1 - \left(-\dfrac{\sqrt{3}}{2}\right)}{2}} = \sqrt{\frac{1 + \dfrac{\sqrt{3}}{2}}{2}} = \sqrt{\frac{2 + \sqrt{3}}{4}} = \frac{\sqrt{2 + \sqrt{3}}}{2}$$

Note that these two answers do not appear to be equal. It is easier to compare the squares of these numbers:

$$\sin^2 105° = \left(\frac{\sqrt{6}+\sqrt{2}}{4}\right)^2 = \frac{6+2\sqrt{12}+2}{16} = \frac{8+4\sqrt{3}}{16} = \frac{2+\sqrt{3}}{4}$$

$$\sin^2 105° = \left(\frac{\sqrt{2+\sqrt{3}}}{2}\right)^2 = \frac{2+\sqrt{3}}{4}$$

Note that the two values are indeed equal.

8.3 The Law of Sines and the Law of Cosines

1.   First find $C = 180° - 80° - 35° = 65°$. Using the law of sines:
$$\frac{\sin 80°}{12} = \frac{\sin 35°}{b}$$
$$b \sin 80° = 12 \sin 35°$$
$$b = \frac{12 \sin 35°}{\sin 80°} \approx 7.0$$
Also using the law of sines:
$$\frac{\sin 80°}{12} = \frac{\sin 65°}{c}$$
$$c \sin 80° = 12 \sin 65°$$
$$c = \frac{12 \sin 65°}{\sin 80°} \approx 11.0$$

3.   First find B using the law of sines:
$$\frac{\sin 72°}{24} = \frac{\sin B}{15}$$
$$24 \sin B = 15 \sin 72°$$
$$\sin B = \frac{15 \sin 72°}{24} \approx 0.5944$$
$$B = \sin^{-1}(0.5944) \approx 36.5°$$
Then $C = 180° - 72° - 36.5° = 71.5°$. Now using the law of sines:
$$\frac{\sin 72°}{24} = \frac{\sin 71.5°}{c}$$
$$c \sin 72° = 24 \sin 71.5°$$
$$c = \frac{24 \sin 71.5°}{\sin 72°} \approx 23.9$$

5.   First find C using the law of cosines:
$$10^2 = 6^2 + 9^2 - 2(6)(9) \cos C$$
$$100 = 36 + 81 - 108 \cos C$$
$$-17 = -108 \cos C$$
$$\cos C = \tfrac{17}{108}$$
$$C = \cos^{-1}\left(\tfrac{17}{108}\right) \approx 80.9°$$

Now find A using the law of sines:
$$\frac{\sin A}{6} = \frac{\sin 80.9°}{10}$$
$$10 \sin A = 6 \sin 80.9°$$
$$\sin A = \frac{6 \sin 80.9°}{10} \approx 0.5925$$
$$A = \sin^{-1}(0.5925) \approx 36.3°$$
Finally $B = 180° - 80.9° - 36.3° = 62.8°$.

7. First find $B = 180° - 110° - 25° = 45°$. Using the law of sines:
$$\frac{\sin 25°}{16} = \frac{\sin 110°}{b}$$
$$b \sin 25° = 16 \sin 110°$$
$$b = \frac{16 \sin 110°}{\sin 25°} \approx 35.6$$
Also using the law of sines:
$$\frac{\sin 25°}{16} = \frac{\sin 45°}{a}$$
$$a \sin 25° = 16 \sin 45°$$
$$a = \frac{16 \sin 45°}{\sin 25°} \approx 26.8$$

9. First find a using the law of cosines:
$$a^2 = 5^2 + 11^2 - 2(5)(11) \cos 138° \approx 227.75$$
$$a \approx 15.1$$
Now find B using the law of sines:
$$\frac{\sin 138°}{15.1} = \frac{\sin B}{5}$$
$$15.1 \sin B = 5 \sin 138°$$
$$\sin B = \frac{5 \sin 138°}{15.1} \approx 0.2217$$
$$B = \sin^{-1}(0.2217) \approx 12.8°$$
Finally $C = 180° - 138° - 12.8° = 29.2°$.

11. Using the law of sines:
$$\frac{\sin 54°}{7} = \frac{\sin B}{10}$$
$$7 \sin B = 10 \sin 54°$$
$$\sin B = \frac{10 \sin 54°}{7} \approx 1.1557$$
Since $-1 \le \sin B \le 1$, no such triangle exists.

13. Using the law of cosines:
$$c^2 = 35^2 + 40^2 - 2(35)(40) \cos 134° \approx 4770.04$$
$$c \approx 69.1$$
One such triangle exists.

15. Let x represent the required distance. Using the law of sines:
$$\frac{\sin 63.8°}{200} = \frac{\sin 84.2°}{x}$$
$$x \sin 63.8° = 200 \sin 84.2°$$
$$x = \frac{200 \sin 84.2°}{\sin 63.8°} \approx 221.8$$
The distance from A to B is 221.8 meters.

17.   First draw the figure:

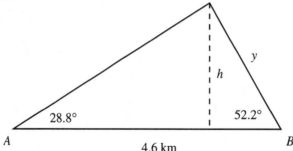

The third angle of the triangle is $180° - 28.8° - 52.2° = 99°$. Now find y using the law of sines:

$$\frac{\sin 99°}{4.6} = \frac{\sin 28.8°}{y}$$
$$y \sin 99° = 4.6 \sin 28.8°$$
$$y = \frac{4.6 \sin 28.8°}{\sin 99°} \approx 2.24$$

Now using the right triangle:

$$\sin 52.2° = \frac{h}{2.24}$$
$$h = 2.24 \sin 52.2° \approx 1.8$$

The altitude of the balloon is 1.8 km.

19.   Let x represent the length of the tunnel. Using the law of cosines:

$$x^2 = 600^2 + 500^2 - 2(600)(500)\cos 76.7° \approx 471970.16$$
$$x \approx 687.0$$

The length of the tunnel is 687.0 meters.

21.   The third angle of $\triangle ABD$ is $180° - 30° - 125° = 25°$. Using the law of sines:

$$\frac{\sin 25°}{4} = \frac{\sin 30°}{BD}$$
$$BD \sin 25° = 4 \sin 30°$$
$$BD = \frac{4 \sin 30°}{\sin 25°} \approx 4.7$$

Now $\angle DBC = 180° - 125° = 55°$. Using the law of sines:

$$\frac{\sin 85°}{4.7} = \frac{\sin 55°}{CD}$$
$$CD \sin 85° = 4.7 \sin 55°$$
$$CD = \frac{4.7 \sin 55°}{\sin 85°} \approx 3.9$$

23.   First find the distance d from first base to third base:

$$d^2 = 90^2 + 90^2 = 16200$$
$$d = \sqrt{16200} \approx 127.3$$

Since the pitcher's mound is midway between these two bases, the distance from the mound to first base and third base is thus $\frac{1}{2}(127.3) \approx 63.6$ feet.

25. Let θ represent the smallest angle. Using the law of cosines:
$$20^2 = 42^2 + 35^2 - 2(42)(35)\cos\theta$$
$$400 = 2989 - 2940\cos\theta$$
$$-2589 = -2940\cos\theta$$
$$\cos\theta = \frac{2589}{2940}$$
$$\theta = \cos^{-1}\left(\frac{2589}{2940}\right) \approx 28.3°$$

27. Note that $\angle QRP = 180° - 38° = 142°$, so $\angle QPR = 180° - 24.2° - 142° = 13.8°$. Using the law of sines:
$$\frac{\sin 13.8°}{200} = \frac{\sin 24.2°}{PR}$$
$$PR\sin 13.8° = 200\sin 24.2°$$
$$PR = \frac{200\sin 24.2°}{\sin 13.8°} \approx 343.70$$
Now using the right triangle:
$$\sin 38° = \frac{PS}{343.70}$$
$$PS = 343.7\sin 38° \approx 211.6$$
The tree is approximately 211.6 feet tall.

29. Drawing the figure:

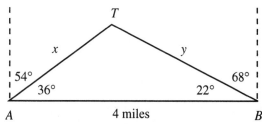

Note that $T = 180° - 36° - 22° = 122°$. Using the law of sines:

$$\frac{\sin 122°}{4} = \frac{\sin 36°}{y} \qquad\qquad \frac{\sin 122°}{4} = \frac{\sin 22°}{x}$$
$$y\sin 122° = 4\sin 36° \qquad\qquad x\sin 122° = 4\sin 22°$$
$$y = \frac{4\sin 36°}{\sin 122°} \approx 2.77 \qquad\qquad x = \frac{4\sin 22°}{\sin 122°} \approx 1.77$$

The range for cannon A is 1.77 miles, and the range for cannon B is 2.77 miles.

31. After 10 minutes $= \frac{1}{6}$ hour, the boat has traveled $15\left(\frac{1}{6}\right) = 2.5$ miles. Drawing the figure:

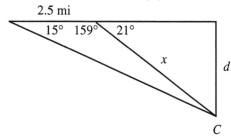

Note that $C = 180° - 15° - 159° = 6°$. Using the law of sines:

$$\frac{\sin 6°}{2.5} = \frac{\sin 15°}{x}$$
$$x \sin 6° = 2.5 \sin 15°$$
$$x = \frac{2.5 \sin 15°}{\sin 6°} \approx 6.19$$

Now using the right triangle:

$$\sin 21° = \frac{d}{6.19}$$
$$d = 6.19 \sin 21° \approx 2.2$$

The boat is 2.2 miles offshore.

33.   Drawing the figure:

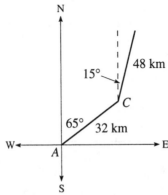

The angle at C is $(90° - 65°) + 90° + 15° = 130°$. Let d represent the required distance. Using the law of cosines:

$$d^2 = 32^2 + 48^2 - 2(32)(48)\cos 130° \approx 5302.64$$
$$d \approx 72.8$$

The ship is 72.8 km from point A.

35.   Drawing the figure:

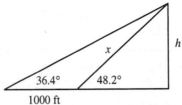

The other two angles in the triangle are $180° - 48.2° = 131.8°$ and $180° - 36.4° - 131.8° = 11.8°$.
Using the law of sines:

$$\frac{\sin 11.8°}{1000} = \frac{\sin 36.4°}{x}$$
$$x \sin 11.8° = 1000 \sin 36.4°$$
$$x = \frac{1000 \sin 36.4°}{\sin 11.8°} \approx 2901.86$$

Using the right triangle:

$$\sin 48.2° = \frac{h}{2901.86}$$
$$h = 2901.86 \sin 48.2° \approx 2163.3$$

The mountain is 2163.3 feet high.

37.   Since 5 minutes = $\frac{1}{12}$ hour, the man has driven $60\left(\frac{1}{12}\right) = 5$ km. Drawing the figure:

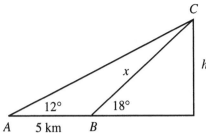

Note that $\angle ABC = 180° - 18° = 162°$, so $\angle ACB = 180° - 12° - 162° = 6°$. Using the law of sines:

$$\frac{\sin 6°}{5} = \frac{\sin 12°}{x}$$
$$x \sin 6° = 5 \sin 12°$$
$$x = \frac{5 \sin 12°}{\sin 6°} \approx 9.95$$

Using the right triangle:

$$\sin 18° = \frac{h}{9.95}$$
$$h = 9.95 \sin 18° \approx 3.1$$

The mountain is 3.1 km tall.

39.   Let d represent the distance between the markers. Drawing the figure:

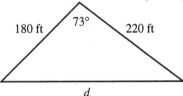

Using the law of cosines:

$$d^2 = 180^2 + 220^2 - 2(180)(220)\cos 73° \approx 57644.16$$
$$d \approx 240.1$$

The two markers are approximately 240.1 feet apart.

41.   Drawing the figure:

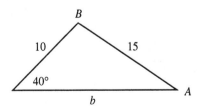

Using the law of sines:

$$\frac{\sin 40°}{15} = \frac{\sin A}{10}$$
$$15 \sin A = 10 \sin 40°$$
$$\sin A = \frac{10 \sin 40°}{15} \approx 0.4285$$
$$A = \sin^{-1}(0.4285) \approx 25.4°$$

Then $B = 180° - 25.4° - 40° = 114.6°$. Using the area formula: $A = \frac{1}{2}(10)(15)\sin 114.6° \approx 68.2$

43.   Using the hint:
$$\tfrac{1}{2}ab\sin C = \tfrac{1}{2}ac\sin B = \tfrac{1}{2}bc\sin A$$
$$\frac{ab\sin C}{abc} = \frac{ac\sin B}{abc} = \frac{bc\sin A}{abc}$$
$$\frac{\sin C}{c} = \frac{\sin B}{b} = \frac{\sin A}{a}$$

45.   **a.**   Using the law of cosines:
$$c^2 = a^2 + b^2 - 2ab\cos C$$
$$c^2 - a^2 - b^2 = -2ab\cos C$$
$$\cos C = \frac{c^2 - a^2 - b^2}{-2ab}$$

Therefore:
$$\tfrac{1}{2}ab(1+\cos C) = \tfrac{1}{2}ab\left(1 + \frac{c^2 - a^2 - b^2}{-2ab}\right)$$
$$= \tfrac{1}{2}ab\left(\frac{-2ab + c^2 - a^2 - b^2}{-2ab}\right)$$
$$= \frac{a^2 + 2ab + b^2 - c^2}{4}$$
$$= \frac{(a+b)^2 - c^2}{4}$$
$$= \frac{(a+b+c)(a+b-c)}{4}$$
$$= \left(\frac{a+b+c}{2}\right)\left(\frac{a+b-c}{2}\right)$$

b.   Proving the relationship:
$$\tfrac{1}{2}ab(1-\cos C) = \tfrac{1}{2}ab\left(1 - \frac{c^2 - a^2 - b^2}{-2ab}\right)$$
$$= \tfrac{1}{2}ab\left(\frac{-2ab - c^2 + a^2 + b^2}{-2ab}\right)$$
$$= \frac{c^2 - a^2 + 2ab - b^2}{4}$$
$$= \frac{c^2 - (a-b)^2}{4}$$
$$= \frac{(c+a-b)(c-a+b)}{4}$$
$$= \left(\frac{a-b+c}{2}\right)\left(\frac{-a+b+c}{2}\right)$$

47.   Deriving Heron's formula:
$$A = \sqrt{\left[\tfrac{1}{2}ab(1+\cos C)\right]\left[\tfrac{1}{2}ab(1-\cos C)\right]} = \sqrt{s(s-c)(s-a)(s-b)} = \sqrt{s(s-a)(s-b)(s-c)}$$

8.4 Vectors

1.   Sketching $2\mathbf{u} = \langle 8, 4 \rangle$:

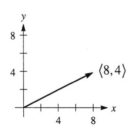

3.   Sketching $\mathbf{u} + \mathbf{v} = \langle 1, 7 \rangle$:

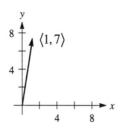

5.   Sketching $\mathbf{u} - \mathbf{v} = \langle 7, -3 \rangle$:

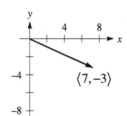

7.   Since $\overrightarrow{OQ} = \langle -1, 2 \rangle$, $\left| \overrightarrow{OQ} \right| = \sqrt{(-1)^2 + 2^2} = \sqrt{5}$. Sketching the vector:

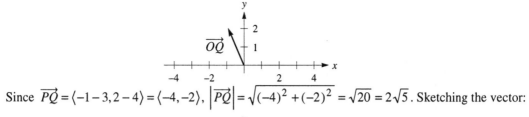

9.   Since $\overrightarrow{PQ} = \langle -1 - 3, 2 - 4 \rangle = \langle -4, -2 \rangle$, $\left| \overrightarrow{PQ} \right| = \sqrt{(-4)^2 + (-2)^2} = \sqrt{20} = 2\sqrt{5}$. Sketching the vector:

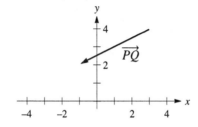

11. Since $\overrightarrow{OP} = \langle 3,4 \rangle$ and $\overrightarrow{OQ} = \langle -1,2 \rangle$, $\overrightarrow{OP} + \overrightarrow{OQ} = \langle 2,6 \rangle$, so $\left| \overrightarrow{OP} + \overrightarrow{OQ} \right| = \sqrt{2^2 + 6^2} = \sqrt{40} = 2\sqrt{10}$.

Sketching the vector:

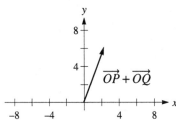

13. Since $\overrightarrow{OR} = \langle 1,5 \rangle$ and $\overrightarrow{OS} = \langle 4,-1 \rangle$, $\overrightarrow{OR} - \overrightarrow{OS} = \langle -3,6 \rangle$, so

$\left| \overrightarrow{OR} - \overrightarrow{OS} \right| = \sqrt{(-3)^2 + 6^2} = \sqrt{45} = 3\sqrt{5}$. Sketching the vector:

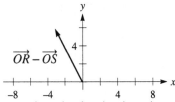

15. Performing the operations: $\mathbf{u} + \mathbf{v} = \langle -3,2 \rangle + \langle 2,4 \rangle = \langle -1,6 \rangle = -\mathbf{i} + 6\mathbf{j}$

17. Performing the operations: $2\mathbf{u} - 3\mathbf{v} = 2\langle -3,2 \rangle - 3\langle 2,4 \rangle = \langle -6,4 \rangle - \langle 6,12 \rangle = \langle -12,-8 \rangle = -12\mathbf{i} - 8\mathbf{j}$

19. Performing the operations:
$$-6\mathbf{v} - \mathbf{u} = -6\langle 2,4 \rangle - \langle -3,2 \rangle = \langle -12,-24 \rangle - \langle -3,2 \rangle = \langle -9,-26 \rangle = -9\mathbf{i} - 26\mathbf{j}$$

21. Since $\mathbf{A} + \mathbf{B} = \langle 3,0 \rangle + \langle 0,2 \rangle = \langle 3,2 \rangle$, $|\mathbf{A} + \mathbf{B}| = \sqrt{3^2 + 2^2} = \sqrt{13}$. Drawing the figure:

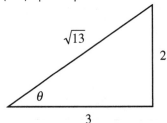

Therefore:
$$\tan \theta = \tfrac{2}{3}$$
$$\theta = \tan^{-1}\left(\tfrac{2}{3}\right) \approx 33.7°$$

23. Since $\mathbf{A} + \mathbf{B} = \langle 12,0 \rangle + \langle 0,-1 \rangle = \langle 12,-1 \rangle$, $|\mathbf{A} + \mathbf{B}| = \sqrt{12^2 + (-1)^2} = \sqrt{145}$. Drawing the figure:

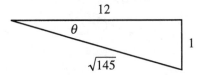

Therefore:
$$\tan \theta = \tfrac{1}{12}$$
$$\theta = \tan^{-1}\left(\tfrac{1}{12}\right) \approx 4.8°$$

25. Note that $|\mathbf{v}| = \sqrt{2^2 + 4^2} = \sqrt{20} = 2\sqrt{5}$ and $|\mathbf{u}| = \sqrt{4^2 + 2^2} = \sqrt{20} = 2\sqrt{5}$. The resultant vector is

$\mathbf{u} + \mathbf{v} = \langle 4, 2 \rangle + \langle 2, 4 \rangle = \langle 6, 6 \rangle$, so its length is $|\mathbf{u} + \mathbf{v}| = \sqrt{6^2 + 6^2} = \sqrt{72} = 6\sqrt{2}$. To find the angle, draw the triangle:

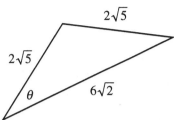

Use the law of cosines to find θ:

$$\left(2\sqrt{5}\right)^2 = \left(2\sqrt{5}\right)^2 + \left(6\sqrt{2}\right)^2 - 2\left(2\sqrt{5}\right)\left(6\sqrt{2}\right)\cos\theta$$
$$20 = 20 + 72 - 24\sqrt{10}\cos\theta$$
$$-72 = -24\sqrt{10}\cos\theta$$
$$\cos\theta = \frac{72}{24\sqrt{10}}$$
$$\theta = \cos^{-1}\left(\frac{3}{\sqrt{10}}\right) \approx 18.4°$$

27. Note that $|\mathbf{v}| = \sqrt{(-1)^2 + (-3)^2} = \sqrt{10}$ and $|\mathbf{w}| = \sqrt{6^2 + 6^2} = \sqrt{72} = 6\sqrt{2}$. The resultant vector is

$\mathbf{v} + \mathbf{w} = \langle -1, -3 \rangle + \langle 6, 6 \rangle = \langle 5, 3 \rangle$, so its length is $|\mathbf{v} + \mathbf{w}| = \sqrt{5^2 + 3^2} = \sqrt{34}$. To find the angle, draw the triangle:

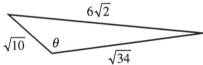

Use the law of cosines to find θ:

$$\left(6\sqrt{2}\right)^2 = \left(\sqrt{10}\right)^2 + \left(\sqrt{34}\right)^2 - 2\left(\sqrt{10}\right)\left(\sqrt{34}\right)\cos\theta$$
$$72 = 10 + 34 - 2\sqrt{340}\cos\theta$$
$$28 = -2\sqrt{340}\cos\theta$$
$$\cos\theta = -\frac{14}{\sqrt{340}}$$
$$\theta = \cos^{-1}\left(-\frac{14}{\sqrt{340}}\right) \approx 139.4°$$

29. Resolving the vector into components:

$$\mathbf{v} = \left\langle |\mathbf{v}|\cos\theta, |\mathbf{v}|\sin\theta \right\rangle = \langle 7\cos 60°, 7\sin 60° \rangle = \left\langle 7 \cdot \frac{1}{2}, 7 \cdot \frac{\sqrt{3}}{2} \right\rangle = \left\langle \frac{7}{2}, \frac{7\sqrt{3}}{2} \right\rangle$$

31. Resolving the vector into components:

$$\mathbf{v} = \left\langle |\mathbf{v}|\cos\theta, |\mathbf{v}|\sin\theta \right\rangle = \langle 100\cos 126°, 100\sin 126° \rangle \approx \langle -58.8, 80.9 \rangle$$

33. Rewriting the vector: $2\mathbf{i} + 7\mathbf{j} = 2\langle 1, 0 \rangle + 7\langle 0, 1 \rangle = \langle 2, 0 \rangle + \langle 0, 7 \rangle = \langle 2, 7 \rangle$

35. Rewriting the vector: $5\langle 1, 0 \rangle - 6\langle 0, 1 \rangle = \langle 5, 0 \rangle - \langle 0, 6 \rangle = \langle 5, -6 \rangle = 5\mathbf{i} - 6\mathbf{j}$

37. Rewriting the vector: $4\langle 2, -1 \rangle + 7\langle -3, 2 \rangle = \langle 8, -4 \rangle + \langle -21, 14 \rangle = \langle -13, 10 \rangle = -13\mathbf{i} + 10\mathbf{j}$

39. Since $|\mathbf{v}| = \sqrt{3^2 + 5^2} = \sqrt{34}$, the unit vector is $\dfrac{1}{\sqrt{34}}\langle 3, 5 \rangle = \left\langle \dfrac{3}{\sqrt{34}}, \dfrac{5}{\sqrt{34}} \right\rangle$.

41. Since $|\mathbf{v}| = \sqrt{(-1)^2 + (-3)^2} = \sqrt{10}$, the unit vector is $\dfrac{1}{\sqrt{10}}\langle -1, -3 \rangle = \left\langle -\dfrac{1}{\sqrt{10}}, -\dfrac{3}{\sqrt{10}} \right\rangle$.

43. Since $\mathbf{v} = -2\langle 1, 0 \rangle - 5\langle 0, 1 \rangle = \langle -2, 0 \rangle - \langle 0, 5 \rangle = \langle -2, -5 \rangle$, $|\mathbf{v}| = \sqrt{(-2)^2 + (-5)^2} = \sqrt{29}$, and the unit

vector is $\dfrac{1}{\sqrt{29}}\langle -2, -5 \rangle = \left\langle -\dfrac{2}{\sqrt{29}}, -\dfrac{5}{\sqrt{29}} \right\rangle$.

45. First draw the figure:

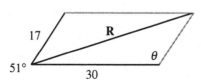

Finding angle θ:
$$2(51°) + 2\theta = 360°$$
$$102° + 2\theta = 360°$$
$$2\theta = 258°$$
$$\theta = 129°$$

Now using the law of cosines:
$$|\mathbf{R}|^2 = 30^2 + 17^2 - 2(30)(17)\cos 129° \approx 1830.91$$
$$|\mathbf{R}| \approx 42.8 \text{ pounds}$$

47. First draw the figure:

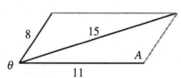

Finding angle A using the law of cosines:
$$15^2 = 11^2 + 8^2 - 2(11)(8)\cos A$$
$$225 = 185 - 176\cos A$$
$$40 = -176\cos A$$
$$\cos A = -\frac{40}{176}$$
$$A = \cos^{-1}\left(-\frac{40}{176}\right) \approx 103.1°$$

Therefore:
$$2\theta + 2(103.1°) = 360°$$
$$2\theta + 206.2° = 360°$$
$$2\theta = 153.8°$$
$$\theta = 76.9°$$

49. First draw the figure:

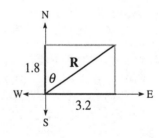

The speed is given by: $|\mathbf{R}| = \sqrt{3.2^2 + 1.8^2} = \sqrt{13.48} \approx 3.7$ mph

Finding θ:

$$\tan\theta = \frac{3.2}{1.8}$$

$$\theta = \tan^{-1}\left(\frac{3.2}{1.8}\right) \approx 60.6°$$

The course of the boat is N 60.6° E.

51. First draw the figure:

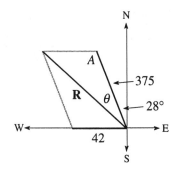

Finding A:

$$2A + 2(62°) = 360°$$
$$2A + 124° = 360°$$
$$2A = 236°$$
$$A = 118°$$

Finding $|\mathbf{R}|$ using the law of cosines:

$$|\mathbf{R}|^2 = 42^2 + 375^2 - 2(42)(375)\cos 118° \approx 157177$$
$$|\mathbf{R}| \approx 396.5 \text{ mph}$$

Finding θ using the law of sines:

$$\frac{\sin\theta}{42} = \frac{\sin 118°}{396.5}$$
$$\sin\theta = \frac{42\sin 118°}{396.5} \approx 0.0935$$
$$\theta \approx 5.4°$$
$$28° + \theta \approx 33.4°$$

The ground speed is 396.5 mph and the direction is N 33.4° W.

53. First draw the figure:

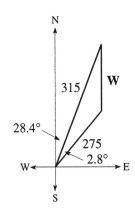

Finding $|\mathbf{W}|$ using the law of cosines:

$$|\mathbf{W}|^2 = 275^2 + 315^2 - 2(275)(315)\cos 2.8° \approx 1806.84$$
$$|\mathbf{W}| \approx 42.5 \text{ mph}$$

The speed of the wind is 42.5 mph.

55. First draw the figure:

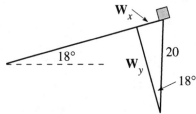

The parallel component of the weight is:

$$\sin 18° = \frac{\mathbf{W}_x}{20}$$
$$\mathbf{W}_x = 20 \sin 18° \approx 6.2 \text{ pounds}$$

The perpendicular component of the weight is:

$$\cos 18° = \frac{\mathbf{W}_y}{20}$$
$$\mathbf{W}_y = 20 \cos 18° \approx 19.0 \text{ pounds}$$

57. First draw the figure:

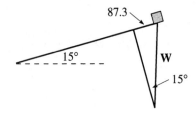

Finding the weight:

$$\sin 15° = \frac{87.3}{|\mathbf{W}|}$$
$$|\mathbf{W}| \sin 15° = 87.3$$
$$|\mathbf{W}| = \frac{87.3}{\sin 15°} \approx 337.3 \text{ pounds}$$

59. First draw the figure:

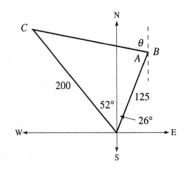

Finding the distance using the law of cosines:
$$BC^2 = 125^2 + 200^2 - 2(125)(200)\cos 78° \approx 45229.42$$
$$BC \approx 212.7 \text{ miles}$$
To find the heading, first find A using the law of sines:
$$\frac{\sin A}{200} = \frac{\sin 78°}{212.7}$$
$$\sin A = \frac{200\sin 78°}{212.7} \approx 0.9199$$
$$A \approx 66.9°$$
Now let θ be the heading, so:
$$\theta + 66.9° + 26° = 180°$$
$$\theta = 87.1°$$
The heading is N 87.1° W.

61. Let $\left|\overrightarrow{OB}\right|$ and $\left|\overrightarrow{OA}\right|$ be the required tensions, and let $\left|\mathbf{W}\right|$ be the weight. The following are the vector components of all three vectors:
$$\left|\overrightarrow{OB}\right| = \left\langle -\left|\overrightarrow{OB}\right|\cos 45°, -\left|\overrightarrow{OB}\right|\sin 45° \right\rangle$$
$$\left|\overrightarrow{OA}\right| = \left\langle \left|\overrightarrow{OA}\right|\cos 30°, -\left|\overrightarrow{OA}\right|\sin 30° \right\rangle$$
$$\left|\mathbf{W}\right| = \left\langle 0, -100 \right\rangle$$
Since the sum of all horizontal components is 0 and the sum of all vertical components is 0, the system of equations is:
$$-\left|\overrightarrow{OB}\right|\cos 45° + \left|\overrightarrow{OA}\right|\cos 30° = 0$$
$$-\left|\overrightarrow{OB}\right|\sin 45° - \left|\overrightarrow{OA}\right|\sin 30° = -100$$
Simplifying the equations:
$$\left|\overrightarrow{OB}\right|\cos 45° - \left|\overrightarrow{OA}\right|\cos 30° = 0$$
$$\left|\overrightarrow{OB}\right|\sin 45° + \left|\overrightarrow{OA}\right|\sin 30° = 100$$
Multiplying the first equation by $\sin 30°$ and the second equation by $\cos 30°$:
$$\left|\overrightarrow{OB}\right|\cos 45° \sin 30° - \left|\overrightarrow{OA}\right|\cos 30° \sin 30° = 0$$
$$\left|\overrightarrow{OB}\right|\sin 45° \cos 30° + \left|\overrightarrow{OA}\right|\sin 30° \cos 30° = 100\cos 30°$$
Adding the two equations:
$$\left|\overrightarrow{OB}\right|(\cos 45° \sin 30° + \sin 45° \cos 30°) = 100\cos 30°$$
$$\left|\overrightarrow{OB}\right|\sin 75° = 100\cos 30°$$
$$\left|\overrightarrow{OB}\right| = \frac{100\cos 30°}{\sin 75°} \approx 89.66 \text{ pounds}$$
Substituting into the first equation:
$$-89.66\cos 45° + \left|\overrightarrow{OA}\right|\cos 30° = 0$$
$$\left|\overrightarrow{OA}\right|\cos 30° = 89.66\cos 45°$$
$$\left|\overrightarrow{OA}\right| = \frac{89.66\cos 45°}{\cos 30°} \approx 73.21 \text{ pounds}$$

63. Finding the dot product: $\mathbf{u} \cdot \mathbf{v} = (-1)(3) + (6)(-4) = -3 - 24 = -27$

65. Finding the dot product: $\mathbf{u} \bullet \mathbf{v} = \left(-\frac{3}{4}\right)\left(\frac{2}{9}\right) + \left(\frac{1}{10}\right)(5) = -\frac{1}{6} + \frac{1}{2} = \frac{1}{3}$

67. Finding the dot product: $\mathbf{A} \bullet (\mathbf{B} - \mathbf{C}) = \langle -3, 4 \rangle \bullet (\langle 5, 2 \rangle - \langle 6, -1 \rangle) = \langle -3, 4 \rangle \bullet \langle -1, 3 \rangle = 3 + 12 = 15$

69. Finding the dot product: $\mathbf{B} \bullet (\mathbf{A} + \mathbf{C}) = \langle 5, 2 \rangle \bullet (\langle -3, 4 \rangle + \langle 6, -1 \rangle) = \langle 5, 2 \rangle \bullet \langle 3, 3 \rangle = 15 + 6 = 21$

71. Finding the dot product:
$$(\mathbf{A} + \mathbf{B}) \bullet (\mathbf{B} - \mathbf{C}) = (\langle -3, 4 \rangle + \langle 5, 2 \rangle) \bullet (\langle 5, 2 \rangle - \langle 6, -1 \rangle) = \langle 2, 6 \rangle \bullet \langle -1, 3 \rangle = -2 + 18 = 16$$

73. The three forces can be represented in component form as:
$$\langle 60 \cos 0°, 60 \sin 0° \rangle, \langle 90 \cos 40°, 90 \sin 40° \rangle, \langle 100 \cos 65°, 100 \sin 65° \rangle$$
The resultant force is therefore:
$$\mathbf{R} = \langle 60 \cos 0° + 90 \cos 40° + 100 \cos 65°, 60 \sin 0° + 90 \sin 40° + 100 \sin 65° \rangle \approx \langle 171.21, 148.48 \rangle$$
The magnitude is: $|\mathbf{R}| = \sqrt{171.21^2 + 148.48^2} \approx 226.6$ pounds
The direction θ can be found by:
$$\tan\theta = \frac{148.48}{171.21}$$
$$\theta = \tan^{-1}\left(\frac{148.48}{171.21}\right) \approx 40.9°$$

75. Since $\vec{PQ} = \langle x_2 - x_1, y_2 - y_1 \rangle$, using the law of cosines:
$$\left|\vec{PQ}\right|^2 = |\mathbf{A}|^2 + |\mathbf{B}|^2 - 2|\mathbf{A}||\mathbf{B}|\cos\theta$$
$$(x_2 - x_1)^2 + (y_2 - y_1)^2 = (x_1^2 + y_1^2) + (x_2^2 + y_2^2) - 2|\mathbf{A}||\mathbf{B}|\cos\theta$$
$$x_2^2 - 2x_1 x_2 + x_1^2 + y_2^2 - 2y_1 y_2 + y_1^2 = x_1^2 + y_1^2 + x_2^2 + y_2^2 - 2|\mathbf{A}||\mathbf{B}|\cos\theta$$
$$-2x_1 x_2 - 2y_1 y_2 = -2|\mathbf{A}||\mathbf{B}|\cos\theta$$
$$x_1 x_2 + y_1 y_2 = |\mathbf{A}||\mathbf{B}|\cos\theta$$
$$\langle x_1, y_1 \rangle \bullet \langle x_2, y_2 \rangle = |\mathbf{A}||\mathbf{B}|\cos\theta$$
$$\mathbf{A} \bullet \mathbf{B} = |\mathbf{A}||\mathbf{B}|\cos\theta$$

8.5 The Trigonometric Form of Complex Numbers and DeMoivre's Theorem

1. Here $r = \sqrt{\left(\sqrt{3}\right)^2 + (-1)^2} = \sqrt{4} = 2$. Thus: $\sqrt{3} - i = 2\left(\frac{\sqrt{3}}{2} - \frac{1}{2}i\right) = 2\left(\cos\frac{11\pi}{6} + i\sin\frac{11\pi}{6}\right)$

3. Converting to rectangular form: $10\left(\cos\frac{7\pi}{4} + i\sin\frac{7\pi}{4}\right) = 10\left(\frac{\sqrt{2}}{2} - \frac{\sqrt{2}}{2}i\right) = 5\sqrt{2} - 5i\sqrt{2}$

5. Here $r = \sqrt{2^2 + (-7)^2} = \sqrt{53}$. Thus:
$$2 - 7i = \sqrt{53}\left(\frac{2}{\sqrt{53}} - \frac{7}{\sqrt{53}}i\right) \approx \sqrt{53}\left(\cos 285.95° + i\sin 285.95°\right)$$

7. Here $r = 6$. Thus: $6 = 6(1 + 0i) = 6(\cos 0 + i\sin 0)$

9. Converting to rectangular form: $2\left(\cos\frac{3\pi}{2} + i\sin\frac{3\pi}{2}\right) = 2(0 - i) = -2i$

11. Here $r = \sqrt{\left(\sqrt{5}\right)^2 + 4^2} = \sqrt{21}$. Thus:

$$\sqrt{5} + 4i = \sqrt{21}\left(\frac{\sqrt{5}}{\sqrt{21}} + \frac{4}{\sqrt{21}}i\right) \approx \sqrt{21}\left(\cos 60.79° + i \sin 60.79°\right)$$

13. Converting to rectangular form: $9\left(\cos 300° + i \sin 300°\right) = 9\left(\frac{1}{2} - \frac{\sqrt{3}}{2}i\right) = \frac{9}{2} - \frac{9\sqrt{3}}{2}i$

15. Here $r = 4$. Thus: $-4 = 4(-1 + 0i) = 4\left(\cos \pi + i \sin \pi\right)$

17. Finding the product and quotient:

$$z_1 z_2 = 6 \cdot 2\left[\cos\left(\frac{\pi}{3} + \frac{\pi}{9}\right) + i \sin\left(\frac{\pi}{3} + \frac{\pi}{9}\right)\right] = 12\left(\cos \frac{4\pi}{9} + i \sin \frac{4\pi}{9}\right)$$

$$\frac{z_1}{z_2} = \frac{6}{2}\left[\cos\left(\frac{\pi}{3} - \frac{\pi}{9}\right) + i \sin\left(\frac{\pi}{3} - \frac{\pi}{9}\right)\right] = 3\left(\cos \frac{2\pi}{9} + i \sin \frac{2\pi}{9}\right)$$

19. Finding the product and quotient:

$$z_1 z_2 = 5 \cdot 1\left[\cos\left(\frac{5\pi}{6} + \frac{\pi}{8}\right) + i \sin\left(\frac{5\pi}{6} + \frac{\pi}{8}\right)\right] = 5\left(\cos \frac{23\pi}{24} + i \sin \frac{23\pi}{24}\right)$$

$$\frac{z_1}{z_2} = \frac{5}{1}\left[\cos\left(\frac{5\pi}{6} - \frac{\pi}{8}\right) + i \sin\left(\frac{5\pi}{6} - \frac{\pi}{8}\right)\right] = 5\left(\cos \frac{17\pi}{24} + i \sin \frac{17\pi}{24}\right)$$

21. Computing in rectangular form: $(3 + 3i)(1 - i) = 3 + 3i - 3i - 3i^2 = 6$

Now convert each number to trigonometric form:

$$3 + 3i = 3\sqrt{2}\left(\frac{1}{\sqrt{2}} + \frac{1}{\sqrt{2}}i\right) = 3\sqrt{2}\left(\cos \frac{\pi}{4} + i \sin \frac{\pi}{4}\right)$$

$$1 - i = \sqrt{2}\left(\frac{1}{\sqrt{2}} - \frac{1}{\sqrt{2}}i\right) = \sqrt{2}\left(\cos \frac{7\pi}{4} + i \sin \frac{7\pi}{4}\right)$$

Therefore:

$$(3 + 3i)(1 - i) = 3\sqrt{2} \cdot \sqrt{2}\left[\cos\left(\frac{\pi}{4} + \frac{7\pi}{4}\right) + i \sin\left(\frac{\pi}{4} + \frac{7\pi}{4}\right)\right]$$
$$= 6\left(\cos 2\pi + i \sin 2\pi\right)$$
$$= 6(1 + 0i)$$
$$= 6$$

23. Computing in rectangular form: $\dfrac{i}{1+i} \cdot \dfrac{1-i}{1-i} = \dfrac{i - i^2}{1 - i^2} = \dfrac{1+i}{2} = \dfrac{1}{2} + \dfrac{1}{2}i$

Now convert each number to trigonometric form:

$$i = 1(0 + 1i) = 1\left(\cos \frac{\pi}{2} + i \sin \frac{\pi}{2}\right)$$

$$1 + i = \sqrt{2}\left(\frac{1}{\sqrt{2}} + \frac{1}{\sqrt{2}}i\right) = \sqrt{2}\left(\cos \frac{\pi}{4} + i \sin \frac{\pi}{4}\right)$$

Therefore:

$$\frac{i}{1+i} = \frac{1}{\sqrt{2}}\left[\cos\left(\frac{\pi}{2} - \frac{\pi}{4}\right) + i\sin\left(\frac{\pi}{2} - \frac{\pi}{4}\right)\right]$$

$$= \frac{1}{\sqrt{2}}\left(\cos\frac{\pi}{4} + i\sin\frac{\pi}{4}\right)$$

$$= \frac{1}{\sqrt{2}}\left(\frac{1}{\sqrt{2}} + \frac{1}{\sqrt{2}}i\right)$$

$$= \tfrac{1}{2} + \tfrac{1}{2}i$$

25. Using DeMoivre's theorem:

$$\left[3\left(\cos\frac{\pi}{6} + i\sin\frac{\pi}{6}\right)\right]^5 = 3^5\left(\cos\frac{5\pi}{6} + i\sin\frac{5\pi}{6}\right) = 243\left(-\frac{\sqrt{3}}{2} + \frac{1}{2}i\right) = -\frac{243\sqrt{3}}{2} + \frac{243}{2}i$$

27. Using DeMoivre's theorem:

$$\left[\sqrt{2}\left(\cos\frac{5\pi}{4} + i\sin\frac{5\pi}{4}\right)\right]^7 = \left(\sqrt{2}\right)^7\left(\cos\frac{35\pi}{4} + i\sin\frac{35\pi}{4}\right) = 8\sqrt{2}\left(-\frac{1}{\sqrt{2}} + \frac{1}{\sqrt{2}}i\right) = -8 + 8i$$

29. Using DeMoivre's theorem:

$$\left[\sqrt{6}\left(\cos 20° + i\sin 20°\right)\right]^6 = 6^3\left(\cos 120° + i\sin 120°\right) = 216\left(-\frac{1}{2} + \frac{\sqrt{3}}{2}i\right) = -108 + 108i\sqrt{3}$$

31. First convert to trigonometric form: $1 + i = \sqrt{2}\left(\frac{1}{\sqrt{2}} + \frac{1}{\sqrt{2}}i\right) = \sqrt{2}\left(\cos\frac{\pi}{4} + i\sin\frac{\pi}{4}\right)$

Using DeMoivre's theorem:

$$(1+i)^8 = \left[\sqrt{2}\left(\cos\frac{\pi}{4} + i\sin\frac{\pi}{4}\right)\right]^8 = 2^4\left(\cos 2\pi + i\sin 2\pi\right) = 16(1 + 0i) = 16$$

33. First convert to trigonometric form: $-3 + 3i = 3\sqrt{2}\left(-\frac{1}{\sqrt{2}} + \frac{1}{\sqrt{2}}i\right) = 3\sqrt{2}\left(\cos\frac{3\pi}{4} + i\sin\frac{3\pi}{4}\right)$

Using DeMoivre's theorem:

$$(-3+3i)^5 = \left[3\sqrt{2}\left(\cos\frac{3\pi}{4} + i\sin\frac{3\pi}{4}\right)\right]^5$$

$$= \left(3\sqrt{2}\right)^5\left(\cos\frac{15\pi}{4} + i\sin\frac{15\pi}{4}\right)$$

$$= 972\sqrt{2}\left(\frac{1}{\sqrt{2}} - \frac{1}{\sqrt{2}}i\right)$$

$$= 972 - 972i$$

35. First convert to trigonometric form:

$$i = 1(1 + 0i) = 1\left(\cos\frac{\pi}{2} + i\sin\frac{\pi}{2}\right)$$

$$1 - i = \sqrt{2}\left(\frac{1}{\sqrt{2}} - \frac{1}{\sqrt{2}}i\right) = \sqrt{2}\left(\cos\frac{7\pi}{4} + i\sin\frac{7\pi}{4}\right)$$

Therefore: $\dfrac{i}{1-i} = \dfrac{1}{\sqrt{2}}\left[\cos\left(\dfrac{\pi}{2} - \dfrac{7\pi}{4}\right) + i\sin\left(\dfrac{\pi}{2} - \dfrac{7\pi}{4}\right)\right] = \dfrac{1}{\sqrt{2}}\left[\cos\left(-\dfrac{5\pi}{4}\right) + i\sin\left(-\dfrac{5\pi}{4}\right)\right]$

Using DeMoivre's theorem:

$$\left(\frac{i}{1-i}\right)^3 = \left(\frac{1}{\sqrt{2}}\right)^3\left[\cos\left(-\frac{15\pi}{4}\right)+i\sin\left(-\frac{15\pi}{4}\right)\right] = \frac{1}{2\sqrt{2}}\left(\frac{1}{\sqrt{2}}+\frac{1}{\sqrt{2}}i\right) = \frac{1}{4}+\frac{1}{4}i$$

37.   First convert to trigonometric form:

$$1-i\sqrt{3} = 2\left(\frac{1}{2}-\frac{\sqrt{3}}{2}i\right) = 2\left(\cos\frac{5\pi}{3}+i\sin\frac{5\pi}{3}\right)$$

$$-1+i = \sqrt{2}\left(-\frac{1}{\sqrt{2}}+\frac{1}{\sqrt{2}}i\right) = \sqrt{2}\left(\cos\frac{3\pi}{4}+i\sin\frac{3\pi}{4}\right)$$

Therefore: $$\frac{1-i\sqrt{3}}{-1+i} = \frac{2}{\sqrt{2}}\left[\cos\left(\frac{5\pi}{3}-\frac{3\pi}{4}\right)+i\sin\left(\frac{5\pi}{3}-\frac{3\pi}{4}\right)\right] = \sqrt{2}\left[\cos\frac{11\pi}{12}+i\sin\frac{11\pi}{12}\right]$$

Using DeMoivre's theorem: $$\left(\frac{1-i\sqrt{3}}{-1+i}\right)^4 = \left(\sqrt{2}\right)^4\left[\cos\frac{11\pi}{3}+i\sin\frac{11\pi}{3}\right] = 4\left(\frac{1}{2}-\frac{\sqrt{3}}{2}i\right) = 2-2i\sqrt{3}$$

39.   First convert to trigonometric form: $$3+3i = 3\sqrt{2}\left(\frac{1}{\sqrt{2}}+\frac{1}{\sqrt{2}}i\right) = 3\sqrt{2}\left(\cos\frac{\pi}{4}+i\sin\frac{\pi}{4}\right)$$

The fourth roots of $3+3i$ are given by:

$$w_k = \sqrt[4]{3\sqrt{2}}\left(\cos\frac{\frac{\pi}{4}+2k\pi}{4}+i\sin\frac{\frac{\pi}{4}+2k\pi}{4}\right) = \sqrt[8]{18}\left[\cos\left(\frac{\pi}{16}+\frac{k\pi}{2}\right)+i\sin\left(\frac{\pi}{16}+\frac{k\pi}{2}\right)\right]$$

Substituting $k = 0, 1, 2, 3$:

$$w_0 = \sqrt[8]{18}\left(\cos\frac{\pi}{16}+i\sin\frac{\pi}{16}\right)$$

$$w_1 = \sqrt[8]{18}\left(\cos\frac{9\pi}{16}+i\sin\frac{9\pi}{16}\right)$$

$$w_2 = \sqrt[8]{18}\left(\cos\frac{17\pi}{16}+i\sin\frac{17\pi}{16}\right)$$

$$w_3 = \sqrt[8]{18}\left(\cos\frac{25\pi}{16}+i\sin\frac{25\pi}{16}\right)$$

41.   First convert to trigonometric form: $1 = 1(\cos 0+i\sin 0)$
The sixth roots of 1 are given by:

$$w_k = \sqrt[6]{1}\left(\cos\frac{0+2k\pi}{6}+i\sin\frac{0+2k\pi}{6}\right) = \cos\left(\frac{k\pi}{3}\right)+i\sin\left(\frac{k\pi}{3}\right)$$

Substituting $k = 0, 1, 2, 3, 4, 5$:

$$w_0 = \cos 0+i\sin 0 = 1$$

$$w_1 = \cos\frac{\pi}{3}+i\sin\frac{\pi}{3} = \frac{1}{2}+\frac{\sqrt{3}}{2}i$$

$$w_2 = \cos\frac{2\pi}{3}+i\sin\frac{2\pi}{3} = -\frac{1}{2}+\frac{\sqrt{3}}{2}i$$

$$w_3 = \cos\pi+i\sin\pi = -1$$

$$w_4 = \cos\frac{4\pi}{3}+i\sin\frac{4\pi}{3} = -\frac{1}{2}-\frac{\sqrt{3}}{2}i$$

$$w_5 = \cos\frac{5\pi}{3}+i\sin\frac{5\pi}{3} = \frac{1}{2}-\frac{\sqrt{3}}{2}i$$

43. First convert to trigonometric form: $-\sqrt{3}+i = 2\left(-\dfrac{\sqrt{3}}{2}+\dfrac{1}{2}i\right) = 2\left(\cos\dfrac{5\pi}{6}+i\sin\dfrac{5\pi}{6}\right)$

The fifth roots of $-\sqrt{3}+i$ are given by:

$$w_k = \sqrt[5]{2}\left(\cos\dfrac{\dfrac{5\pi}{6}+2k\pi}{5}+i\sin\dfrac{\dfrac{5\pi}{6}+2k\pi}{5}\right) = \sqrt[5]{2}\left[\cos\left(\dfrac{\pi}{6}+\dfrac{2k\pi}{5}\right)+i\sin\left(\dfrac{\pi}{6}+\dfrac{2k\pi}{5}\right)\right]$$

Substituting $k = 0, 1, 2, 3, 4$:

$$w_0 = \sqrt[5]{2}\left(\cos\dfrac{\pi}{6}+i\sin\dfrac{\pi}{6}\right)$$

$$w_1 = \sqrt[5]{2}\left(\cos\dfrac{17\pi}{30}+i\sin\dfrac{17\pi}{30}\right)$$

$$w_2 = \sqrt[5]{2}\left(\cos\dfrac{29\pi}{30}+i\sin\dfrac{29\pi}{30}\right)$$

$$w_3 = \sqrt[5]{2}\left(\cos\dfrac{41\pi}{30}+i\sin\dfrac{41\pi}{30}\right)$$

$$w_4 = \sqrt[5]{2}\left(\cos\dfrac{53\pi}{30}+i\sin\dfrac{53\pi}{30}\right)$$

45. First convert to trigonometric form: $1+i\sqrt{3} = 2\left(\dfrac{1}{2}+\dfrac{\sqrt{3}}{2}i\right) = 2\left(\cos\dfrac{\pi}{3}+i\sin\dfrac{\pi}{3}\right)$

The cube roots of $1+i\sqrt{3}$ are given by:

$$w_k = \sqrt[3]{2}\left(\cos\dfrac{\dfrac{\pi}{3}+2k\pi}{3}+i\sin\dfrac{\dfrac{\pi}{3}+2k\pi}{3}\right) = \sqrt[3]{2}\left[\cos\left(\dfrac{\pi}{9}+\dfrac{2k\pi}{3}\right)+i\sin\left(\dfrac{\pi}{9}+\dfrac{2k\pi}{3}\right)\right]$$

Substituting $k = 0, 1, 2$:

$$w_0 = \sqrt[3]{2}\left(\cos\dfrac{\pi}{9}+i\sin\dfrac{\pi}{9}\right) \approx 1.18+0.43i$$

$$w_1 = \sqrt[3]{2}\left(\cos\dfrac{7\pi}{9}+i\sin\dfrac{7\pi}{9}\right) \approx -0.97+0.81i$$

$$w_2 = \sqrt[3]{2}\left(\cos\dfrac{13\pi}{9}+i\sin\dfrac{13\pi}{9}\right) \approx -0.22-1.24i$$

47. First convert to trigonometric form: $i = 1(0+1i) = 1\left(\cos\dfrac{\pi}{2}+i\sin\dfrac{\pi}{2}\right)$

The fourth roots of i are given by:

$$w_k = \sqrt[4]{1}\left(\cos\dfrac{\dfrac{\pi}{2}+2k\pi}{4}+i\sin\dfrac{\dfrac{\pi}{2}+2k\pi}{4}\right) = \cos\left(\dfrac{\pi}{8}+\dfrac{k\pi}{2}\right)+i\sin\left(\dfrac{\pi}{8}+\dfrac{k\pi}{2}\right)$$

Substituting $k = 0, 1, 2, 3$:

$$w_0 = \cos\frac{\pi}{8} + i\sin\frac{\pi}{8}$$

$$w_1 = \cos\frac{5\pi}{8} + i\sin\frac{5\pi}{8}$$

$$w_2 = \cos\frac{9\pi}{8} + i\sin\frac{9\pi}{8}$$

$$w_3 = \cos\frac{13\pi}{8} + i\sin\frac{13\pi}{8}$$

For each of these values, use the half-angle formula (note the signs due to quadrants):

$$\cos\frac{\pi}{8} = \sqrt{\frac{1+\cos\frac{\pi}{4}}{2}} = \sqrt{\frac{1+\frac{\sqrt{2}}{2}}{2}} = \sqrt{\frac{2+\sqrt{2}}{4}} = \frac{\sqrt{2+\sqrt{2}}}{2}$$

$$\sin\frac{\pi}{8} = \sqrt{\frac{1-\cos\frac{\pi}{4}}{2}} = \sqrt{\frac{1-\frac{\sqrt{2}}{2}}{2}} = \sqrt{\frac{2-\sqrt{2}}{4}} = \frac{\sqrt{2-\sqrt{2}}}{2}$$

$$\cos\frac{5\pi}{8} = -\sqrt{\frac{1+\cos\frac{5\pi}{4}}{2}} = -\sqrt{\frac{1-\frac{\sqrt{2}}{2}}{2}} = -\sqrt{\frac{2-\sqrt{2}}{4}} = -\frac{\sqrt{2-\sqrt{2}}}{2}$$

$$\sin\frac{5\pi}{8} = \sqrt{\frac{1-\cos\frac{5\pi}{4}}{2}} = \sqrt{\frac{1+\frac{\sqrt{2}}{2}}{2}} = \sqrt{\frac{2+\sqrt{2}}{4}} = \frac{\sqrt{2+\sqrt{2}}}{2}$$

$$\cos\frac{9\pi}{8} = -\sqrt{\frac{1+\cos\frac{9\pi}{4}}{2}} = -\sqrt{\frac{1+\frac{\sqrt{2}}{2}}{2}} = -\sqrt{\frac{2+\sqrt{2}}{4}} = -\frac{\sqrt{2+\sqrt{2}}}{2}$$

$$\sin\frac{9\pi}{8} = -\sqrt{\frac{1-\cos\frac{9\pi}{4}}{2}} = -\sqrt{\frac{1-\frac{\sqrt{2}}{2}}{2}} = -\sqrt{\frac{2-\sqrt{2}}{4}} = -\frac{\sqrt{2-\sqrt{2}}}{2}$$

$$\cos\frac{13\pi}{8} = \sqrt{\frac{1+\cos\frac{13\pi}{4}}{2}} = \sqrt{\frac{1-\frac{\sqrt{2}}{2}}{2}} = \sqrt{\frac{2-\sqrt{2}}{4}} = \frac{\sqrt{2-\sqrt{2}}}{2}$$

$$\sin\frac{13\pi}{8} = -\sqrt{\frac{1-\cos\frac{13\pi}{4}}{2}} = -\sqrt{\frac{1+\frac{\sqrt{2}}{2}}{2}} = -\sqrt{\frac{2+\sqrt{2}}{4}} = -\frac{\sqrt{2+\sqrt{2}}}{2}$$

Therefore:

$$w_0 = \cos\frac{\pi}{8} + i\sin\frac{\pi}{8} = \frac{\sqrt{2+\sqrt{2}}}{2} + \frac{\sqrt{2-\sqrt{2}}}{2}i$$

$$w_1 = \cos\frac{5\pi}{8} + i\sin\frac{5\pi}{8} = -\frac{\sqrt{2-\sqrt{2}}}{2} + \frac{\sqrt{2+\sqrt{2}}}{2}i$$

$$w_2 = \cos\frac{9\pi}{8} + i\sin\frac{9\pi}{8} = -\frac{\sqrt{2+\sqrt{2}}}{2} - \frac{\sqrt{2-\sqrt{2}}}{2}i$$

$$w_3 = \cos\frac{13\pi}{8} + i\sin\frac{13\pi}{8} = \frac{\sqrt{2-\sqrt{2}}}{2} - \frac{\sqrt{2+\sqrt{2}}}{2}i$$

49. Since $1 = 1(\cos 0 + i\sin 0)$, the cube roots of 1 are given by:

$$w_k = \sqrt[3]{1}\left(\cos\frac{0+2k\pi}{3} + i\sin\frac{0+2k\pi}{3}\right) = \cos\frac{2\pi k}{3} + i\sin\frac{2\pi k}{3}$$

Substituting $k = 0, 1, 2$:

$$w_0 = \cos 0 + i\sin 0 = 1$$

$$w_1 = \cos\frac{2\pi}{3} + i\sin\frac{2\pi}{3} = -\frac{1}{2} + \frac{\sqrt{3}}{2}i$$

$$w_2 = \cos\frac{4\pi}{3} + i\sin\frac{4\pi}{3} = -\frac{1}{2} - \frac{\sqrt{3}}{2}i$$

Therefore: $w_0 + w_1 + w_2 = 1 + \left(-\frac{1}{2} + \frac{\sqrt{3}}{2}i\right) + \left(-\frac{1}{2} - \frac{\sqrt{3}}{2}i\right) = 0$

51. Using the hint:

$$\frac{r_1(\cos\theta_1 + i\sin\theta_1)}{r_2(\cos\theta_2 + i\sin\theta_2)} \cdot \frac{\cos\theta_2 - i\sin\theta_2}{\cos\theta_2 - i\sin\theta_2}$$

$$= \frac{r_1(\cos\theta_1\cos\theta_2 + i\sin\theta_1\cos\theta_2 - i\cos\theta_1\sin\theta_2 + \sin\theta_1\sin\theta_2)}{r_2(\cos^2\theta_2 + \sin^2\theta_2)}$$

$$= \frac{r_1}{r_2}\left[(\cos\theta_1\cos\theta_2 + \sin\theta_1\sin\theta_2) + i(\sin\theta_1\cos\theta_2 - \cos\theta_1\sin\theta_2)\right]$$

$$= \frac{r_1}{r_2}\left[\cos(\theta_1 - \theta_2) + i\sin(\theta_1 - \theta_2)\right]$$

53. If $z = a + bi$, $|z| = \sqrt{a^2 + b^2}$. If $z = r(\cos\theta + i\sin\theta)$, then $|z| = r$. Thus the trigonometric form does not require any computation.

8.6 Polar Coordinates

1. Plotting the point:

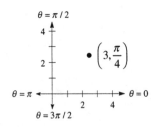

3.   Plotting the point:

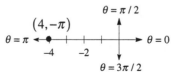

5.   Plotting the point:

7.   Two additional points are $\left(5,-\dfrac{5\pi}{3}\right)$ and $\left(-5,\dfrac{4\pi}{3}\right)$.

9.   Two additional points are $\left(2,\dfrac{3\pi}{2}\right)$ and $\left(-2,-\dfrac{3\pi}{2}\right)$.

11.   The rectangular point is: $\left(2\cos\dfrac{2\pi}{3}, 2\sin\dfrac{2\pi}{3}\right) = \left(2\cdot\left(-\dfrac{1}{2}\right), 2\cdot\dfrac{\sqrt{3}}{2}\right) = \left(-1,\sqrt{3}\right)$

13.   Since $r = \sqrt{\left(-4\sqrt{3}\right)^2 + (-4)^2} = \sqrt{64} = 8$ and $\theta = \dfrac{7\pi}{6}$, the polar point is $\left(8,\dfrac{7\pi}{6}\right)$.

15.   The rectangular point is: $\left(3\cos\dfrac{\pi}{7}, 3\sin\dfrac{\pi}{7}\right) \approx (2.70, 1.30)$

17.   Since $r = \sqrt{(-6)^2} = \sqrt{36} = 6$ and $\theta = \pi$, the polar point is $(6,\pi)$.

19.   The rectangular point is: $(2\cos 3, 2\sin 3) \approx (-1.98, 0.28)$

21.   Since $r = \sqrt{(-1)^2 + \left(\dfrac{\pi}{4}\right)^2} = \sqrt{1+\dfrac{\pi^2}{16}} = \dfrac{\sqrt{\pi^2+16}}{6}$ and $\theta = \pi + \tan^{-1}\left(-\dfrac{\pi}{4}\right) \approx 2.48$, the polar

point is $\left(\dfrac{\sqrt{\pi^2+16}}{6}, 2.48\right)$.

23.   Converting to rectangular form:
$$r = 5\cos\theta$$
$$r^2 = 5r\cos\theta$$
$$x^2 + y^2 = 5x$$
This is the equation of a circle.

25.   Converting to rectangular form:
$$\theta = \dfrac{\pi}{3}$$
$$\tan\theta = \tan\dfrac{\pi}{3}$$
$$\dfrac{y}{x} = \sqrt{3}$$
$$y = x\sqrt{3}$$
This is the equation of a line.

27.   Converting to rectangular form:
$$r\sin\theta = -1$$
$$y = -1$$
This is the equation of a line.

29.   Converting to rectangular form:
$$r = \tan\theta$$
$$\sqrt{x^2 + y^2} = \frac{y}{x}$$
$$x^2 + y^2 = \frac{y^2}{x^2}$$
$$x^4 + x^2 y^2 = y^2$$
$$x^4 = y^2\left(1 - x^2\right)$$
$$y^2 = \frac{x^4}{1 - x^2}$$

31.   Converting to rectangular form:
$$r = \frac{2}{1 - \sin\theta}$$
$$r - r\sin\theta = 2$$
$$r = 2 + r\sin\theta$$
$$\sqrt{x^2 + y^2} = 2 + y$$
$$x^2 + y^2 = y^2 + 4y + 4$$
$$x^2 = 4y + 4$$
$$4y = x^2 - 4$$
$$y = \tfrac{1}{4}x^2 - 1$$
This is the equation of a parabola.

33.   Converting to polar form:
$$x^2 + y^2 = 16$$
$$r^2 = 16$$
$$r = 4$$

35.   Converting to polar form:
$$2xy = 1$$
$$2(r\cos\theta)(r\sin\theta) = 1$$
$$r^2 \sin 2\theta = 1$$
$$r^2 = \csc 2\theta$$

37.   Converting to polar form:
$$x^2 + 2x + y^2 + 2y = 0$$
$$r^2 \cos^2\theta + 2r\cos\theta + r^2 \sin^2\theta + 2r\sin\theta = 0$$
$$r^2\left(\cos^2\theta + \sin^2\theta\right) + 2r(\sin\theta + \cos\theta) = 0$$
$$r + 2(\sin\theta + \cos\theta) = 0$$
$$r = -2(\sin\theta + \cos\theta)$$

39.   Converting to polar form:
$$y = 4$$
$$r\sin\theta = 4$$
$$r = 4\csc\theta$$

41. Graphing the curve:

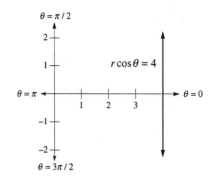

43. Graphing the curve:

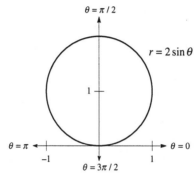

45. Graphing the curve:

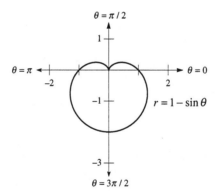

47. Graphing the curve:

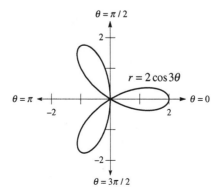

49 . Graphing the curve:

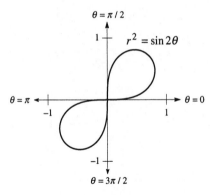

51 . Graphing the curve:

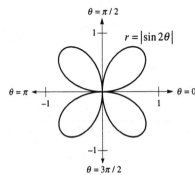

53 . Graphing the curve:

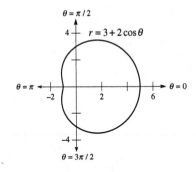

55 . Graphing the curve:

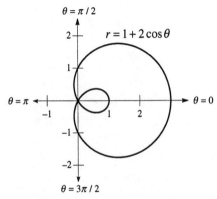

57. Graphing the curve:

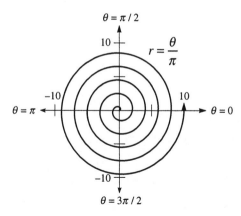

59. Graphing the curve:

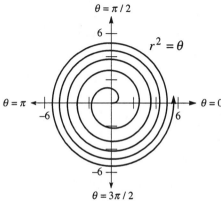

61. The polar coordinate system is a completely different graphing system from the rectangular coordinate system, so there is no reason $y = 4\sin x$ would look like $y = 4\sin\theta$.

63. There is no vertical line test for the graph in polar coordinates.

Chapter 8 Review Exercises

1. Using the addition formula:
$$\cos 15° = \cos(45° - 30°) = \cos 45° \cos 30° + \sin 45° \sin 30° = \frac{\sqrt{2}}{2} \cdot \frac{\sqrt{3}}{2} + \frac{\sqrt{2}}{2} \cdot \frac{1}{2} = \frac{\sqrt{6} + \sqrt{2}}{4}$$

2. Using the addition formula:
$$\sin \frac{5\pi}{12} = \sin\left(\frac{\pi}{4} + \frac{\pi}{6}\right) = \sin \frac{\pi}{4} \cos \frac{\pi}{6} + \cos \frac{\pi}{4} \sin \frac{\pi}{6} = \frac{\sqrt{2}}{2} \cdot \frac{\sqrt{3}}{2} + \frac{\sqrt{2}}{2} \cdot \frac{1}{2} = \frac{\sqrt{6} + \sqrt{2}}{4}$$

3. Using the addition formula: $\tan \dfrac{7\pi}{12} = \tan\left(\dfrac{\pi}{3} + \dfrac{\pi}{4}\right) = \dfrac{\tan \dfrac{\pi}{3} + \tan \dfrac{\pi}{4}}{1 - \tan \dfrac{\pi}{3} \tan \dfrac{\pi}{4}} = \dfrac{\sqrt{3} + 1}{1 - \sqrt{3} \cdot 1} = \dfrac{1 + \sqrt{3}}{1 - \sqrt{3}}$

4.   Using the addition formula:
$$\cos 195° = \cos(135° + 60°)$$
$$= \cos 135° \cos 60° - \sin 135° \sin 60°$$
$$= -\frac{\sqrt{2}}{2} \cdot \frac{1}{2} - \frac{\sqrt{2}}{2} \cdot \frac{\sqrt{3}}{2}$$
$$= \frac{-\sqrt{2} - \sqrt{6}}{4}$$

5.   Verifying the identity: $\sin(3\pi - x) = \sin 3\pi \cos x - \cos 3\pi \sin x = 0 \bullet \cos x - (-1) \bullet \sin x = \sin x$

6.   Verifying the identity: $\cos(x + \pi) = \cos x \cos \pi - \sin x \sin \pi = \cos x \bullet (-1) - \sin x \bullet 0 = -\cos x$

7.   Verifying the identity: $\tan\left(\theta - \dfrac{\pi}{4}\right) = \dfrac{\tan\theta - \tan\dfrac{\pi}{4}}{1 + \tan\theta \tan\dfrac{\pi}{4}} = \dfrac{\tan\theta - 1}{1 + \tan\theta \bullet 1} = \dfrac{\tan\theta - 1}{\tan\theta + 1}$

8.   Verifying the identity:
$$\sin(A + B)\sin(A - B) = (\sin A \cos B + \cos A \sin B)(\sin A \cos B - \cos A \sin B)$$
$$= \sin^2 A \cos^2 B - \cos^2 A \sin^2 B$$
$$= \sin^2 A(1 - \sin^2 B) - (1 - \sin^2 A)\sin^2 B$$
$$= \sin^2 A - \sin^2 A \sin^2 B - \sin^2 B + \sin^2 A \sin^2 B$$
$$= \sin^2 A - \sin^2 B$$

9.   Verifying the identity:
$$\sin\left(\frac{\pi}{6} + x\right) = \sin\frac{\pi}{6}\cos x + \cos\frac{\pi}{6}\sin x = \tfrac{1}{2}\cos x + \frac{\sqrt{3}}{2}\sin x = \tfrac{1}{2}\left(\cos x + \sqrt{3}\sin x\right)$$

10.   Verifying the identity:
$$\sin\left(\frac{\pi}{2} + \theta\right) + \cos(\pi - \theta) = \sin\frac{\pi}{2}\cos\theta + \cos\frac{\pi}{2}\sin\theta + \cos\pi\cos\theta + \sin\pi\sin\theta$$
$$= 1 \bullet \cos\theta + 0 \bullet \sin\theta + (-1) \bullet \cos\theta + 0 \bullet \sin\theta$$
$$= \cos\theta - \cos\theta$$
$$= 1$$

11.   Verifying the identity:
$$\sin(\alpha + \beta)\cos(\alpha - \beta)$$
$$= (\sin\alpha\cos\beta + \cos\alpha\sin\beta)(\cos\alpha\cos\beta + \sin\alpha\sin\beta)$$
$$= \sin\alpha\cos\alpha\cos^2\beta + \sin\beta\cos\beta\cos^2\alpha + \sin^2\alpha\sin\beta\cos\beta + \sin^2\beta\sin\alpha\cos\alpha$$
$$= \sin\alpha\cos\alpha(\cos^2\beta + \sin^2\beta) + \sin\beta\cos\beta(\cos^2\alpha + \sin^2\alpha)$$
$$= \sin\alpha\cos\alpha + \sin\beta\cos\beta$$

12.   Verifying the identity: $\dfrac{1 + \cos 2x}{\sin^2 2x} = \dfrac{1 + 2\cos^2 x - 1}{(2\sin x \cos x)^2} = \dfrac{2\cos^2 x}{4\sin^2 x \cos^2 x} = \dfrac{1}{2\sin^2 x} = \dfrac{\csc^2 x}{2}$

13.   Verifying the identity: $\dfrac{1 - \cos 2\theta}{\sin 2\theta} = \dfrac{1 - (1 - 2\sin^2\theta)}{2\sin\theta\cos\theta} = \dfrac{2\sin^2\theta}{2\sin\theta\cos\theta} = \dfrac{\sin\theta}{\cos\theta} = \tan\theta$

14.   Verifying the identity: $\dfrac{\sin 2\theta}{1 + \cos 2\theta} = \dfrac{2\sin\theta\cos\theta}{1 + 2\cos^2\theta - 1} = \dfrac{2\sin\theta\cos\theta}{2\cos^2\theta} = \dfrac{\sin\theta}{\cos\theta} = \tan\theta$

15. Verifying the identity:

$$\tan 2A - \tan A = \frac{2\tan A}{1-\tan^2 A} - \tan A$$

$$= \frac{2\tan A - \tan A + \tan^3 A}{1-\tan^2 A}$$

$$= \frac{\tan A + \tan^3 A}{1-\tan^2 A}$$

$$= \frac{\tan A\left(1+\tan^2 A\right)}{1-\tan^2 A}$$

$$= \frac{\tan A \sec^2 A}{1-\tan^2 A}$$

$$\tan A \sec 2A = \frac{\tan A}{\cos 2A} = \frac{\tan A}{\cos^2 A - \sin^2 A} \div \frac{\cos^2 A}{\cos^2 A} = \frac{\dfrac{\tan A}{\cos^2 A}}{1-\dfrac{\sin^2 A}{\cos^2 A}} = \frac{\tan A \sec^2 A}{1-\tan^2 A}$$

Since both sides simplify to the same quantity, the identity is verified.

16. Verifying the identity: $\dfrac{2}{\cot \beta - \tan \beta} = \dfrac{2}{\dfrac{1}{\tan \beta} - \tan \beta} \bullet \dfrac{\tan \beta}{\tan \beta} = \dfrac{2\tan \beta}{1-\tan^2 \beta} = \tan 2\beta$

17. Solving the equation:

$$\cos 2x = \sin x$$
$$1 - 2\sin^2 x = \sin x$$
$$2\sin^2 x + \sin x - 1 = 0$$
$$(2\sin x - 1)(\sin x + 1) = 0$$
$$\sin x = -1, \tfrac{1}{2}$$
$$x = \frac{\pi}{6}, \frac{5\pi}{6}, \frac{3\pi}{2}$$

18. Solving the equation:

$$2\cos^2 x + \cos 2x = 0$$
$$2\cos^2 x + 2\cos^2 x - 1 = 0$$
$$4\cos^2 x = 1$$
$$\cos^2 x = \tfrac{1}{4}$$
$$\cos x = \pm \tfrac{1}{2}$$
$$x = \frac{\pi}{3}, \frac{2\pi}{3}, \frac{4\pi}{3}, \frac{5\pi}{3}$$

19. Solving the equation:
$$\cos 2x = 2\sin^2 x$$
$$1 - 2\sin^2 x = 2\sin^2 x$$
$$4\sin^2 x = 1$$
$$\sin^2 x = \tfrac{1}{4}$$
$$\sin x = \pm\tfrac{1}{2}$$
$$x = \frac{\pi}{6}, \frac{5\pi}{6}, \frac{7\pi}{6}, \frac{11\pi}{6}$$

20. Solving the equation:
$$\cos 2x + 2 = 3\cos x$$
$$2\cos^2 x - 1 + 2 = 3\cos x$$
$$2\cos^2 x - 3\cos x + 1 = 0$$
$$(2\cos x - 1)(\cos x - 1) = 0$$
$$\cos x = \tfrac{1}{2}, 1$$
$$x = 0, \frac{\pi}{3}, \frac{5\pi}{3}$$

21. Solving the equation:
$$\sin 4x = \cos 2x$$
$$2\sin 2x \cos 2x = \cos 2x$$
$$2\sin 2x \cos 2x - \cos 2x = 0$$
$$\cos 2x(2\sin 2x - 1) = 0$$
$$\cos 2x = 0, \sin 2x = \tfrac{1}{2}$$
$$2x = \frac{\pi}{6}, \frac{\pi}{2}, \frac{5\pi}{6}, \frac{3\pi}{2}, \frac{13\pi}{6}, \frac{5\pi}{2}, \frac{17\pi}{6}, \frac{7\pi}{2}$$
$$x = \frac{\pi}{12}, \frac{\pi}{4}, \frac{5\pi}{12}, \frac{3\pi}{4}, \frac{13\pi}{12}, \frac{5\pi}{4}, \frac{17\pi}{12}, \frac{7\pi}{4}$$

22. Solving the equation:
$$\sin 2x = 2\cos x$$
$$2\sin x \cos x = 2\cos x$$
$$2\sin x \cos x - 2\cos x = 0$$
$$2\cos x(\sin x - 1) = 0$$
$$\cos x = 0, \sin x = 1$$
$$x = \frac{\pi}{2}, \frac{3\pi}{2}$$

23. Solving the equation:

$$\cos x = \cos \frac{x}{2}$$

$$\cos^2 x = \cos^2 \frac{x}{2}$$

$$\cos^2 x = \frac{1 + \cos x}{2}$$

$$2\cos^2 x = 1 + \cos x$$

$$2\cos^2 x - \cos x - 1 = 0$$

$$(2\cos x + 1)(\cos x - 1) = 0$$

$$\cos x = -\tfrac{1}{2}, 1$$

$$x = 0, \frac{2\pi}{3}, \frac{4\pi}{3}$$

Upon checking, $x = 0, \dfrac{4\pi}{3}$ are the solutions.

24. Solving the equation:

$$\cos 4x - 7\cos 2x = 8$$

$$2\cos^2 2x - 1 - 7\cos 2x = 8$$

$$2\cos^2 2x - 7\cos 2x - 9 = 0$$

$$(2\cos 2x - 9)(\cos 2x + 1) = 0$$

$$\cos 2x = -1 \qquad \left(\cos 2x = \tfrac{9}{2} \text{ is impossible}\right)$$

$$2x = \pi, 3\pi$$

$$x = \frac{\pi}{2}, \frac{3\pi}{2}$$

25. Solving the equation:

$$2 - \sin^2 x = 2\cos^2 \frac{x}{2}$$

$$2 - \left(1 - \cos^2 x\right) = 2 \cdot \frac{1 + \cos x}{2}$$

$$1 + \cos^2 x = 1 + \cos x$$

$$\cos^2 x - \cos x = 0$$

$$\cos x(\cos x - 1) = 0$$

$$\cos x = 0, 1$$

$$x = 0, \frac{\pi}{2}, \frac{3\pi}{2}$$

26. Solving the equation:

$$\sin 2x + \sqrt{2}\sin x = 0$$

$$2\sin x \cos x + \sqrt{2}\sin x = 0$$

$$\sin x\left(2\cos x + \sqrt{2}\right) = 0$$

$$\sin x = 0, \cos x = -\frac{\sqrt{2}}{2}$$

$$x = 0, \frac{3\pi}{4}, \pi, \frac{5\pi}{4}$$

27. Since $\sin\theta = -\frac{2}{5}$ and θ is in quadrant III, $\cos\theta = -\frac{\sqrt{21}}{5}$. Using the double-angle formula:

$$\sin 2\theta = 2\sin\theta\cos\theta = 2\left(-\frac{2}{5}\right)\left(-\frac{\sqrt{21}}{5}\right) = \frac{4\sqrt{21}}{25}$$

28. Since $\cos\theta = \frac{3}{7}$ and θ is in quadrant IV, $\sin\theta = -\frac{\sqrt{40}}{7}$. Using the double-angle formula:

$$\cos 2\theta = \cos^2\theta - \sin^2\theta = \left(\frac{3}{7}\right)^2 - \left(-\frac{\sqrt{40}}{7}\right)^2 = \frac{9}{49} - \frac{40}{49} = -\frac{31}{49}$$

29. Using the double-angle formula: $\tan 2\theta = \dfrac{2\tan\theta}{1-\tan^2\theta} = \dfrac{2 \cdot \frac{4}{9}}{1-\left(\frac{4}{9}\right)^2} = \dfrac{\frac{8}{9}}{1-\frac{16}{81}} \cdot \dfrac{81}{81} = \dfrac{72}{81-16} = \dfrac{72}{65}$

30. Since $\tan\theta = -\frac{1}{4}$ and $\frac{\pi}{2} < \theta < \pi$, $\sin\theta = \frac{1}{\sqrt{17}}$ and $\cos\theta = -\frac{4}{\sqrt{17}}$.

   a. Using the double-angle formula: $\sin 2\theta = 2\sin\theta\cos\theta = 2\left(\frac{1}{\sqrt{17}}\right)\left(-\frac{4}{\sqrt{17}}\right) = -\frac{8}{17}$

   b. Using the double-angle formula:

$$\cos 2\theta = \cos^2\theta - \sin^2\theta = \left(-\frac{4}{\sqrt{17}}\right)^2 - \left(\frac{1}{\sqrt{17}}\right)^2 = \frac{16}{17} - \frac{1}{17} = \frac{15}{17}$$

31. Since $\sin\theta = -\frac{3}{5}$ and $\frac{3\pi}{2} < \theta < 2\pi$, $\cos\theta = \frac{4}{5}$. Also note that $\frac{3\pi}{4} < \frac{\theta}{2} < \pi$, so $\sin\frac{\theta}{2} > 0$ while $\cos\frac{\theta}{2} < 0$.

   a. Using the half-angle formula: $\sin\dfrac{\theta}{2} = \sqrt{\dfrac{1-\cos\theta}{2}} = \sqrt{\dfrac{1-\frac{4}{5}}{2}} = \sqrt{\dfrac{1}{10}} = \dfrac{1}{\sqrt{10}}$

   b. Using the half-angle formula: $\cos\dfrac{\theta}{2} = -\sqrt{\dfrac{1+\cos\theta}{2}} = -\sqrt{\dfrac{1+\frac{4}{5}}{2}} = -\sqrt{\dfrac{9}{10}} = -\dfrac{3}{\sqrt{10}}$

   c. Computing the value: $\tan\dfrac{\theta}{2} = \dfrac{\sin\frac{\theta}{2}}{\cos\frac{\theta}{2}} = \dfrac{\frac{1}{\sqrt{10}}}{-\frac{3}{\sqrt{10}}} = -\dfrac{1}{3}$

32. The other side of the triangle is $\sqrt{25-x^2}$. Using the double-angle formula:

$$\sin 2\theta = 2\sin\theta\cos\theta = 2 \cdot \frac{x}{5} \cdot \frac{\sqrt{25-x^2}}{5} = \frac{2x\sqrt{25-x^2}}{25}$$

33. The other side of the triangle is $\sqrt{25-x^2}$. Using the double-angle formula:

$$\cos 2\theta = \cos^2\theta - \sin^2\theta = \left(\frac{\sqrt{25-x^2}}{5}\right)^2 - \left(\frac{x}{5}\right)^2 = \frac{25-x^2-x^2}{25} = \frac{25-2x^2}{25}$$

34. Using the double-angle formula: $\cos 2\theta = 1-2\sin^2\theta = 1-2\left(\frac{a}{7}\right)^2 = 1-\frac{2a^2}{49} = \frac{49-2a^2}{49}$

35. Since $\cos\theta = \dfrac{a}{5}$, $\sin\theta = \dfrac{\sqrt{25-a^2}}{5}$ and thus $\tan\theta = \dfrac{\sqrt{25-a^2}}{a}$. Using the double-angle formula:

$$\tan 2\theta = \frac{2\tan\theta}{1-\tan^2\theta} = \frac{2\cdot\dfrac{\sqrt{25-a^2}}{a}}{1-\left(\dfrac{\sqrt{25-a^2}}{a}\right)^2}\cdot\frac{a^2}{a^2} = \frac{2a\sqrt{25-a^2}}{a^2-\left(25-a^2\right)} = \frac{2a\sqrt{25-a^2}}{2a^2-25}$$

36. Using the addition and double-angle formulas:

$$\cos 3\theta = \cos(2\theta+\theta)$$
$$= \cos 2\theta\cos\theta - \sin 2\theta\sin\theta$$
$$= \left(2\cos^2\theta-1\right)\cos\theta - 2\sin^2\theta\cos\theta$$
$$= 2\cos^3\theta - \cos\theta - 2\left(1-\cos^2\theta\right)\cos\theta$$
$$= 2\cos^3\theta - \cos\theta - 2\cos\theta + 2\cos^3\theta$$
$$= 4\cos^3\theta - 3\cos\theta$$

37. First note that $C = 180° - 70° - 32° = 78°$. Using the law of sines:

$$\frac{\sin 70°}{14} = \frac{\sin 32°}{b} \qquad\qquad \frac{\sin 70°}{14} = \frac{\sin 78°}{c}$$
$$b\sin 70° = 14\sin 32° \qquad\qquad c\sin 70° = 14\sin 78°$$
$$b = \frac{14\sin 32°}{\sin 70°} \approx 7.9 \qquad\qquad c = \frac{14\sin 78°}{\sin 70°} \approx 14.6$$

38. Using the law of cosines:

$$a^2 = 8^2 + 15^2 - 2(8)(15)\cos 105° \approx 351.12$$
$$a \approx 18.7$$

Using the law of sines:

$$\frac{\sin 105°}{18.7} = \frac{\sin B}{8}$$
$$18.7\sin B = 8\sin 105°$$
$$\sin B = \frac{8\sin 105°}{18.7} \approx 0.4124$$
$$B \approx 24.4°$$

Finally $C = 180° - 105° - 24.4° = 50.6°$.

39. Using the law of cosines:

$$b^2 = 9^2 + 16^2 - 2(9)(16)\cos 39° \approx 113.18$$
$$b \approx 10.6$$

Using the law of sines:

$$\frac{\sin 39°}{10.6} = \frac{\sin A}{9}$$
$$10.6\sin A = 9\sin 39°$$
$$\sin A = \frac{9\sin 39°}{10.6} \approx 0.5343$$
$$A \approx 32.3°$$

Finally $C = 180° - 39° - 32.3° = 108.7°$.

40. First note that $A = 180° - 41° - 24° = 115°$. Using the law of sines:

$$\frac{\sin 24°}{5} = \frac{\sin 41°}{b} \qquad\qquad \frac{\sin 24°}{5} = \frac{\sin 115°}{a}$$

$$b \sin 24° = 5 \sin 41° \qquad\qquad a \sin 24° = 5 \sin 115°$$

$$b = \frac{5 \sin 41°}{\sin 24°} \approx 8.1 \qquad\qquad a = \frac{5 \sin 115°}{\sin 24°} \approx 11.1$$

41. Using the law of cosines to find C:

$$9^2 = 3^2 + 7^2 - 2(3)(7)\cos C$$

$$81 = 58 - 42 \cos C$$

$$23 = -42 \cos C$$

$$\cos C = -\tfrac{23}{42}$$

$$C = \cos^{-1}\left(-\tfrac{23}{42}\right) \approx 123.2°$$

Using the law of sines:

$$\frac{\sin 123.2°}{9} = \frac{\sin B}{7}$$

$$9 \sin B = 7 \sin 123.2°$$

$$\sin B = \frac{7 \sin 123.2°}{9} \approx 0.6508$$

$$B \approx 40.6°$$

Finally $A = 180° - 123.2° - 40.6° = 16.2°$.

42. Using the law of sines:

$$\frac{\sin 115°}{40} = \frac{\sin B}{25}$$

$$40 \sin B = 25 \sin 115°$$

$$\sin B = \frac{25 \sin 115°}{40} \approx 0.5664$$

$$B \approx 34.5°$$

Thus $A = 180° - 115° - 34.5° = 30.5°$. Using the law of sines:

$$\frac{\sin 115°}{40} = \frac{\sin 30.5°}{a}$$

$$a \sin 115° = 40 \sin 30.5°$$

$$a = \frac{40 \sin 30.5°}{\sin 115°} \approx 22.4$$

43. Using the law of sines:

$$\frac{\sin 52°}{8} = \frac{\sin B}{11}$$

$$8 \sin B = 11 \sin 52°$$

$$\sin B = \frac{11 \sin 52°}{8} \approx 1.0835$$

Since $-1 \le \sin B \le 1$, no such triangle exists.

44. Using the law of sines:

$$\frac{\sin 71°}{30} = \frac{\sin B}{27}$$

$$30 \sin B = 27 \sin 71°$$

$$\sin B = \frac{27 \sin 71°}{30} \approx 0.8510$$

$$B \approx 58.3°$$

One such triangle exists.

45. Using the law of cosines:
$$c^2 = 65^2 + 70^2 - 2(65)(70)\cos 130° \approx 14974.37$$
$$c \approx 122.37$$
One such triangle exists.

46. Using the law of sines:
$$\frac{\sin 47°}{10} = \frac{\sin A}{12}$$
$$10\sin A = 12\sin 47°$$
$$\sin A = \frac{12\sin 47°}{10} \approx 0.8776$$
$$A \approx 61.4°, 118.6°$$
Two such triangles exist.

47. Since the trains have traveled for 1.25 hours, their distances traveled are $90(1.25) = 112.5$ km and $110(1.25) = 137.5$ km. Let d represent the distance between the trains. Using the law of cosines:
$$d^2 = 137.5^2 + 112.5^2 - 2(137.5)(112.5)\cos 106° \approx 40090.03$$
$$d \approx 200.2$$
The two trains are 200.2 km apart.

48. First draw a figure:

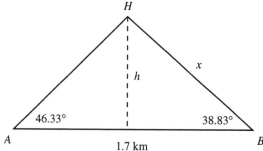

Note that $H = 180° - 46.33° - 38.83° = 94.84°$. Finding x using the law of sines:
$$\frac{\sin 94.84°}{1.7} = \frac{\sin 46.33°}{x}$$
$$x\sin 94.84° = 1.7\sin 46.33°$$
$$x = \frac{1.7\sin 46.33°}{\sin 94.84°} \approx 1.23$$
Now finding h:
$$\sin 38.83° = \frac{h}{1.23}$$
$$h = 1.23\sin 38.83° \approx 0.77$$
The altitude of the helicopter is approximately 0.8 km.

49. First draw the figure:

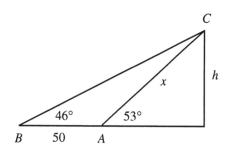

Note that $\angle BAC = 180° - 53° = 127°$, and thus $\angle BCA = 180° - 127° - 46° = 7°$. Using the law of sines:

$$\frac{\sin 7°}{50} = \frac{\sin 46°}{x}$$
$$x \sin 7° = 50 \sin 46°$$
$$x = \frac{50 \sin 46°}{\sin 7°} \approx 295.13$$

Now finding h:

$$\sin 53° = \frac{h}{295.13}$$
$$h = 295.13 \sin 53° \approx 235.7$$

The height of the tree is approximately 235.7 feet.

50. First draw the figure:

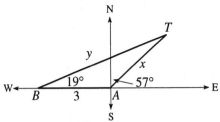

Note that $T = 180° - 19° - 147° = 14°$. Using the law of sines:

$$\frac{\sin 14°}{3} = \frac{\sin 19°}{x} \qquad\qquad \frac{\sin 14°}{3} = \frac{\sin 147°}{y}$$
$$x \sin 14° = 3 \sin 19° \qquad\qquad y \sin 14° = 3 \sin 147°$$
$$x = \frac{3 \sin 19°}{\sin 14°} \approx 4.0 \qquad\qquad y = \frac{3 \sin 147°}{\sin 14°} \approx 6.8$$

The target is 4.0 miles from A and 6.8 miles from B.

51. Since $\mathbf{u} + \mathbf{v} = \langle 7, 10 \rangle$, $|\mathbf{u} + \mathbf{v}| = \sqrt{7^2 + 10^2} = \sqrt{149}$. Drawing the vectors:

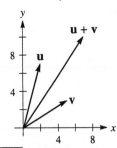

52. Since $\mathbf{u} + \mathbf{v} = \langle 1, 1 \rangle$, $|\mathbf{u} + \mathbf{v}| = \sqrt{1^2 + 1^2} = \sqrt{2}$. Drawing the vectors:

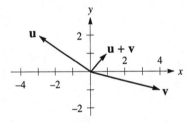

53. Since $\mathbf{u}+\mathbf{v}=5i$, $|\mathbf{u}+\mathbf{v}|=\sqrt{5^2}=5$. Drawing the vectors:

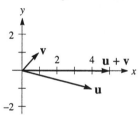

54. Since $\mathbf{u}+\mathbf{v}=-2i-7j$, $|\mathbf{u}+\mathbf{v}|=\sqrt{(-2)^2+(-7)^2}=\sqrt{53}$. Drawing the vectors:

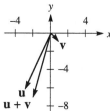

55. Simplifying: $\mathbf{u}+\mathbf{v}=\langle 2,-5\rangle+\langle 4,3\rangle=\langle 6,-2\rangle=6\mathbf{i}-2\mathbf{j}$

56. Simplifying: $\mathbf{u}-5\mathbf{v}=\langle 2,-5\rangle-5\langle 4,3\rangle=\langle 2,-5\rangle-\langle 20,15\rangle=\langle -18,-20\rangle=-18\mathbf{i}-20\mathbf{j}$

57. Simplifying: $2\mathbf{u}+4\mathbf{v}=2\langle 2,-5\rangle+4\langle 4,3\rangle=\langle 4,-10\rangle+\langle 16,12\rangle=\langle 20,2\rangle=20\mathbf{i}+2\mathbf{j}$

58. Simplifying: $-3\mathbf{u}+5\mathbf{v}=-3\langle 2,-5\rangle+5\langle 4,3\rangle=\langle -6,15\rangle+\langle 20,15\rangle=\langle 14,30\rangle=14\mathbf{i}+30\mathbf{j}$

59. Resolving the vector: $\mathbf{v}=\langle 3\cos 45°,3\sin 45°\rangle=\left\langle 3\cdot\dfrac{\sqrt{2}}{2},3\cdot\dfrac{\sqrt{2}}{2}\right\rangle=\left\langle\dfrac{3\sqrt{2}}{2},\dfrac{3\sqrt{2}}{2}\right\rangle$

60. Resolving the vector: $\mathbf{v}=\langle 8\cos 30°,8\sin 30°\rangle=\left\langle 8\cdot\dfrac{\sqrt{3}}{2},8\cdot\dfrac{1}{2}\right\rangle=\left\langle 4\sqrt{3},4\right\rangle$

61. Resolving the vector: $\mathbf{v}=\langle 10\cos 118°,10\sin 118°\rangle\approx\langle -4.7,8.8\rangle$

62. Resolving the vector: $\mathbf{v}=\langle 5\cos 25°,5\sin 25°\rangle\approx\langle 4.5,2.1\rangle$

63. Since $|\mathbf{v}|=\sqrt{5^2+4^2}=\sqrt{41}$, a unit vector is $\dfrac{1}{\sqrt{41}}\langle 5,4\rangle=\left\langle\dfrac{5}{\sqrt{41}},\dfrac{4}{\sqrt{41}}\right\rangle$.

64. Since $|\mathbf{v}|=\sqrt{3^2+2^2}=\sqrt{13}$, a unit vector is $\dfrac{1}{\sqrt{13}}(3\mathbf{i}+2\mathbf{j})=\dfrac{3}{\sqrt{13}}\mathbf{i}+\dfrac{2}{\sqrt{13}}\mathbf{j}$.

65. Since $\mathbf{v}=4\langle 3,5\rangle+\langle 2,1\rangle=\langle 12,20\rangle+\langle 2,1\rangle=\langle 14,21\rangle=7\langle 2,3\rangle$: $|\mathbf{v}|=7\sqrt{2^2+3^2}=7\sqrt{13}$

So a unit vector is $\dfrac{1}{7\sqrt{13}}\cdot 7\langle 2,3\rangle=\dfrac{1}{\sqrt{13}}\langle 2,3\rangle=\left\langle\dfrac{2}{\sqrt{13}},\dfrac{3}{\sqrt{13}}\right\rangle$.

66. Since $|\mathbf{v}|=\sqrt{\dfrac{1}{9}+\dfrac{9}{16}}=\dfrac{\sqrt{97}}{12}$, a unit vector is: $\dfrac{12}{\sqrt{97}}\left(\dfrac{1}{3}\mathbf{i}-\dfrac{3}{4}\mathbf{j}\right)=\dfrac{1}{\sqrt{97}}(4\mathbf{i}-9\mathbf{j})=\dfrac{4}{\sqrt{97}}\mathbf{i}-\dfrac{9}{\sqrt{97}}\mathbf{j}$

67.   Since $\mathbf{A} + \mathbf{B} = \langle 4,0 \rangle + \langle 0.5 \rangle = \langle 4,5 \rangle$, $|\mathbf{A} + \mathbf{B}| = \sqrt{4^2 + 5^2} = \sqrt{41}$. Drawing the figure:

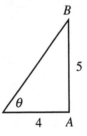

Therefore:
$$\tan \theta = \tfrac{5}{4}$$
$$\theta = \tan^{-1}\left(\tfrac{5}{4}\right) \approx 51.3°$$

68.   Since $\mathbf{A} + \mathbf{B} = \langle 1,3 \rangle + \langle 2,4 \rangle = \langle 3,7 \rangle$, $|\mathbf{A} + \mathbf{B}| = \sqrt{3^2 + 7^2} = \sqrt{58}$. Note that $|\mathbf{A}| = \sqrt{1^2 + 3^2} = \sqrt{10}$ and $|\mathbf{B}| = \sqrt{2^2 + 4^2} = \sqrt{20} = 2\sqrt{5}$. Drawing the figure:

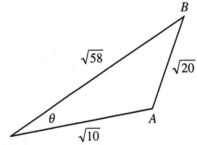

Using the law of cosines to find θ:
$$\left(\sqrt{20}\right)^2 = \left(\sqrt{10}\right)^2 + \left(\sqrt{58}\right)^2 - 2\sqrt{10}\sqrt{58}\cos\theta$$
$$20 = 68 - 2\sqrt{580}\cos\theta$$
$$-48 = -2\sqrt{580}\cos\theta$$
$$\cos\theta = \frac{24}{\sqrt{580}}$$
$$\theta \approx 4.8°$$

69.   Since $\mathbf{A} + \mathbf{B} = \langle 6,2 \rangle + \langle 1,8 \rangle = \langle 7,10 \rangle$, $|\mathbf{A} + \mathbf{B}| = \sqrt{7^2 + 10^2} = \sqrt{149}$. Note that $|\mathbf{A}| = \sqrt{6^2 + 2^2} = \sqrt{40} = 2\sqrt{5}$ and $|\mathbf{B}| = \sqrt{1^2 + 8^2} = \sqrt{65}$. Drawing the figure:

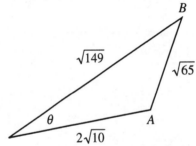

Using the law of cosines to find θ:
$$\left(\sqrt{65}\right)^2 = \left(2\sqrt{10}\right)^2 + \left(\sqrt{149}\right)^2 - 4\sqrt{10}\sqrt{149}\cos\theta$$
$$65 = 189 - 4\sqrt{1490}\cos\theta$$
$$-124 = -4\sqrt{1490}\cos\theta$$
$$\cos\theta = \frac{31}{\sqrt{1490}}$$
$$\theta \approx 36.6°$$

70. Since $\mathbf{A} + \mathbf{B} = \langle 10,0\rangle + \langle 0,7\rangle = \langle 10,7\rangle$, $|\mathbf{A} + \mathbf{B}| = \sqrt{10^2 + 7^2} = \sqrt{149}$. Drawing the figure:

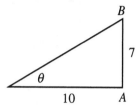

Therefore:
$$\tan\theta = \tfrac{7}{10}$$
$$\theta = \tan^{-1}\left(\tfrac{7}{10}\right) \approx 35.0°$$

71. First draw the figure:

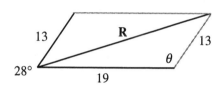

Finding angle θ:
$$2(28°) + 2\theta = 360°$$
$$56° + 2\theta = 360°$$
$$2\theta = 304°$$
$$\theta = 152°$$
Now using the law of cosines:
$$|\mathbf{R}|^2 = 19^2 + 13^2 - 2(19)(13)\cos 152° \approx 966.18$$
$$|\mathbf{R}| \approx 31.1 \text{ pounds}$$

72. First draw the figure:

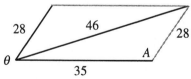

Finding angle A using the law of cosines:
$$46^2 = 35^2 + 28^2 - 2(35)(28)\cos A$$
$$2116 = 2009 - 1960\cos A$$
$$107 = -1960\cos A$$
$$\cos A = -\frac{107}{1960}$$
$$A = \cos^{-1}\left(-\frac{107}{1960}\right) \approx 93.1°$$

Therefore:
$$2\theta + 2(93.1°) = 360°$$
$$2\theta + 186.2° = 360°$$
$$2\theta = 173.8°$$
$$\theta = 86.9°$$

73. First draw the figure:

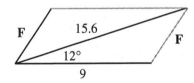

Using the law of cosines:
$$|\mathbf{F}|^2 = 15.6^2 + 9^2 - 2(15.6)(9)\cos 12° \approx 49.70$$
$$|\mathbf{F}| \approx 7.0 \text{ pounds}$$

74. First draw the figure:

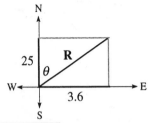

The speed of the boat is $|\mathbf{R}| = \sqrt{25^2 + 3.6^2} \approx 25.3$ mph. Finding θ:
$$\tan\theta = \frac{3.6}{25}$$
$$\theta = \tan^{-1}\left(\frac{3.6}{25}\right) \approx 8.2°$$
The course of the boat is N 8.2° E.

75. First draw the figure:

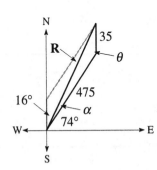

Finding angle θ:
$$2\theta + 2(16°) = 360°$$
$$2\theta + 32° = 360°$$
$$2\theta = 328°$$
$$\theta = 164°$$

Using the law of cosines to find $|\mathbf{R}|$:
$$|\mathbf{R}|^2 = 475^2 + 35^2 - 2(475)(35)\cos 164° \approx 258811.95$$
$$|\mathbf{R}| \approx 508.7 \text{ mph}$$

The groundspeed of the plane is 508.7 mph. Using the law of sines:

$$\frac{\sin 164°}{508.7} = \frac{\sin \alpha}{35}$$

$$508.7 \sin \alpha = 35 \sin 164°$$

$$\sin \alpha = \frac{35 \sin 164°}{508.7} \approx 0.0190$$

$$\alpha \approx 1.1°$$

$$16° - \alpha = 14.9°$$

The course of the plane is N 14.9° E.

76. Drawing the figure:

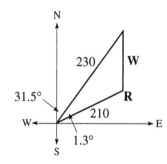

Using the law of cosines:

$$|\mathbf{W}|^2 = 230^2 + 210^2 - 2(230)(210)\cos 1.3° \approx 424.86$$

$$|\mathbf{W}| \approx 20.6 \text{ mph}$$

The speed of the wind is 20.6 mph.

77. First draw the figure:

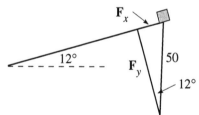

The parallel component of the weight is:

$$\sin 12° = \frac{F_x}{50}$$

$$F_x = 50 \sin 12° \approx 10.4 \text{ pounds}$$

The perpendicular component of the weight is:

$$\cos 12° = \frac{F_y}{50}$$

$$F_y = 50 \cos 12° \approx 48.9 \text{ pounds}$$

78. First draw the figure:

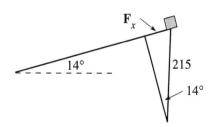

The force required is:

$$\sin 14° = \frac{F_x}{215}$$

$$F_x = 215\sin 14° \approx 52.0 \text{ pounds}$$

79.   First draw the figure:

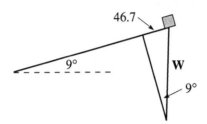

Finding the weight:

$$\sin 9° = \frac{46.7}{|W|}$$

$$|W|\sin 9° = 46.7$$

$$|W| = \frac{46.7}{\sin 9°} \approx 298.5 \text{ pounds}$$

80.   First draw the figure:

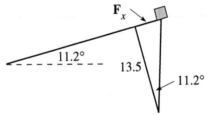

Finding the weight parallel to the ramp:

$$\tan 11.2° = \frac{F_x}{13.5}$$

$$F_x = 13.5\tan 11.2° \approx 2.7 \text{ pounds}$$

81.   Here $r = \sqrt{\left(\sqrt{2}\right)^2 + (-1)^2} = \sqrt{3}$. Thus:

$$\sqrt{2} - i = \sqrt{3}\left(\frac{\sqrt{2}}{\sqrt{3}} - \frac{1}{\sqrt{3}}i\right) = \sqrt{3}(\cos 324.74° + i\sin 324.74°)$$

82.   Converting to rectangular form: $6\left(\cos\frac{7\pi}{6} + i\sin\frac{7\pi}{6}\right) = 6\left(-\frac{\sqrt{3}}{2} - \frac{1}{2}i\right) = -3\sqrt{3} - 3i$

83.   Converting to rectangular form: $8\left(\cos\frac{3\pi}{4} + i\sin\frac{3\pi}{4}\right) = 8\left(-\frac{\sqrt{2}}{2} + \frac{\sqrt{2}}{2}i\right) = -4\sqrt{2} + 4i\sqrt{2}$

84.   Here $r = \sqrt{3^2 + 3^2} = \sqrt{18} = 3\sqrt{2}$. Thus: $3 + 3i = 3\sqrt{2}\left(\frac{1}{\sqrt{2}} + \frac{1}{\sqrt{2}}i\right) = 3\sqrt{2}\left(\cos\frac{\pi}{4} + i\sin\frac{\pi}{4}\right)$

85. Finding the product and quotient:

$$z_1 z_2 = 4 \cdot 3 \left[\cos\left(\frac{\pi}{3} + \frac{\pi}{5}\right) + i \sin\left(\frac{\pi}{3} + \frac{\pi}{5}\right) \right] = 12\left(\cos\frac{8\pi}{15} + i \sin\frac{8\pi}{15} \right)$$

$$\frac{z_1}{z_2} = \frac{4}{3}\left[\cos\left(\frac{\pi}{3} - \frac{\pi}{5}\right) + i \sin\left(\frac{\pi}{3} - \frac{\pi}{5}\right) \right] = \frac{4}{3}\left(\cos\frac{2\pi}{15} + i \sin\frac{2\pi}{15} \right)$$

86. Finding the product and quotient:

$$z_1 z_2 = 9 \cdot 3\left[\cos(120° + 210°) + i \sin(120° + 210°) \right]$$

$$= 27(\cos 330° + i \sin 330°)$$

$$= 27\left(\frac{\sqrt{3}}{2} - \frac{1}{2}i \right)$$

$$= \frac{27\sqrt{3}}{2} - \frac{27}{2}i$$

$$\frac{z_1}{z_2} = \frac{9}{3}\left[\cos(120° - 210°) + i \sin(120° - 210°) \right] = 3(\cos(-90°) + i \sin(-90°)) = 3(0 - i) = -3i$$

87. Using DeMoivre's theorem:

$$\left[4\left(\cos\frac{\pi}{3} + i \sin\frac{\pi}{3} \right) \right]^4 = 4^4\left(\cos\frac{4\pi}{3} + i \sin\frac{4\pi}{3} \right) = 256\left(-\frac{1}{2} - \frac{\sqrt{3}}{2}i \right) = -128 - 128i\sqrt{3}$$

88. Using DeMoivre's theorem:

$$\left[\sqrt[3]{4}\left(\cos\frac{5\pi}{6} + i \sin\frac{5\pi}{6} \right) \right]^6 = \left(\sqrt[3]{4}\right)^6 (\cos 5\pi + i \sin 5\pi) = 16(-1 + 0i) = -16$$

89. First convert to trigonometric form: $1 - i = \sqrt{2}\left(\frac{1}{\sqrt{2}} - \frac{1}{\sqrt{2}}i \right) = \sqrt{2}\left(\cos\frac{7\pi}{4} + i \sin\frac{7\pi}{4} \right)$

Using DeMoivre's theorem: $(1 - i)^6 = \left(\sqrt{2}\right)^6\left(\cos\frac{21\pi}{2} + i \sin\frac{21\pi}{2} \right) = 8(0 + i) = 8i$

90. First convert to trigonometric form: $\sqrt{3} + i = 2\left(\frac{\sqrt{3}}{2} + \frac{1}{2}i \right) = 2\left(\cos\frac{\pi}{6} + i \sin\frac{\pi}{6} \right)$

Using DeMoivre's theorem:

$$\left(\sqrt{3} + i\right)^8 = 2^8\left(\cos\frac{4\pi}{3} + i \sin\frac{4\pi}{3} \right) = 256\left(-\frac{1}{2} - \frac{\sqrt{3}}{2}i \right) = -128 - 128i\sqrt{3}$$

91. First convert to trigonometric form: $-2 - 2i = 2\sqrt{2}\left(-\frac{1}{\sqrt{2}} - \frac{1}{\sqrt{2}}i \right) = 2\sqrt{2}\left(\cos\frac{5\pi}{4} + i \sin\frac{5\pi}{4} \right)$

Using DeMoivre's theorem:

$$(-2 - 2i)^5 = \left(2\sqrt{2}\right)^5\left(\cos\frac{25\pi}{4} + i \sin\frac{25\pi}{4} \right) = 128\sqrt{2}\left(\frac{1}{\sqrt{2}} + \frac{1}{\sqrt{2}}i \right) = 128 + 128i$$

92. First convert to trigonometric form: $-\frac{1}{2} + \frac{\sqrt{3}}{2}i = 1\left(\cos\frac{2\pi}{3} + i \sin\frac{2\pi}{3} \right)$

Using DeMoivre's theorem: $\left(-\frac{1}{2} + \frac{\sqrt{3}}{2}i \right)^6 = (1)^6(\cos 4\pi + i \sin 4\pi) = 1 + 0i = 1$

93. First convert to trigonometric form: $2\sqrt{3} - 2i = 4\left(\dfrac{\sqrt{3}}{2} - \dfrac{1}{2}i\right) = 4\left(\cos\dfrac{11\pi}{6} + i\sin\dfrac{11\pi}{6}\right)$

The fourth roots of $2\sqrt{3} - 2i$ are given by:

$$w_k = \sqrt[4]{4}\left(\cos\dfrac{\dfrac{11\pi}{6} + 2k\pi}{4} + i\sin\dfrac{\dfrac{11\pi}{6} + 2k\pi}{4}\right) = \sqrt{2}\left[\cos\left(\dfrac{11\pi}{24} + \dfrac{k\pi}{2}\right) + i\sin\left(\dfrac{11\pi}{24} + \dfrac{k\pi}{2}\right)\right]$$

Substituting $k = 0, 1, 2, 3$:

$$w_0 = \sqrt{2}\left(\cos\dfrac{11\pi}{24} + i\sin\dfrac{11\pi}{24}\right)$$

$$w_1 = \sqrt{2}\left(\cos\dfrac{23\pi}{24} + i\sin\dfrac{23\pi}{24}\right)$$

$$w_2 = \sqrt{2}\left(\cos\dfrac{35\pi}{24} + i\sin\dfrac{35\pi}{24}\right)$$

$$w_3 = \sqrt{2}\left(\cos\dfrac{47\pi}{24} + i\sin\dfrac{47\pi}{24}\right)$$

94. First convert to trigonometric form: $1 + i = \sqrt{2}\left(\dfrac{1}{\sqrt{2}} + \dfrac{1}{\sqrt{2}}i\right) = \sqrt{2}\left(\cos\dfrac{\pi}{4} + i\sin\dfrac{\pi}{4}\right)$

The cube roots of $1 + i$ are given by:

$$w_k = \sqrt[3]{\sqrt{2}}\left(\cos\dfrac{\dfrac{\pi}{4} + 2k\pi}{3} + i\sin\dfrac{\dfrac{\pi}{4} + 2k\pi}{3}\right) = \sqrt[6]{2}\left[\cos\left(\dfrac{\pi}{12} + \dfrac{2k\pi}{3}\right) + i\sin\left(\dfrac{\pi}{12} + \dfrac{2k\pi}{3}\right)\right]$$

Substituting $k = 0, 1, 2$:

$$w_0 = \sqrt[6]{2}\left(\cos\dfrac{\pi}{12} + i\sin\dfrac{\pi}{12}\right)$$

$$w_1 = \sqrt[6]{2}\left(\cos\dfrac{9\pi}{12} + i\sin\dfrac{9\pi}{12}\right) = \sqrt[6]{2}\left(\cos\dfrac{3\pi}{4} + i\sin\dfrac{3\pi}{4}\right)$$

$$w_2 = \sqrt[6]{2}\left(\cos\dfrac{17\pi}{12} + i\sin\dfrac{17\pi}{12}\right)$$

95. First convert to trigonometric form: $1 = 1(1 + 0i) = 1(\cos 0 + i\sin 0)$

The fifth roots of 1 are given by:

$$w_k = \sqrt[5]{1}\left(\cos\dfrac{0 + 2k\pi}{5} + i\sin\dfrac{0 + 2k\pi}{5}\right) = \cos\dfrac{2k\pi}{5} + i\sin\dfrac{2k\pi}{5}$$

Substituting $k = 0, 1, 2, 3, 4, 5$:

$$w_0 = \cos 0 + i\sin 0 = 1$$

$$w_1 = \cos\dfrac{2\pi}{5} + i\sin\dfrac{2\pi}{5}$$

$$w_2 = \cos\dfrac{4\pi}{5} + i\sin\dfrac{4\pi}{5}$$

$$w_3 = \cos\dfrac{6\pi}{5} + i\sin\dfrac{6\pi}{5}$$

$$w_4 = \cos\dfrac{8\pi}{5} + i\sin\dfrac{8\pi}{5}$$

96. First convert to trigonometric form: $-i = 1(0-1i) = 1\left(\cos\dfrac{3\pi}{2} + i\sin\dfrac{3\pi}{2}\right)$

The square roots of i are given by:

$$w_k = \sqrt{1}\left(\cos\dfrac{\dfrac{3\pi}{2}+2k\pi}{2} + i\sin\dfrac{\dfrac{3\pi}{2}+2k\pi}{2}\right) = \cos\left(\dfrac{3\pi}{4}+k\pi\right) + i\sin\left(\dfrac{3\pi}{4}+k\pi\right)$$

Substituting $k = 0, 1$:

$$w_0 = \cos\dfrac{3\pi}{4} + i\sin\dfrac{3\pi}{4} = -\dfrac{1}{\sqrt{2}} + \dfrac{1}{\sqrt{2}}i$$

$$w_1 = \cos\dfrac{7\pi}{4} + i\sin\dfrac{7\pi}{4} = \dfrac{1}{\sqrt{2}} - \dfrac{1}{\sqrt{2}}i$$

97. Two additional points are $\left(3, -\dfrac{7\pi}{4}\right)$ and $\left(-3, \dfrac{5\pi}{4}\right)$.

98. Two additional points are $\left(5, \dfrac{4\pi}{3}\right)$ and $\left(-5, \dfrac{\pi}{3}\right)$.

99. Two additional points are $(1, 0)$ and $(-1, -\pi)$.

100. Two additional points are $\left(2, \dfrac{\pi}{6}\right)$ and $\left(-2, \dfrac{7\pi}{6}\right)$.

101. The rectangular point is $\left(3\cos\dfrac{3\pi}{4}, 3\sin\dfrac{3\pi}{4}\right) = \left(3\cdot\left(-\dfrac{1}{\sqrt{2}}\right), 3\cdot\dfrac{1}{\sqrt{2}}\right) = \left(-\dfrac{3}{\sqrt{2}}, \dfrac{3}{\sqrt{2}}\right)$.

102. Since $r = \sqrt{2^2 + 2^2} = \sqrt{8} = 2\sqrt{2}$ and $\theta = \dfrac{\pi}{4}$, the polar point is $\left(2\sqrt{2}, \dfrac{\pi}{4}\right)$.

103. Since $r = \sqrt{\left(-5\sqrt{3}\right)^2 + (-5)^2} = \sqrt{100} = 10$ and $\theta = \dfrac{7\pi}{6}$, the polar point is $\left(10, \dfrac{7\pi}{6}\right)$.

104. The rectangular point is $\left(-4\cos\left(-\dfrac{5\pi}{6}\right), -4\sin\left(-\dfrac{5\pi}{6}\right)\right) = \left(-4\cdot\left(-\dfrac{\sqrt{3}}{2}\right), -4\cdot\left(-\dfrac{1}{2}\right)\right) = \left(2\sqrt{3}, 2\right)$.

105. Converting to rectangular form:
$$r = 6\sin\theta$$
$$r^2 = 6r\sin\theta$$
$$x^2 + y^2 = 6y$$
This is the equation of a circle.

106. Converting to rectangular form:
$$r = 8$$
$$r^2 = 64$$
$$x^2 + y^2 = 64$$
This is the equation of a circle.

107. Converting to rectangular form:
$$\theta = \frac{\pi}{4}$$
$$\tan \theta = \tan \frac{\pi}{4}$$
$$\frac{y}{x} = 1$$
$$y = x$$
This is the equation of a line.

108. Converting to rectangular form:
$$r = 4 \sec \theta$$
$$r = \frac{4}{\cos \theta}$$
$$r \cos \theta = 4$$
$$x = 4$$
This is the equation of a line.

109. Converting to rectangular form:
$$r \sin \theta = 4$$
$$y = 4$$
This is the equation of a line.

110. Converting to rectangular form:
$$r \cos \theta = -2$$
$$x = -2$$
This is the equation of a line.

111. Converting to rectangular form:
$$r = \frac{1}{1 - \cos \theta}$$
$$r - r \cos \theta = 1$$
$$r = 1 + r \cos \theta$$
$$\sqrt{x^2 + y^2} = 1 + x$$
$$x^2 + y^2 = 1 + 2x + x^2$$
$$y^2 = 1 + 2x$$
$$y^2 - 1 = 2x$$
$$x = \tfrac{1}{2} y^2 - \tfrac{1}{2}$$
This is the equation of a parabola.

112. Converting to rectangular form:
$$r = 2 \tan \theta$$
$$\sqrt{x^2 + y^2} = 2 \cdot \frac{y}{x}$$
$$x^2 + y^2 = \frac{4y^2}{x^2}$$
$$x^4 + x^2 y^2 = 4y^2$$
$$x^4 = 4y^2 - x^2 y^2$$
$$x^4 = y^2 \left(4 - x^2 \right)$$
$$y^2 = \frac{x^4}{4 - x^2}$$

113. Graphing the curve:

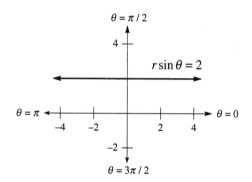

114. Graphing the curve:

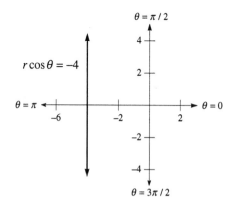

115. Graphing the curve:

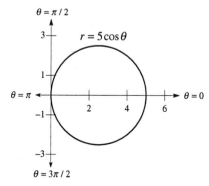

116. Graphing the curve:

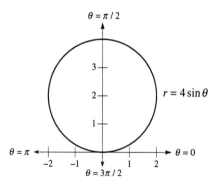

117. Graphing the curve:

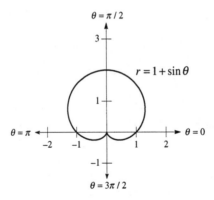

118. Graphing the curve:

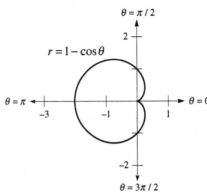

119. Graphing the curve:

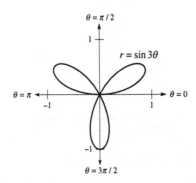

120. Graphing the curve:

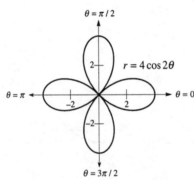

121. Graphing the curve:

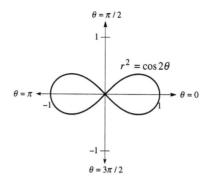

122. Graphing the curve:

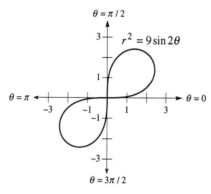

123. Graphing the curve:

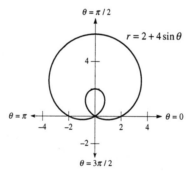

124. Graphing the curve:

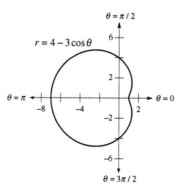

Chapter 8 Practice Test

1. Using the addition formula:
$$\cos\frac{7\pi}{12} = \cos\left(\frac{\pi}{3}+\frac{\pi}{4}\right) = \cos\frac{\pi}{3}\cos\frac{\pi}{4} - \sin\frac{\pi}{3}\sin\frac{\pi}{4} = \frac{1}{2}\cdot\frac{\sqrt{2}}{2} - \frac{\sqrt{3}}{2}\cdot\frac{\sqrt{2}}{2} = \frac{\sqrt{2}-\sqrt{6}}{4}$$

2. a. Verifying the identity: $\sin(\theta+\pi) = \sin\theta\cos\pi + \cos\theta\sin\pi = \sin\theta\cdot(-1) + \cos\theta\cdot 0 = -\sin\theta$

   b. Verifying the identity: $\dfrac{1-\cos 2x}{1+\cos 2x} = \dfrac{1-\left(1-2\sin^2 x\right)}{1+\left(2\cos^2 x - 1\right)} = \dfrac{2\sin^2 x}{2\cos^2 x} = \left(\dfrac{\sin x}{\cos x}\right)^2 = \tan^2 x$

3. Solving the equation:
$$3 - \cos^2\theta = 2\sin^2\frac{\theta}{2}$$
$$3 - \cos^2\theta = 2\cdot\frac{1-\cos\theta}{2}$$
$$3 - \cos^2\theta = 1 - \cos\theta$$
$$\cos^2\theta - \cos\theta - 2 = 0$$
$$(\cos\theta - 2)(\cos\theta + 1) = 0$$
$$\cos\theta = -1 \qquad (\cos\theta = 2 \text{ is impossible})$$
$$\theta = \pi$$

4. Since $\tan\theta = \frac{1}{6}$ and $\pi < \theta < \dfrac{3\pi}{2}$, $\sin\theta = -\dfrac{1}{\sqrt{37}}$ and $\cos\theta = -\dfrac{6}{\sqrt{37}}$.

   a. Using the double-angle formula: $\sin 2\theta = 2\sin\theta\cos\theta = 2\left(-\dfrac{1}{\sqrt{37}}\right)\left(-\dfrac{6}{\sqrt{37}}\right) = \frac{12}{37}$

   b. Using the double-angle formula:
$$\cos 2\theta = \cos^2\theta - \sin^2\theta = \left(-\dfrac{6}{\sqrt{37}}\right)^2 - \left(-\dfrac{1}{\sqrt{37}}\right)^2 = \tfrac{36}{37} - \tfrac{1}{37} = \tfrac{35}{37}$$

5. Since $\cos\theta = \dfrac{2a}{3}$, $\sin\theta = \dfrac{\sqrt{9-4a^2}}{3}$ and thus $\tan\theta = \dfrac{\sqrt{9-4a^2}}{2a}$. Using the double-angle formula:
$$\tan 2\theta = \frac{2\tan\theta}{1-\tan^2\theta} = \frac{2\cdot\dfrac{\sqrt{9-4a^2}}{2a}}{1-\left(\dfrac{\sqrt{9-4a^2}}{2a}\right)^2}\cdot\frac{4a^2}{4a^2} = \frac{4a\sqrt{9-4a^2}}{4a^2-\left(9-4a^2\right)} = \frac{4a\sqrt{9-4a^2}}{8a^2-9}$$

6. a. Note that $C = 180° - 75° - 43° = 62°$. Using the law of sines:

   $\dfrac{\sin 75°}{20} = \dfrac{\sin 43°}{b}$ $\dfrac{\sin 75°}{20} = \dfrac{\sin 62°}{c}$

   $b\sin 75° = 20\sin 43°$ $c\sin 75° = 20\sin 62°$

   $b = \dfrac{20\sin 43°}{\sin 75°} \approx 14.1$ $c = \dfrac{20\sin 62°}{\sin 75°} \approx 18.3$

b. Using the law of cosines:
$$b^2 = 10^2 + 20^2 - 2(10)(20)\cos 105° \approx 603.53$$
$$b \approx 24.6$$
Using the law of sines:
$$\frac{\sin 105°}{24.6} = \frac{\sin C}{20}$$
$$24.6 \sin C = 20 \sin 105°$$
$$\sin C = \frac{20 \sin 105°}{24.6} \approx 0.7864$$
$$C = \sin^{-1}(0.7864) \approx 51.8°$$
Thus $A = 180° - 105° - 51.8° = 23.2°$.

7. Drawing the figure:

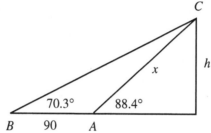

Note that $\angle BAC = 180° - 88.4° = 91.6°$, thus $\angle BCA = 180° - 91.6° - 70.3° = 18.1°$. Using the law of sines:
$$\frac{\sin 18.1°}{90} = \frac{\sin 70.3°}{x}$$
$$x \sin 18.1° = 90 \sin 70.3°$$
$$x = \frac{90 \sin 70.3°}{\sin 18.1°} \approx 272.74$$
Therefore:
$$\sin 88.4° = \frac{h}{272.74}$$
$$h = 272.74 \sin 88.4° \approx 272.6$$
The height of the bridge is 272.6 feet.

8. **a.** Since $\mathbf{u} + \mathbf{v} = \langle 3, -1 \rangle + \langle 2, 4 \rangle = \langle 5, 3 \rangle$, $|\mathbf{u} + \mathbf{v}| = \sqrt{5^2 + 3^2} = \sqrt{34}$. Drawing the vectors:

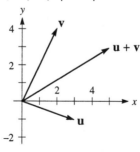

b.   Since $\mathbf{u} + \mathbf{v} = 6\mathbf{i}$, $|\mathbf{u} + \mathbf{v}| = \sqrt{6^2} = \sqrt{36} = 6$. Drawing the vectors:

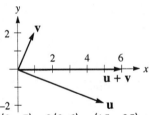

9.   **a.**   Simplifying: $5\mathbf{u} + 3\mathbf{v} = 5\langle 3, -7 \rangle + 3\langle 2, 6 \rangle = \langle 15, -35 \rangle + \langle 6, 18 \rangle = \langle 21, -17 \rangle$

  b.   Simplifying: $-4\mathbf{u} + 3\mathbf{v} = -4\langle 3, -7 \rangle + 3\langle 2, 6 \rangle = \langle -12, 28 \rangle + \langle 6, 18 \rangle = \langle -6, 46 \rangle = -6\mathbf{i} + 46\mathbf{j}$

10.   Resolving the components: $\mathbf{v} = \langle 8\cos 56°, 8\sin 56° \rangle \approx \langle 4.5, 6.6 \rangle$

11.   Since $|\langle 4, 5 \rangle| = \sqrt{4^2 + 5^2} = \sqrt{41}$, a unit vector is $\dfrac{1}{\sqrt{41}}\langle 4, 5 \rangle = \left(\dfrac{4}{\sqrt{41}}, \dfrac{5}{\sqrt{41}} \right)$.

12.   Since $\overrightarrow{OP} + \overrightarrow{OQ} = \langle 1, 4 \rangle + \langle -5, 2 \rangle = \langle -4, 6 \rangle$, the lengths are $\left| \overrightarrow{OP} \right| = \sqrt{1^2 + 4^2} = \sqrt{17}$,

  $\left| \overrightarrow{OQ} \right| = \sqrt{(-5)^2 + 2^2} = \sqrt{29}$, and $\left| \overrightarrow{OP} + \overrightarrow{OQ} \right| = \sqrt{(-4)^2 + 6^2} = \sqrt{52} = 2\sqrt{13}$. Drawing the figure:

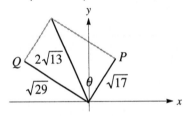

Using the law of cosines:

$$\left(\sqrt{29} \right)^2 = \left(2\sqrt{13} \right)^2 + \left(\sqrt{17} \right)^2 - 2\left(2\sqrt{13} \right)\left(\sqrt{17} \right)\cos\theta$$
$$29 = 69 - 4\sqrt{221}\cos\theta$$
$$-40 = -4\sqrt{221}\cos\theta$$
$$\cos\theta = \frac{10}{\sqrt{221}}$$
$$\theta = \cos^{-1}\left(\frac{10}{\sqrt{221}} \right) \approx 47.7°$$

13.   First draw the figure:

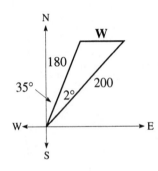

Using the law of cosines:
$$|\mathbf{W}|^2 = 180^2 + 200^2 - 2(180)(200)\cos 2° \approx 443.86$$
$$|\mathbf{W}| \approx 21.1$$
The speed of the wind is approximately 21.1 mph.

14. First draw the figure:

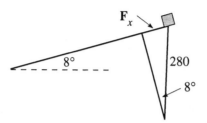

Therefore:
$$\sin 8° = \frac{\mathbf{F}_x}{280}$$
$$\mathbf{F}_x = 280 \sin 8° \approx 39.0$$
Any force greater than or equal to 39.0 pounds will be enough to roll the barrel up the ramp.

15. Here $r = \sqrt{(-1)^2 + \left(-\sqrt{3}\right)^2} = \sqrt{4} = 2$. Thus: $-1 - i\sqrt{3} = 2\left(-\dfrac{1}{2} - \dfrac{\sqrt{3}}{2}i\right) = 2\left(\cos\dfrac{4\pi}{3} + i\sin\dfrac{4\pi}{3}\right)$

16. First convert to trigonometric form: $1 + i = \sqrt{2}\left(\dfrac{1}{\sqrt{2}} + \dfrac{1}{\sqrt{2}}i\right) = \sqrt{2}\left(\cos\dfrac{\pi}{4} + i\sin\dfrac{\pi}{4}\right)$

Using DeMoivre's theorem: $(1 + i)^8 = \left(\sqrt{2}\right)^8 (\cos 2\pi + i\sin 2\pi) = 16(1 + 0i) = 16$

17. Two additional points are $\left(5, -\dfrac{5\pi}{6}\right)$ and $\left(-5, \dfrac{\pi}{6}\right)$.

18. Graphing the curve:

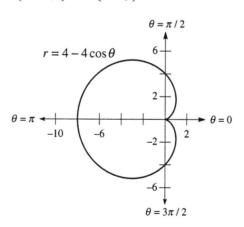

19. First convert to trigonometric form: $27i = 27(0 + 1i) = 27\left(\cos\dfrac{\pi}{2} + i\sin\dfrac{\pi}{2}\right)$

The cube roots of $27i$ are given by:

$$w_k = \sqrt[3]{27}\left(\cos\dfrac{\dfrac{\pi}{2} + 2k\pi}{3} + i\sin\dfrac{\dfrac{\pi}{2} + 2k\pi}{3}\right) = 3\left[\cos\left(\dfrac{\pi}{6} + \dfrac{2k\pi}{3}\right) + i\sin\left(\dfrac{\pi}{6} + \dfrac{2k\pi}{3}\right)\right]$$

Substituting $k = 0, 1, 2$:

$$w_0 = 3\left(\cos\dfrac{\pi}{6} + i\sin\dfrac{\pi}{6}\right) = 3\left(\dfrac{\sqrt{3}}{2} + \dfrac{1}{2}i\right) = \dfrac{3\sqrt{3}}{2} + \dfrac{3}{2}i$$

$$w_1 = 3\left(\cos\dfrac{5\pi}{6} + i\sin\dfrac{5\pi}{6}\right) = 3\left(-\dfrac{\sqrt{3}}{2} + \dfrac{1}{2}i\right) = -\dfrac{3\sqrt{3}}{2} + \dfrac{3}{2}i$$

$$w_2 = 3\left(\cos\dfrac{3\pi}{2} + i\sin\dfrac{3\pi}{2}\right) = 3(0 - 1i) = -3i$$

Chapter 9
Systems of Linear Equations and Inequalities

9.1 Elimination and Substitution: 2 x 2 Linear Systems

1. Adding the two equations:
$$5x = 20$$
$$x = 4$$
Substituting into the first equation:
$$2(4) + y = 12$$
$$8 + y = 12$$
$$y = 4$$
The solution is (4,4).

3. Multiplying the second equation by 2:
$$3x - 2y = 15$$
$$4x + 2y = 20$$
Adding the two equations:
$$7x = 35$$
$$x = 5$$
Substituting into the second equation:
$$2(5) + y = 10$$
$$10 + y = 10$$
$$y = 0$$
The solution is (5,0).

5. Multiplying the second equation by –5:
$$5x - 2y = 1$$
$$-5x + 25y = 160$$
Adding the two equations:
$$23y = 161$$
$$y = 7$$
Substituting into the second equation:
$$x - 5(7) = -32$$
$$x - 35 = -32$$
$$x = 3$$
The solution is (3,7).

7. Substituting into the second equation:
$$8x - 2(3x - 15) = 10$$
$$8x - 6x + 30 = 10$$
$$2x + 30 = 10$$
$$2x = -20$$
$$x = -10$$
Substituting into the first equation: $y = 3(-10) - 15 = -45$
The solution is $(-10, -45)$.

9. Multiplying the first equation by 6 and the second equation by 12 to clear fractions:
$$6\left(\tfrac{2}{3}y + \tfrac{3}{2}x\right) = 6(2) \qquad\qquad 12\left(\tfrac{3}{4}x + \tfrac{1}{12}y\right) = 12(1)$$
$$4y + 9x = 12 \qquad\qquad\qquad 9x + y = 12$$
The system of equations becomes:
$$9x + 4y = 12$$
$$9x + y = 12$$
Multiplying the second equation by -1:
$$9x + 4y = 12$$
$$-9x - y = -12$$
Adding the two equations:
$$3y = 0$$
$$y = 0$$
Substituting into the second equation:
$$9x + 0 = 12$$
$$x = \tfrac{4}{3}$$
The solution is $\left(\tfrac{4}{3}, 0\right)$.

11. Multiplying the first equation by 24 and the second equation by 12 to clear fractions:
$$24\left(\frac{a}{6} + \frac{b}{8}\right) = 24\left(\tfrac{3}{4}\right) \qquad\qquad 12\left(\frac{a}{4} + \frac{b}{3}\right) = 12\left(\tfrac{7}{12}\right)$$
$$4a + 3b = 18 \qquad\qquad\qquad 3a + 4b = 7$$
The system of equations becomes:
$$4a + 3b = 18$$
$$3a + 4b = 7$$
Multiplying the first equation by -3 and the second equation by 4:
$$-12a - 9b = -54$$
$$12a + 16b = 68$$
Adding the two equations:
$$7b = 14$$
$$b = 2$$
Substituting into the first equation:
$$4a + 3(2) = 18$$
$$4a + 6 = 18$$
$$4a = 12$$
$$a = 3$$
The solution is $(3, 2)$.

13. Multiplying the first equation by 6 and the second equation by 6 to clear fractions:
$$6\left(\frac{x+3}{2} + \frac{y-4}{3}\right) = 6\left(\tfrac{10}{6}\right) \qquad\qquad 6\left(\frac{x-2}{3} + \frac{y-2}{2}\right) = 6(2)$$
$$3x + 9 + 2y - 8 = 10 \qquad\qquad 2x - 4 + 3y - 6 = 12$$
$$3x + 2y = 9 \qquad\qquad\qquad 2x + 3y = 22$$

The system of equations becomes:

$$3x + 2y = 9$$
$$2x + 3y = 22$$

Multiplying the first equation by –3 and the second equation by 2:

$$-9x - 6y = -27$$
$$4x + 6y = 44$$

Adding the two equations:

$$-5x = 17$$
$$x = -\tfrac{17}{5}$$

Substituting into the first equation:

$$3\left(-\tfrac{17}{5}\right) + 2y = 9$$
$$-\tfrac{51}{5} + 2y = 9$$
$$2y = \tfrac{96}{5}$$
$$y = \tfrac{48}{5}$$

The solution is $\left(-\tfrac{17}{5}, \tfrac{48}{5}\right)$.

15. Solving the first equation for x:

$$\frac{x}{2} - 0.03y = 0.6$$
$$\frac{x}{2} = 0.03y + 0.6$$
$$x = 0.06y + 1.2$$

Substituting into the second equation:

$$0.0004y + \frac{0.06y + 1.2}{2} = 0.2$$
$$0.0004y + 0.03y + 0.6 = 0.2$$
$$0.0304y = -0.4$$
$$y = -\tfrac{250}{19}$$

Substituting into the first equation: $x = \tfrac{3}{50}\left(-\tfrac{250}{19}\right) + \tfrac{6}{5} = \tfrac{39}{95}$

The solution is $\left(\tfrac{39}{95}, -\tfrac{250}{19}\right)$.

17. Let x and y represent the amounts of 30% and 45% alcohol solutions, respectively. The system of equations is:

$$x + y = 30$$
$$0.30x + 0.45y = 0.40(30)$$

Solving the first equation for y yields $y = 30 - x$. Substituting into the second equation:

$$0.30x + 0.45(30 - x) = 12$$
$$0.30x + 13.5 - 0.45x = 12$$
$$-0.15x + 13.5 = 12$$
$$-0.15x = -1.5$$
$$x = 10$$
$$y = 30 - 10 = 20$$

So 10 liters of 30% solution and 20 liters of 45% solution should be mixed.

19. Let x and y represent the amounts invested at 8% and 9.5%, respectively. The system of equations is:
$$x + y = 20,000$$
$$0.08x + 0.095y = 1690$$
Solving the first equation for y yields $y = 20,000 - x$. Substituting into the second equation:
$$0.08x + 0.095(20,000 - x) = 1690$$
$$0.08x + 1900 - 0.095x = 1690$$
$$-0.015x + 1900 = 1690$$
$$-0.015x = -210$$
$$x = 14,000$$
$$y = 20,000 - 14,000 = 6,000$$
Carol should invest $14,000 at 8% and $6,000 at 9.5% interest.

21. Let F represent the flat fee and m represent the mileage charge. The system of equations is:
$$F + 90m = 58.30$$
$$F + 140m = 74.30$$
Multiplying the first equation by -1:
$$-F - 90m = -58.30$$
$$F + 140m = 74.30$$
Adding the two equations:
$$50m = 16$$
$$m = 0.32$$
Substituting into the first equation:
$$F + 90(0.32) = 58.30$$
$$F + 28.80 = 58.30$$
$$F = 29.50$$
The flat fee is $29.50/day and the mileage fee is 32 cents/mile.

23. Let s represent the speed of the plane and w represent the speed of the wind. The system of equations is:
$$4(s + w) = 2520$$
$$4(s - w) = 2280$$
Simplifying the equations:
$$s + w = 630$$
$$s - w = 570$$
Adding the two equations:
$$2s = 1200$$
$$s = 600$$
Substituting into the simplified first equation:
$$600 + w = 630$$
$$w = 30$$
The speed of the plane is 600 mph and the speed of the wind is 30 mph.

25. Let m and l represent the more and less expensive models, respectively. The system of equations is:
$$1m + \tfrac{3}{4}l = 150$$
$$\tfrac{1}{2}m + \tfrac{1}{4}l = 60$$
Multiplying the first equation by 4 and the second equation by 4 to clear fractions:

$$4\left(1m + \tfrac{3}{4}l\right) = 4(150) \qquad\qquad 4\left(\tfrac{1}{2}m + \tfrac{1}{4}l\right) = 4(60)$$
$$4m + 3l = 600 \qquad\qquad\qquad\qquad 2m + l = 240$$
The system of equations is:
$$4m + 3l = 600$$
$$2m + l = 240$$

Multiplying the second equation by -2:
$$4m + 3l = 600$$
$$-4m - 2l = -480$$
Adding the two equations yields $l = 120$. Substituting into the second equation:
$$2m + 120 = 240$$
$$2m = 120$$
$$m = 60$$
The company can produce 60 of the more expensive model and 120 of the less expensive model.

27. Let m represent the number of miles driven during the day. The total charges for each company are:

Goodman: $18 + 0.28m$
Hirsch: $14 + 0.32m$

Finding when these are equal:
$$18 + 0.28m = 14 + 0.32m$$
$$18 = 14 + 0.04m$$
$$4 = 0.04m$$
$$m = 100$$
For mileage under 100 miles, the Hirsch company will be cheaper. For mileage over 100 miles, the Goodman company will be cheaper.

29. Let t represent the air time. The charges for each plan are:

plan A: $0.90t$
plan B: $20 + 0.70t$

Finding when these are equal:
$$20 + 0.70t = 0.90t$$
$$20 = 0.20t$$
$$t = 100$$
Plan A is cheaper for air time less than 100 minutes, while plan B is cheaper for air time greater than 100 minutes.

31. Substituting each pair of values results in the equations:
$$64 = -16(1)^2 + v_0(1) + s_0$$
$$96 = -16(2)^2 + v_0(2) + s_0$$
Simplifying these equations:
$$v_0 + s_0 = 80$$
$$2v_0 + s_0 = 160$$
Multiplying the first equation by -1:
$$-v_0 - s_0 = -80$$
$$2v_0 + s_0 = 160$$
Adding the two equations yields $v_0 = 80$. Substituting into the simplified first equation:
$$80 + s_0 = 80$$
$$s_0 = 0$$
The initial velocity and height are $v_0 = 80$ ft/sec and $s_0 = 0$ ft.

33. Multiplying the equation by $(x - 2)(2x + 1)$:
$$10 = A(2x + 1) + B(x - 2)$$
$$10 = 2Ax + A + Bx - 2B$$
$$10 = (2A + B)x + (A - 2B)$$
Setting coefficients equal results in the equations:
$$2A + B = 0$$
$$A - 2B = 10$$

Multiplying the first equation by 2:
$$4A + 2B = 0$$
$$A - 2B = 10$$
Adding the two equations:
$$5A = 10$$
$$A = 2$$
Substituting into the first equation:
$$2(2) + B = 0$$
$$B = -4$$
So $A = 2$, $B = -4$.

35. Multiplying the equation by $(x+1)^2$:
$$2x - 1 = A(x+1) + B$$
$$2x - 1 = Ax + (A + B)$$
Setting coefficients equal results in the equations:
$$A = 2$$
$$A + B = -1$$
Substituting into the second equation:
$$2 + B = -1$$
$$B = -3$$
So $A = 2$, $B = -3$.

37. Using the slope formula, the equations represented are:

$$\frac{y-2}{x+1} = 3 \qquad\qquad \frac{y+1}{x-2} = 1$$
$$y - 2 = 3x + 3 \qquad\qquad y + 1 = x - 2$$
$$y = 3x + 5 \qquad\qquad\qquad y = x - 3$$

Solving the system by substitution:
$$3x + 5 = x - 3$$
$$2x = -8$$
$$x = -4$$
$$y = -4 - 3 = -7$$
The point is $(-4, -7)$.

39. There are an infinite number of solutions, all lying on the line $y = 2x - 5$.

41. Geometrically, it means that the two lines (represented by the equations) are parallel.

43. Dividing the second equation by 2 yields $y = -5$. Substituting into the first equation:
$$5x - 3(-5) = 8$$
$$5x + 15 = 8$$
$$5x = -7$$
$$x = -\frac{7}{5}$$

The solution is $\left(-\frac{7}{5}, -5\right)$. No elimination is necessary.

9.2 Elimination and Gaussian Elimination: 3 x 3 Linear Systems

1. Adding equations 1 and 2 results in $4x + 2y = 12$. Adding equations 1 and 3 results in $3x + 2y = 10$.
    We have the 2 x 2 system:
    $$4x + 2y = 12$$
    $$3x + 2y = 10$$
    Multiplying the second equation by –1:
    $$4x + 2y = 12$$
    $$-3x - 2y = -10$$
    Adding yields $x = 2$. Substituting to find y:
    $$4(2) + 2y = 12$$
    $$8 + 2y = 12$$
    $$2y = 4$$
    $$y = 2$$
    Substituting into the original first equation to find z:
    $$2 + 2 + z = 6$$
    $$z = 2$$
    The solution is (2,2,2).

3. Adding equations 1 and 3 results in:
    $$2x = 4$$
    $$x = 2$$
    Multiply equation 3 by 2:
    $$2x + 2y + 2z = 0$$
    $$2x - 2y + 2z = 6$$
    Adding:
    $$4x + 4z = 6$$
    $$4(2) + 4z = 6$$
    $$4z = -2$$
    $$z = -\tfrac{1}{2}$$
    Substituting into equation 2:
    $$2(2) + 2y + 2\left(-\tfrac{1}{2}\right) = 0$$
    $$4 + 2y - 1 = 0$$
    $$2y = -3$$
    $$y = -\tfrac{3}{2}$$
    The solution is $\left(2, -\tfrac{3}{2}, -\tfrac{1}{2}\right)$.

5. Multiply equation 1 by –2:
    $$-2x - 2y - 2z = -12$$
    $$2x + 2y + 2z = 4$$
    Adding yields $0 = -8$, which is impossible. There is no solution to the system (inconsistent).

7. Adding equations 2 and 3 results in:
    $$1.3a + 1.2c = 6.86$$
    $$13a + 12c = 68.6$$
    Multiply equation 2 by 0.2:
    $$0.01a + 0.2b + c = 0.45$$
    $$0.06a - 0.2b + 0.04c = 0.212$$
    Adding:
    $$0.07a + 1.04c = 0.662$$
    $$7a + 104c = 66.2$$

We have the 2 x 2 system:
$$13a + 12c = 68.6$$
$$7a + 104c = 66.2$$
Multiplying the first equation by –7 and the second equation is 13:
$$-91a - 84c = -480.2$$
$$91a + 1352c = 860.6$$
Adding:
$$1268c = 380.4$$
$$c = 0.3$$
Substituting to find a:
$$7a + 104(0.3) = 66.2$$
$$7a = 35$$
$$a = 5$$
Substituting into the original third equation:
$$5 + b + 0.3 = 5.8$$
$$b = 0.5$$
The solution is (5,0.5,0.3).

9.   First simplify the system by multiplying equation 1 by 12, equation 2 by 6, and equation 3 by 8:

$$12\left(\frac{x}{4} + \frac{y}{6} - \frac{z}{3}\right) = 12(1) \qquad 6\left(\frac{x}{2} + \frac{y}{3} + z\right) = 6(-3) \qquad 8\left(\frac{x}{8} + \frac{y}{4} - z\right) = 8(5)$$

$$3x + 2y - 4z = 12 \qquad\qquad 3x + 2y + 6z = -18 \qquad\qquad x + 2y - 8z = 40$$

We have the system:
$$3x + 2y - 4z = 12$$
$$3x + 2y + 6z = -18$$
$$x + 2y - 8z = 40$$
Multiply equation 1 by –1:
$$-3x - 2y + 4z = -12$$
$$3x + 2y + 6z = -18$$
Adding yields $10z = -30$, so $z = -3$. Multiply equation 3 by –1:
$$3x + 2y + 6z = -18$$
$$-x - 2y + 8z = -40$$
Adding yields $2x + 14z = -58$. Substituting $z = -3$:
$$2x + 14(-3) = -58$$
$$2x - 42 = -58$$
$$2x = -16$$
$$x = -8$$
Substituting into the first equation:
$$3(-8) + 2y - 4(-3) = 12$$
$$2y - 12 = 12$$
$$2y = 24$$
$$y = 12$$
The solution is (–8,12,–3).

11.   Multiplying equation 1 by –1 and adding to equation 2:
$$-x - y + z = -7$$
$$x + 2y + 2z = 4$$
Adding yields $y + 3z = -3$. Multiplying equation 1 by –2 and adding to equation 3:
$$-2x - 2y + 2z = -14$$
$$2x - y + z = -1$$

Adding yields $-3y + 3z = -15$, or $-y + z = -5$. So we have the system:
$$x + y - z = 7$$
$$y + 3z = -3$$
$$-y + z = -5$$
Adding equations 2 and 3 yields $4z = -8$, so the system becomes:
$$x + y - z = 7$$
$$y + 3z = -3$$
$$4z = -8$$
Solving equation 3 for z yields $z = -2$. Substituting into equation 2:
$$y + 3(-2) = -3$$
$$y - 6 = -3$$
$$y = 3$$
Substituting into equation 1:
$$x + 3 - (-2) = 7$$
$$x + 5 = 7$$
$$x = 2$$
The solution is $(2, 3, -2)$.

13.   Multiplying equation 1 by -2 and adding to equation 2:
$$-2x - 2y + 4z = -6$$
$$2x + y + z = -4$$
Adding yields $-y + 5z = -10$. Multiplying equation 1 by -1 and adding to equation 3:
$$-x - y + 2z = -3$$
$$x + 2y - 3z = 5$$
Adding yields $y - z = 2$. So we have the system:
$$x + y - 2z = 3$$
$$-y + 5z = -10$$
$$y - z = 2$$
Adding equations 2 and 3 yields $z = -2$. Substituting into equation 2:
$$-y + 5(-2) = -10$$
$$-y - 10 = -10$$
$$-y = 0$$
$$y = 0$$
Substituting into equation 1:
$$x + 0 - 2(-2) = 3$$
$$x + 4 = 3$$
$$x = -1$$
The solution is $(-1, 0, -2)$.

15.   Using $x - y - z = 5$ as equation 1, multiply by -2 and add to equation 2:
$$-2x + 2y + 2z = -10$$
$$2x + y + z = 4$$
Adding yields $3y + 3z = -6$, or $y + z = -2$. Multiplying equation 1 by -2 and adding to equation 3:
$$-2x + 2y + 2z = -10$$
$$2x - 2y - 2z = 1$$
Adding yields $0 = -9$, which is false. There is no solution to the system (inconsistent).

17.   Let b, c, and s represent the amount invested in the bank, certificate, and stock, respectively. The system of equations is:
$$b + c + s = 19750$$
$$0.06b + 0.08c + 0.105s = 1680$$
$$0.105s = 0.06b + 0.08c$$

Rewriting the system:
$$b + c + s = 19750$$
$$0.06b + 0.08c + 0.105s = 1680$$
$$-0.06b - 0.08c + 0.105s = 0$$
Adding equation 2 and equation 3:
$$0.21s = 1680$$
$$s = 8000$$
Substituting into the first and second equations:

$b + c + 8000 = 19750$ $\qquad\qquad$ $0.06b + 0.08c + 0.105(8000) = 1680$

$\qquad b + c = 11750$ $\qquad\qquad\qquad\qquad$ $0.06b + 0.08c = 840$

We have the 2 x 2 system:
$$b + c = 11750$$
$$0.06b + 0.08c = 840$$
Multiplying the first equation by –0.06:
$$-0.06b - 0.06c = -750$$
$$0.06b + 0.08c = 840$$
Adding yields:
$$0.02c = 135$$
$$c = 6750$$
Substituting into the first equation:
$$b + 6750 = 11750$$
$$b = 5000$$
Elaine invested $5000 in the bank, $6750 in the certificate of deposit, and $8000 in the stock.

19. The system of equations is:
$$0.2A + 0.3B + 0.6C = 147$$
$$0.4A + 0.5B + 0.8C = 227$$
$$1.2A + 1.4B + 1.8C = 586$$
Multiplying the first equation by –2 and adding it to the second equation:
$$-0.4A - 0.6B - 1.2C = -294$$
$$0.4A + 0.5B + 0.8C = 227$$
Adding:
$$-0.1B - 0.4C = -67$$
$$0.1B + 0.4C = 67$$
Multiplying the first equation by –6 and adding it to the third equation:
$$-1.2A - 1.8B - 3.6C = -882$$
$$1.2A + 1.4B + 1.8C = 586$$
Adding:
$$-0.4B - 1.8C = -296$$
$$0.4B + 1.8C = 296$$
The system of equations is:
$$0.2A + 0.3B + 0.6C = 147$$
$$0.1B + 0.4C = 67$$
$$0.4B + 1.8C = 296$$
Multiplying the second equation by –4 and adding it to the third equation:
$$-0.4B - 1.6C = -268$$
$$0.4B + 1.8C = 296$$
Adding yields $0.2C = 28$. So we have the system:
$$0.2A + 0.3B + 0.6C = 147$$
$$0.1B + 0.4C = 67$$
$$0.2C = 28$$

Solving the third equation for C yields $C = 140$. Substituting into the second equation:
$$0.1B + 0.4(140) = 67$$
$$0.1B + 56 = 67$$
$$0.1B = 11$$
$$B = 110$$
Substituting into the first equation:
$$0.2A + 0.3(110) + 0.6(140) = 147$$
$$0.2A + 117 = 147$$
$$0.2A = 30$$
$$A = 150$$
The company can produce 150 of model A, 110 of model B, and 140 of model C.

21. Let x, y, z represent the quantities of 10%, 25%, and 40% acid to use. The system of equations is:
$$x + y + z = 100$$
$$0.10x + 0.25y + 0.40z = 0.30(100)$$
$$y = x + 20$$
Substituting $y = x + 20$ into the first equation:
$$x + x + 20 + z = 100$$
$$2x + z = 80$$
Substituting $y = x + 20$ into the second equation:
$$0.10x + 0.25(x + 20) + 0.40z = 30$$
$$0.10x + 0.25x + 5 + 0.40z = 30$$
$$0.35x + 0.40z = 25$$
We have the 2 x 2 system:
$$2x + z = 80$$
$$0.35x + 0.40z = 25$$
Multiplying the first equation by -0.40:
$$-0.80x - 0.40z = -32$$
$$0.35x + 0.40z = 25$$
Adding yields $-0.45x = -7$, so $x = \frac{140}{9} = 15\frac{5}{9}$. Substituting to find z:
$$2\left(\frac{140}{9}\right) + z = 80$$
$$\frac{280}{9} + z = 80$$
$$z = \frac{440}{9} = 48\frac{8}{9}$$
Finally: $y = 15\frac{5}{9} + 20 = 35\frac{5}{9}$

The chemist will need $15\frac{5}{9}$ liters of 10% acid, $35\frac{5}{9}$ liters of 25% acid, and $48\frac{8}{9}$ liters of 40% acid.

23. Substituting $f(1) = 2$, $f(-1) = 6$, and $f(2) = 9$:

$$a(1)^2 + b(1) + c = 2 \qquad a(-1)^2 + b(-1) + c = 6 \qquad a(2)^2 + b(2) + c = 9$$
$$a + b + c = 2 \qquad\qquad a - b + c = 6 \qquad\qquad 4a + 2b + c = 9$$

The system of equations is:
$$a + b + c = 2$$
$$a - b + c = 6$$
$$4a + 2b + c = 9$$
Multiplying the first equation by -1 and adding it to the second equation:
$$-a - b - c = -2$$
$$a - b + c = 6$$

Adding yields $-2b = 4$, so $b = -2$. Multiplying the first equation by -4 and adding it to the third equation:

$$-4a - 4b - 4c = -8$$
$$4a + 2b + c = 9$$

Adding yields $-2b - 3c = 1$. Substituting $b = -2$:

$$-2(-2) - 3c = 1$$
$$4 - 3c = 1$$
$$-3c = -3$$
$$c = 1$$

Substituting into the first equation:

$$a + (-2) + 1 = 2$$
$$a - 1 = 2$$
$$a = 3$$

Thus the function is $f(x) = 3x^2 - 2x + 1$.

25. Multiplying the equation by $x(x-1)^2$:

$$2x + 5 = A(x-1)^2 + Bx(x-1) + Cx$$
$$2x + 5 = Ax^2 - 2Ax + A + Bx^2 - Bx + Cx$$
$$2x + 5 = (A + B)x^2 + (-2A - B + C)x + A$$

Setting coefficients equal results in the equations:

$$A + B = 0$$
$$-2A - B + C = 2$$
$$A = 5$$

Substituting into the first equation:

$$5 + B = 0$$
$$B = -5$$

Substituting into the second equation:

$$-2(5) - (-5) + C = 2$$
$$-5 + C = 2$$
$$C = 7$$

So $A = 5$, $B = -5$, $C = 7$.

27. Multiplying the equation by $\left(x^2 + 1\right)^2$:

$$x^3 + 2x^2 + 3x + 5 = (Ax + B)\left(x^2 + 1\right) + Cx + D$$
$$x^3 + 2x^2 + 3x + 5 = Ax^3 + Bx^2 + Ax + B + Cx + D$$
$$x^3 + 2x^2 + 3x + 5 = Ax^3 + Bx^2 + (A + C)x + (B + D)$$

Setting coefficients equal results in the equations:

$$A = 1$$
$$B = 2$$
$$A + C = 3$$
$$B + D = 5$$

Substituting $A = 1$ into the third equation results in $1 + C = 3$, so $C = 2$. Substituting $B = 2$ into the fourth equation results in $2 + D = 5$, so $D = 3$. So $A = 1$, $B = 2$, $C = 2$, $D = 3$.

9.3 Solving Linear Systems Using Augmented Matrices

1. The augmented matrix is $\begin{bmatrix} 7 & 1 & \vdots & 1 \\ 1 & -3 & \vdots & 4 \end{bmatrix}$.

3. The augmented matrix is $\begin{bmatrix} 3 & -2 & 1 & \vdots & 15 \\ 2 & 1 & -3 & \vdots & 10 \\ 5 & -3 & 1 & \vdots & -2 \end{bmatrix}$.

5. Performing the row operation $-5R_1 + R_2 \rightarrow R_2$: $\begin{bmatrix} 1 & 3 & \vdots & 2 \\ 0 & -11 & \vdots & -10 \end{bmatrix}$

7. Performing the row operation $-2R_1 + R_2 \rightarrow R_2$: $\begin{bmatrix} 1 & 2 & 1 & \vdots & 0 \\ 0 & 2 & 1 & \vdots & 1 \\ 3 & 8 & 7 & \vdots & -8 \end{bmatrix}$

   Performing the row operation $-3R_1 + R_3 \rightarrow R_3$: $\begin{bmatrix} 1 & 2 & 1 & \vdots & 0 \\ 0 & 2 & 1 & \vdots & 1 \\ 0 & 2 & 4 & \vdots & -8 \end{bmatrix}$

   Performing the row operation $-R_2 + R_3 \rightarrow R_3$: $\begin{bmatrix} 1 & 2 & 1 & \vdots & 0 \\ 0 & 2 & 1 & \vdots & 1 \\ 0 & 0 & 3 & \vdots & -9 \end{bmatrix}$

9. Form the augmented matrix: $\begin{bmatrix} 1 & -1 & \vdots & 4 \\ 2 & 1 & \vdots & 11 \end{bmatrix}$

   Using the row operation $-2R_1 + R_2 \rightarrow R_2$: $\begin{bmatrix} 1 & -1 & \vdots & 4 \\ 0 & 3 & \vdots & 3 \end{bmatrix}$

   The system of equations is:
   $$x - y = 4$$
   $$3y = 3$$
   Since $3y = 3$, $y = 1$. Substituting:
   $$x - 1 = 4$$
   $$x = 5$$
   The solution is (5,1).

11. Form the augmented matrix: $\begin{bmatrix} 1 & 1 & \vdots & -1 \\ 3 & -2 & \vdots & 12 \end{bmatrix}$

   Using the row operation $-3R_1 + R_2 \rightarrow R_2$: $\begin{bmatrix} 1 & 1 & \vdots & -1 \\ 0 & -5 & \vdots & 15 \end{bmatrix}$

   The system of equations is:
   $$x + y = -1$$
   $$-5y = 15$$
   Since $-5y = 15$, $y = -3$. Substituting:
   $$x - 3 = -1$$
   $$x = 2$$
   The solution is (2,–3).

13. Form the augmented matrix: $\begin{bmatrix} 5 & -2 & | & 1 \\ 1 & -5 & | & -32 \end{bmatrix}$

Using the row operation $R_1 \leftrightarrow R_2$: $\begin{bmatrix} 1 & -5 & | & -32 \\ 5 & -2 & | & 1 \end{bmatrix}$

Using the row operation $-5R_1 + R_2 \rightarrow R_2$: $\begin{bmatrix} 1 & -5 & | & -32 \\ 0 & 23 & | & 161 \end{bmatrix}$

The system of equations is:
$$x - 5y = -32$$
$$23y = 161$$
Since $23y = 161$, $y = 7$. Substituting:
$$x - 5(7) = -32$$
$$x - 35 = -32$$
$$x = 3$$
The solution is (3,7).

15. Write the second equation as $-3x + y = -15$. Form the augmented matrix: $\begin{bmatrix} 8 & -2 & | & 10 \\ -3 & 1 & | & -15 \end{bmatrix}$

Using the row operation $\frac{1}{8}R_1 \rightarrow R_1$: $\begin{bmatrix} 1 & -\frac{1}{4} & | & \frac{5}{4} \\ -3 & 1 & | & -15 \end{bmatrix}$

Using the row operation $3R_1 + R_2 \rightarrow R_2$: $\begin{bmatrix} 1 & -\frac{1}{4} & | & \frac{5}{4} \\ 0 & \frac{1}{4} & | & -\frac{45}{4} \end{bmatrix}$

The system of equations is:
$$x - \frac{1}{4}y = \frac{5}{4}$$
$$\frac{1}{4}y = -\frac{45}{4}$$
Since $\frac{1}{4}y = -\frac{45}{4}$, $y = -45$. Substituting:
$$x - \frac{1}{4}(-45) = \frac{5}{4}$$
$$x + \frac{45}{4} = \frac{5}{4}$$
$$x = -10$$
The solution is (–10,–45).

17. Form the augmented matrix: $\begin{bmatrix} 3 & 2 & | & 2 \\ 3 & 1 & | & 1 \end{bmatrix}$

Using the row operation $\frac{1}{3}R_1 \rightarrow R_1$: $\begin{bmatrix} 1 & \frac{2}{3} & | & \frac{2}{3} \\ 3 & 1 & | & 1 \end{bmatrix}$

Using the row operation $-3R_1 + R_2 \rightarrow R_2$: $\begin{bmatrix} 1 & \frac{2}{3} & | & \frac{2}{3} \\ 0 & -1 & | & -1 \end{bmatrix}$

The system of equations is:
$$x + \frac{2}{3}y = \frac{2}{3}$$
$$-y = -1$$
Since $-y = -1$, $y = 1$. Substituting:
$$x + \frac{2}{3}(1) = \frac{2}{3}$$
$$x = 0$$
The solution is (0,1).

19. Form the augmented matrix:
$$\begin{bmatrix} 1 & 1 & 1 & | & 6 \\ 2 & 1 & -1 & | & 4 \\ 3 & 1 & -1 & | & 6 \end{bmatrix}$$

Using the row operations $-2R_1 + R_2 \to R_2$ and $-3R_1 + R_3 \to R_3$:
$$\begin{bmatrix} 1 & 1 & 1 & | & 6 \\ 0 & -1 & -3 & | & -8 \\ 0 & -2 & -4 & | & -12 \end{bmatrix}$$

Using the row operations $-R_2 \to R_2$ and $-R_3 \to R_3$:
$$\begin{bmatrix} 1 & 1 & 1 & | & 6 \\ 0 & 1 & 3 & | & 8 \\ 0 & 2 & 4 & | & 12 \end{bmatrix}$$

Using the row operation $-2R_2 + R_3 \to R_3$:
$$\begin{bmatrix} 1 & 1 & 1 & | & 6 \\ 0 & 1 & 3 & | & 8 \\ 0 & 0 & -2 & | & -4 \end{bmatrix}$$

The system of equations is:
$$x + y + z = 6$$
$$y + 3z = 8$$
$$-2z = -4$$

Since $-2z = -4$, $z = 2$. Substituting into the second equation:
$$y + 3(2) = 8$$
$$y + 6 = 8$$
$$y = 2$$

Substituting into the first equation:
$$x + 2 + 2 = 6$$
$$x + 4 = 6$$
$$x = 2$$

The solution is $(2,2,2)$.

21. Form the augmented matrix:
$$\begin{bmatrix} 1 & 1 & -2 & | & -5 \\ 1 & -1 & 2 & | & 9 \\ 3 & 1 & -1 & | & 2 \end{bmatrix}$$

Using the row operations $-R_1 + R_2 \to R_2$ and $-3R_1 + R_3 \to R_3$:
$$\begin{bmatrix} 1 & 1 & -2 & | & -5 \\ 0 & -2 & 4 & | & 14 \\ 0 & -2 & 5 & | & 17 \end{bmatrix}$$

Using the row operation $-R_2 + R_3 \to R_3$:
$$\begin{bmatrix} 1 & 1 & -2 & | & -5 \\ 0 & -2 & 4 & | & 14 \\ 0 & 0 & 1 & | & 3 \end{bmatrix}$$

The system of equations is:
$$x + y - 2z = -5$$
$$-2y + 4z = 14$$
$$z = 3$$

Substituting into the second equation:
$$-2y + 4(3) = 14$$
$$-2y + 12 = 14$$
$$-2y = 2$$
$$y = -1$$

Substituting into the first equation:
$$x - 1 - 2(3) = -5$$
$$x - 7 = -5$$
$$x = 2$$
The solution is $(2,-1,3)$.

23. Form the augmented matrix: $\begin{bmatrix} 1 & 1 & 1 & \vdots & 6 \\ 3 & 1 & -1 & \vdots & 4 \\ 2 & 2 & 2 & \vdots & 4 \end{bmatrix}$

Using the row operations $-3R_1 + R_2 \rightarrow R_2$ and $-2R_1 + R_3 \rightarrow R_3$: $\begin{bmatrix} 1 & 1 & 1 & \vdots & 6 \\ 0 & -2 & -4 & \vdots & -14 \\ 0 & 0 & 0 & \vdots & -8 \end{bmatrix}$

The system of equations is:
$$x + y + z = 6$$
$$-2y - 4z = -14$$
$$0 = -8$$
Since this last statement is false, there is no solution (inconsistent).

25. Form the augmented matrix: $\begin{bmatrix} 2 & -2 & 1 & \vdots & 0 \\ 4 & -1 & 2 & \vdots & 6 \\ 2 & 3 & -1 & \vdots & 4 \end{bmatrix}$

Using the row operations $-2R_1 + R_2 \rightarrow R_2$ and $-R_1 + R_3 \rightarrow R_3$: $\begin{bmatrix} 2 & -2 & 1 & \vdots & 0 \\ 0 & 3 & 0 & \vdots & 6 \\ 0 & 5 & -2 & \vdots & 4 \end{bmatrix}$

Using the row operation $\frac{1}{3}R_2 \rightarrow R_2$: $\begin{bmatrix} 2 & -2 & 1 & \vdots & 0 \\ 0 & 1 & 0 & \vdots & 2 \\ 0 & 5 & -2 & \vdots & 4 \end{bmatrix}$

Using the row operation $-5R_2 + R_3 \rightarrow R_3$: $\begin{bmatrix} 2 & -2 & 1 & \vdots & 0 \\ 0 & 1 & 0 & \vdots & 2 \\ 0 & 0 & -2 & \vdots & -6 \end{bmatrix}$

The system of equations is:
$$2x - 2y + z = 0$$
$$y = 2$$
$$-2z = -6$$
Since $-2z = -6$, $z = 3$. Substituting into the first equation:
$$2x - 2(2) + 3 = 0$$
$$2x - 1 = 0$$
$$2x = 1$$
$$x = \frac{1}{2}$$
The solution is $\left(\frac{1}{2}, 2, 3\right)$.

27. Form the augmented matrix: $\begin{bmatrix} 1 & 1 & 1 & -1 & \vdots & 4 \\ 1 & 2 & 2 & 1 & \vdots & 3 \\ 2 & 1 & 3 & -2 & \vdots & 7 \\ 3 & 1 & -1 & 2 & \vdots & 5 \end{bmatrix}$

Using the row operations $-R_1 + R_2 \rightarrow R_2$, $-2R_1 + R_3 \rightarrow R_3$, and $-3R_1 + R_4 \rightarrow R_4$:

$\begin{bmatrix} 1 & 1 & 1 & -1 & \vdots & 4 \\ 0 & 1 & 1 & 2 & \vdots & -1 \\ 0 & -1 & 1 & 0 & \vdots & -1 \\ 0 & -2 & -4 & 5 & \vdots & -7 \end{bmatrix}$

Using the row operations $R_2 + R_3 \rightarrow R_3$ and $2R_2 + R_4 \rightarrow R_4$: $\begin{bmatrix} 1 & 1 & 1 & -1 & \vdots & 4 \\ 0 & 1 & 1 & 2 & \vdots & -1 \\ 0 & 0 & 2 & 2 & \vdots & -2 \\ 0 & 0 & -2 & 9 & \vdots & -9 \end{bmatrix}$

Using the row operation $R_3 + R_4 \rightarrow R_4$: $\begin{bmatrix} 1 & 1 & 1 & -1 & \vdots & 4 \\ 0 & 1 & 1 & 2 & \vdots & -1 \\ 0 & 0 & 2 & 2 & \vdots & -2 \\ 0 & 0 & 0 & 11 & \vdots & -11 \end{bmatrix}$

The system of equations is:
$$w + x + y - z = 4$$
$$x + y + 2z = -1$$
$$2y + 2z = -2$$
$$11z = -11$$
Since $11z = -11$, $z = -1$. Substituting into the third equation:
$$2y + 2(-1) = -2$$
$$2y - 2 = -2$$
$$2y = 0$$
$$y = 0$$
Substituting into the second equation:
$$w + 1 + 0 - (-1) = 4$$
$$w + 2 = 4$$
$$w = 2$$
The solution is $(2,1,0,-1)$.

29. Using the row operation $\frac{1}{3}R_1 \rightarrow R_1$: $\begin{bmatrix} 1 & 2 & \vdots & 0 \\ 3 & 1 & \vdots & 5 \end{bmatrix}$

Using the row operation $-3R_1 + R_2 \rightarrow R_2$: $\begin{bmatrix} 1 & 2 & \vdots & 0 \\ 0 & -5 & \vdots & 5 \end{bmatrix}$

Using the row operation $-\frac{1}{5}R_2 \rightarrow R_2$: $\begin{bmatrix} 1 & 2 & \vdots & 0 \\ 0 & 1 & \vdots & -1 \end{bmatrix}$

Using the row operation $-2R_2 + R_1 \rightarrow R_1$: $\begin{bmatrix} 1 & 0 & \vdots & 2 \\ 0 & 1 & \vdots & -1 \end{bmatrix}$

31. Using the row operation $-2R_1 + R_2 \rightarrow R_2$: $\begin{bmatrix} 1 & 4 & | & 0 \\ 0 & -5 & | & -5 \end{bmatrix}$

Using the row operation $-\frac{1}{5}R_2 \rightarrow R_2$: $\begin{bmatrix} 1 & 4 & | & 0 \\ 0 & 1 & | & 1 \end{bmatrix}$

Using the row operation $-4R_2 + R_1 \rightarrow R_1$: $\begin{bmatrix} 1 & 0 & | & -4 \\ 0 & 1 & | & 1 \end{bmatrix}$

33. Using the row operations $-2R_1 + R_2 \rightarrow R_2$ and $-3R_1 + R_3 \rightarrow R_3$: $\begin{bmatrix} 1 & 1 & 2 & | & 0 \\ 0 & 2 & -2 & | & 6 \\ 0 & -2 & -4 & | & 6 \end{bmatrix}$

Using the row operation $\frac{1}{2}R_2 \rightarrow R_2$: $\begin{bmatrix} 1 & 1 & 2 & | & 0 \\ 0 & 1 & -1 & | & 3 \\ 0 & -2 & -4 & | & 6 \end{bmatrix}$

Using the row operation $2R_2 + R_3 \rightarrow R_3$: $\begin{bmatrix} 1 & 1 & 2 & | & 0 \\ 0 & 1 & -1 & | & 3 \\ 0 & 0 & -6 & | & 12 \end{bmatrix}$

Using the row operation $-\frac{1}{6}R_3 \rightarrow R_3$: $\begin{bmatrix} 1 & 1 & 2 & | & 0 \\ 0 & 1 & -1 & | & 3 \\ 0 & 0 & 1 & | & -2 \end{bmatrix}$

Using the row operations $-2R_3 + R_1 \rightarrow R_1$ and $R_3 + R_2 \rightarrow R_2$: $\begin{bmatrix} 1 & 1 & 0 & | & 4 \\ 0 & 1 & 0 & | & 1 \\ 0 & 0 & 1 & | & -2 \end{bmatrix}$

Using the row operation $-R_2 + R_1 \rightarrow R_1$: $\begin{bmatrix} 1 & 0 & 0 & | & 3 \\ 0 & 1 & 0 & | & 1 \\ 0 & 0 & 1 & | & -2 \end{bmatrix}$

35. Form the augmented matrix: $\begin{bmatrix} 1 & -2 & | & 4 \\ 3 & 5 & | & 1 \end{bmatrix}$

Using the row operation $-3R_1 + R_2 \rightarrow R_2$: $\begin{bmatrix} 1 & -2 & | & 4 \\ 0 & 11 & | & -11 \end{bmatrix}$

Using the row operation $\frac{1}{11}R_2 \rightarrow R_2$: $\begin{bmatrix} 1 & -2 & | & 4 \\ 0 & 1 & | & -1 \end{bmatrix}$

Using the row operation $2R_2 + R_1 \rightarrow R_1$: $\begin{bmatrix} 1 & 0 & | & 2 \\ 0 & 1 & | & -1 \end{bmatrix}$

The solution is $(2,-1)$.

37. Form the augmented matrix: $\begin{bmatrix} 1 & 2 & 1 & \vdots & 8 \\ 1 & 4 & -1 & \vdots & 12 \\ 1 & -2 & 1 & \vdots & -4 \end{bmatrix}$

Using the row operations $-R_1 + R_2 \to R_2$ and $-R_1 + R_3 \to R_3$: $\begin{bmatrix} 1 & 2 & 1 & \vdots & 8 \\ 0 & 2 & -2 & \vdots & 4 \\ 0 & -4 & 0 & \vdots & -12 \end{bmatrix}$

Using the row operation $\frac{1}{2}R_2 \to R_2$: $\begin{bmatrix} 1 & 2 & 1 & \vdots & 8 \\ 0 & 1 & -1 & \vdots & 2 \\ 0 & -4 & 0 & \vdots & -12 \end{bmatrix}$

Using the row operation $4R_2 + R_3 \to R_3$: $\begin{bmatrix} 1 & 2 & 1 & \vdots & 8 \\ 0 & 1 & -1 & \vdots & 2 \\ 0 & 0 & -4 & \vdots & -4 \end{bmatrix}$

Using the row operation $-\frac{1}{4}R_3 \to R_3$: $\begin{bmatrix} 1 & 2 & 1 & \vdots & 8 \\ 0 & 1 & -1 & \vdots & 2 \\ 0 & 0 & 1 & \vdots & 1 \end{bmatrix}$

Using the row operations $R_3 + R_2 \to R_2$ and $-R_3 + R_1 \to R_1$: $\begin{bmatrix} 1 & 2 & 0 & \vdots & 7 \\ 0 & 1 & 0 & \vdots & 3 \\ 0 & 0 & 1 & \vdots & 1 \end{bmatrix}$

Using the row operation $-2R_2 + R_1 \to R_1$: $\begin{bmatrix} 1 & 0 & 0 & \vdots & 1 \\ 0 & 1 & 0 & \vdots & 3 \\ 0 & 0 & 1 & \vdots & 1 \end{bmatrix}$

The solution is $(1,3,1)$.

39. Form the augmented matrix: $\begin{bmatrix} 1 & 1 & 2 & \vdots & 1 \\ 2 & 4 & 2 & \vdots & 6 \\ 3 & 1 & 2 & \vdots & -4 \end{bmatrix}$

Using the row operations $-2R_1 + R_2 \to R_2$ and $-3R_1 + R_3 \to R_3$: $\begin{bmatrix} 1 & 1 & 2 & \vdots & 1 \\ 0 & 2 & -2 & \vdots & 4 \\ 0 & -2 & -4 & \vdots & -7 \end{bmatrix}$

Using the row operation $\frac{1}{2}R_2 \to R_2$: $\begin{bmatrix} 1 & 1 & 2 & \vdots & 1 \\ 0 & 1 & -1 & \vdots & 2 \\ 0 & -2 & -4 & \vdots & -7 \end{bmatrix}$

Using the row operation $2R_2 + R_3 \to R_3$: $\begin{bmatrix} 1 & 1 & 2 & \vdots & 1 \\ 0 & 1 & -1 & \vdots & 2 \\ 0 & 0 & -6 & \vdots & -3 \end{bmatrix}$

Using the row operation $-\frac{1}{6}R_3 \to R_3$: $\begin{bmatrix} 1 & 1 & 2 & \vdots & 1 \\ 0 & 1 & -1 & \vdots & 2 \\ 0 & 0 & 1 & \vdots & \frac{1}{2} \end{bmatrix}$

Using the row operations $R_3 + R_2 \rightarrow R_2$ and $-2R_3 + R_1 \rightarrow R_1$:
$$\begin{bmatrix} 1 & 1 & 0 & | & 0 \\ 0 & 1 & 0 & | & \frac{5}{2} \\ 0 & 0 & 1 & | & \frac{1}{2} \end{bmatrix}$$

Using the row operation $-R_2 + R_1 \rightarrow R_1$:
$$\begin{bmatrix} 1 & 0 & 0 & | & -\frac{5}{2} \\ 0 & 1 & 0 & | & \frac{5}{2} \\ 0 & 0 & 1 & | & \frac{1}{2} \end{bmatrix}$$

The solution is $\left(-\frac{5}{2}, \frac{5}{2}, \frac{1}{2} \right)$.

41. Multiplying the equation by $\left(x^2 + 5 \right)\left(x^2 + 4 \right)$:
$$5x^3 + 3x^2 + 2 = (Ax + B)\left(x^2 + 4 \right) + (Cx + D)\left(x^2 + 5 \right)$$
$$5x^3 + 3x^2 + 2 = Ax^3 + Bx^2 + 4Ax + 4B + Cx^3 + Dx^2 + 5Cx + 5D$$
$$5x^3 + 3x^2 + 2 = (A + C)x^3 + (B + D)x^2 + (4A + 5C)x + (4B + 5D)$$
Setting coefficients equal results in the equations:
$$A + C = 5$$
$$B + D = 3$$
$$4A + 5C = 0$$
$$4B + 5D = 2$$

Form the augmented matrix:
$$\begin{bmatrix} 1 & 0 & 1 & 0 & | & 5 \\ 0 & 1 & 0 & 1 & | & 3 \\ 4 & 0 & 5 & 0 & | & 0 \\ 0 & 4 & 0 & 5 & | & 2 \end{bmatrix}$$

Using the row operations $-4R_1 + R_3 \rightarrow R_3$ and $-4R_2 + R_4 \rightarrow R_4$:
$$\begin{bmatrix} 1 & 0 & 1 & 0 & | & 5 \\ 0 & 1 & 0 & 1 & | & 3 \\ 0 & 0 & 1 & 0 & | & -20 \\ 0 & 0 & 0 & 1 & | & -10 \end{bmatrix}$$

Using the row operations $-R_4 + R_2 \rightarrow R_2$ and $-R_3 + R_1 \rightarrow R_1$:
$$\begin{bmatrix} 1 & 0 & 0 & 0 & | & 25 \\ 0 & 1 & 0 & 0 & | & 13 \\ 0 & 0 & 1 & 0 & | & -20 \\ 0 & 0 & 0 & 1 & | & -10 \end{bmatrix}$$

So $A = 25$, $B = 13$, $C = -20$, $D = -10$.

43. Re-solving Exercise 21 by forming the augmented matrix:
$$\begin{bmatrix} 1 & 1 & -2 & | & -5 \\ 1 & -1 & 2 & | & 9 \\ 3 & 1 & -1 & | & 2 \end{bmatrix}$$

Using the row operations $2R_3 + R_2 \rightarrow R_2$ and $-2R_3 + R_1 \rightarrow R_1$:
$$\begin{bmatrix} -5 & -1 & 0 & | & -9 \\ 7 & 1 & 0 & | & 13 \\ 3 & 1 & -1 & | & 2 \end{bmatrix}$$

Using the row operation $R_2 + R_1 \rightarrow R_1$:
$$\begin{bmatrix} 2 & 0 & 0 & | & 4 \\ 7 & 1 & 0 & | & 13 \\ 3 & 1 & -1 & | & 2 \end{bmatrix}$$

The system of equations is:
$$2x = 4$$
$$7x + y = 13$$
$$3x + y - z = 2$$
Since $2x = 4$, $x = 2$. Substituting into equation 2:
$$7(2) + y = 13$$
$$14 + y = 13$$
$$y = -1$$
Substituting into equation 3:
$$3(2) + (-1) - z = 2$$
$$5 - z = 2$$
$$-z = -3$$
$$z = 3$$
The solution is $(2, -1, 3)$.

Re-solving Exercise 22 by forming the augmented matrix: $\begin{bmatrix} 1 & -2 & 1 & \vdots & 4 \\ 1 & -1 & -2 & \vdots & -3 \\ 3 & 2 & 1 & \vdots & 0 \end{bmatrix}$

Using the row operations $2R_3 + R_2 \rightarrow R_2$ and $-R_3 + R_1 \rightarrow R_1$: $\begin{bmatrix} -2 & -4 & 0 & \vdots & 4 \\ 7 & 3 & 0 & \vdots & -3 \\ 3 & 2 & 1 & \vdots & 0 \end{bmatrix}$

Using the row operation $\frac{1}{3}R_2 \rightarrow R_2$: $\begin{bmatrix} -2 & -4 & 0 & \vdots & 4 \\ \frac{7}{3} & 1 & 0 & \vdots & -1 \\ 3 & 2 & 1 & \vdots & 0 \end{bmatrix}$

Using the row operation $4R_2 + R_1 \rightarrow R_1$: $\begin{bmatrix} \frac{22}{3} & 0 & 0 & \vdots & 0 \\ \frac{7}{3} & 1 & 0 & \vdots & -1 \\ 3 & 2 & 1 & \vdots & 0 \end{bmatrix}$

The system of equations is:
$$\tfrac{22}{3}x = 0$$
$$\tfrac{7}{3}x + y = -1$$
$$3x + 2y + z = 0$$
Since $\frac{22}{3}x = 0$, $x = 0$. Substituting into equation 2:
$$\tfrac{7}{3}(0) + y = -1$$
$$y = -1$$
Substituting into equation 3:
$$3(0) + 2(-1) + z = 0$$
$$-2 + z = 0$$
$$z = 2$$
The solution is $(0, -1, 2)$.

9.4 The Algebra of Matrices

1.   Performing the operations: $\begin{bmatrix} 1 & -3 \\ 5 & 2 \end{bmatrix} + \begin{bmatrix} 2 & 5 \\ 3 & 0 \end{bmatrix} = \begin{bmatrix} 3 & 2 \\ 8 & 2 \end{bmatrix}$

3.   Performing the operations: $\begin{bmatrix} 1 & 3 \\ -5 & 0 \\ 2 & -7 \end{bmatrix} + \begin{bmatrix} -1 & -3 \\ 5 & 0 \\ -2 & 7 \end{bmatrix} = \begin{bmatrix} 0 & 0 \\ 0 & 0 \\ 0 & 0 \end{bmatrix}$

5.   The operation is not defined, since these matrices are different sizes.

7.   Performing the operations: $3 \begin{bmatrix} 0 & 3 \\ 17 & 0 \end{bmatrix} = \begin{bmatrix} 0 & 9 \\ 51 & 0 \end{bmatrix}$

9.   Computing the matrix: $-A = -1 \begin{bmatrix} 2 & -3 \\ -1 & 0 \end{bmatrix} = \begin{bmatrix} -2 & 3 \\ 1 & 0 \end{bmatrix}$

11.   Computing the matrix: $\frac{2}{3}B = \frac{2}{3} \begin{bmatrix} 5 & 1 \\ -2 & 3 \end{bmatrix} = \begin{bmatrix} \frac{10}{3} & \frac{2}{3} \\ -\frac{4}{3} & 2 \end{bmatrix}$

13.   Computing the matrix: $A - B = \begin{bmatrix} 2 & -3 \\ -1 & 0 \end{bmatrix} - \begin{bmatrix} 5 & 1 \\ -2 & 3 \end{bmatrix} = \begin{bmatrix} -3 & -4 \\ 1 & -3 \end{bmatrix}$

15.   Computing the matrix: $2A - B = 2 \begin{bmatrix} 2 & -3 \\ -1 & 0 \end{bmatrix} - \begin{bmatrix} 5 & 1 \\ -2 & 3 \end{bmatrix} = \begin{bmatrix} 4 & -6 \\ -2 & 0 \end{bmatrix} - \begin{bmatrix} 5 & 1 \\ -2 & 3 \end{bmatrix} = \begin{bmatrix} -1 & -7 \\ 0 & -3 \end{bmatrix}$

17.   Solving for X:
$$2X = B$$
$$2X = \begin{bmatrix} 5 & 1 \\ -2 & 3 \end{bmatrix}$$
$$X = \frac{1}{2} \begin{bmatrix} 5 & 1 \\ -2 & 3 \end{bmatrix} = \begin{bmatrix} \frac{5}{2} & \frac{1}{2} \\ -1 & \frac{3}{2} \end{bmatrix}$$

19.   Solving for X:
$$2X - B = 2A$$
$$2X = 2A + B$$
$$X = \frac{1}{2}(2A + B)$$
$$X = \frac{1}{2}\left(\begin{bmatrix} 4 & -6 \\ -2 & 0 \end{bmatrix} + \begin{bmatrix} 5 & 1 \\ -2 & 3 \end{bmatrix} \right) = \frac{1}{2} \begin{bmatrix} 9 & -5 \\ -4 & 3 \end{bmatrix} = \begin{bmatrix} \frac{9}{2} & -\frac{5}{2} \\ -2 & \frac{3}{2} \end{bmatrix}$$

21.   Computing the product: $\begin{bmatrix} 3 & -3 & 1 \end{bmatrix} \begin{bmatrix} 5 \\ 1 \\ -2 \end{bmatrix} = [15 - 3 - 2] = [10]$

23.   Computing the product: $\begin{bmatrix} 0 & -1 & 1 & 2 \end{bmatrix} \begin{bmatrix} 1 \\ -1 \\ -2 \\ 2 \end{bmatrix} = [0 + 1 - 2 + 4] = [3]$

25.   Computing the product: $\begin{bmatrix} 0 & -1 \\ 3 & 2 \end{bmatrix} \begin{bmatrix} -1 & 3 \\ 0 & -2 \end{bmatrix} = \begin{bmatrix} 0 & 2 \\ -3 & 5 \end{bmatrix}$

27. This multiplication is not defined.

29. Computing the product: $\begin{bmatrix} 0 & 1 & -5 \\ -3 & 1 & 6 \end{bmatrix}\begin{bmatrix} -1 & 3 \\ 0 & -2 \\ 6 & 0 \end{bmatrix} = \begin{bmatrix} -30 & -2 \\ 39 & -11 \end{bmatrix}$

31. Computing the product: $\begin{bmatrix} 1 & 0 & -2 \\ 3 & 1 & -1 \\ 2 & 0 & 1 \end{bmatrix}\begin{bmatrix} 1 & -3 & 0 \\ 0 & 2 & -2 \\ -6 & 0 & 5 \end{bmatrix} = \begin{bmatrix} 13 & -3 & -10 \\ 9 & -7 & -7 \\ -4 & -6 & 5 \end{bmatrix}$

33. Computing the product: $\begin{bmatrix} 2 & 0 & -2 & 1 \\ -2 & 3 & -1 & 1 \end{bmatrix}\begin{bmatrix} 1 & 3 \\ 1 & 0 \\ -1 & 5 \\ 1 & -3 \end{bmatrix} = \begin{bmatrix} 5 & -7 \\ 3 & -14 \end{bmatrix}$

35. Computing the product: $\begin{bmatrix} 1 & 0 & 0 \\ 0 & 1 & 0 \\ 0 & 0 & 1 \end{bmatrix}\begin{bmatrix} 3 & 1 & 0 \\ -2 & 4 & 6 \\ 3 & 8 & -10 \end{bmatrix} = \begin{bmatrix} 3 & 1 & 0 \\ -2 & 4 & 6 \\ 3 & 8 & -10 \end{bmatrix}$

37. Computing each product:

$$CB = \begin{bmatrix} 2 & -2 \\ 4 & 1 \end{bmatrix}\begin{bmatrix} 3 & 2 \\ -5 & 0 \end{bmatrix} = \begin{bmatrix} 16 & 4 \\ 7 & 8 \end{bmatrix}$$

$$BC = \begin{bmatrix} 3 & 2 \\ -5 & 0 \end{bmatrix}\begin{bmatrix} 2 & -2 \\ 4 & 1 \end{bmatrix} = \begin{bmatrix} 14 & -4 \\ -10 & 10 \end{bmatrix}$$

Since $CB \neq BC$, the statement is false.

39. Computing each product:

$$(B+C)A = \left(\begin{bmatrix} 3 & 2 \\ -5 & 0 \end{bmatrix} + \begin{bmatrix} 2 & -2 \\ 4 & 1 \end{bmatrix} \right)\begin{bmatrix} 2 & 1 & 0 \\ 3 & 1 & -2 \end{bmatrix} = \begin{bmatrix} 5 & 0 \\ -1 & 1 \end{bmatrix}\begin{bmatrix} 2 & 1 & 0 \\ 3 & 1 & -2 \end{bmatrix} = \begin{bmatrix} 10 & 5 & 0 \\ 1 & 0 & -2 \end{bmatrix}$$

$$BA + CA = \begin{bmatrix} 3 & 2 \\ -5 & 0 \end{bmatrix}\begin{bmatrix} 2 & 1 & 0 \\ 3 & 1 & -2 \end{bmatrix} + \begin{bmatrix} 2 & -2 \\ 4 & 1 \end{bmatrix}\begin{bmatrix} 2 & 1 & 0 \\ 3 & 1 & -2 \end{bmatrix}$$

$$= \begin{bmatrix} 12 & 5 & -4 \\ -10 & -5 & 0 \end{bmatrix} + \begin{bmatrix} -2 & 0 & 4 \\ 11 & 5 & -2 \end{bmatrix}$$

$$= \begin{bmatrix} 10 & 5 & 0 \\ 1 & 0 & -2 \end{bmatrix}$$

Since $(B+C)A = BA + CA$, the statement is true.

41. Computing each product:

$$\left(\frac{1}{2}\begin{bmatrix} 3 & 5 \\ 2 & 4 \end{bmatrix}\begin{bmatrix} 0 & 2 \\ 1 & -1 \end{bmatrix} \right) = \begin{bmatrix} \frac{3}{2} & \frac{5}{2} \\ 1 & 2 \end{bmatrix}\begin{bmatrix} 0 & 2 \\ 1 & -1 \end{bmatrix} = \begin{bmatrix} \frac{5}{2} & \frac{1}{2} \\ 2 & 0 \end{bmatrix}$$

$$\frac{1}{2}\left(\begin{bmatrix} 3 & 5 \\ 2 & 4 \end{bmatrix}\begin{bmatrix} 0 & 2 \\ 1 & -1 \end{bmatrix} \right) = \frac{1}{2}\begin{bmatrix} 5 & 1 \\ 4 & 0 \end{bmatrix} = \begin{bmatrix} \frac{5}{2} & \frac{1}{2} \\ 2 & 0 \end{bmatrix}$$

The two products are equal.

43. AB is undefined, and BA is 3 x 4

45. Let $A = \begin{bmatrix} a_{11} & a_{12} & a_{13} \\ a_{21} & a_{22} & a_{23} \\ a_{31} & a_{32} & a_{33} \end{bmatrix}$ and $B = \begin{bmatrix} b_{11} & b_{12} & b_{13} \\ b_{21} & b_{22} & b_{23} \\ b_{31} & b_{32} & b_{33} \end{bmatrix}$. Then:

$$A + B = \begin{bmatrix} a_{11}+b_{11} & a_{12}+b_{12} & a_{13}+b_{13} \\ a_{21}+b_{21} & a_{22}+b_{22} & a_{23}+b_{23} \\ a_{31}+b_{31} & a_{32}+b_{32} & a_{33}+b_{33} \end{bmatrix} = \begin{bmatrix} b_{11}+a_{11} & b_{12}+a_{12} & b_{13}+a_{13} \\ b_{21}+a_{21} & b_{22}+a_{22} & b_{23}+a_{23} \\ b_{31}+a_{31} & b_{32}+a_{32} & b_{33}+a_{33} \end{bmatrix} = B + A$$

47. Yes. The additive inverse of A is $-A = (-1)A$.

49. Let a_{ij} be an arbitrary element of A and b_{ij} be an arbitrary element of B. Then:

$$c\left(a_{ij} + b_{ij}\right) = ca_{ij} + cb_{ij}$$

Since a_{ij} and b_{ij} are arbitrary elements, $c(A + B) = cA + cB$.

51. Use the matrix $\begin{bmatrix} 12 & 18 & 14 \\ 14 & 6 & 12 \\ 18 & 12 & 5 \end{bmatrix}$ to represent the tubing required. Therefore:

George's: $\begin{bmatrix} 1 & 1.5 & 1.8 \end{bmatrix} \begin{bmatrix} 12 & 18 & 14 \\ 14 & 6 & 12 \\ 18 & 12 & 5 \end{bmatrix} = \begin{bmatrix} 65.4 & 48.6 & 41 \end{bmatrix}$ (155 total)

Raju's: $\begin{bmatrix} 1.2 & 1.4 & 1.6 \end{bmatrix} \begin{bmatrix} 12 & 18 & 14 \\ 14 & 6 & 12 \\ 18 & 12 & 5 \end{bmatrix} = \begin{bmatrix} 62.8 & 49.2 & 41.6 \end{bmatrix}$ (153.6 total)

Brand A should be ordered from Raju (62.8 cents), Brand B should be ordered from George (48.6 cents), and Brand C should be ordered from George (41 cents). If all three brands must be ordered from the same company, then Raju is the least expensive (153.6 cents).

53. The inner product is undefined, while the matrix product is $\begin{bmatrix} 1 \\ 0 \\ 4 \end{bmatrix} \begin{bmatrix} 2 & -3 & 1 \end{bmatrix} = \begin{bmatrix} 2 & -3 & 1 \\ 0 & 0 & 0 \\ 8 & -12 & 4 \end{bmatrix}$. The inner product is undefined, while the matrix product is a 3 x 3 matrix.

9.5 Solving Linear Systems Using Matrix Inverses

1. Finding the product: $\begin{bmatrix} 3 & -1 \\ 3 & -2 \end{bmatrix} \begin{bmatrix} 1 & 0 \\ 0 & 1 \end{bmatrix} = \begin{bmatrix} 3 & -1 \\ 3 & -2 \end{bmatrix}$

3. Let $A = \begin{bmatrix} a_{11} & a_{12} & a_{13} \\ a_{21} & a_{22} & a_{23} \\ a_{31} & a_{32} & a_{33} \end{bmatrix}$, then:

$$AI_3 = \begin{bmatrix} a_{11} & a_{12} & a_{13} \\ a_{21} & a_{22} & a_{23} \\ a_{31} & a_{32} & a_{33} \end{bmatrix} \begin{bmatrix} 1 & 0 & 0 \\ 0 & 1 & 0 \\ 0 & 0 & 1 \end{bmatrix} = \begin{bmatrix} a_{11} & a_{12} & a_{13} \\ a_{21} & a_{22} & a_{23} \\ a_{31} & a_{32} & a_{33} \end{bmatrix} = A$$

$$I_3 A = \begin{bmatrix} 1 & 0 & 0 \\ 0 & 1 & 0 \\ 0 & 0 & 1 \end{bmatrix} \begin{bmatrix} a_{11} & a_{12} & a_{13} \\ a_{21} & a_{22} & a_{23} \\ a_{31} & a_{32} & a_{33} \end{bmatrix} = \begin{bmatrix} a_{11} & a_{12} & a_{13} \\ a_{21} & a_{22} & a_{23} \\ a_{31} & a_{32} & a_{33} \end{bmatrix} = A$$

5.   Form the grafted matrix: $\begin{bmatrix} 1 & 2 & \vdots & 1 & 0 \\ 2 & 0 & \vdots & 0 & 1 \end{bmatrix}$

Using the row operation $-2R_1 + R_2 \rightarrow R_2$: $\begin{bmatrix} 1 & 2 & \vdots & 1 & 0 \\ 0 & -4 & \vdots & -2 & 1 \end{bmatrix}$

Using the row operation $-\frac{1}{4}R_2 \rightarrow R_2$: $\begin{bmatrix} 1 & 2 & \vdots & 1 & 0 \\ 0 & 1 & \vdots & \frac{1}{2} & -\frac{1}{4} \end{bmatrix}$

Using the row operation $-2R_2 + R_1 \rightarrow R_1$: $\begin{bmatrix} 1 & 0 & \vdots & 0 & \frac{1}{2} \\ 0 & 1 & \vdots & \frac{1}{2} & -\frac{1}{4} \end{bmatrix}$

The inverse is $\begin{bmatrix} 0 & \frac{1}{2} \\ \frac{1}{2} & -\frac{1}{4} \end{bmatrix}$.

7.   Form the grafted matrix: $\begin{bmatrix} -1 & 5 & \vdots & 1 & 0 \\ 3 & 1 & \vdots & 0 & 1 \end{bmatrix}$

Using the row operation $-R_1 \rightarrow R_1$: $\begin{bmatrix} 1 & -5 & \vdots & -1 & 0 \\ 3 & 1 & \vdots & 0 & 1 \end{bmatrix}$

Using the row operation $-3R_1 + R_2 \rightarrow R_2$: $\begin{bmatrix} 1 & -5 & \vdots & -1 & 0 \\ 0 & 16 & \vdots & 3 & 1 \end{bmatrix}$

Using the row operation $\frac{1}{16}R_2 \rightarrow R_2$: $\begin{bmatrix} 1 & -5 & \vdots & -1 & 0 \\ 0 & 1 & \vdots & \frac{3}{16} & \frac{1}{16} \end{bmatrix}$

Using the row operation $5R_2 + R_1 \rightarrow R_1$: $\begin{bmatrix} 1 & 0 & \vdots & -\frac{1}{16} & \frac{5}{16} \\ 0 & 1 & \vdots & \frac{3}{16} & \frac{1}{16} \end{bmatrix}$

The inverse is $\begin{bmatrix} -\frac{1}{16} & \frac{5}{16} \\ \frac{3}{16} & \frac{1}{16} \end{bmatrix} = \frac{1}{16} \begin{bmatrix} -1 & 5 \\ 3 & 1 \end{bmatrix}$.

9.   Form the grafted matrix: $\begin{bmatrix} 4 & 2 & \vdots & 1 & 0 \\ 2 & 1 & \vdots & 0 & 1 \end{bmatrix}$

Using the row operation $\frac{1}{4}R_1 \rightarrow R_1$: $\begin{bmatrix} 1 & \frac{1}{2} & \vdots & \frac{1}{4} & 0 \\ 2 & 1 & \vdots & 0 & 1 \end{bmatrix}$

Using the row operation $-2R_1 + R_2 \rightarrow R_2$: $\begin{bmatrix} 1 & \frac{1}{2} & \vdots & \frac{1}{4} & 0 \\ 0 & 0 & \vdots & -\frac{1}{2} & 1 \end{bmatrix}$

The inverse does not exist.

11. Form the grafted matrix: $\begin{bmatrix} -1 & 5 & | & 1 & 0 \\ -2 & 10 & | & 0 & 1 \end{bmatrix}$

Using the row operation $-R_1 \to R_1$: $\begin{bmatrix} 1 & -5 & | & -1 & 0 \\ -2 & 10 & | & 0 & 1 \end{bmatrix}$

Using the row operation $2R_1 + R_2 \to R_2$: $\begin{bmatrix} 1 & -5 & | & -1 & 0 \\ 0 & 0 & | & -2 & 1 \end{bmatrix}$

The inverse does not exist.

13. Form the grafted matrix: $\begin{bmatrix} 1 & 1 & 2 & | & 1 & 0 & 0 \\ 2 & 0 & 1 & | & 0 & 1 & 0 \\ 3 & -1 & 1 & | & 0 & 0 & 1 \end{bmatrix}$

Using the row operations $-2R_1 + R_2 \to R_2$ and $-3R_1 + R_3 \to R_3$: $\begin{bmatrix} 1 & 1 & 2 & | & 1 & 0 & 0 \\ 0 & -2 & -3 & | & -2 & 1 & 0 \\ 0 & -4 & -5 & | & -3 & 0 & 1 \end{bmatrix}$

Using the row operation $-\frac{1}{2}R_2 \to R_2$: $\begin{bmatrix} 1 & 1 & 2 & | & 1 & 0 & 0 \\ 0 & 1 & \frac{3}{2} & | & 1 & -\frac{1}{2} & 0 \\ 0 & -4 & -5 & | & -3 & 0 & 1 \end{bmatrix}$

Using the row operation $4R_2 + R_3 \to R_3$: $\begin{bmatrix} 1 & 1 & 2 & | & 1 & 0 & 0 \\ 0 & 1 & \frac{3}{2} & | & 1 & -\frac{1}{2} & 0 \\ 0 & 0 & 1 & | & 1 & -2 & 1 \end{bmatrix}$

Using the row operations $-\frac{3}{2}R_3 + R_2 \to R_2$ and $-2R_3 + R_1 \to R_1$: $\begin{bmatrix} 1 & 1 & 0 & | & -1 & 4 & -2 \\ 0 & 1 & 0 & | & -\frac{1}{2} & \frac{5}{2} & -\frac{3}{2} \\ 0 & 0 & 1 & | & 1 & -2 & 1 \end{bmatrix}$

Using the row operation $-R_2 + R_1 \to R_1$: $\begin{bmatrix} 1 & 0 & 0 & | & -\frac{1}{2} & \frac{3}{2} & -\frac{1}{2} \\ 0 & 1 & 0 & | & -\frac{1}{2} & \frac{5}{2} & -\frac{3}{2} \\ 0 & 0 & 1 & | & 1 & -2 & 1 \end{bmatrix}$

The inverse is $\begin{bmatrix} -\frac{1}{2} & \frac{3}{2} & -\frac{1}{2} \\ -\frac{1}{2} & \frac{5}{2} & -\frac{3}{2} \\ 1 & -2 & 1 \end{bmatrix} = \frac{1}{2}\begin{bmatrix} -1 & 3 & -1 \\ -1 & 5 & -3 \\ 2 & -4 & 2 \end{bmatrix}$.

15. Form the grafted matrix: $\begin{bmatrix} 1 & 0 & 1 & | & 1 & 0 & 0 \\ -2 & 3 & 2 & | & 0 & 1 & 0 \\ 2 & -2 & 0 & | & 0 & 0 & 1 \end{bmatrix}$

Using the row operations $2R_1 + R_2 \to R_2$ and $-2R_1 + R_3 \to R_3$: $\begin{bmatrix} 1 & 0 & 1 & | & 1 & 0 & 0 \\ 0 & 3 & 4 & | & 2 & 1 & 0 \\ 0 & -2 & -2 & | & -2 & 0 & 1 \end{bmatrix}$

Using the row operation $R_2 \leftrightarrow R_3$: $\begin{bmatrix} 1 & 0 & 1 & | & 1 & 0 & 0 \\ 0 & -2 & -2 & | & -2 & 0 & 1 \\ 0 & 3 & 4 & | & 2 & 1 & 0 \end{bmatrix}$

Using the row operation $-\frac{1}{2}R_2 \rightarrow R_2$: $\begin{bmatrix} 1 & 0 & 1 & \vdots & 1 & 0 & 0 \\ 0 & 1 & 1 & \vdots & 1 & 0 & -\frac{1}{2} \\ 0 & 3 & 4 & \vdots & 2 & 1 & 0 \end{bmatrix}$

Using the row operation $-3R_2 + R_3 \rightarrow R_3$: $\begin{bmatrix} 1 & 0 & 1 & \vdots & 1 & 0 & 0 \\ 0 & 1 & 1 & \vdots & 1 & 0 & -\frac{1}{2} \\ 0 & 0 & 1 & \vdots & -1 & 1 & \frac{3}{2} \end{bmatrix}$

Using the row operations $-R_3 + R_2 \rightarrow R_2$ and $-R_3 + R_1 \rightarrow R_1$: $\begin{bmatrix} 1 & 0 & 0 & \vdots & 2 & -1 & -\frac{3}{2} \\ 0 & 1 & 0 & \vdots & 2 & -1 & -2 \\ 0 & 0 & 1 & \vdots & -1 & 1 & \frac{3}{2} \end{bmatrix}$

The inverse is $\begin{bmatrix} 2 & -1 & -\frac{3}{2} \\ 2 & -1 & -2 \\ -1 & 1 & \frac{3}{2} \end{bmatrix} = \frac{1}{2}\begin{bmatrix} 4 & -2 & -3 \\ 4 & -2 & -4 \\ -2 & 2 & 3 \end{bmatrix}$.

17. Form the grafted matrix: $\begin{bmatrix} 2 & 1 & 3 & \vdots & 1 & 0 & 0 \\ 1 & 0 & 1 & \vdots & 0 & 1 & 0 \\ 4 & -2 & 2 & \vdots & 0 & 0 & 1 \end{bmatrix}$

Using the row operation $R_1 \leftrightarrow R_2$: $\begin{bmatrix} 1 & 0 & 1 & \vdots & 0 & 1 & 0 \\ 2 & 1 & 3 & \vdots & 1 & 0 & 0 \\ 4 & -2 & 2 & \vdots & 0 & 0 & 1 \end{bmatrix}$

Using the row operations $-2R_1 + R_2 \rightarrow R_2$ and $-4R_1 + R_3 \rightarrow R_3$: $\begin{bmatrix} 1 & 0 & 1 & \vdots & 0 & 1 & 0 \\ 0 & 1 & 1 & \vdots & 1 & -2 & 0 \\ 0 & -2 & -2 & \vdots & 0 & -4 & 1 \end{bmatrix}$

Using the row operation $2R_2 + R_3 \rightarrow R_3$: $\begin{bmatrix} 1 & 0 & 1 & \vdots & 0 & 1 & 0 \\ 0 & 1 & 1 & \vdots & 1 & -2 & 0 \\ 0 & 0 & 0 & \vdots & 2 & -8 & 1 \end{bmatrix}$

The inverse does not exist.

19. Forming the grafted matrix: $\begin{bmatrix} 1 & -1 & 2 & \vdots & 1 & 0 & 0 \\ 2 & -2 & 1 & \vdots & 0 & 1 & 0 \\ 3 & -1 & 0 & \vdots & 0 & 0 & 1 \end{bmatrix}$

Using the row operations $-2R_1 + R_2 \rightarrow R_2$ and $-3R_1 + R_3 \rightarrow R_3$: $\begin{bmatrix} 1 & -1 & 2 & \vdots & 1 & 0 & 0 \\ 0 & 0 & -3 & \vdots & -2 & 1 & 0 \\ 0 & 2 & -6 & \vdots & -3 & 0 & 1 \end{bmatrix}$

Using the row operation $R_2 \leftrightarrow R_3$: $\begin{bmatrix} 1 & -1 & 2 & \vdots & 1 & 0 & 0 \\ 0 & 2 & -6 & \vdots & -3 & 0 & 1 \\ 0 & 0 & -3 & \vdots & -2 & 1 & 0 \end{bmatrix}$

Using the row operations $\frac{1}{2}R_2 \rightarrow R_2$ and $-\frac{1}{3}R_3 \rightarrow R_3$: $\begin{bmatrix} 1 & -1 & 2 & \vdots & 1 & 0 & 0 \\ 0 & 1 & -3 & \vdots & -\frac{3}{2} & 0 & \frac{1}{2} \\ 0 & 0 & 1 & \vdots & \frac{2}{3} & -\frac{1}{3} & 0 \end{bmatrix}$

Using the row operations $3R_3 + R_2 \rightarrow R_2$ and $-2R_3 + R_1 \rightarrow R_1$:
$$\begin{bmatrix} 1 & -1 & 0 & \vdots & -\frac{1}{3} & \frac{2}{3} & 0 \\ 0 & 1 & 0 & \vdots & \frac{1}{2} & -1 & \frac{1}{2} \\ 0 & 0 & 1 & \vdots & \frac{2}{3} & -\frac{1}{3} & 0 \end{bmatrix}$$

Using the row operation $R_2 + R_1 \rightarrow R_1$:
$$\begin{bmatrix} 1 & 0 & 0 & \vdots & \frac{1}{6} & -\frac{1}{3} & \frac{1}{2} \\ 0 & 1 & 0 & \vdots & \frac{1}{2} & -1 & \frac{1}{2} \\ 0 & 0 & 1 & \vdots & \frac{2}{3} & -\frac{1}{3} & 0 \end{bmatrix}$$

The inverse is $\begin{bmatrix} \frac{1}{6} & -\frac{1}{3} & \frac{1}{2} \\ \frac{1}{2} & -1 & \frac{1}{2} \\ \frac{2}{3} & -\frac{1}{3} & 0 \end{bmatrix} = \frac{1}{6}\begin{bmatrix} 1 & -2 & 3 \\ 3 & -6 & 3 \\ 4 & -2 & 0 \end{bmatrix}$.

21. Let $A = \begin{bmatrix} 1 & 2 \\ 1 & -1 \end{bmatrix}$. Form the grafted matrix: $\begin{bmatrix} 1 & 2 & \vdots & 1 & 0 \\ 1 & -1 & \vdots & 0 & 1 \end{bmatrix}$

Using the row operation $-R_1 + R_2 \rightarrow R_2$: $\begin{bmatrix} 1 & 2 & \vdots & 1 & 0 \\ 0 & -3 & \vdots & -1 & 1 \end{bmatrix}$

Using the row operation $-\frac{1}{3}R_2 \rightarrow R_2$: $\begin{bmatrix} 1 & 2 & \vdots & 1 & 0 \\ 0 & 1 & \vdots & \frac{1}{3} & -\frac{1}{3} \end{bmatrix}$

Using the row operation $-2R_2 + R_1 \rightarrow R_1$: $\begin{bmatrix} 1 & 0 & \vdots & \frac{1}{3} & \frac{2}{3} \\ 0 & 1 & \vdots & \frac{1}{3} & -\frac{1}{3} \end{bmatrix}$

So $A^{-1} = \begin{bmatrix} \frac{1}{3} & \frac{2}{3} \\ \frac{1}{3} & -\frac{1}{3} \end{bmatrix}$. Therefore: $\begin{bmatrix} x \\ y \end{bmatrix} = \begin{bmatrix} \frac{1}{3} & \frac{2}{3} \\ \frac{1}{3} & -\frac{1}{3} \end{bmatrix}\begin{bmatrix} 3 \\ 6 \end{bmatrix} = \begin{bmatrix} 5 \\ -1 \end{bmatrix}$

The solution is $(5,-1)$.

23. Let $A = \begin{bmatrix} 1 & -2 \\ 2 & 1 \end{bmatrix}$. Form the grafted matrix: $\begin{bmatrix} 1 & -2 & \vdots & 1 & 0 \\ 2 & 1 & \vdots & 0 & 1 \end{bmatrix}$

Using the row operation $-2R_1 + R_2 \rightarrow R_2$: $\begin{bmatrix} 1 & -2 & \vdots & 1 & 0 \\ 0 & 5 & \vdots & -2 & 1 \end{bmatrix}$

Using the row operation $\frac{1}{5}R_2 \rightarrow R_2$: $\begin{bmatrix} 1 & -2 & \vdots & 1 & 0 \\ 0 & 1 & \vdots & -\frac{2}{5} & \frac{1}{5} \end{bmatrix}$

Using the row operation $2R_2 + R_1 \rightarrow R_1$: $\begin{bmatrix} 1 & 0 & \vdots & \frac{1}{5} & \frac{2}{5} \\ 0 & 1 & \vdots & -\frac{2}{5} & \frac{1}{5} \end{bmatrix}$

So $A^{-1} = \begin{bmatrix} \frac{1}{5} & \frac{2}{5} \\ -\frac{2}{5} & \frac{1}{5} \end{bmatrix}$. Therefore: $\begin{bmatrix} x \\ y \end{bmatrix} = \begin{bmatrix} \frac{1}{5} & \frac{2}{5} \\ -\frac{2}{5} & \frac{1}{5} \end{bmatrix}\begin{bmatrix} 4 \\ 13 \end{bmatrix} = \begin{bmatrix} 6 \\ 1 \end{bmatrix}$

The solution is $(6,1)$.

25. Let $A = \begin{bmatrix} 5 & -2 \\ 1 & -5 \end{bmatrix}$. Form the grafted matrix: $\begin{bmatrix} 5 & -2 & \vdots & 1 & 0 \\ 1 & -5 & \vdots & 0 & 1 \end{bmatrix}$

Using the row operation $R_1 \leftrightarrow R_2$: $\begin{bmatrix} 1 & -5 & \vdots & 0 & 1 \\ 5 & -2 & \vdots & 1 & 0 \end{bmatrix}$

Using the row operation $-5R_1 + R_2 \rightarrow R_2$: $\begin{bmatrix} 1 & -5 & | & 0 & 1 \\ 0 & 23 & | & 1 & -5 \end{bmatrix}$

Using the row operation $\frac{1}{23}R_2 \rightarrow R_2$: $\begin{bmatrix} 1 & -5 & | & 0 & 1 \\ 0 & 1 & | & \frac{1}{23} & -\frac{5}{23} \end{bmatrix}$

Using the row operation $5R_2 + R_1 \rightarrow R_1$: $\begin{bmatrix} 1 & 0 & | & \frac{5}{23} & -\frac{2}{23} \\ 0 & 1 & | & \frac{1}{23} & -\frac{5}{23} \end{bmatrix}$

So $A^{-1} = \begin{bmatrix} \frac{5}{23} & -\frac{2}{23} \\ \frac{1}{23} & -\frac{5}{23} \end{bmatrix}$. Therefore: $\begin{bmatrix} x \\ y \end{bmatrix} = \begin{bmatrix} \frac{5}{23} & -\frac{2}{23} \\ \frac{1}{23} & -\frac{5}{23} \end{bmatrix} \begin{bmatrix} 11 \\ -7 \end{bmatrix} = \begin{bmatrix} 3 \\ 2 \end{bmatrix}$

The solution is (3,2).

27. First write the system as:
$$-3x + y = -15$$
$$8x - 2y = 10$$

Let $A = \begin{bmatrix} -3 & 1 \\ 8 & -2 \end{bmatrix}$. Form the grafted matrix: $\begin{bmatrix} -3 & 1 & | & 1 & 0 \\ 8 & -2 & | & 0 & 1 \end{bmatrix}$

Using the row operation $-\frac{1}{3}R_1 \rightarrow R_1$: $\begin{bmatrix} 1 & -\frac{1}{3} & | & -\frac{1}{3} & 0 \\ 8 & -2 & | & 0 & 1 \end{bmatrix}$

Using the row operation $-8R_1 + R_2 \rightarrow R_2$: $\begin{bmatrix} 1 & -\frac{1}{3} & | & -\frac{1}{3} & 0 \\ 0 & \frac{2}{3} & | & \frac{8}{3} & 1 \end{bmatrix}$

Using the row operation $\frac{3}{2}R_2 \rightarrow R_2$: $\begin{bmatrix} 1 & -\frac{1}{3} & | & -\frac{1}{3} & 0 \\ 0 & 1 & | & 4 & \frac{3}{2} \end{bmatrix}$

Using the row operation $\frac{1}{3}R_2 + R_1 \rightarrow R_1$: $\begin{bmatrix} 1 & 0 & | & 1 & \frac{1}{2} \\ 0 & 1 & | & 4 & \frac{3}{2} \end{bmatrix}$

So $A^{-1} = \begin{bmatrix} 1 & \frac{1}{2} \\ 4 & \frac{3}{2} \end{bmatrix}$. Therefore: $\begin{bmatrix} x \\ y \end{bmatrix} = \begin{bmatrix} 1 & \frac{1}{2} \\ 4 & \frac{3}{2} \end{bmatrix} \begin{bmatrix} -15 \\ 10 \end{bmatrix} = \begin{bmatrix} -10 \\ -45 \end{bmatrix}$

The solution is (−10,−45).

29. Let $A = \begin{bmatrix} \frac{3}{2} & \frac{1}{3} \\ 1 & \frac{1}{12} \end{bmatrix}$. Form the grafted matrix: $\begin{bmatrix} \frac{3}{2} & \frac{1}{3} & | & 1 & 0 \\ 1 & \frac{1}{12} & | & 0 & 1 \end{bmatrix}$

Using the row operation $R_1 \leftrightarrow R_2$: $\begin{bmatrix} 1 & \frac{1}{12} & | & 0 & 1 \\ \frac{3}{2} & \frac{1}{3} & | & 1 & 0 \end{bmatrix}$

Using the row operation $-\frac{3}{2}R_1 + R_2 \rightarrow R_2$: $\begin{bmatrix} 1 & \frac{1}{12} & | & 0 & 1 \\ 0 & \frac{5}{24} & | & 1 & -\frac{3}{2} \end{bmatrix}$

Using the row operation $\frac{24}{5}R_2 \rightarrow R_2$: $\begin{bmatrix} 1 & \frac{1}{12} & | & 0 & 1 \\ 0 & 1 & | & \frac{24}{5} & -\frac{36}{5} \end{bmatrix}$

Using the row operation $-\frac{1}{12}R_2 + R_1 \rightarrow R_1$: $\begin{bmatrix} 1 & 0 & | & -\frac{2}{5} & \frac{8}{5} \\ 0 & 1 & | & \frac{24}{5} & -\frac{36}{5} \end{bmatrix}$

So $A^{-1} = \begin{bmatrix} -\frac{2}{5} & \frac{8}{5} \\ \frac{24}{5} & -\frac{36}{5} \end{bmatrix}$. Therefore: $\begin{bmatrix} x \\ y \end{bmatrix} = \begin{bmatrix} -\frac{2}{5} & \frac{8}{5} \\ \frac{24}{5} & -\frac{36}{5} \end{bmatrix} \begin{bmatrix} 1 \\ 4 \end{bmatrix} = \begin{bmatrix} 0 \\ 3 \end{bmatrix}$

The solution is (0,3).

31. Let $A = \begin{bmatrix} 1 & 1 & 1 \\ 3 & 1 & -1 \\ 2 & 1 & -1 \end{bmatrix}$. Form the grafted matrix: $\left[\begin{array}{ccc|ccc} 1 & 1 & 1 & 1 & 0 & 0 \\ 3 & 1 & -1 & 0 & 1 & 0 \\ 2 & 1 & -1 & 0 & 0 & 1 \end{array}\right]$

Using the row operations $-3R_1 + R_2 \rightarrow R_2$ and $-2R_1 + R_3 \rightarrow R_3$: $\left[\begin{array}{ccc|ccc} 1 & 1 & 1 & 1 & 0 & 0 \\ 0 & -2 & -4 & -3 & 1 & 0 \\ 0 & -1 & -3 & -2 & 0 & 1 \end{array}\right]$

Using the row operation $R_2 \leftrightarrow R_3$: $\left[\begin{array}{ccc|ccc} 1 & 1 & 1 & 1 & 0 & 0 \\ 0 & -1 & -3 & -2 & 0 & 1 \\ 0 & -2 & -4 & -3 & 1 & 0 \end{array}\right]$

Using the row operation $-R_2 \rightarrow R_2$: $\left[\begin{array}{ccc|ccc} 1 & 1 & 1 & 1 & 0 & 0 \\ 0 & 1 & 3 & 2 & 0 & -1 \\ 0 & -2 & -4 & -3 & 1 & 0 \end{array}\right]$

Using the row operation $2R_2 + R_3 \rightarrow R_3$: $\left[\begin{array}{ccc|ccc} 1 & 1 & 1 & 1 & 0 & 0 \\ 0 & 1 & 3 & 2 & 0 & -1 \\ 0 & 0 & 2 & 1 & 1 & -2 \end{array}\right]$

Using the row operation $\frac{1}{2}R_3 \rightarrow R_3$: $\left[\begin{array}{ccc|ccc} 1 & 1 & 1 & 1 & 0 & 0 \\ 0 & 1 & 3 & 2 & 0 & -1 \\ 0 & 0 & 1 & \frac{1}{2} & \frac{1}{2} & -1 \end{array}\right]$

Using the row operations $-3R_3 + R_2 \rightarrow R_2$ and $-R_3 + R_1 \rightarrow R_1$: $\left[\begin{array}{ccc|ccc} 1 & 1 & 0 & \frac{1}{2} & -\frac{1}{2} & 1 \\ 0 & 1 & 0 & \frac{1}{2} & -\frac{3}{2} & 2 \\ 0 & 0 & 1 & \frac{1}{2} & \frac{1}{2} & -1 \end{array}\right]$

Using the row operation $-R_2 + R_1 \rightarrow R_1$: $\left[\begin{array}{ccc|ccc} 1 & 0 & 0 & 0 & 1 & -1 \\ 0 & 1 & 0 & \frac{1}{2} & -\frac{3}{2} & 2 \\ 0 & 0 & 1 & \frac{1}{2} & \frac{1}{2} & -1 \end{array}\right]$

So $A^{-1} = \begin{bmatrix} 0 & 1 & -1 \\ \frac{1}{2} & -\frac{3}{2} & 2 \\ \frac{1}{2} & \frac{1}{2} & -1 \end{bmatrix}$. Therefore: $\begin{bmatrix} x \\ y \\ z \end{bmatrix} = \begin{bmatrix} 0 & 1 & -1 \\ \frac{1}{2} & -\frac{3}{2} & 2 \\ \frac{1}{2} & \frac{1}{2} & -1 \end{bmatrix} \begin{bmatrix} 6 \\ 6 \\ 4 \end{bmatrix} = \begin{bmatrix} 2 \\ 2 \\ 2 \end{bmatrix}$

The solution is (2,2,2).

33. Let $A = \begin{bmatrix} 1 & -3 & 1 \\ 3 & -1 & 1 \\ 1 & 2 & -1 \end{bmatrix}$. Form the grafted matrix: $\begin{bmatrix} 1 & -3 & 1 & | & 1 & 0 & 0 \\ 3 & -1 & 1 & | & 0 & 1 & 0 \\ 1 & 2 & -1 & | & 0 & 0 & 1 \end{bmatrix}$

Using the row operations $-3R_1 + R_2 \rightarrow R_2$ and $-R_1 + R_3 \rightarrow R_3$: $\begin{bmatrix} 1 & -3 & 1 & | & 1 & 0 & 0 \\ 0 & 8 & -2 & | & -3 & 1 & 0 \\ 0 & 5 & -2 & | & -1 & 0 & 1 \end{bmatrix}$

Using the row operation $\frac{1}{8}R_2 \rightarrow R_2$: $\begin{bmatrix} 1 & -3 & 1 & | & 1 & 0 & 0 \\ 0 & 1 & -\frac{1}{4} & | & -\frac{3}{8} & \frac{1}{8} & 0 \\ 0 & 5 & -2 & | & -1 & 0 & 1 \end{bmatrix}$

Using the row operation $-5R_2 + R_3 \rightarrow R_3$: $\begin{bmatrix} 1 & -3 & 1 & | & 1 & 0 & 0 \\ 0 & 1 & -\frac{1}{4} & | & -\frac{3}{8} & \frac{1}{8} & 0 \\ 0 & 0 & -\frac{3}{4} & | & \frac{7}{8} & -\frac{5}{8} & 1 \end{bmatrix}$

Using the row operation $-\frac{4}{3}R_3 \rightarrow R_3$: $\begin{bmatrix} 1 & -3 & 1 & | & 1 & 0 & 0 \\ 0 & 1 & -\frac{1}{4} & | & -\frac{3}{8} & \frac{1}{8} & 0 \\ 0 & 0 & 1 & | & -\frac{7}{6} & \frac{5}{6} & -\frac{4}{3} \end{bmatrix}$

Using the row operations $\frac{1}{4}R_3 + R_2 \rightarrow R_2$ and $-R_3 + R_1 \rightarrow R_1$: $\begin{bmatrix} 1 & -3 & 0 & | & \frac{13}{6} & -\frac{5}{6} & \frac{4}{3} \\ 0 & 1 & 0 & | & -\frac{2}{3} & \frac{1}{3} & -\frac{1}{3} \\ 0 & 0 & 1 & | & -\frac{7}{6} & \frac{5}{6} & -\frac{4}{3} \end{bmatrix}$

Using the row operation $3R_2 + R_1 \rightarrow R_1$: $\begin{bmatrix} 1 & 0 & 0 & | & \frac{1}{6} & \frac{1}{6} & \frac{1}{3} \\ 0 & 1 & 0 & | & -\frac{2}{3} & \frac{1}{3} & -\frac{1}{3} \\ 0 & 0 & 1 & | & -\frac{7}{6} & \frac{5}{6} & -\frac{4}{3} \end{bmatrix}$

So $A^{-1} = \begin{bmatrix} \frac{1}{6} & \frac{1}{6} & \frac{1}{3} \\ -\frac{2}{3} & \frac{1}{3} & -\frac{1}{3} \\ -\frac{7}{6} & \frac{5}{6} & -\frac{4}{3} \end{bmatrix}$. Therefore: $\begin{bmatrix} x \\ y \\ z \end{bmatrix} = \begin{bmatrix} \frac{1}{6} & \frac{1}{6} & \frac{1}{3} \\ -\frac{2}{3} & \frac{1}{3} & -\frac{1}{3} \\ -\frac{7}{6} & \frac{5}{6} & -\frac{4}{3} \end{bmatrix}\begin{bmatrix} 0 \\ 2 \\ 2 \end{bmatrix} = \begin{bmatrix} 1 \\ 0 \\ -1 \end{bmatrix}$

The solution is $(1, 0, -1)$.

35. Let $A = \begin{bmatrix} 1 & 1 & 1 \\ 3 & 1 & -1 \\ 2 & 2 & 2 \end{bmatrix}$. Form the grafted matrix: $\begin{bmatrix} 1 & 1 & 1 & | & 1 & 0 & 0 \\ 3 & 1 & -1 & | & 0 & 1 & 0 \\ 2 & 2 & 2 & | & 0 & 0 & 1 \end{bmatrix}$

Using the row operations $-3R_1 + R_2 \rightarrow R_2$ and $-2R_1 + R_3 \rightarrow R_3$: $\begin{bmatrix} 1 & 1 & 1 & | & 1 & 0 & 0 \\ 0 & -2 & -4 & | & -3 & 1 & 0 \\ 0 & 0 & 0 & | & -2 & 0 & 1 \end{bmatrix}$

So the inverse does not exist. Multiplying the first equation by -2:
$$-2x - 2y - 2z = -12$$
$$2x + 2y + 2z = 4$$
Adding yields $0 = -8$, so there is no solution (inconsistent).

37. Let $A = \begin{bmatrix} 1 & 0.2 & 1 \\ 0.3 & -1 & 0.2 \\ 1 & 1 & 1 \end{bmatrix}$. Form the grafted matrix: $\begin{bmatrix} 1 & 0.2 & 1 & | & 1 & 0 & 0 \\ 0.3 & -1 & 0.2 & | & 0 & 1 & 0 \\ 1 & 1 & 1 & | & 0 & 0 & 1 \end{bmatrix}$

Using the row operation $R_1 \leftrightarrow R_3$: $\begin{bmatrix} 1 & 1 & 1 & | & 0 & 0 & 1 \\ 0.3 & -1 & 0.2 & | & 0 & 1 & 0 \\ 1 & 0.2 & 1 & | & 1 & 0 & 0 \end{bmatrix}$

Using the row operations $-0.3R_1 + R_2 \rightarrow R_2$ and $-R_1 + R_3 \rightarrow R_3$: $\begin{bmatrix} 1 & 1 & 1 & | & 0 & 0 & 1 \\ 0 & -1.3 & -0.1 & | & 0 & 1 & -0.3 \\ 0 & -0.8 & 0 & | & 1 & 0 & -1 \end{bmatrix}$

Using the row operation $R_2 \leftrightarrow R_3$: $\begin{bmatrix} 1 & 1 & 1 & | & 0 & 0 & 1 \\ 0 & -0.8 & 0 & | & 1 & 0 & -1 \\ 0 & -1.3 & -0.1 & | & 0 & 1 & -0.3 \end{bmatrix}$

Using the row operation $-\dfrac{1}{0.8}R_2 \rightarrow R_2$: $\begin{bmatrix} 1 & 1 & 1 & | & 0 & 0 & 1 \\ 0 & 1 & 0 & | & -1.25 & 0 & 1.25 \\ 0 & -1.3 & -0.1 & | & 0 & 1 & -0.3 \end{bmatrix}$

Using the row operations $-R_2 + R_1 \rightarrow R_1$ and $1.3R_2 + R_3 \rightarrow R_3$:
$\begin{bmatrix} 1 & 0 & 1 & | & 1.25 & 0 & -0.25 \\ 0 & 1 & 0 & | & -1.25 & 0 & 1.25 \\ 0 & 0 & -0.1 & | & -1.625 & 1 & 1.325 \end{bmatrix}$

Using the row operation $-10R_3 \rightarrow R_3$: $\begin{bmatrix} 1 & 0 & 1 & | & 1.25 & 0 & -0.25 \\ 0 & 1 & 0 & | & -1.25 & 0 & 1.25 \\ 0 & 0 & 1 & | & 16.25 & -10 & -13.25 \end{bmatrix}$

Using the row operation $-R_3 + R_1 \rightarrow R_1$: $\begin{bmatrix} 1 & 0 & 0 & | & -15 & 10 & 13 \\ 0 & 1 & 0 & | & -1.25 & 0 & 1.25 \\ 0 & 0 & 1 & | & 16.25 & -10 & -13.25 \end{bmatrix}$

So $A^{-1} = \begin{bmatrix} -15 & 10 & 13 \\ -1.25 & 0 & 1.25 \\ 16.25 & -10 & -13.25 \end{bmatrix}$. Therefore: $\begin{bmatrix} x \\ y \\ z \end{bmatrix} = \begin{bmatrix} -15 & 10 & 13 \\ -1.25 & 0 & 1.25 \\ 16.25 & -10 & -13.25 \end{bmatrix} \begin{bmatrix} 0.02 \\ 0 \\ 0.1 \end{bmatrix} = \begin{bmatrix} 1 \\ 0.1 \\ -1 \end{bmatrix}$

The solution is $(1, 0.1, -1)$.

39. Let r represent the interest rate. The system of equations is:
$$x + y = 20,000$$
$$0.06x + 0.18y = r(20,000)$$

Let $A = \begin{bmatrix} 1 & 1 \\ 0.06 & 0.18 \end{bmatrix}$. Form the grafted matrix: $\begin{bmatrix} 1 & 1 & | & 1 & 0 \\ 0.06 & 0.18 & | & 0 & 1 \end{bmatrix}$

Using the row operation $-0.06R_1 + R_2 \rightarrow R_2$: $\begin{bmatrix} 1 & 1 & | & 1 & 0 \\ 0 & 0.12 & | & -0.06 & 1 \end{bmatrix}$

Using the row operation $\dfrac{1}{0.12}R_2 \rightarrow R_2$: $\begin{bmatrix} 1 & 1 & | & 1 & 0 \\ 0 & 1 & | & -\frac{1}{2} & \frac{25}{3} \end{bmatrix}$

Using the row operation $-R_2 + R_1 \rightarrow R_1$: $\begin{bmatrix} 1 & 0 & | & \frac{3}{2} & -\frac{25}{3} \\ 0 & 1 & | & -\frac{1}{2} & \frac{25}{3} \end{bmatrix}$

So $A^{-1} = \begin{bmatrix} \frac{3}{2} & -\frac{25}{3} \\ -\frac{1}{2} & \frac{25}{3} \end{bmatrix}$.

a. Since $0.08(20000) = 1600$, the solution is: $\begin{bmatrix} x \\ y \end{bmatrix} = \begin{bmatrix} \frac{3}{2} & -\frac{25}{3} \\ -\frac{1}{2} & \frac{25}{3} \end{bmatrix} \begin{bmatrix} 20000 \\ 1600 \end{bmatrix} = \begin{bmatrix} 16,666.67 \\ 3,333.33 \end{bmatrix}$

Jenna should invest $16,666.67 at 6% and $3,333.33 at 18%.

b. Since $0.10(20000) = 2000$, the solution is: $\begin{bmatrix} x \\ y \end{bmatrix} = \begin{bmatrix} \frac{3}{2} & -\frac{25}{3} \\ -\frac{1}{2} & \frac{25}{3} \end{bmatrix} \begin{bmatrix} 20000 \\ 2000 \end{bmatrix} = \begin{bmatrix} 13,333.33 \\ 6,666.67 \end{bmatrix}$

Jenna should invest $13,333.33 at 6% and $6,666.67 at 18%.

c. Since $0.12(20000) = 2400$, the solution is: $\begin{bmatrix} x \\ y \end{bmatrix} = \begin{bmatrix} \frac{3}{2} & -\frac{25}{3} \\ -\frac{1}{2} & \frac{25}{3} \end{bmatrix} \begin{bmatrix} 20000 \\ 2400 \end{bmatrix} = \begin{bmatrix} 10,000 \\ 10,000 \end{bmatrix}$

Jenna should invest $10,000 at 6% and $10,000 at 18%.

41. The matrix representing the nutritional needs is: $\begin{bmatrix} 8 & 4 & 15 \\ 8 & 16 & 10 \\ 6 & 10 & 25 \end{bmatrix}$

Form the grafted matrix: $\left[\begin{array}{ccc|ccc} 8 & 4 & 15 & 1 & 0 & 0 \\ 8 & 16 & 10 & 0 & 1 & 0 \\ 6 & 10 & 25 & 0 & 0 & 1 \end{array}\right]$

Using the row operation $\frac{1}{8} R_1 \rightarrow R_1$: $\left[\begin{array}{ccc|ccc} 1 & \frac{1}{2} & \frac{15}{8} & \frac{1}{8} & 0 & 0 \\ 8 & 16 & 10 & 0 & 1 & 0 \\ 6 & 10 & 25 & 0 & 0 & 1 \end{array}\right]$

Using the row operations $-8R_1 + R_2 \rightarrow R_2$ and $-6R_1 + R_3 \rightarrow R_3$: $\left[\begin{array}{ccc|ccc} 1 & \frac{1}{2} & \frac{15}{8} & \frac{1}{8} & 0 & 0 \\ 0 & 12 & -5 & -1 & 1 & 0 \\ 0 & 7 & \frac{55}{4} & -\frac{3}{4} & 0 & 1 \end{array}\right]$

Using the row operation $\frac{1}{12} R_2 \rightarrow R_2$: $\left[\begin{array}{ccc|ccc} 1 & \frac{1}{2} & \frac{15}{8} & \frac{1}{8} & 0 & 0 \\ 0 & 1 & -\frac{5}{12} & -\frac{1}{12} & \frac{1}{12} & 0 \\ 0 & 7 & \frac{55}{4} & -\frac{3}{4} & 0 & 1 \end{array}\right]$

Using the row operation $-7R_2 + R_3 \rightarrow R_3$: $\left[\begin{array}{ccc|ccc} 1 & \frac{1}{2} & \frac{15}{8} & \frac{1}{8} & 0 & 0 \\ 0 & 1 & -\frac{5}{12} & -\frac{1}{12} & \frac{1}{12} & 0 \\ 0 & 0 & \frac{50}{3} & -\frac{1}{6} & -\frac{7}{12} & 1 \end{array}\right]$

Using the row operation $\frac{3}{50} R_3 \rightarrow R_3$: $\left[\begin{array}{ccc|ccc} 1 & \frac{1}{2} & \frac{15}{8} & \frac{1}{8} & 0 & 0 \\ 0 & 1 & -\frac{5}{12} & -\frac{1}{12} & \frac{1}{12} & 0 \\ 0 & 0 & 1 & -\frac{1}{100} & -\frac{7}{200} & \frac{3}{50} \end{array}\right]$

Using the row operations $-\frac{15}{8}R_3 + R_1 \to R_1$ and $\frac{5}{12}R_3 + R_2 \to R_2$:

$$\begin{bmatrix} 1 & \frac{1}{2} & 0 & | & \frac{23}{160} & \frac{21}{320} & -\frac{9}{80} \\ 0 & 1 & 0 & | & -\frac{7}{80} & \frac{11}{160} & \frac{1}{40} \\ 0 & 0 & 1 & | & -\frac{1}{100} & -\frac{7}{200} & \frac{3}{50} \end{bmatrix}$$

Using the row operation $-\frac{1}{2}R_2 + R_1 \to R_1$: $\begin{bmatrix} 1 & 0 & 0 & | & \frac{3}{16} & \frac{1}{32} & -\frac{1}{8} \\ 0 & 1 & 0 & | & -\frac{7}{80} & \frac{11}{160} & \frac{1}{40} \\ 0 & 0 & 1 & | & -\frac{1}{100} & -\frac{7}{200} & \frac{3}{50} \end{bmatrix}$

So the inverse is $\begin{bmatrix} \frac{3}{16} & \frac{1}{32} & -\frac{1}{8} \\ -\frac{7}{80} & \frac{11}{160} & \frac{1}{40} \\ -\frac{1}{100} & -\frac{7}{200} & \frac{3}{50} \end{bmatrix}$. Solving the system: $\begin{bmatrix} \frac{3}{16} & \frac{1}{32} & -\frac{1}{8} \\ -\frac{7}{80} & \frac{11}{160} & \frac{1}{40} \\ -\frac{1}{100} & -\frac{7}{200} & \frac{3}{50} \end{bmatrix}\begin{bmatrix} 300 \\ 280 \\ 400 \end{bmatrix} = \begin{bmatrix} 15 \\ 3 \\ 11.2 \end{bmatrix}$

To feed the mice, 15 mg = 0.015 g of A, 3 mg = 0.003 g of B, and 11.2 mg = 0.0112 g of C should be mixed. Solving the system:

$$\begin{bmatrix} \frac{3}{16} & \frac{1}{32} & -\frac{1}{8} \\ -\frac{7}{80} & \frac{11}{160} & \frac{1}{40} \\ -\frac{1}{100} & -\frac{7}{200} & \frac{3}{50} \end{bmatrix}\begin{bmatrix} 450 \\ 400 \\ 600 \end{bmatrix} = \begin{bmatrix} 21\frac{7}{8} \\ 3\frac{1}{8} \\ 17\frac{1}{2} \end{bmatrix}$$

To feed the rats, $21\frac{7}{8}$ mg = 0.021875 g of A, $3\frac{1}{8}$ mg = 0.008125 g of B, and $17\frac{1}{2}$ mg = 0.0175 g of C should be mixed.

9.6 Determinants and Cramer's Rule: 2 x 2 and 3 x 3 Systems

1.   Evaluating the determinant: $\begin{vmatrix} 2 & 3 \\ 1 & 4 \end{vmatrix} = (2)(4) - (1)(3) = 8 - 3 = 5$

3.   Evaluating the determinant: $\begin{vmatrix} 1 & -5 \\ 2 & 4 \end{vmatrix} = (1)(4) - (-5)(2) = 4 + 10 = 14$

5.   Evaluating the determinant: $\begin{vmatrix} 4 & -3 \\ -8 & 6 \end{vmatrix} = (4)(6) - (-3)(-8) = 24 - 24 = 0$

7.   Expanding along the first row:

$$\begin{vmatrix} 2 & 3 & -1 \\ 1 & 4 & 1 \\ 2 & 0 & 1 \end{vmatrix} = 2\begin{vmatrix} 4 & 1 \\ 0 & 1 \end{vmatrix} - 3\begin{vmatrix} 1 & 1 \\ 2 & 1 \end{vmatrix} + (-1)\begin{vmatrix} 1 & 4 \\ 2 & 0 \end{vmatrix} = 2(4-0) - 3(1-2) - 1(0-8) = 8 + 3 + 8 = 19$$

9.   Expanding along the second column: $\begin{vmatrix} 1 & 3 & -1 \\ 1 & 0 & 1 \\ 2 & 0 & 5 \end{vmatrix} = -3\begin{vmatrix} 1 & 1 \\ 2 & 5 \end{vmatrix} + 0 + 0 = -3(5-2) = -9$

11. Expanding along the first row:

$$\begin{vmatrix} 2 & 3 & 1 \\ 2 & 0 & 2 \\ 4 & 6 & 2 \end{vmatrix} = 2\begin{vmatrix} 0 & 2 \\ 6 & 2 \end{vmatrix} - 3\begin{vmatrix} 2 & 2 \\ 4 & 2 \end{vmatrix} + 1\begin{vmatrix} 2 & 0 \\ 4 & 6 \end{vmatrix}$$

$$= 2(0-12) - 3(4-8) + 1(12-0)$$
$$= -24 + 12 + 12$$
$$= 0$$

13. Finding the required determinants:

$$D = \begin{vmatrix} 1 & -2 \\ 3 & 5 \end{vmatrix} = (1)(5) - (-2)(3) = 5 + 6 = 11$$

$$D_x = \begin{vmatrix} 4 & -2 \\ 1 & 5 \end{vmatrix} = (4)(5) - (-2)(1) = 20 + 2 = 22$$

$$D_y = \begin{vmatrix} 1 & 4 \\ 3 & 1 \end{vmatrix} = (1)(1) - (4)(3) = 1 - 12 = -11$$

Now using Cramer's rule:

$$x = \frac{D_x}{D} = \frac{22}{11} = 2 \qquad\qquad y = \frac{D_y}{D} = \frac{-11}{11} = -1$$

The solution is $(2,-1)$.

15. Finding the required determinants:

$$D = \begin{vmatrix} 3 & -1 \\ 3 & 5 \end{vmatrix} = (3)(5) - (-1)(3) = 15 + 3 = 18$$

$$D_x = \begin{vmatrix} 7 & -1 \\ -1 & 5 \end{vmatrix} = (7)(5) - (-1)(-1) = 35 - 1 = 34$$

$$D_y = \begin{vmatrix} 3 & 7 \\ 3 & -1 \end{vmatrix} = (3)(-1) - (7)(3) = -3 - 21 = -24$$

Now using Cramer's rule:

$$x = \frac{D_x}{D} = \frac{34}{18} = \frac{17}{9} \qquad\qquad y = \frac{D_y}{D} = \frac{-24}{18} = -\frac{4}{3}$$

The solution is $\left(\frac{17}{9}, -\frac{4}{3}\right)$.

17. Finding the required determinants:

$$D = \begin{vmatrix} 3 & 4 \\ 5 & -5 \end{vmatrix} = (3)(-5) - (4)(5) = -15 - 20 = -35$$

$$D_x = \begin{vmatrix} 4 & 4 \\ 1 & -5 \end{vmatrix} = (4)(-5) - (4)(1) = -20 - 4 = -24$$

$$D_y = \begin{vmatrix} 3 & 4 \\ 5 & 1 \end{vmatrix} = (3)(1) - (4)(5) = 3 - 20 = -17$$

Now using Cramer's rule:

$$x = \frac{D_x}{D} = \frac{-24}{-35} = \frac{24}{35} \qquad\qquad y = \frac{D_y}{D} = \frac{-17}{-35} = \frac{17}{35}$$

The solution is $\left(\frac{24}{35}, \frac{17}{35}\right)$.

19. Finding the required determinant: $D = \begin{vmatrix} 2 & 4 \\ 1 & 2 \end{vmatrix} = (2)(2) - (4)(1) = 4 - 4 = 0$

Since $D = 0$, Cramer's rule cannot be applied. Since the equations are multiples of each other:
$$x + 2y = 2$$
$$x = 2 - 2y$$

The solutions are $(2 - 2a, a)$, where a is any real number.

21. Finding the required determinant: $D = \begin{vmatrix} 5 & -10 \\ 1 & -2 \end{vmatrix} = (5)(-2) - (-10)(1) = -10 + 10 = 0$

Since $D = 0$, Cramer's rule cannot be applied. Multiplying the second equation by -5:
$$5x - 10y = 4$$
$$-5x + 10y = -5$$

Adding yields $0 = -1$, which is false. There is no solution to the system (inconsistent).

23. Finding the required determinants:

$$D = \begin{vmatrix} 1 & 2 & 1 \\ 1 & 4 & -1 \\ 1 & -2 & 1 \end{vmatrix} = 1 \begin{vmatrix} 4 & -1 \\ -2 & 1 \end{vmatrix} - 1 \begin{vmatrix} 2 & 1 \\ -2 & 1 \end{vmatrix} + 1 \begin{vmatrix} 2 & 1 \\ 4 & -1 \end{vmatrix} = 1(2) - 1(4) + 1(-6) = -8$$

$$D_x = \begin{vmatrix} 8 & 2 & 1 \\ 12 & 4 & -1 \\ -4 & -2 & 1 \end{vmatrix} = 8 \begin{vmatrix} 4 & -1 \\ -2 & 1 \end{vmatrix} - 12 \begin{vmatrix} 2 & 1 \\ -2 & 1 \end{vmatrix} - 4 \begin{vmatrix} 2 & 1 \\ 4 & -1 \end{vmatrix} = 8(2) - 12(4) - 4(-6) = -8$$

$$D_y = \begin{vmatrix} 1 & 8 & 1 \\ 1 & 12 & -1 \\ 1 & -4 & 1 \end{vmatrix} = 1 \begin{vmatrix} 12 & -1 \\ -4 & 1 \end{vmatrix} - 1 \begin{vmatrix} 8 & 1 \\ -4 & 1 \end{vmatrix} + 1 \begin{vmatrix} 8 & 1 \\ 12 & -1 \end{vmatrix} = 1(8) - 1(12) + 1(-20) = -24$$

$$D_z = \begin{vmatrix} 1 & 2 & 8 \\ 1 & 4 & 12 \\ 1 & -2 & -4 \end{vmatrix} = 1 \begin{vmatrix} 4 & 12 \\ -2 & -4 \end{vmatrix} - 1 \begin{vmatrix} 2 & 8 \\ -2 & -4 \end{vmatrix} + 1 \begin{vmatrix} 2 & 8 \\ 4 & 12 \end{vmatrix} = 1(8) - 1(8) + 1(-8) = -8$$

Now using Cramer's rule:
$$x = \frac{D_x}{D} = \frac{-8}{-8} = 1 \qquad y = \frac{D_y}{D} = \frac{-24}{-8} = 3 \qquad z = \frac{D_z}{D} = \frac{-8}{-8} = 1$$

The solution is $(1, 3, 1)$.

25. Finding the required determinants:

$$D = \begin{vmatrix} 3 & 2 & 1 \\ 1 & -2 & -1 \\ 1 & -2 & 1 \end{vmatrix} = 3 \begin{vmatrix} -2 & -1 \\ -2 & 1 \end{vmatrix} - 1 \begin{vmatrix} 2 & 1 \\ -2 & 1 \end{vmatrix} + 1 \begin{vmatrix} 2 & 1 \\ -2 & -1 \end{vmatrix} = 3(-4) - 1(4) + 1(0) = -16$$

$$D_x = \begin{vmatrix} 1 & 2 & 1 \\ 3 & -2 & -1 \\ 3 & -2 & 1 \end{vmatrix} = 1 \begin{vmatrix} -2 & -1 \\ -2 & 1 \end{vmatrix} - 3 \begin{vmatrix} 2 & 1 \\ -2 & 1 \end{vmatrix} + 3 \begin{vmatrix} 2 & 1 \\ -2 & -1 \end{vmatrix} = 1(-4) - 3(4) + 3(0) = -16$$

$$D_y = \begin{vmatrix} 3 & 1 & 1 \\ 1 & 3 & -1 \\ 1 & 3 & 1 \end{vmatrix} = 3 \begin{vmatrix} 3 & -1 \\ 3 & 1 \end{vmatrix} - 1 \begin{vmatrix} 1 & 1 \\ 3 & 1 \end{vmatrix} + 1 \begin{vmatrix} 1 & 1 \\ 3 & -1 \end{vmatrix} = 3(6) - 1(-2) + 1(-4) = 16$$

$$D_z = \begin{vmatrix} 3 & 2 & 1 \\ 1 & -2 & 3 \\ 1 & -2 & 3 \end{vmatrix} = 3 \begin{vmatrix} -2 & 3 \\ -2 & 3 \end{vmatrix} - 1 \begin{vmatrix} 2 & 1 \\ -2 & 3 \end{vmatrix} + 1 \begin{vmatrix} 2 & 1 \\ -2 & 3 \end{vmatrix} = 3(0) - 1(8) + 1(8) = 0$$

Now using Cramer's rule:

$$x = \frac{D_x}{D} = \frac{-16}{-16} = 1 \qquad y = \frac{D_y}{D} = \frac{16}{-16} = -1 \qquad z = \frac{D_z}{D} = \frac{0}{-16} = 0$$

The solution is $(1,-1,0)$.

27. Finding the required determinants:

$$D = \begin{vmatrix} 2 & -2 & 1 \\ 1 & -2 & -1 \\ 1 & -2 & 1 \end{vmatrix} = 2\begin{vmatrix} -2 & -1 \\ -2 & 1 \end{vmatrix} - 1\begin{vmatrix} -2 & 1 \\ -2 & 1 \end{vmatrix} + 1\begin{vmatrix} -2 & 1 \\ -2 & -1 \end{vmatrix} = 2(-4) - 1(0) + 1(4) = -4$$

$$D_x = \begin{vmatrix} 8 & -2 & 1 \\ 6 & -2 & -1 \\ 4 & -2 & 1 \end{vmatrix} = 8\begin{vmatrix} -2 & -1 \\ -2 & 1 \end{vmatrix} - 6\begin{vmatrix} -2 & 1 \\ -2 & 1 \end{vmatrix} + 4\begin{vmatrix} -2 & 1 \\ -2 & -1 \end{vmatrix} = 8(-4) - 6(0) + 4(4) = -16$$

$$D_y = \begin{vmatrix} 2 & 8 & 1 \\ 1 & 6 & -1 \\ 1 & 4 & 1 \end{vmatrix} = 2\begin{vmatrix} 6 & -1 \\ 4 & 1 \end{vmatrix} - 1\begin{vmatrix} 8 & 1 \\ 4 & 1 \end{vmatrix} + 1\begin{vmatrix} 8 & 1 \\ 6 & -1 \end{vmatrix} = 2(10) - 1(4) + 1(-14) = 2$$

$$D_z = \begin{vmatrix} 2 & -2 & 8 \\ 1 & -2 & 6 \\ 1 & -2 & 4 \end{vmatrix} = 2\begin{vmatrix} -2 & 6 \\ -2 & 4 \end{vmatrix} - 1\begin{vmatrix} -2 & 8 \\ -2 & 4 \end{vmatrix} + 1\begin{vmatrix} -2 & 8 \\ -2 & 6 \end{vmatrix} = 2(4) - 1(8) + 1(4) = 4$$

Now using Cramer's rule:

$$x = \frac{D_x}{D} = \frac{-16}{-4} = 4 \qquad y = \frac{D_y}{D} = \frac{2}{-4} = -\frac{1}{2} \qquad z = \frac{D_z}{D} = \frac{4}{-4} = -1$$

The solution is $\left(4, -\frac{1}{2}, -1\right)$.

29. Finding the required determinants:

$$D = \begin{vmatrix} 1 & 2 & 1 \\ 2 & 4 & 1 \\ 1 & -2 & 1 \end{vmatrix} = 1\begin{vmatrix} 4 & 1 \\ -2 & 1 \end{vmatrix} - 2\begin{vmatrix} 2 & 1 \\ -2 & 1 \end{vmatrix} + 1\begin{vmatrix} 2 & 1 \\ 4 & 1 \end{vmatrix} = 1(6) - 2(4) + 1(-2) = -4$$

$$D_x = \begin{vmatrix} 1 & 2 & 1 \\ 2 & 4 & 1 \\ -4 & -2 & 1 \end{vmatrix} = 1\begin{vmatrix} 4 & 1 \\ -2 & 1 \end{vmatrix} - 2\begin{vmatrix} 2 & 1 \\ -2 & 1 \end{vmatrix} - 4\begin{vmatrix} 2 & 1 \\ 4 & 1 \end{vmatrix} = 1(6) - 2(4) - 4(-2) = 6$$

$$D_y = \begin{vmatrix} 1 & 1 & 1 \\ 2 & 2 & 1 \\ 1 & -4 & 1 \end{vmatrix} = 1\begin{vmatrix} 2 & 1 \\ -4 & 1 \end{vmatrix} - 2\begin{vmatrix} 1 & 1 \\ -4 & 1 \end{vmatrix} + 1\begin{vmatrix} 1 & 1 \\ 2 & 1 \end{vmatrix} = 1(6) - 2(5) + 1(-1) = -5$$

$$D_z = \begin{vmatrix} 1 & 2 & 1 \\ 2 & 4 & 2 \\ 1 & -2 & -4 \end{vmatrix} = 1\begin{vmatrix} 4 & 2 \\ -2 & -4 \end{vmatrix} - 2\begin{vmatrix} 2 & 1 \\ -2 & -4 \end{vmatrix} + 1\begin{vmatrix} 2 & 1 \\ 4 & 2 \end{vmatrix} = 1(-12) - 2(-6) + 1(0) = 0$$

Now using Cramer's rule:

$$x = \frac{D_x}{D} = \frac{6}{-4} = -\frac{3}{2} \qquad y = \frac{D_y}{D} = \frac{-5}{-4} = \frac{5}{4} \qquad z = \frac{D_z}{D} = \frac{0}{-4} = 0$$

The solution is $\left(-\frac{3}{2}, \frac{5}{4}, 0\right)$.

31.   Finding the required determinants:

$$D = \begin{vmatrix} 1 & 0 & 1 \\ 0 & 2 & -1 \\ 1 & 0 & -1 \end{vmatrix} = 1\begin{vmatrix} 2 & -1 \\ 0 & -1 \end{vmatrix} - 0 + 1\begin{vmatrix} 0 & 1 \\ 2 & -1 \end{vmatrix} = 1(-2) + 1(-2) = -4$$

$$D_x = \begin{vmatrix} 1 & 0 & 1 \\ 2 & 2 & -1 \\ 4 & 0 & -1 \end{vmatrix} = 2\begin{vmatrix} 1 & 1 \\ 4 & -1 \end{vmatrix} = 2(-5) = -10$$

$$D_y = \begin{vmatrix} 1 & 1 & 1 \\ 0 & 2 & -1 \\ 1 & 4 & -1 \end{vmatrix} = 1\begin{vmatrix} 2 & -1 \\ 4 & -1 \end{vmatrix} - 0 + 1\begin{vmatrix} 1 & 1 \\ 2 & -1 \end{vmatrix} = 1(2) + 1(-3) = -1$$

$$D_z = \begin{vmatrix} 1 & 0 & 1 \\ 0 & 2 & 2 \\ 1 & 0 & 4 \end{vmatrix} = 2\begin{vmatrix} 1 & 1 \\ 1 & 4 \end{vmatrix} = 2(3) = 6$$

Now using Cramer's rule:

$$x = \frac{D_x}{D} = \frac{-10}{-4} = \frac{5}{2} \qquad y = \frac{D_y}{D} = \frac{-1}{-4} = \frac{1}{4} \qquad z = \frac{D_z}{D} = \frac{6}{-4} = -\frac{3}{2}$$

The solution is $\left(\frac{5}{2}, \frac{1}{4}, -\frac{3}{2}\right)$.

33.   Let x and y represent the amounts invested in each account. The system of equations is:

$$x + y = 10,000$$
$$0.06x + 0.085y = 0.07(10,000) = 700$$

Finding the required determinants:

$$D = \begin{vmatrix} 1 & 1 \\ 0.06 & 0.085 \end{vmatrix} = (1)(0.085) - (1)(0.06) = 0.025$$

$$D_x = \begin{vmatrix} 10000 & 1 \\ 700 & 0.085 \end{vmatrix} = (10000)(0.085) - (1)(700) = 150$$

$$D_y = \begin{vmatrix} 1 & 10000 \\ 0.06 & 700 \end{vmatrix} = (1)(700) - (0.06)(10000) = 100$$

Now using Cramer's rule:

$$x = \frac{D_x}{D} = \frac{150}{0.025} = 6000 \qquad y = \frac{D_y}{D} = \frac{100}{0.025} = 4000$$

Brian invested $6000 at 6% and $4000 at 8.5%.

35.   Let a and b represent the amounts of model A and model B, respectively. The system of equations is:

$$1.5a + 1b = 100$$
$$1a + 0.5b = 59$$

Finding the required determinants:

$$D = \begin{vmatrix} 1.5 & 1 \\ 1 & 0.5 \end{vmatrix} = (1.5)(0.5) - (1)(1) = -0.25$$

$$D_a = \begin{vmatrix} 100 & 1 \\ 59 & 0.5 \end{vmatrix} = (100)(0.5) - (1)(59) = -9$$

$$D_b = \begin{vmatrix} 1.5 & 100 \\ 1 & 59 \end{vmatrix} = (1.5)(59) - (100)(1) = -11.5$$

Now using Cramer's rule:

$$a = \frac{D_a}{D} = \frac{-9}{-0.25} = 36 \qquad\qquad b = \frac{D_b}{D} = \frac{-11.5}{-0.25} = 46$$

The manufacturer can produce 36 units of model A and 46 units of model B.

37. The system of equations is:

$$0.8A + 0.7B + 0.4C = 123$$
$$0.9A + 0.5B + 0.8C = 145$$
$$9.8A + 6.2B + 3.1C = 1280$$

Finding the required determinants:

$$D = \begin{vmatrix} 0.8 & 0.7 & 0.4 \\ 0.9 & 0.5 & 0.8 \\ 9.8 & 6.2 & 3.1 \end{vmatrix}$$

$$= 0.8 \begin{vmatrix} 0.5 & 0.8 \\ 6.2 & 3.1 \end{vmatrix} - 0.9 \begin{vmatrix} 0.7 & 0.4 \\ 6.2 & 3.1 \end{vmatrix} + 9.8 \begin{vmatrix} 0.7 & 0.4 \\ 0.5 & 0.8 \end{vmatrix}$$

$$= 0.8(-3.41) - 0.9(-0.31) + 9.8(0.36)$$

$$= 1.079$$

$$D_A = \begin{vmatrix} 123 & 0.7 & 0.4 \\ 145 & 0.5 & 0.8 \\ 1280 & 6.2 & 3.1 \end{vmatrix}$$

$$= 123 \begin{vmatrix} 0.5 & 0.8 \\ 6.2 & 3.1 \end{vmatrix} - 145 \begin{vmatrix} 0.7 & 0.4 \\ 6.2 & 3.1 \end{vmatrix} + 1280 \begin{vmatrix} 0.7 & 0.4 \\ 0.5 & 0.8 \end{vmatrix}$$

$$= 123(-3.41) - 145(-0.31) + 1280(0.36)$$

$$= 86.32$$

$$D_B = \begin{vmatrix} 0.8 & 123 & 0.4 \\ 0.9 & 145 & 0.8 \\ 9.8 & 1280 & 3.1 \end{vmatrix}$$

$$= 0.8 \begin{vmatrix} 145 & 0.8 \\ 1280 & 3.1 \end{vmatrix} - 0.9 \begin{vmatrix} 123 & 0.4 \\ 1280 & 3.1 \end{vmatrix} + 9.8 \begin{vmatrix} 123 & 0.4 \\ 145 & 0.8 \end{vmatrix}$$

$$= 0.8(-574.5) - 0.9(-130.7) + 9.8(40.4)$$

$$= 53.95$$

$$D_C = \begin{vmatrix} 0.8 & 0.7 & 123 \\ 0.9 & 0.5 & 145 \\ 9.8 & 6.2 & 1280 \end{vmatrix}$$

$$= 0.8 \begin{vmatrix} 0.5 & 145 \\ 6.2 & 1280 \end{vmatrix} - 0.9 \begin{vmatrix} 0.7 & 123 \\ 6.2 & 1280 \end{vmatrix} + 9.8 \begin{vmatrix} 0.7 & 123 \\ 0.5 & 145 \end{vmatrix}$$

$$= 0.8(-259) - 0.9(133.4) + 9.8(40)$$

$$= 64.74$$

Now using Cramer's rule:

$$A = \frac{D_A}{D} = \frac{86.32}{1.079} = 80 \qquad B = \frac{D_B}{D} = \frac{53.95}{1.079} = 50 \qquad C = \frac{D_C}{D} = \frac{64.74}{1.079} = 60$$

The manufacturer can produce 80 units of model A, 50 units of model B, and 60 units of model C.

39. Since $\begin{vmatrix} 5 & 0 \\ -1 & 2 \end{vmatrix} = (5)(2) - (0)(-1) = 10$, $\begin{vmatrix} 5 & 0 \\ -1 & 2 \end{vmatrix}$ is a real number. But $\begin{bmatrix} 5 & 0 \\ -1 & 2 \end{bmatrix}$ is a 2 x 2 matrix.

9.7 Properties of Determinants

1.   The minor of 6 is $\begin{vmatrix} 2 & 5 & -1 \\ 7 & 4 & -3 \\ -2 & 1 & 9 \end{vmatrix}$.

3.   The cofactor of 0 is $-\begin{vmatrix} 2 & 5 & -1 \\ -2 & 1 & 9 \\ -6 & -4 & -5 \end{vmatrix}$.

5.   Here $A_{24} = \begin{vmatrix} 2 & 5 & 3 \\ -2 & 1 & 8 \\ -6 & -4 & 6 \end{vmatrix}$.

7.   **a.**   Computing each determinant:

$$\begin{vmatrix} 4 & 1 \\ -2 & 3 \end{vmatrix} = (4)(3) - (1)(-2) = 12 + 2 = 14$$

$$\begin{vmatrix} -2 & 3 \\ 4 & 1 \end{vmatrix} = (-2)(1) - (3)(4) = -2 - 12 = -14$$

Note that interchanging rows (or columns) changes the sign of the determinant.

b.   Multiplying row 1 by the constant k:

$$\begin{vmatrix} 4k & k \\ -2 & 3 \end{vmatrix} = (4k)(3) - (k)(-2) = 12k + 2k = 14k$$

Note that multiplying a row (or column) by a constant k multiplies the determinant by that same constant.

c.   Adding a multiple (k) of the second row to the first row:

$$\begin{vmatrix} 4 - 2k & 1 + 3k \\ -2 & 3 \end{vmatrix} = (4 - 2k)(3) - (1 + 3k)(-2) = 12 - 6k + 2 + 6k = 14$$

Note that the determinant is unchanged.

9.   Computing each determinant:

$$\begin{vmatrix} a & b \\ c & d \end{vmatrix} = ad - bc$$

$$\begin{vmatrix} c & d \\ a & b \end{vmatrix} = bc - ad$$

Note that interchanging rows (or columns) changes the sign of the determinant. Multiplying column 1 by the constant k:

$$\begin{vmatrix} ak & b \\ ck & d \end{vmatrix} = adk - bck = k(ad - bc)$$

Note that multiplying a column (or row) by a constant k multiplies the determinant by that same constant. Adding a multiple (k) of the second row to the first row:

$$\begin{vmatrix} a + kc & b + kd \\ c & d \end{vmatrix} = ad + kcd - bc - kcd = ad - bc$$

Note that the determinant is unchanged.

11. Switching the first and third columns: $\begin{vmatrix} 3 & -3 & 1 \\ 5 & 2 & 1 \\ 5 & 2 & 4 \end{vmatrix} = -\begin{vmatrix} 1 & -3 & 3 \\ 1 & 2 & 5 \\ 4 & 2 & 5 \end{vmatrix}$

Adding -1 times row 1 to row 2, and -4 times row 1 to row 3: $-\begin{vmatrix} 1 & -3 & 3 \\ 1 & 2 & 5 \\ 4 & 2 & 5 \end{vmatrix} = -\begin{vmatrix} 1 & -3 & 3 \\ 0 & 5 & 2 \\ 0 & 14 & -7 \end{vmatrix}$

Factoring 5 out of the second row: $-\begin{vmatrix} 1 & -3 & 3 \\ 0 & 5 & 2 \\ 0 & 14 & -7 \end{vmatrix} = -5\begin{vmatrix} 1 & -3 & 3 \\ 0 & 1 & \frac{2}{5} \\ 0 & 14 & -7 \end{vmatrix}$

Adding -14 times row 2 to row 3: $-5\begin{vmatrix} 1 & -3 & 3 \\ 0 & 1 & \frac{2}{5} \\ 0 & 14 & -7 \end{vmatrix} = -5\begin{vmatrix} 1 & -3 & 3 \\ 0 & 1 & \frac{2}{5} \\ 0 & 0 & -\frac{63}{5} \end{vmatrix}$

The determinant is therefore: $-5(1)(1)\left(-\frac{63}{5}\right) = 63$

13. Adding -5 times row 1 to row 3: $\begin{vmatrix} 1 & 2 & -1 \\ 0 & 3 & 4 \\ 5 & 1 & 2 \end{vmatrix} = \begin{vmatrix} 1 & 2 & -1 \\ 0 & 3 & 4 \\ 0 & -9 & 7 \end{vmatrix} = 1\begin{vmatrix} 3 & 4 \\ -9 & 7 \end{vmatrix} = 21 + 36 = 57$

15. Subtracting row 1 from row 2 and expanding along the third column:
$-3\begin{vmatrix} 3 & 1 & 2 \\ 3 & 1 & 4 \\ 2 & 2 & 5 \end{vmatrix} = -3\begin{vmatrix} 3 & 1 & 2 \\ 0 & 0 & 2 \\ 2 & 2 & 5 \end{vmatrix} = (-3)(-2)\begin{vmatrix} 3 & 1 \\ 2 & 2 \end{vmatrix} = 6(6-2) = 24$

17. Adding -3 times row 3 to row 1 and -4 times row 3 to row 2:
$\begin{vmatrix} 2 & 1 & 3 & 1 \\ -1 & 2 & 4 & 0 \\ 3 & 1 & 1 & -2 \\ 2 & -1 & 0 & 5 \end{vmatrix} = \begin{vmatrix} -7 & -2 & 0 & 7 \\ -13 & -2 & 0 & 8 \\ 3 & 1 & 1 & -2 \\ 2 & -1 & 0 & 5 \end{vmatrix} = 1\begin{vmatrix} -7 & -2 & 7 \\ -13 & -2 & 8 \\ 2 & -1 & 5 \end{vmatrix}$

Adding -2 times row 3 to row 1 and -2 times row 3 to row 2:
$\begin{vmatrix} -7 & -2 & 7 \\ -13 & -2 & 8 \\ 2 & -1 & 5 \end{vmatrix} = \begin{vmatrix} -11 & 0 & -3 \\ -17 & 0 & -2 \\ 2 & -1 & 5 \end{vmatrix} = -(-1)\begin{vmatrix} -11 & -3 \\ -17 & -2 \end{vmatrix} = 22 - 51 = -29$

19. Since the third column is all zeros, the determinant is 0.

21. Given the system:
$$a_{11}x_1 + a_{12}x_2 + a_{13}x_3 + a_{14}x_4 = k_1$$
$$a_{21}x_1 + a_{22}x_2 + a_{23}x_3 + a_{24}x_4 = k_2$$
$$a_{31}x_1 + a_{32}x_2 + a_{33}x_3 + a_{34}x_4 = k_3$$
$$a_{41}x_1 + a_{42}x_2 + a_{43}x_3 + a_{44}x_4 = k_4$$

The solutions to the system are given by $x_1 = \dfrac{D_{x_1}}{D}, x_2 = \dfrac{D_{x_2}}{D}, x_3 = \dfrac{D_{x_3}}{D}, x_4 = \dfrac{D_{x_4}}{D}$, where $D, D_{x_1}, D_{x_2}, D_{x_3}$, and D_{x_4} are defined by:

$$D = \begin{vmatrix} a_{11} & a_{12} & a_{13} & a_{14} \\ a_{12} & a_{22} & a_{23} & a_{24} \\ a_{13} & a_{32} & a_{33} & a_{34} \\ a_{14} & a_{42} & a_{43} & a_{44} \end{vmatrix} \qquad D_{x_1} = \begin{vmatrix} k_1 & a_{12} & a_{13} & a_{14} \\ k_2 & a_{22} & a_{23} & a_{24} \\ k_3 & a_{32} & a_{33} & a_{34} \\ k_4 & a_{42} & a_{43} & a_{44} \end{vmatrix}$$

$$D_{x_2} = \begin{vmatrix} a_{11} & k_1 & a_{13} & a_{14} \\ a_{21} & k_2 & a_{23} & a_{24} \\ a_{31} & k_3 & a_{33} & a_{34} \\ a_{41} & k_4 & a_{43} & a_{44} \end{vmatrix} \qquad D_{x_3} = \begin{vmatrix} a_{11} & a_{12} & k_1 & a_{14} \\ a_{21} & a_{22} & k_2 & a_{24} \\ a_{31} & a_{32} & k_3 & a_{34} \\ a_{41} & a_{42} & k_4 & a_{44} \end{vmatrix}$$

$$D_{x_4} = \begin{vmatrix} a_{11} & a_{12} & a_{13} & k_1 \\ a_{21} & a_{22} & a_{23} & k_2 \\ a_{31} & a_{32} & a_{33} & k_3 \\ a_{41} & a_{42} & a_{43} & k_4 \end{vmatrix}$$

23. Finding the required determinants:

$$D = \begin{vmatrix} 1 & 1 & 0 & 0 \\ 1 & 0 & 1 & 1 \\ 0 & 1 & 1 & 0 \\ 0 & 0 & 1 & 1 \end{vmatrix} = \begin{vmatrix} 1 & 1 & 0 & 0 \\ 0 & -1 & 1 & 1 \\ 0 & 1 & 1 & 0 \\ 0 & 0 & 1 & 1 \end{vmatrix} = 1\begin{vmatrix} -1 & 1 & 1 \\ 1 & 1 & 0 \\ 0 & 1 & 1 \end{vmatrix} = \begin{vmatrix} -1 & 1 & 1 \\ 0 & 2 & 1 \\ 0 & 1 & 1 \end{vmatrix} = -1\begin{vmatrix} 2 & 1 \\ 1 & 1 \end{vmatrix} = -1$$

$$D_w = \begin{vmatrix} -1 & 1 & 0 & 0 \\ 2 & 0 & 1 & 1 \\ 1 & 1 & 1 & 0 \\ 3 & 0 & 1 & 1 \end{vmatrix} = \begin{vmatrix} -1 & 1 & 0 & 0 \\ 2 & 0 & 1 & 1 \\ 1 & 1 & 1 & 0 \\ 1 & 0 & 0 & 0 \end{vmatrix} = -1\begin{vmatrix} 1 & 0 & 0 \\ 0 & 1 & 1 \\ 1 & 1 & 0 \end{vmatrix} = -1\begin{vmatrix} 1 & 1 \\ 1 & 0 \end{vmatrix} = -1$$

$$D_x = \begin{vmatrix} 1 & -1 & 0 & 0 \\ 1 & 2 & 1 & 1 \\ 0 & 1 & 1 & 0 \\ 0 & 3 & 1 & 1 \end{vmatrix} = \begin{vmatrix} 1 & 0 & 0 & 0 \\ 1 & 3 & 1 & 1 \\ 0 & 1 & 1 & 0 \\ 0 & 3 & 1 & 1 \end{vmatrix} = 1\begin{vmatrix} 3 & 1 & 1 \\ 1 & 1 & 0 \\ 3 & 1 & 1 \end{vmatrix} = 1\begin{vmatrix} 3 & 1 & 1 \\ 1 & 1 & 0 \\ 0 & 0 & 0 \end{vmatrix} = 0$$

$$D_y = \begin{vmatrix} 1 & 1 & -1 & 0 \\ 1 & 0 & 2 & 1 \\ 0 & 1 & 1 & 0 \\ 0 & 0 & 3 & 1 \end{vmatrix} = \begin{vmatrix} 1 & 1 & -1 & 0 \\ 0 & -1 & 3 & 1 \\ 0 & 1 & 1 & 0 \\ 0 & 0 & 3 & 1 \end{vmatrix} = 1\begin{vmatrix} -1 & 3 & 1 \\ 1 & 1 & 0 \\ 0 & 3 & 1 \end{vmatrix} = 1\begin{vmatrix} -1 & 3 & 1 \\ 0 & 4 & 1 \\ 0 & 3 & 1 \end{vmatrix} = -1\begin{vmatrix} 4 & 1 \\ 3 & 1 \end{vmatrix} = -1$$

$$D_z = \begin{vmatrix} 1 & 1 & 0 & -1 \\ 1 & 0 & 1 & 2 \\ 0 & 1 & 1 & 1 \\ 0 & 0 & 1 & 3 \end{vmatrix} = \begin{vmatrix} 1 & 1 & 0 & -1 \\ 0 & -1 & 1 & 3 \\ 0 & 1 & 1 & 1 \\ 0 & 0 & 1 & 3 \end{vmatrix} = 1\begin{vmatrix} -1 & 1 & 3 \\ 1 & 1 & 1 \\ 0 & 1 & 3 \end{vmatrix} = 1\begin{vmatrix} -1 & 1 & 3 \\ 0 & 2 & 4 \\ 0 & 1 & 3 \end{vmatrix} = -1\begin{vmatrix} 2 & 4 \\ 1 & 3 \end{vmatrix} = -2$$

Now using Cramer's rule:

$$w = \frac{D_w}{D} = \frac{-1}{-1} = 1 \qquad x = \frac{D_x}{D} = \frac{0}{-1} = 0 \qquad y = \frac{D_y}{D} = \frac{-1}{-1} = 1 \qquad z = \frac{D_z}{D} = \frac{-2}{-1} = 2$$

The solution is $(1,0,1,2)$.

25. Finding the required determinants:

$$D = \begin{vmatrix} 1 & 1 & 2 & -1 \\ 0 & 2 & 3 & 1 \\ 1 & 0 & 1 & -1 \\ 1 & 1 & -1 & 0 \end{vmatrix} = \begin{vmatrix} 1 & 1 & 2 & -1 \\ 0 & 2 & 3 & 1 \\ 0 & -1 & -1 & 0 \\ 0 & 0 & -3 & 1 \end{vmatrix} = 1 \begin{vmatrix} 2 & 3 & 1 \\ -1 & -1 & 0 \\ 0 & -3 & 1 \end{vmatrix} = \begin{vmatrix} -1 & 3 & 1 \\ 0 & -1 & 0 \\ 3 & -3 & 1 \end{vmatrix} = -1 \begin{vmatrix} -1 & 1 \\ 3 & 1 \end{vmatrix} = 4$$

$$D_w = \begin{vmatrix} 1 & 1 & 2 & -1 \\ 1 & 2 & 3 & 1 \\ 0 & 0 & 1 & -1 \\ 4 & 1 & -1 & 0 \end{vmatrix} = \begin{vmatrix} 1 & 1 & 2 & 1 \\ 1 & 2 & 3 & 4 \\ 0 & 0 & 1 & 0 \\ 4 & 1 & -1 & -1 \end{vmatrix} = 1 \begin{vmatrix} 1 & 1 & 1 \\ 1 & 2 & 4 \\ 4 & 1 & -1 \end{vmatrix} = \begin{vmatrix} 1 & 0 & 0 \\ 1 & 1 & 3 \\ 4 & -3 & -5 \end{vmatrix} = 1 \begin{vmatrix} 1 & 3 \\ -3 & -5 \end{vmatrix} = 4$$

$$D_x = \begin{vmatrix} 1 & 1 & 2 & -1 \\ 0 & 1 & 3 & 1 \\ 1 & 0 & 1 & -1 \\ 1 & 4 & -1 & 0 \end{vmatrix} = \begin{vmatrix} 1 & 1 & 2 & -1 \\ 0 & 1 & 3 & 1 \\ 0 & -1 & -1 & 0 \\ 0 & 3 & -3 & 1 \end{vmatrix} = 1 \begin{vmatrix} 1 & 3 & 1 \\ -1 & -1 & 0 \\ 3 & -3 & 1 \end{vmatrix} = \begin{vmatrix} -2 & 3 & 1 \\ 0 & -1 & 0 \\ 6 & -3 & 1 \end{vmatrix} = -1 \begin{vmatrix} -2 & 1 \\ 6 & 1 \end{vmatrix} = 8$$

$$D_y = \begin{vmatrix} 1 & 1 & 1 & -1 \\ 0 & 2 & 1 & 1 \\ 1 & 0 & 0 & -1 \\ 1 & 1 & 4 & 0 \end{vmatrix} = \begin{vmatrix} 1 & 1 & 1 & 0 \\ 0 & 2 & 1 & 1 \\ 1 & 0 & 0 & 0 \\ 1 & 1 & 4 & 1 \end{vmatrix} = 1 \begin{vmatrix} 1 & 1 & 0 \\ 2 & 1 & 1 \\ 1 & 4 & 1 \end{vmatrix} = \begin{vmatrix} 1 & 0 & 0 \\ 2 & -1 & 1 \\ 1 & 3 & 1 \end{vmatrix} = 1 \begin{vmatrix} -1 & 1 \\ 3 & 1 \end{vmatrix} = -4$$

$$D_z = \begin{vmatrix} 1 & 1 & 2 & 1 \\ 0 & 2 & 3 & 1 \\ 1 & 0 & 1 & 0 \\ 1 & 1 & -1 & 4 \end{vmatrix} = \begin{vmatrix} 1 & 1 & 1 & 1 \\ 0 & 2 & 3 & 1 \\ 1 & 0 & 0 & 0 \\ 1 & 1 & -2 & 4 \end{vmatrix} = 1 \begin{vmatrix} 1 & 1 & 1 \\ 2 & 3 & 1 \\ 1 & -2 & 4 \end{vmatrix} = \begin{vmatrix} 1 & 0 & 0 \\ 2 & 1 & -1 \\ 1 & -3 & 3 \end{vmatrix} = 1 \begin{vmatrix} 1 & -1 \\ -3 & 3 \end{vmatrix} = 0$$

Now using Cramer's rule:

$$w = \frac{D_w}{D} = \frac{4}{4} = 1 \qquad x = \frac{D_x}{D} = \frac{8}{4} = 2 \qquad y = \frac{D_y}{D} = \frac{-4}{4} = -1 \qquad z = \frac{D_z}{D} = \frac{0}{4} = 0$$

The solution is $(1,2,-1,0)$.

27. Finding the required determinant:

$$D = \begin{vmatrix} 1 & 1 & 1 & 1 \\ 1 & 1 & 1 & 0 \\ 2 & 2 & 2 & 0 \\ 0 & 2 & -1 & 1 \end{vmatrix} = \begin{vmatrix} 1 & 1 & 1 & 1 \\ 0 & 0 & 0 & -1 \\ 2 & 2 & 2 & 0 \\ 0 & 2 & -1 & 1 \end{vmatrix} = -1 \begin{vmatrix} 1 & 1 & 1 \\ 2 & 2 & 2 \\ 0 & 2 & -1 \end{vmatrix} = -1 \begin{vmatrix} 1 & 1 & 1 \\ 0 & 0 & 0 \\ 0 & 2 & -1 \end{vmatrix} = 0$$

Since $D = 0$, Cramer's rule cannot be applied. Subtracting the second equation from the first yields $z = 0$. From the fourth equation:

$$2x - y = 0$$
$$y = 2x$$

Let $x = a$, so $y = 2a$. Substituting into the first equation:

$$w + a + 2a + 0 = 0$$
$$w = -3a$$

The solution is $(-3a, a, 2a, 0)$, where a is any real number.

29. The value of $D = 0$. The system will be dependent, and must be solved by other methods (elimination).

9.8 Systems of Linear Inequalities

1.   Sketching the graph:

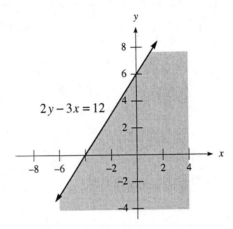

$$2y - 3x = 12$$

3.   Sketching the graph:

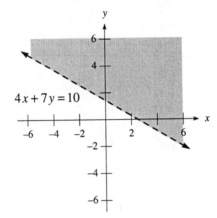

$$4x + 7y = 10$$

5.   Sketching the graph:

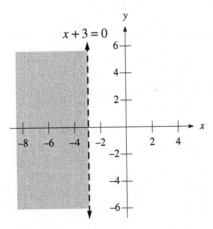

$$x + 3 = 0$$

7. Sketching the graph:

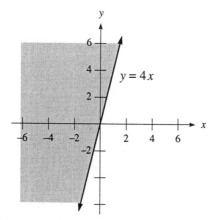

9. Sketching the solution set:

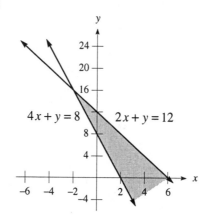

11. Sketching the solution set:

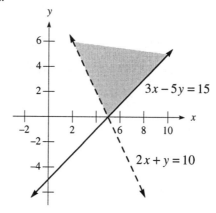

13 .   Sketching the solution set:

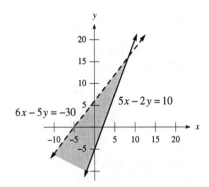

$6x - 5y = -30$ $5x - 2y = 10$

15 .   Sketching the solution set:

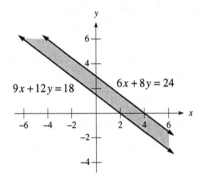

$9x + 12y = 18$ $6x + 8y = 24$

17 .   Sketching the solution set:

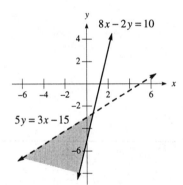

$8x - 2y = 10$ $5y = 3x - 15$

19 .   Sketching the solution set:

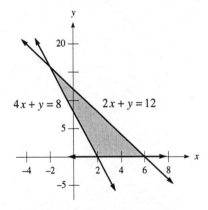

$4x + y = 8$ $2x + y = 12$

21.   Sketching the solution set:

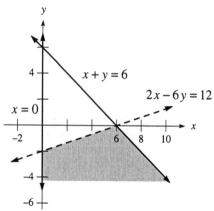

23.   Sketching the solution set:

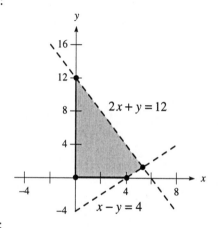

25.   Sketching the solution set:

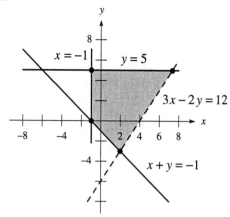

27. Sketching the solution set:

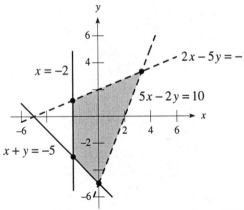

29. Let v represent the number of VCRs and t represent the number of TVs. The system of inequalities is:

$$200v + 380t \geq 6000$$
$$2v + 6t \leq 120$$
$$t \geq 0$$
$$v \leq 0$$

Sketching the graph:

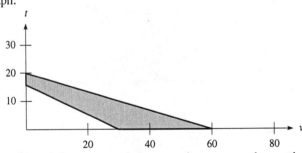

31. Let c represent the number of custom-made boxes and s represent the number of standard models. The system of inequalities is:

$$3c + s \leq 20$$
$$80c + 20s \geq 200$$
$$c \geq 0$$
$$s \geq 0$$

Sketching the graph:

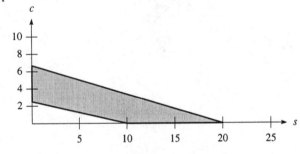

33. Let x represent the amount invested in the six-month certificate and y represent the amount invested in the 1-year certificate. The system of inequalities is:

$$x + y \leq 60,000$$
$$0.05x + 0.065y \geq 3300$$
$$x \geq 0$$
$$y \geq 0$$

Sketching the graph:

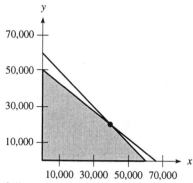

35. Let d represent the number of divorce cases and m represent the number of malpractice cases. The system of inequalities is:

$$12d + 22m \leq 60$$
$$20d + 18m \geq 40$$
$$d \geq 0$$
$$m \geq 0$$

Sketching the graph:

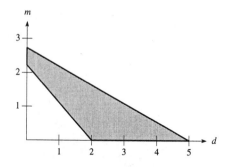

9.9 An Introduction to Linear Programming: Geometric Solutions

1. First sketch the constraints:

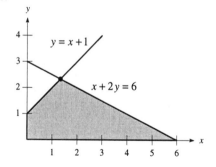

Evaluating f at these vertices:
$$f(0,0) = 3(0) + 2(0) = 0$$
$$f(0,1) = 3(0) + 2(1) = 2$$
$$f(6,0) = 3(6) + 2(0) = 18$$
$$f\left(\tfrac{4}{3}, \tfrac{7}{3}\right) = 3\left(\tfrac{4}{3}\right) + 2\left(\tfrac{7}{3}\right) = \tfrac{26}{3}$$

The maximum value is 18 which occurs at (6,0), and the minimum value is 0 which occurs at (0,0).

3.   First sketch the constraints:

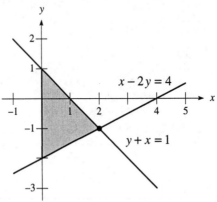

Evaluating f at these vertices:
$$f(0,1) = 0.3(0) - 1 = -1$$
$$f(0,-2) = 0.3(0) - (-2) = 2$$
$$f(2,-1) = 0.3(2) - (-1) = 1.6$$

The maximum value is 2 which occurs at (0,–2), and the minimum value is –1 which occurs at (0,1).

5.   First sketch the constraints:

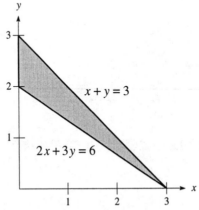

Evaluating f at these vertices:
$$f(0,2) = 2(0) + 5(2) - 1 = 9$$
$$f(0,3) = 2(0) + 5(3) - 1 = 14$$
$$f(3,0) = 2(3) + 5(0) - 1 = 5$$

The maximum value is 14 which occurs at (0,3), and the minimum value is 5 which occurs at (3,0).

7.   Let h represent the high-risk investments and c represent the conservative investments. The system of inequalities is:

$$h + c \leq 75$$
$$h \leq 0.3(h + c)$$
$$h \geq 0$$
$$c \geq 0$$

Now $P = 0.14h + 0.10c$. Sketching the constraints:

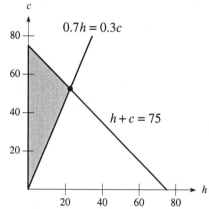

Evaluating P at these vertices:

$$P(0,0) = 0.14(0) + 0.10(0) = 0$$
$$P(0,75) = 0.14(0) + 0.10(75) = 7.5$$
$$P(22.5, 52.5) = 0.14(22.5) + 0.10(52.5) = 8.4$$

The maximum profit is $8.4 million with $22.5 million invested in high-risk and $52.5 million invested in conservative investments.

9.   Let x and y represent the quantities of model X and model Y, respectively. The system of inequalities is:

$$400x + 500y \leq 50,000$$
$$16x + 12y \leq 1600$$
$$x \geq 0$$
$$y \geq 0$$

Now $P = 90x + 100y$. Sketching the constraints:

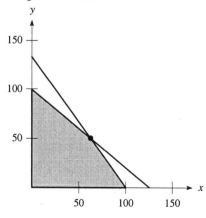

Evaluating P at these vertices:
$$P(0,100) = 90(0) + 100(100) = 10,000$$
$$P(100,0) = 90(100) + 100(0) = 9,000$$
$$P(63,49) = 90(63) + 100(49) = 10,570$$

The maximum profit is $10,570 if she sells 63 of model X and 49 of model Y.

11. Let s represent the number of scientific calculators and g represent the number of graphics calculators. The system of inequalities is:
$$8s + 25g \leq 15,000$$
$$s + \tfrac{1}{2}g \leq 1000$$
$$s \geq 0$$
$$g \geq 0$$

Now $P = 10s + 30g$. Sketching the constraints:

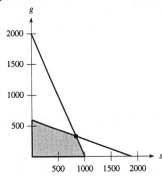

Evaluating P at these vertices:
$$P(0,600) = 10(0) + 30(600) = 18,000$$
$$P(0,0) = 10(0) + 30(0) = 0$$
$$P(1000,0) = 10(1000) + 30(0) = 10,000$$
$$P(833,333) = 10(833) + 30(333) = 18,320$$

The maximum profit is $18,230 when 833 scientific calculators and 333 graphics calculators are produced.

13. The system of inequalities is:
$$8X + 12Y \geq 24$$
$$0.25X + 0.34Y \leq 1$$
$$X \geq 0$$
$$Y \geq 0$$

Sketching the graph:

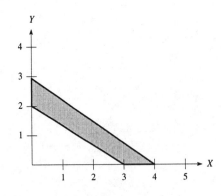

Now $C = 25X + 30Y$. Evaluating C at these vertices:

$C(0,2) = 25(0) + 30(2) = 60$

$C(0,2.94) = 25(0) + 30(2.94) = 88.2$

$C(3,0) = 25(3) + 30(0) = 75$

$C(4,0) = 25(4) + 30(0) = 100$

The least expensive combination is 60 cents for 0 oz of cereal X and 2 oz of cereal Y.

15. Let d represent the number of divorce cases and m represent the number of malpractice cases. The system of inequalities is:

$12d + 22m \leq 60$

$20d + 18m \leq 80$

$d \geq 0$

$m \geq 0$

Now $I = 6000d + 8000m$. Sketching the constraints:

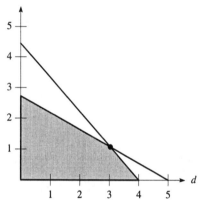

Evaluating I at these vertices:

$I(0,0) = 6000(0) + 8000(0) = 0$

$I(4,0) = 6000(4) + 8000(0) = 24,000$

$I(0,2) = 6000(0) + 8000(2) = 16,000$

$I(3,1) = 6000(3) + 8000(1) = 26,000$

She will maximize her income of $26,000 with 3 divorce cases and 1 malpractice case.

17. Let x and y represent the quantities of Zomine X and Zomine Y, respectively. The system of inequalities is:

$2x + 1y \leq 6$

$4x + 1y \leq 8$

$x + y \geq 1$

Let $R = 5x + 2y$. Evaluating R at these vertices:

$R(0,1) = 5(0) + 2(1) = 2$

$R(0,6) = 5(0) + 2(6) = 12$

$R(1,0) = 5(1) + 2(0) = 5$

$R(2,0) = 5(2) + 20(1) = 10$

$R(1,4) = 5(1) + 2(4) = 13$

The maximum pain relief is 13 mg when 1 g of Zomine X and 4 g of Zomine Y are administered.

Chapter 9 Review Exercises

1. Adding the two equations:
$$5x = 20$$
$$x = 4$$
Substituting into the first equation:
$$2(4) + y = 9$$
$$8 + y = 9$$
$$y = 1$$
The solution is (4,1).

2. Adding the two equations yields $y = 3$. Substituting into the second equation:
$$x - 3 = -5$$
$$x = -2$$
The solution is (–2,3).

3. Multiplying the second equation by 2:
$$3x - 2y = 10$$
$$4x + 2y = -10$$
Adding the two equations:
$$7x = 0$$
$$x = 0$$
Substituting into the second equation:
$$2(0) + y = -5$$
$$y = -5$$
The solution is (0,–5).

4. Multiplying the second equation by –2:
$$5x + 2y = 34$$
$$-6x - 2y = -40$$
Adding the two equations:
$$-x = -6$$
$$x = 6$$
Substituting into the second equation:
$$3(6) + y = 20$$
$$18 + y = 20$$
$$y = 2$$
The solution is (6,2).

5. Multiplying the second equation by –5:
$$5x - 2y = 9$$
$$-5x + 25y = 60$$
Adding the two equations:
$$23y = 69$$
$$y = 3$$
Substituting into the second equation:
$$x - 5(3) = -12$$
$$x - 15 = -12$$
$$x = 3$$
The solution is (3,3).

6. Substituting into the second equation:
$$2(3y-1)-5y=1$$
$$6y-2-5y=1$$
$$y-2=1$$
$$y=3$$
Substituting into the first equation: $x=3(3)-1=8$
The solution is (8,3).

7. Adding equations 1 and 2 results in $4x+2y=8$. Adding equations 1 and 3 results in $3x+2y=7$.
We have the 2 x 2 system:
$$4x+2y=8$$
$$3x+2y=7$$
Multiplying the second equation by –1:
$$4x+2y=8$$
$$-3x-2y=-7$$
Adding yields $x=1$. Substituting to find y:
$$4(1)+2y=8$$
$$4+2y=8$$
$$2y=4$$
$$y=2$$
Substituting into the original first equation to find z:
$$1+2+z=6$$
$$z=3$$
The solution is (1,2,3).

8. Adding equations 1 and 3 results in:
$$2x=8$$
$$x=4$$
Adding equations 2 and 3 results in:
$$4x+y=16$$
$$4(4)+y=16$$
$$16+y=16$$
$$y=0$$
Substituting into equation 1:
$$4-2(0)+z=6$$
$$4+z=6$$
$$z=2$$
The solution is (4,0,2).

9. Adding equations 1 and 3 results in:
$$2x=17$$
$$x=\frac{17}{2}$$
Multiply equation 3 by 2:
$$2x+2y+2z=10$$
$$2x-2y+2z=30$$
Adding:
$$4x+4z=40$$
$$4\left(\frac{17}{2}\right)+4z=40$$
$$34+4z=40$$
$$4z=6$$
$$z=\frac{3}{2}$$

Substituting into equation 1:
$$\frac{17}{2} + y - \frac{3}{2} = 2$$
$$y + 7 = 2$$
$$y = -5$$

The solution is $\left(\frac{17}{2}, -5, \frac{3}{2}\right)$.

10. Adding equations 2 and 3 results in $2x - y = 2$. Multiplying equation 2 by 2:
$$3x + y + 2z = 1$$
$$2x - 4y - 2z = 2$$

Adding yields $5x - 3y = 3$. We have the 2 x 2 system:
$$2x - y = 2$$
$$5x - 3y = 3$$

Multiplying the first equation by -3:
$$-6x + 3y = -6$$
$$5x - 3y = 3$$

Adding yields $-x = -3$, so $x = 3$. Substituting to find y:
$$2(3) - y = 2$$
$$6 - y = 2$$
$$-y = -4$$
$$y = 4$$

Substituting into the original third equation to find z:
$$3 + 4 + z = 1$$
$$7 + z = 1$$
$$z = -6$$

The solution is $(3, 4, -6)$.

11. Multiplying the first equation by 10 and rewrite the system as:
$$a + 2b + c = 8$$
$$a - 0.5b + 2c = 0.5$$
$$2a + b - 10c = 7$$

Adding -1 times equation 1 to equation 2:
$$-a - 2b - c = -8$$
$$a - 0.5b + 2c = 0.5$$

Adding yields $-2.5b + 1.5c = -7.5$. Adding -2 times equation 1 to equation 3:
$$-2a - 4b - 2c = -16$$
$$2a + b - 10c = 7$$

Adding yields $-3b - 12c = -9$, or $b + 4c = 3$. We have the 2 x 2 system:
$$b + 4c = 3$$
$$-2.5b + 1.5c = -7.5$$

Multiplying the first equation by 2.5:
$$2.5b + 10c = 7.5$$
$$-2.5b + 1.5c = -7.5$$

Adding yields $11.5c = 0$, so $c = 0$. Substituting:
$$b + 4(0) = 3$$
$$b = 3$$

Substituting:
$$a + 2(3) + 0 = 8$$
$$a + 6 = 8$$
$$a = 2$$

The solution is $(2, 3, 0)$.

12. First rewrite the system of equations as:
$$x - 2y - 2z = 0$$
$$x - 4y = -1$$
$$2x - 3y - 3z = 0$$
Adding -1 times equation 1 to equation 2:
$$-x + 2y + 2z = 0$$
$$x - 4y = -1$$
Adding yields $-2y + 2z = -1$. Adding -2 times equation 1 to equation 3:
$$-2x + 4y + 4z = 0$$
$$2x - 3y - 3z = 0$$
Adding yields $y + z = 0$. We have the 2 x 2 system:
$$y + z = 0$$
$$-2y + 2z = -1$$
Multiplying the first equation by 2:
$$2y + 2z = 0$$
$$-2y + 2z = -1$$
Adding yields $4z = -1$, so $z = -\frac{1}{4}$. So $y - \frac{1}{4} = 0$, thus $y = \frac{1}{4}$. Substituting into the original first equation:
$$x - 4\left(\tfrac{1}{4}\right) = -1$$
$$x - 1 = -1$$
$$x = 0$$
The solution is $\left(0, \frac{1}{4}, -\frac{1}{4}\right)$.

13. Using the row operation $-R_1 \to R_1$: $\begin{bmatrix} 1 & -2 & \vdots & -2 \\ 4 & 6 & \vdots & 6 \end{bmatrix}$

Using the row operation $-4R_1 + R_2 \to R_2$: $\begin{bmatrix} 1 & -2 & \vdots & -2 \\ 0 & 14 & \vdots & 14 \end{bmatrix}$

Using the row operation $\frac{1}{14}R_2 \to R_2$: $\begin{bmatrix} 1 & -2 & \vdots & -2 \\ 0 & 1 & \vdots & 1 \end{bmatrix}$

Using the row operation $2R_2 + R_1 \to R_1$: $\begin{bmatrix} 1 & 0 & \vdots & 0 \\ 0 & 1 & \vdots & 1 \end{bmatrix}$

14. Using the row operation $\frac{1}{3}R_1 \to R_1$: $\begin{bmatrix} 1 & 0 & \vdots & 2 \\ -6 & 1 & \vdots & 3 \end{bmatrix}$

Using the row operation $6R_1 + R_2 \to R_2$: $\begin{bmatrix} 1 & 0 & \vdots & 2 \\ 0 & 1 & \vdots & 15 \end{bmatrix}$

15. Using the row operations $-2R_1 + R_2 \to R_2$ and $-3R_1 + R_3 \to R_3$: $\begin{bmatrix} 1 & 1 & 2 & \vdots & 2 \\ 0 & -2 & -2 & \vdots & -4 \\ 0 & -4 & -5 & \vdots & -5 \end{bmatrix}$

Using the row operation $-\frac{1}{2}R_2 \to R_2$: $\begin{bmatrix} 1 & 1 & 2 & \vdots & 2 \\ 0 & 1 & 1 & \vdots & 2 \\ 0 & -4 & -5 & \vdots & -5 \end{bmatrix}$

Using the row operation $4R_2 + R_3 \rightarrow R_3$: $\begin{bmatrix} 1 & 1 & 2 & | & 2 \\ 0 & 1 & 1 & | & 2 \\ 0 & 0 & -1 & | & 3 \end{bmatrix}$

Using the row operation $-R_3 \rightarrow R_3$: $\begin{bmatrix} 1 & 1 & 2 & | & 2 \\ 0 & 1 & 1 & | & 2 \\ 0 & 0 & 1 & | & -3 \end{bmatrix}$

Using the row operations $-R_3 + R_2 \rightarrow R_2$ and $-2R_3 + R_1 \rightarrow R_1$: $\begin{bmatrix} 1 & 1 & 0 & | & 8 \\ 0 & 1 & 0 & | & 5 \\ 0 & 0 & 1 & | & -3 \end{bmatrix}$

Using the row operation $-R_2 + R_1 \rightarrow R_1$: $\begin{bmatrix} 1 & 0 & 0 & | & 3 \\ 0 & 1 & 0 & | & 5 \\ 0 & 0 & 1 & | & -3 \end{bmatrix}$

16. Using the row operations $2R_1 + R_2 \rightarrow R_2$ and $-4R_1 + R_3 \rightarrow R_3$: $\begin{bmatrix} 1 & -1 & 0 & | & 5 \\ 0 & 0 & 1 & | & 9 \\ 0 & 3 & 0 & | & -18 \end{bmatrix}$

Using the row operation $R_2 \leftrightarrow R_3$: $\begin{bmatrix} 1 & -1 & 0 & | & 5 \\ 0 & 3 & 0 & | & -18 \\ 0 & 0 & 1 & | & 9 \end{bmatrix}$

Using the row operation $\frac{1}{3}R_2 \rightarrow R_2$: $\begin{bmatrix} 1 & -1 & 0 & | & 5 \\ 0 & 1 & 0 & | & -6 \\ 0 & 0 & 1 & | & 9 \end{bmatrix}$

Using the row operation $R_2 + R_1 \rightarrow R_1$: $\begin{bmatrix} 1 & 0 & 0 & | & -1 \\ 0 & 1 & 0 & | & -6 \\ 0 & 0 & 1 & | & 9 \end{bmatrix}$

17. Form the augmented matrix: $\begin{bmatrix} 1 & 1 & | & 5 \\ 3 & -1 & | & 8 \end{bmatrix}$

Using the row operation $-3R_1 + R_2 \rightarrow R_2$: $\begin{bmatrix} 1 & 1 & | & 5 \\ 0 & -4 & | & -7 \end{bmatrix}$

The system of equations is:
$$x + y = 5$$
$$-4y = -7$$

Since $-4y = -7$, $y = \frac{7}{4}$. Substituting into the first equation:
$$x + \frac{7}{4} = 5$$
$$x = \frac{13}{4}$$

The solution is $\left(\frac{13}{4}, \frac{7}{4} \right)$.

18. Form the augmented matrix: $\begin{bmatrix} 1 & 2 & | & 9 \\ 1 & -1 & | & -6 \end{bmatrix}$

Using the row operation $-R_1 + R_2 \rightarrow R_2$: $\begin{bmatrix} 1 & 2 & | & 9 \\ 0 & -3 & | & -15 \end{bmatrix}$

The system of equations is:
$$x + 2y = 9$$
$$-3y = -15$$

Since $-3y = -15$, $y = 5$. Substituting into the first equation:
$$x + 2(5) = 9$$
$$x + 10 = 9$$
$$x = -1$$

The solution is $(-1, 5)$.

19. Form the augmented matrix: $\begin{bmatrix} 1 & 1 & 1 & | & -3 \\ 3 & 1 & -1 & | & 11 \\ 2 & 1 & -1 & | & 9 \end{bmatrix}$

Using the row operations $-3R_1 + R_2 \rightarrow R_2$ and $-2R_1 + R_3 \rightarrow R_3$: $\begin{bmatrix} 1 & 1 & 1 & | & -3 \\ 0 & -2 & -4 & | & 20 \\ 0 & -1 & -3 & | & 15 \end{bmatrix}$

Using the row operation $-\frac{1}{2}R_2 \rightarrow R_2$: $\begin{bmatrix} 1 & 1 & 1 & | & -3 \\ 0 & 1 & 2 & | & -10 \\ 0 & -1 & -3 & | & 15 \end{bmatrix}$

Using the row operation $R_2 + R_3 \rightarrow R_3$: $\begin{bmatrix} 1 & 1 & 1 & | & -3 \\ 0 & 1 & 2 & | & -10 \\ 0 & 0 & -1 & | & 5 \end{bmatrix}$

The system of equations is:
$$x + y + z = -3$$
$$y + 2z = -10$$
$$-z = 5$$

Since $-z = 5$, $z = -5$. Substituting into the second equation:
$$y + 2(-5) = -10$$
$$y - 10 = -10$$
$$y = 0$$

Substituting into the first equation:
$$x + 0 - 5 = -3$$
$$x = 2$$

The solution is $(2, 0, -5)$.

20. Form the augmented matrix: $\begin{bmatrix} 1 & 1 & -1 & | & 1 \\ 2 & 2 & 2 & | & 1 \\ 1 & -1 & 1 & | & 3 \end{bmatrix}$

Using the row operations $-2R_1 + R_2 \rightarrow R_2$ and $-R_1 + R_3 \rightarrow R_3$: $\begin{bmatrix} 1 & 1 & -1 & | & 1 \\ 0 & 0 & 4 & | & -1 \\ 0 & -2 & 2 & | & 2 \end{bmatrix}$

Using the row operation $R_2 \leftrightarrow R_3$: $\begin{bmatrix} 1 & 1 & -1 & | & 1 \\ 0 & -2 & 2 & | & 2 \\ 0 & 0 & 4 & | & -1 \end{bmatrix}$

The system of equations is:
$$x + y - z = 1$$
$$-2y + 2z = 2$$
$$4z = -1$$

Since $4z = -1$, $z = -\frac{1}{4}$. Substituting into the second equation:

$$-2y + 2\left(-\frac{1}{4}\right) = 2$$
$$-2y - \frac{1}{2} = 2$$
$$-2y = \frac{5}{2}$$
$$y = -\frac{5}{4}$$

Substituting into the first equation:
$$x - \frac{5}{4} + \frac{1}{4} = 1$$
$$x - 1 = 1$$
$$x = 2$$

The solution is $\left(2, -\frac{5}{4}, \frac{1}{4}\right)$.

21. Form the augmented matrix: $\begin{bmatrix} 1 & -2 & | & 4 \\ 2 & 1 & | & 13 \end{bmatrix}$

Using the row operation $-2R_1 + R_2 \rightarrow R_2$: $\begin{bmatrix} 1 & -2 & | & 4 \\ 0 & 5 & | & 5 \end{bmatrix}$

Using the row operation $\frac{1}{5}R_2 \rightarrow R_2$: $\begin{bmatrix} 1 & -2 & | & 4 \\ 0 & 1 & | & 1 \end{bmatrix}$

Using the row operation $2R_2 + R_1 \rightarrow R_1$: $\begin{bmatrix} 1 & 0 & | & 6 \\ 0 & 1 & | & 1 \end{bmatrix}$

The solution is $(6,1)$.

22. Form the augmented matrix: $\begin{bmatrix} 3 & 2 & | & 3 \\ 3 & 1 & | & 0 \end{bmatrix}$

Using the row operation $\frac{1}{3}R_1 \rightarrow R_1$: $\begin{bmatrix} 1 & \frac{2}{3} & | & 1 \\ 3 & 1 & | & 0 \end{bmatrix}$

Using the row operation $-3R_1 + R_2 \rightarrow R_2$: $\begin{bmatrix} 1 & \frac{2}{3} & | & 1 \\ 0 & -1 & | & -3 \end{bmatrix}$

Using the row operation $-R_2 \rightarrow R_2$: $\begin{bmatrix} 1 & \frac{2}{3} & | & 1 \\ 0 & 1 & | & 3 \end{bmatrix}$

Using the row operation $-\frac{2}{3}R_2 + R_1 \rightarrow R_1$: $\begin{bmatrix} 1 & 0 & | & -1 \\ 0 & 1 & | & 3 \end{bmatrix}$

The solution is $(-1,3)$.

23. Form the augmented matrix: $\begin{bmatrix} 1 & -3 & 1 & \vdots & 0 \\ 1 & -1 & 1 & \vdots & 0 \\ 1 & 2 & -1 & \vdots & 2 \end{bmatrix}$

Using the row operations $-R_1 + R_2 \rightarrow R_2$ and $-R_1 + R_3 \rightarrow R_3$: $\begin{bmatrix} 1 & -3 & 1 & \vdots & 0 \\ 0 & 2 & 0 & \vdots & 0 \\ 0 & 5 & -2 & \vdots & 2 \end{bmatrix}$

Using the row operation $\frac{1}{2}R_2 \rightarrow R_2$: $\begin{bmatrix} 1 & -3 & 1 & \vdots & 0 \\ 0 & 1 & 0 & \vdots & 0 \\ 0 & 5 & -2 & \vdots & 2 \end{bmatrix}$

Using the row operation $-5R_2 + R_3 \rightarrow R_3$: $\begin{bmatrix} 1 & -3 & 1 & \vdots & 0 \\ 0 & 1 & 0 & \vdots & 0 \\ 0 & 0 & -2 & \vdots & 2 \end{bmatrix}$

Using the row operation $-\frac{1}{2}R_3 \rightarrow R_3$: $\begin{bmatrix} 1 & -3 & 1 & \vdots & 0 \\ 0 & 1 & 0 & \vdots & 0 \\ 0 & 0 & 1 & \vdots & -1 \end{bmatrix}$

Using the row operation $-R_3 + R_1 \rightarrow R_1$: $\begin{bmatrix} 1 & -3 & 0 & \vdots & 1 \\ 0 & 1 & 0 & \vdots & 0 \\ 0 & 0 & 1 & \vdots & -1 \end{bmatrix}$

Using the row operation $3R_2 + R_1 \rightarrow R_1$: $\begin{bmatrix} 1 & 0 & 0 & \vdots & 1 \\ 0 & 1 & 0 & \vdots & 0 \\ 0 & 0 & 1 & \vdots & -1 \end{bmatrix}$

The solution is $(1,0,-1)$.

24. Form the augmented matrix: $\begin{bmatrix} 1 & 1 & 2 & \vdots & 7 \\ 1 & -2 & -1 & \vdots & 1 \\ 1 & 1 & 1 & \vdots & 4 \end{bmatrix}$

Using the row operations $-R_1 + R_2 \rightarrow R_2$ and $-R_1 + R_3 \rightarrow R_3$: $\begin{bmatrix} 1 & 1 & 2 & \vdots & 7 \\ 0 & -3 & -3 & \vdots & -6 \\ 0 & 0 & -1 & \vdots & -3 \end{bmatrix}$

Using the row operations $-\frac{1}{3}R_2 \rightarrow R_2$ and $-R_3 \rightarrow R_3$: $\begin{bmatrix} 1 & 1 & 2 & \vdots & 7 \\ 0 & 1 & 1 & \vdots & 2 \\ 0 & 0 & 1 & \vdots & 3 \end{bmatrix}$

Using the row operations $-R_3 + R_2 \rightarrow R_2$ and $-2R_3 + R_1 \rightarrow R_1$: $\begin{bmatrix} 1 & 1 & 0 & \vdots & 1 \\ 0 & 1 & 0 & \vdots & -1 \\ 0 & 0 & 1 & \vdots & 3 \end{bmatrix}$

Using the row operation $-R_2 + R_1 \rightarrow R_1$: $\begin{bmatrix} 1 & 0 & 0 & \vdots & 2 \\ 0 & 1 & 0 & \vdots & -1 \\ 0 & 0 & 1 & \vdots & 3 \end{bmatrix}$

The solution is $(2,-1,3)$.

25. Performing the operations: $\begin{bmatrix} 3 & -2 \\ 5 & -8 \end{bmatrix} + \begin{bmatrix} 7 & -1 \\ 4 & 0 \end{bmatrix} = \begin{bmatrix} 10 & -3 \\ 9 & -8 \end{bmatrix}$

26. Performing the operations: $-2\begin{bmatrix} 2 & -3 & 8 \\ 5 & -1 & 0 \\ -2 & 7 & 18 \end{bmatrix} = \begin{bmatrix} -4 & 6 & -16 \\ -10 & 2 & 0 \\ 4 & -14 & -36 \end{bmatrix}$

27. Performing the operations: $2\begin{bmatrix} 1 & -3 \\ 5 & 2 \end{bmatrix} - \begin{bmatrix} 2 & 5 \\ 3 & 0 \end{bmatrix} = \begin{bmatrix} 2 & -6 \\ 10 & 4 \end{bmatrix} - \begin{bmatrix} 2 & 5 \\ 3 & 0 \end{bmatrix} = \begin{bmatrix} 0 & -11 \\ 7 & 4 \end{bmatrix}$

28. The operation is not defined, since the matrices are not the same size.

29. Performing the operations: $\begin{bmatrix} 1 & 0 & 2 & 1 \end{bmatrix}\begin{bmatrix} 1 \\ 0 \\ 2 \\ 1 \end{bmatrix} = [6]$

30. The operation is not defined, since the first matrix is 4 x 1 and the second matrix is 2 x 4.

31. Performing the operations: $\begin{bmatrix} 1 & 0 \\ 0 & 1 \end{bmatrix}\begin{bmatrix} 5 & 8 \\ -4 & 1 \end{bmatrix} = \begin{bmatrix} 5 & 8 \\ -4 & 1 \end{bmatrix}$

32. Performing the operations: $\begin{bmatrix} -1 & 0 & 3 \\ 2 & 2 & 5 \end{bmatrix}\begin{bmatrix} 1 & 0 \\ -4 & 3 \\ 2 & -1 \end{bmatrix} = \begin{bmatrix} 5 & -3 \\ 4 & 1 \end{bmatrix}$

33. Performing the operations: $\begin{bmatrix} 1 & -1 & 0 \\ 2 & 3 & 1 \\ 5 & 1 & 3 \end{bmatrix}\begin{bmatrix} 1 & -2 & 1 \\ 2 & 0 & 1 \\ 3 & -1 & 4 \end{bmatrix} = \begin{bmatrix} -1 & -2 & 0 \\ 11 & -5 & 9 \\ 16 & -13 & 18 \end{bmatrix}$

34. Performing the operations: $\begin{bmatrix} 1 & -2 & 1 \\ 2 & 0 & 1 \\ 3 & -1 & 4 \end{bmatrix}\begin{bmatrix} 1 & -1 & 0 \\ 2 & 3 & 1 \\ 5 & 1 & 3 \end{bmatrix} = \begin{bmatrix} 2 & -6 & 1 \\ 7 & -1 & 3 \\ 21 & -2 & 11 \end{bmatrix}$

35. Form the grafted matrix: $\begin{bmatrix} 1 & 3 & | & 1 & 0 \\ 2 & 2 & | & 0 & 1 \end{bmatrix}$

Using the row operation $-2R_1 + R_2 \rightarrow R_2$: $\begin{bmatrix} 1 & 3 & | & 1 & 0 \\ 0 & -4 & | & -2 & 1 \end{bmatrix}$

Using the row operation $-\frac{1}{4}R_2 \rightarrow R_2$: $\begin{bmatrix} 1 & 3 & | & 1 & 0 \\ 0 & 1 & | & \frac{1}{2} & -\frac{1}{4} \end{bmatrix}$

Using the row operation $-3R_2 + R_1 \rightarrow R_1$: $\begin{bmatrix} 1 & 0 & | & -\frac{1}{2} & \frac{3}{4} \\ 0 & 1 & | & \frac{1}{2} & -\frac{1}{4} \end{bmatrix}$

The inverse is $\begin{bmatrix} -\frac{1}{2} & \frac{3}{4} \\ \frac{1}{2} & -\frac{1}{4} \end{bmatrix} = \frac{1}{4}\begin{bmatrix} -2 & 3 \\ 2 & -1 \end{bmatrix}$.

36. Form the grafted matrix: $\begin{bmatrix} 1 & -1 & 1 & \vdots & 1 & 0 & 0 \\ 2 & -2 & 1 & \vdots & 0 & 1 & 0 \\ 1 & 1 & 0 & \vdots & 0 & 0 & 1 \end{bmatrix}$

Using the row operations $-2R_1 + R_2 \rightarrow R_2$ and $-R_1 + R_3 \rightarrow R_3$: $\begin{bmatrix} 1 & -1 & 1 & \vdots & 1 & 0 & 0 \\ 0 & 0 & -1 & \vdots & -2 & 1 & 0 \\ 0 & 2 & -1 & \vdots & -1 & 0 & 1 \end{bmatrix}$

Using the row operation $R_2 \leftrightarrow R_3$: $\begin{bmatrix} 1 & -1 & 1 & \vdots & 1 & 0 & 0 \\ 0 & 2 & -1 & \vdots & -1 & 0 & 1 \\ 0 & 0 & -1 & \vdots & -2 & 1 & 0 \end{bmatrix}$

Using the row operations $\frac{1}{2}R_2 \rightarrow R_2$ and $-R_3 \rightarrow R_3$: $\begin{bmatrix} 1 & -1 & 1 & \vdots & 1 & 0 & 0 \\ 0 & 1 & -\frac{1}{2} & \vdots & -\frac{1}{2} & 0 & \frac{1}{2} \\ 0 & 0 & 1 & \vdots & 2 & -1 & 0 \end{bmatrix}$

Using the row operations $\frac{1}{2}R_3 + R_2 \rightarrow R_2$ and $-R_3 + R_1 \rightarrow R_1$: $\begin{bmatrix} 1 & -1 & 0 & \vdots & -1 & 1 & 0 \\ 0 & 1 & 0 & \vdots & \frac{1}{2} & -\frac{1}{2} & \frac{1}{2} \\ 0 & 0 & 1 & \vdots & 2 & -1 & 0 \end{bmatrix}$

Using the row operation $R_2 + R_1 \rightarrow R_1$: $\begin{bmatrix} 1 & 0 & 0 & \vdots & -\frac{1}{2} & \frac{1}{2} & \frac{1}{2} \\ 0 & 1 & 0 & \vdots & \frac{1}{2} & -\frac{1}{2} & \frac{1}{2} \\ 0 & 0 & 1 & \vdots & 2 & -1 & 0 \end{bmatrix}$

The inverse is $\begin{bmatrix} -\frac{1}{2} & \frac{1}{2} & \frac{1}{2} \\ \frac{1}{2} & -\frac{1}{2} & \frac{1}{2} \\ 2 & -1 & 0 \end{bmatrix} = \frac{1}{2}\begin{bmatrix} -1 & 1 & 1 \\ 1 & -1 & 1 \\ 4 & -2 & 0 \end{bmatrix}$.

37. Let $A = \begin{bmatrix} 1 & -2 \\ 2 & 1 \end{bmatrix}$. Form the grafted matrix: $\begin{bmatrix} 1 & -2 & \vdots & 1 & 0 \\ 2 & 1 & \vdots & 0 & 1 \end{bmatrix}$

Using the row operation $-2R_1 + R_2 \rightarrow R_2$: $\begin{bmatrix} 1 & -2 & \vdots & 1 & 0 \\ 0 & 5 & \vdots & -2 & 1 \end{bmatrix}$

Using the row operation $\frac{1}{5}R_2 \rightarrow R_2$: $\begin{bmatrix} 1 & -2 & \vdots & 1 & 0 \\ 0 & 1 & \vdots & -\frac{2}{5} & \frac{1}{5} \end{bmatrix}$

Using the row operation $2R_2 + R_1 \rightarrow R_1$: $\begin{bmatrix} 1 & 0 & \vdots & \frac{1}{5} & \frac{2}{5} \\ 0 & 1 & \vdots & -\frac{2}{5} & \frac{1}{5} \end{bmatrix}$

So $A^{-1} = \begin{bmatrix} \frac{1}{5} & \frac{2}{5} \\ -\frac{2}{5} & \frac{1}{5} \end{bmatrix}$. Therefore: $\begin{bmatrix} x \\ y \end{bmatrix} = \begin{bmatrix} \frac{1}{5} & \frac{2}{5} \\ -\frac{2}{5} & \frac{1}{5} \end{bmatrix}\begin{bmatrix} -5 \\ 5 \end{bmatrix} = \begin{bmatrix} 1 \\ 3 \end{bmatrix}$

The solution is (1,3).

38. Let $A = \begin{bmatrix} 3 & 2 \\ 3 & 1 \end{bmatrix}$. Form the grafted matrix: $\begin{bmatrix} 3 & 2 & | & 1 & 0 \\ 3 & 1 & | & 0 & 1 \end{bmatrix}$

Using the row operation $\frac{1}{3}R_1 \to R_1$: $\begin{bmatrix} 1 & \frac{2}{3} & | & \frac{1}{3} & 0 \\ 3 & 1 & | & 0 & 1 \end{bmatrix}$

Using the row operation $-3R_1 + R_2 \to R_2$: $\begin{bmatrix} 1 & \frac{2}{3} & | & \frac{1}{3} & 0 \\ 0 & -1 & | & -1 & 1 \end{bmatrix}$

Using the row operation $-R_2 \to R_2$: $\begin{bmatrix} 1 & \frac{2}{3} & | & \frac{1}{3} & 0 \\ 0 & 1 & | & 1 & -1 \end{bmatrix}$

Using the row operation $-\frac{2}{3}R_2 + R_1 \to R_1$: $\begin{bmatrix} 1 & 0 & | & -\frac{1}{3} & \frac{2}{3} \\ 0 & 1 & | & 1 & -1 \end{bmatrix}$

So $A^{-1} = \begin{bmatrix} -\frac{1}{3} & \frac{2}{3} \\ 1 & -1 \end{bmatrix}$. Therefore: $\begin{bmatrix} x \\ y \end{bmatrix} = \begin{bmatrix} -\frac{1}{3} & \frac{2}{3} \\ 1 & -1 \end{bmatrix}\begin{bmatrix} -4 \\ -2 \end{bmatrix} = \begin{bmatrix} 0 \\ -2 \end{bmatrix}$

The solution is $(0,-2)$.

39. Let $A = \begin{bmatrix} 1 & -1 & 1 \\ 1 & -2 & 1 \\ 1 & 2 & -1 \end{bmatrix}$. Form the grafted matrix: $\begin{bmatrix} 1 & -1 & 1 & | & 1 & 0 & 0 \\ 1 & -2 & 1 & | & 0 & 1 & 0 \\ 1 & 2 & -1 & | & 0 & 0 & 1 \end{bmatrix}$

Using the row operations $-R_1 + R_2 \to R_2$ and $-R_1 + R_3 \to R_3$: $\begin{bmatrix} 1 & -1 & 1 & | & 1 & 0 & 0 \\ 0 & -1 & 0 & | & -1 & 1 & 0 \\ 0 & 3 & -2 & | & -1 & 0 & 1 \end{bmatrix}$

Using the row operation $-R_2 \to R_2$: $\begin{bmatrix} 1 & -1 & 1 & | & 1 & 0 & 0 \\ 0 & 1 & 0 & | & 1 & -1 & 0 \\ 0 & 3 & -2 & | & -1 & 0 & 1 \end{bmatrix}$

Using the row operation $-3R_2 + R_3 \to R_3$: $\begin{bmatrix} 1 & -1 & 1 & | & 1 & 0 & 0 \\ 0 & 1 & 0 & | & 1 & -1 & 0 \\ 0 & 0 & -2 & | & -4 & 3 & 1 \end{bmatrix}$

Using the row operation $-\frac{1}{2}R_3 \to R_3$: $\begin{bmatrix} 1 & -1 & 1 & | & 1 & 0 & 0 \\ 0 & 1 & 0 & | & 1 & -1 & 0 \\ 0 & 0 & 1 & | & 2 & -\frac{3}{2} & -\frac{1}{2} \end{bmatrix}$

Using the row operation $-R_3 + R_1 \to R_1$: $\begin{bmatrix} 1 & -1 & 0 & | & -1 & \frac{3}{2} & \frac{1}{2} \\ 0 & 1 & 0 & | & 1 & -1 & 0 \\ 0 & 0 & 1 & | & 2 & -\frac{3}{2} & -\frac{1}{2} \end{bmatrix}$

Using the row operation $R_2 + R_1 \to R_1$: $\begin{bmatrix} 1 & 0 & 0 & | & 0 & \frac{1}{2} & \frac{1}{2} \\ 0 & 1 & 0 & | & 1 & -1 & 0 \\ 0 & 0 & 1 & | & 2 & -\frac{3}{2} & -\frac{1}{2} \end{bmatrix}$

So $A^{-1} = \begin{bmatrix} 0 & \frac{1}{2} & \frac{1}{2} \\ 1 & -1 & 0 \\ 2 & -\frac{3}{2} & -\frac{1}{2} \end{bmatrix}$. Therefore: $\begin{bmatrix} x \\ y \\ z \end{bmatrix} = \begin{bmatrix} 0 & \frac{1}{2} & \frac{1}{2} \\ 1 & -1 & 0 \\ 2 & -\frac{3}{2} & -\frac{1}{2} \end{bmatrix} \begin{bmatrix} 5 \\ 7 \\ -3 \end{bmatrix} = \begin{bmatrix} 2 \\ -2 \\ 1 \end{bmatrix}$

The solution is (2,–2,1).

40. Let $A = \begin{bmatrix} 1 & -1 & 2 \\ 1 & -2 & 1 \\ 1 & -1 & 1 \end{bmatrix}$. Form the grafted matrix: $\begin{bmatrix} 1 & -1 & 2 & | & 1 & 0 & 0 \\ 1 & -2 & 1 & | & 0 & 1 & 0 \\ 1 & -1 & 1 & | & 0 & 0 & 1 \end{bmatrix}$

Using the row operations $-R_1 + R_2 \to R_2$ and $-R_1 + R_3 \to R_3$: $\begin{bmatrix} 1 & -1 & 2 & | & 1 & 0 & 0 \\ 0 & -1 & -1 & | & -1 & 1 & 0 \\ 0 & 0 & -1 & | & -1 & 0 & 1 \end{bmatrix}$

Using the row operations $-R_2 \to R_2$ and $-R_3 \to R_3$: $\begin{bmatrix} 1 & -1 & 2 & | & 1 & 0 & 0 \\ 0 & 1 & 1 & | & 1 & -1 & 0 \\ 0 & 0 & 1 & | & 1 & 0 & -1 \end{bmatrix}$

Using the row operations $-R_3 + R_2 \to R_2$ and $-2R_3 + R_1 \to R_1$: $\begin{bmatrix} 1 & -1 & 0 & | & -1 & 0 & 2 \\ 0 & 1 & 0 & | & 0 & -1 & 1 \\ 0 & 0 & 1 & | & 1 & 0 & -1 \end{bmatrix}$

Using the row operation $R_2 + R_1 \to R_1$: $\begin{bmatrix} 1 & 0 & 0 & | & -1 & -1 & 3 \\ 0 & 1 & 0 & | & 0 & -1 & 1 \\ 0 & 0 & 1 & | & 1 & 0 & -1 \end{bmatrix}$

So $A^{-1} = \begin{bmatrix} -1 & -1 & 3 \\ 0 & -1 & 1 \\ 1 & 0 & -1 \end{bmatrix}$. Therefore: $\begin{bmatrix} x \\ y \\ z \end{bmatrix} = \begin{bmatrix} -1 & -1 & 3 \\ 0 & -1 & 1 \\ 1 & 0 & -1 \end{bmatrix} \begin{bmatrix} 1 \\ 2 \\ 2 \end{bmatrix} = \begin{bmatrix} 3 \\ 0 \\ -1 \end{bmatrix}$

The solution is (3,0,–1).

41. Evaluating the determinant: $\begin{vmatrix} 1 & 0 \\ 5 & -2 \end{vmatrix} = (1)(-2) - (0)(5) = -2 - 0 = -2$

42. Evaluating the determinant: $\begin{vmatrix} 3 & 3 \\ 1 & 2 \end{vmatrix} = (3)(2) - (3)(1) = 6 - 3 = 3$

43. Expanding along the first row: $\begin{vmatrix} 1 & 0 & 2 \\ 5 & -2 & 1 \\ 3 & -1 & 0 \end{vmatrix} = 1 \begin{vmatrix} -2 & 1 \\ -1 & 0 \end{vmatrix} - 0 \begin{vmatrix} 5 & 1 \\ 3 & 0 \end{vmatrix} + 2 \begin{vmatrix} 5 & -2 \\ 3 & -1 \end{vmatrix} = 1(1) + 2(1) = 3$

44. Expanding along the first row: $\begin{vmatrix} 1 & 0 & 2 \\ 3 & -1 & 1 \\ 2 & -1 & 5 \end{vmatrix} = 1 \begin{vmatrix} -1 & 1 \\ -1 & 5 \end{vmatrix} - 0 \begin{vmatrix} 3 & 1 \\ 1 & 5 \end{vmatrix} + 2 \begin{vmatrix} 3 & -1 \\ 2 & -1 \end{vmatrix} = 1(-4) + 2(-1) = -6$

45. Finding the required determinants:

$$D = \begin{vmatrix} 1 & -2 \\ 3 & 5 \end{vmatrix} = (1)(5) - (-2)(3) = 5 + 6 = 11$$

$$D_x = \begin{vmatrix} -5 & -2 \\ 7 & 5 \end{vmatrix} = (-5)(5) - (-2)(7) = -25 + 14 = -11$$

$$D_y = \begin{vmatrix} 1 & -5 \\ 3 & 7 \end{vmatrix} = (1)(7) - (-5)(3) = 7 + 15 = 22$$

Now using Cramer's rule:

$$x = \frac{D_x}{D} = \frac{-11}{11} = -1 \qquad y = \frac{D_y}{D} = \frac{22}{11} = 2$$

The solution is $(-1, 2)$.

46. Finding the required determinants:

$$D = \begin{vmatrix} 1 & 2 \\ 6 & -2 \end{vmatrix} = (1)(-2) - (2)(6) = -2 - 12 = -14$$

$$D_x = \begin{vmatrix} 5 & 2 \\ 23 & -2 \end{vmatrix} = (5)(-2) - (2)(23) = -10 - 46 = -56$$

$$D_y = \begin{vmatrix} 1 & 5 \\ 6 & 23 \end{vmatrix} = (1)(23) - (5)(6) = 23 - 30 = -7$$

Now using Cramer's rule:

$$x = \frac{D_x}{D} = \frac{-56}{-14} = 4 \qquad y = \frac{D_y}{D} = \frac{-7}{-14} = \tfrac{1}{2}$$

The solution is $\left(4, \tfrac{1}{2}\right)$.

47. Finding the required determinants:

$$D = \begin{vmatrix} 1 & 2 & 1 \\ 1 & 4 & -1 \\ 1 & -2 & 1 \end{vmatrix} = \begin{vmatrix} 1 & 2 & 1 \\ 0 & 2 & -2 \\ 0 & -4 & 0 \end{vmatrix} = 1 \begin{vmatrix} 2 & -2 \\ -4 & 0 \end{vmatrix} = -8$$

$$D_x = \begin{vmatrix} 5 & 2 & 1 \\ -1 & 4 & -1 \\ 5 & -2 & 1 \end{vmatrix} = \begin{vmatrix} 5 & 2 & 1 \\ 4 & 6 & 0 \\ 0 & -4 & 0 \end{vmatrix} = 1 \begin{vmatrix} 4 & 6 \\ 0 & -4 \end{vmatrix} = -16$$

$$D_y = \begin{vmatrix} 1 & 5 & 1 \\ 1 & -1 & -1 \\ 1 & 5 & 1 \end{vmatrix} = \begin{vmatrix} 1 & 5 & 1 \\ 0 & -6 & -2 \\ 0 & 0 & 0 \end{vmatrix} = 0$$

$$D_z = \begin{vmatrix} 1 & 2 & 5 \\ 1 & 4 & -1 \\ 1 & -2 & 5 \end{vmatrix} = \begin{vmatrix} 1 & 2 & 5 \\ 0 & 2 & -6 \\ 0 & -4 & 0 \end{vmatrix} = 1 \begin{vmatrix} 2 & -6 \\ -4 & 0 \end{vmatrix} = -24$$

Now using Cramer's rule:

$$x = \frac{D_x}{D} = \frac{-16}{-8} = 2 \qquad y = \frac{D_y}{D} = \frac{0}{-8} = 0 \qquad z = \frac{D_z}{D} = \frac{-24}{-8} = 3$$

The solution is $(2, 0, 3)$.

48. Finding the required determinants:

$$D = \begin{vmatrix} 1 & -1 & 1 \\ 2 & -2 & -1 \\ 3 & -2 & 1 \end{vmatrix} = \begin{vmatrix} 1 & -1 & 1 \\ 3 & -3 & 0 \\ 2 & -1 & 0 \end{vmatrix} = 1\begin{vmatrix} 3 & -3 \\ 2 & -1 \end{vmatrix} = 3$$

$$D_x = \begin{vmatrix} 2 & -1 & 1 \\ -11 & -2 & -1 \\ -2 & -2 & 1 \end{vmatrix} = \begin{vmatrix} 2 & -1 & 1 \\ -9 & -3 & 0 \\ -4 & -1 & 0 \end{vmatrix} = 1\begin{vmatrix} -9 & -3 \\ -4 & -1 \end{vmatrix} = -3$$

$$D_y = \begin{vmatrix} 1 & 2 & 1 \\ 2 & -11 & -1 \\ 3 & -2 & 1 \end{vmatrix} = \begin{vmatrix} 1 & 2 & 1 \\ 3 & -9 & 0 \\ 2 & -4 & 0 \end{vmatrix} = 1\begin{vmatrix} 3 & -9 \\ 2 & -4 \end{vmatrix} = 6$$

$$D_z = \begin{vmatrix} 1 & -1 & 2 \\ 2 & -2 & -11 \\ 3 & -2 & -2 \end{vmatrix} = \begin{vmatrix} 1 & 0 & 2 \\ 2 & 0 & -11 \\ 3 & 1 & -2 \end{vmatrix} = -1\begin{vmatrix} 1 & 2 \\ 2 & -11 \end{vmatrix} = 15$$

Now using Cramer's rule:

$$x = \frac{D_x}{D} = \frac{-3}{3} = -1 \qquad y = \frac{D_y}{D} = \frac{6}{3} = 2 \qquad z = \frac{D_z}{D} = \frac{15}{3} = 5$$

The solution is $(-1, 2, 5)$.

49. Let x and y represent the quantities of more expensive and less expensive cameras to produce, respectively. The system of equations is:

$$2x + 1y = 280$$
$$\tfrac{1}{3}x + \tfrac{1}{4}y = 60$$

Multiplying the second equation by -4:

$$2x + 1y = 280$$
$$-\tfrac{4}{3}x - y = -240$$

Adding yields $\frac{2}{3}x = 40$, so $x = 60$. Substituting into the first equation:

$$2(60) + y = 280$$
$$120 + y = 280$$
$$y = 160$$

They can produce 60 of the more expensive cameras and 160 of the less expensive cameras.

50. Let g and t represent the hourly wage for grading and tutoring, respectively. The system of equations is:

$$20g + 40t = 855$$
$$40g + 20t = 630$$

Multiplying the first equation by -2:

$$-40g - 80t = -1710$$
$$40g + 20t = 630$$

Adding yields $-60t = -1080$, so $t = 18$. Substituting into the first equation:

$$20g + 40(18) = 855$$
$$20g + 720 = 855$$
$$20g = 135$$
$$g = 6.75$$

The department pays \$18 per hour for tutoring and \$6.75 per hour for grading.

51. Let x and y represent the amounts invested at 6% and 8%, respectively. The system of equations is:
$$x + y = 10000$$
$$0.06x + 0.08y = 750$$
Multiplying the first equation by –0.06:
$$-0.06x - 0.06y = -600$$
$$0.06x + 0.08y = 750$$
Adding yields $0.02y = 150$, so $y = 7500$. Substituting:
$$x + 7500 = 10000$$
$$x = 2500$$
Brian invested $2500 at 6% and $7500 at 8%.

52. Let l and a represent the lawyer and assistant time, respectively. The system of equations is:
$$110l + 50a = 2650$$
$$a = l + 5$$
Substituting into the first equation:
$$110l + 50(l + 5) = 2650$$
$$110l + 50l + 250 = 2650$$
$$160l = 2400$$
$$l = 15$$
$$a = 15 + 5 = 20$$
The lawyer spent 15 hours and the assistant spent 20 hours.

53. Let b, c, and s represent the three investments. The system of equations is:
$$b + c + s = 40,000$$
$$0.05b + 0.065c + 0.09s = 2900$$
$$s = c + 6000$$

Form the augmented matrix: $\begin{bmatrix} 1 & 1 & 1 & \vdots & 40000 \\ 0.05 & 0.065 & 0.09 & \vdots & 2900 \\ 0 & -1 & 1 & \vdots & 6000 \end{bmatrix}$

Using the row operations $-0.05R_1 + R_2 \rightarrow R_2$ and $-R_3 \rightarrow R_3$: $\begin{bmatrix} 1 & 1 & 1 & \vdots & 40000 \\ 0 & 0.015 & 0.04 & \vdots & 900 \\ 0 & 1 & -1 & \vdots & -6000 \end{bmatrix}$

Using the row operation $R_2 \leftrightarrow R_3$: $\begin{bmatrix} 1 & 1 & 1 & \vdots & 40000 \\ 0 & 1 & -1 & \vdots & -6000 \\ 0 & 0.015 & 0.04 & \vdots & 900 \end{bmatrix}$

Using the row operation $-0.015R_2 + R_3 \rightarrow R_3$: $\begin{bmatrix} 1 & 1 & 1 & \vdots & 40000 \\ 0 & 1 & -1 & \vdots & -6000 \\ 0 & 0 & 0.055 & \vdots & 990 \end{bmatrix}$

Using the row operation $\dfrac{1}{0.055}R_3 \rightarrow R_3$: $\begin{bmatrix} 1 & 1 & 1 & \vdots & 40000 \\ 0 & 1 & -1 & \vdots & -6000 \\ 0 & 0 & 1 & \vdots & 18000 \end{bmatrix}$

Using the row operation $R_3 + R_2 \rightarrow R_2$ and $-R_3 + R_1 \rightarrow R_1$: $\begin{bmatrix} 1 & 1 & 0 & \vdots & 22000 \\ 0 & 1 & 0 & \vdots & 12000 \\ 0 & 0 & 1 & \vdots & 18000 \end{bmatrix}$

Using the row operation $-R_2 + R_1 \to R_1$:
$$\begin{bmatrix} 1 & 0 & 0 & \vdots & 10000 \\ 0 & 1 & 0 & \vdots & 12000 \\ 0 & 0 & 1 & \vdots & 18000 \end{bmatrix}$$

Josh invested $10,000 in the bank account, $12,000 in the certificate of deposit, and $18,000 in the stock.

54. The system of equations is:
$$0.8A + 0.7B + 0.4C = 90$$
$$0.9A + 0.5B + 0.8C = 87$$
$$9.8A + 6.2B + 4.2C = 895$$

Multiplying each equation by 10:
$$8A + 7B + 4C = 900$$
$$9A + 5B + 8C = 870$$
$$98A + 62B + 42C = 8950$$

We'll use Cramer's rule. Finding the required determinants:
$$D = \begin{vmatrix} 8 & 7 & 4 \\ 9 & 5 & 8 \\ 98 & 62 & 42 \end{vmatrix}$$
$$= 8\begin{vmatrix} 5 & 8 \\ 62 & 42 \end{vmatrix} - 7\begin{vmatrix} 9 & 8 \\ 98 & 42 \end{vmatrix} + 4\begin{vmatrix} 9 & 5 \\ 98 & 62 \end{vmatrix}$$
$$= 8(-286) - 7(-406) + 4(68)$$
$$= 826$$

$$D_A = \begin{vmatrix} 900 & 7 & 4 \\ 870 & 5 & 8 \\ 8950 & 62 & 42 \end{vmatrix}$$
$$= 900\begin{vmatrix} 5 & 8 \\ 62 & 42 \end{vmatrix} - 870\begin{vmatrix} 7 & 4 \\ 62 & 42 \end{vmatrix} + 8950\begin{vmatrix} 7 & 4 \\ 5 & 8 \end{vmatrix}$$
$$= 900(-286) - 870(46) + 8950(36)$$
$$= 24780$$

$$D_B = \begin{vmatrix} 8 & 900 & 4 \\ 9 & 870 & 8 \\ 98 & 8950 & 42 \end{vmatrix}$$
$$= 8\begin{vmatrix} 870 & 8 \\ 8950 & 42 \end{vmatrix} - 9\begin{vmatrix} 900 & 4 \\ 8950 & 42 \end{vmatrix} + 98\begin{vmatrix} 900 & 4 \\ 870 & 8 \end{vmatrix}$$
$$= 8(-35060) - 9(2000) + 98(3720)$$
$$= 66080$$

$$D_C = \begin{vmatrix} 8 & 7 & 900 \\ 9 & 5 & 870 \\ 98 & 62 & 8950 \end{vmatrix}$$
$$= 8\begin{vmatrix} 5 & 870 \\ 62 & 8950 \end{vmatrix} - 7\begin{vmatrix} 9 & 870 \\ 98 & 8950 \end{vmatrix} + 900\begin{vmatrix} 9 & 5 \\ 98 & 62 \end{vmatrix}$$
$$= 8(-9190) - 7(-4710) + 900(68)$$
$$= 20650$$

Now using Cramer's rule:

$$A = \frac{D_A}{D} = \frac{24780}{826} = 30 \qquad B = \frac{D_B}{D} = \frac{66080}{826} = 80 \qquad C = \frac{D_C}{D} = \frac{20650}{826} = 25$$

The company can produce 30 of model A, 80 of model B, and 25 of model C.

55. Sketching the graph:

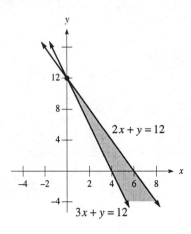

56. Sketching the graph:

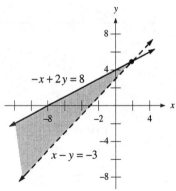

57. Sketching the graph:

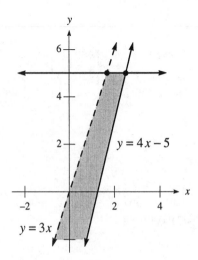

58. Sketching the graph:

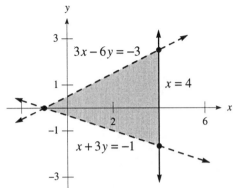

59. Let v represent the number of VCRs and t represent the number of TVs. The system of inequalities is:

$$200v + 250t \geq 10{,}000$$
$$2v + 6t \leq 500$$
$$v \geq 0$$
$$t \geq 0$$

Sketching the graph:

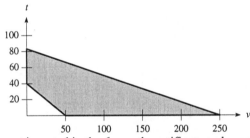

60. Let x represent the amount invested in the 6-month certificate and y represent the amount invested in the 1-year certificate. The system of inequalities is:

$$x + y \leq 20{,}000$$
$$y \geq 2x$$
$$2000 \leq x \leq 4000$$

Sketching the graph:

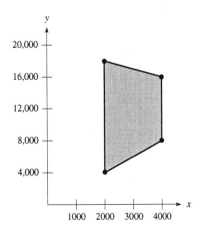

61. Let x represent the number of civil complaints and y represent the number of criminal complaints.
The system of inequalities is:

$$x + \tfrac{3}{4}y \leq 35$$
$$x + \tfrac{3}{4}y \geq 20$$
$$y \geq x$$
$$x \geq 0$$
$$y \geq 0$$

Sketching the graph:

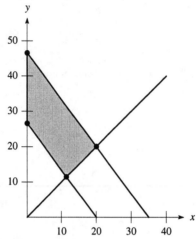

62. The system of inequalities is:

$$6A + 4B \leq 800$$
$$14A + 16B \geq 1000$$
$$A \geq 0$$
$$B \geq 0$$

Sketching the graph:

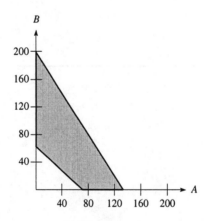

63 . Let h represent the high-risk investments and c represent the conservative investments. The system of inequalities is:
$$h+c \leq 80$$
$$h \leq 0.25(h+c)$$
$$h \geq 0$$
$$c \geq 0$$

Now $P = 0.14h + 0.08c$. Sketching the constraints:

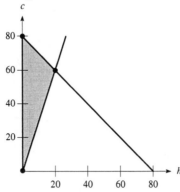

Evaluating P at these vertices:
$$P(0,80) = 0.14(0) + 0.08(80) = 6.4$$
$$P(0,0) = 0.14(0) + 0.08(0) = 0$$
$$P(20,60) = 0.14(20) + 0.08(60) = 7.6$$

The maximum profit is $7.6 million with $20 million invested in high-risk investments and $60 million invested in conservative investments.

64 . The system of inequalities is:
$$300X + 400Y \geq 48,000$$
$$12X + 6Y \leq 1080$$
$$X \geq 0$$
$$Y \geq 0$$

Now $P = 225X + 100Y$. Sketching the constraints:

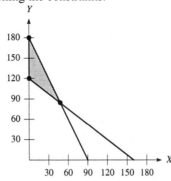

Evaluating P at these vertices:
$$P(0,180) = 225(0) + 100(180) = 18,000$$
$$P(0,120) = 225(0) + 100(120) = 12,000$$
$$P(48,84) = 225(48) + 100(84) = 19,200$$

The maximum profit is $19,200 if she purchases 48 of model X and 84 of model Y.

Chapter 9 Practice Test

1. **a.** Adding the two equations yields $5x = 20$, so $x = 4$. Substituting into the first equation:
$$2(4) + y = 12$$
$$8 + y = 12$$
$$y = 4$$
The solution is $(4,4)$.

b. Multiplying the first equation by 5:
$$-5x + 10y = -20$$
$$5x + 3y = 7$$
Adding yields $13y = -13$, so $y = -1$. Substituting into the second equation:
$$5x + 3(-1) = 7$$
$$5x - 3 = 7$$
$$5x = 10$$
$$x = 2$$
The solution is $(2,-1)$.

c. Adding -1 times equation 1 to equation 2:
$$-x + y + z = -4$$
$$x + 2y + 2z = -2$$
Adding yields $3y + 3z = -6$, or $y + z = -2$. Adding -1 times equation 1 to equation 3:
$$-x + y + z = -4$$
$$x - y + z = 3$$
Adding yields $2z = -1$, so $z = -\frac{1}{2}$. Substituting:
$$y - \frac{1}{2} = -2$$
$$y = -\frac{3}{2}$$
Substituting into the second equation:
$$x + 2\left(-\frac{3}{2}\right) + 2\left(-\frac{1}{2}\right) = -2$$
$$x - 4 = -2$$
$$x = 2$$
The solution is $\left(2, -\frac{3}{2}, -\frac{1}{2}\right)$.

2. Form the augmented matrix:
$$\begin{bmatrix} 1 & 1 & -2 & \vdots & -7 \\ 2 & 2 & 2 & \vdots & 10 \\ -1 & 1 & 1 & \vdots & -1 \end{bmatrix}$$

Using the row operations $-2R_1 + R_2 \to R_2$ and $R_1 + R_3 \to R_3$:
$$\begin{bmatrix} 1 & 1 & -2 & \vdots & -7 \\ 0 & 0 & 6 & \vdots & 24 \\ 0 & 2 & -1 & \vdots & -8 \end{bmatrix}$$

Using the row operation $R_2 \leftrightarrow R_3$:
$$\begin{bmatrix} 1 & 1 & -2 & \vdots & -7 \\ 0 & 2 & -1 & \vdots & -8 \\ 0 & 0 & 6 & \vdots & 24 \end{bmatrix}$$

Using the row operations $\frac{1}{2}R_2 \to R_2$ and $\frac{1}{6}R_3 \to R_3$:
$$\begin{bmatrix} 1 & 1 & -2 & \vdots & -7 \\ 0 & 1 & -\frac{1}{2} & \vdots & -4 \\ 0 & 0 & 1 & \vdots & 4 \end{bmatrix}$$

Using the row operations $\frac{1}{2}R_3 + R_2 \rightarrow R_2$ and $2R_3 + R_1 \rightarrow R_1$:
$\begin{bmatrix} 1 & 1 & 0 & | & 1 \\ 0 & 1 & 0 & | & -2 \\ 0 & 0 & 1 & | & 4 \end{bmatrix}$

Using the row operation $-R_2 + R_1 \rightarrow R_1$:
$\begin{bmatrix} 1 & 0 & 0 & | & 3 \\ 0 & 1 & 0 & | & -2 \\ 0 & 0 & 1 & | & 4 \end{bmatrix}$

The solution is $(3, -2, 4)$.

3. **a.**   Performing the operations: $3\begin{bmatrix} 1 & 0 \\ -2 & 5 \end{bmatrix} - \begin{bmatrix} 2 & 5 \\ 1 & 3 \end{bmatrix} = \begin{bmatrix} 3 & 0 \\ -6 & 15 \end{bmatrix} - \begin{bmatrix} 2 & 5 \\ 1 & 3 \end{bmatrix} = \begin{bmatrix} 1 & -5 \\ -7 & 12 \end{bmatrix}$

   b.   Performing the operations: $\begin{bmatrix} 3 & 5 & 1 \\ -4 & 6 & 0 \\ 2 & 1 & -1 \end{bmatrix}\begin{bmatrix} 1 & 0 & 4 \\ 0 & 1 & -1 \\ 2 & 1 & -2 \end{bmatrix} = \begin{bmatrix} 5 & 6 & 5 \\ -4 & 6 & -22 \\ 0 & 0 & 9 \end{bmatrix}$

4.   Form the grafted matrix: $\begin{bmatrix} 2 & 8 & | & 1 & 0 \\ 1 & -4 & | & 0 & 1 \end{bmatrix}$

   Using the row operation $R_1 \leftrightarrow R_2$: $\begin{bmatrix} 1 & -4 & | & 0 & 1 \\ 2 & 8 & | & 1 & 0 \end{bmatrix}$

   Using the row operation $-2R_1 + R_2 \rightarrow R_2$: $\begin{bmatrix} 1 & -4 & | & 0 & 1 \\ 0 & 16 & | & 1 & -2 \end{bmatrix}$

   Using the row operation $\frac{1}{16}R_2 \rightarrow R_2$: $\begin{bmatrix} 1 & -4 & | & 0 & 1 \\ 0 & 1 & | & \frac{1}{16} & -\frac{1}{8} \end{bmatrix}$

   Using the row operation $4R_2 + R_1 \rightarrow R_1$: $\begin{bmatrix} 1 & 0 & | & \frac{1}{4} & \frac{1}{2} \\ 0 & 1 & | & \frac{1}{16} & -\frac{1}{8} \end{bmatrix}$

   The inverse is $\begin{bmatrix} \frac{1}{4} & \frac{1}{2} \\ \frac{1}{16} & -\frac{1}{8} \end{bmatrix} = \frac{1}{16}\begin{bmatrix} 4 & 8 \\ 1 & -2 \end{bmatrix}$.

5.   Let $A = \begin{bmatrix} 1 & -2 \\ 2 & 1 \end{bmatrix}$. Form the grafted matrix: $\begin{bmatrix} 1 & -2 & | & 1 & 0 \\ 2 & 1 & | & 0 & 1 \end{bmatrix}$

   Using the row operation $-2R_1 + R_2 \rightarrow R_2$: $\begin{bmatrix} 1 & -2 & | & 1 & 0 \\ 0 & 5 & | & -2 & 1 \end{bmatrix}$

   Using the row operation $\frac{1}{5}R_2 \rightarrow R_2$: $\begin{bmatrix} 1 & -2 & | & 1 & 0 \\ 0 & 1 & | & -\frac{2}{5} & \frac{1}{5} \end{bmatrix}$

   Using the row operation $2R_2 + R_1 \rightarrow R_1$: $\begin{bmatrix} 1 & 0 & | & \frac{1}{5} & \frac{2}{5} \\ 0 & 1 & | & -\frac{2}{5} & \frac{1}{5} \end{bmatrix}$

   The inverse is $\begin{bmatrix} \frac{1}{5} & \frac{2}{5} \\ -\frac{2}{5} & \frac{1}{5} \end{bmatrix}$. Therefore: $\begin{bmatrix} x \\ y \end{bmatrix} = \begin{bmatrix} \frac{1}{5} & \frac{2}{5} \\ -\frac{2}{5} & \frac{1}{5} \end{bmatrix}\begin{bmatrix} 4 \\ 13 \end{bmatrix} = \begin{bmatrix} 6 \\ 1 \end{bmatrix}$

   The solution is $(6, 1)$.

6. **a.**   Evaluating the determinant: $\begin{vmatrix} 2 & 5 \\ -8 & 1 \end{vmatrix} = (2)(1)-(5)(-8) = 2+40 = 42$

  b.   Evaluating the determinant: $\begin{vmatrix} 3 & 0 & 2 \\ 2 & -1 & 4 \\ 1 & 2 & -3 \end{vmatrix} = 3\begin{vmatrix} -1 & 4 \\ 2 & -3 \end{vmatrix} - 0 + 2\begin{vmatrix} 2 & -1 \\ 1 & 2 \end{vmatrix} = 3(-5)+2(5) = -5$

7.   Finding the required determinants:

$$D = \begin{vmatrix} 1 & -2 & 3 \\ -1 & 1 & -2 \\ 2 & 1 & -1 \end{vmatrix} = \begin{vmatrix} 1 & -2 & 3 \\ 0 & -1 & 1 \\ 0 & 5 & -7 \end{vmatrix} = 1\begin{vmatrix} -1 & 1 \\ 5 & -7 \end{vmatrix} = 2$$

$$D_x = \begin{vmatrix} 6 & -2 & 3 \\ -4 & 1 & -2 \\ -1 & 1 & -1 \end{vmatrix} = \begin{vmatrix} 6 & -2 & 3 \\ -4 & -3 & 2 \\ -1 & 0 & 0 \end{vmatrix} = -1\begin{vmatrix} 4 & -3 \\ -3 & 2 \end{vmatrix} = 1$$

$$D_y = \begin{vmatrix} 1 & 6 & 3 \\ -1 & -4 & -2 \\ 2 & -1 & -1 \end{vmatrix} = \begin{vmatrix} 1 & 6 & 3 \\ 0 & 2 & 1 \\ 0 & -13 & -7 \end{vmatrix} = 1\begin{vmatrix} 2 & 1 \\ -13 & -7 \end{vmatrix} = -1$$

$$D_z = \begin{vmatrix} 1 & -2 & 6 \\ -1 & 1 & -4 \\ 2 & 1 & -1 \end{vmatrix} = \begin{vmatrix} 1 & -2 & 6 \\ 0 & -1 & 2 \\ 0 & 5 & -13 \end{vmatrix} = 1\begin{vmatrix} -1 & 2 \\ 5 & -13 \end{vmatrix} = 3$$

Now using Cramer's rule:

$$x = \frac{D_x}{D} = \frac{1}{2} \qquad\qquad y = \frac{D_y}{D} = -\frac{1}{2} \qquad\qquad z = \frac{D_z}{D} = \frac{3}{2}$$

The solution is $\left(\frac{1}{2},-\frac{1}{2},\frac{3}{2}\right)$.

8.   Let f represent the flat fee and m represent the mileage charge. The system of equations is:
$$f + 120m = 65.70$$
$$f + 80m = 51.30$$
Subtracting the two equations yields $40m = 14.4$, so $m = 0.36$. Substituting into the first equation:
$$f + 120(0.36) = 65.70$$
$$f + 43.20 = 65.70$$
$$f = 22.50$$
The flat fee is \$22.50 and the mileage fee is 36 cents per mile.

9.   Let x, y, z represent the three types of fertilizers. The system of equations is:
$$x + y + z = 2000$$
$$0.10x + 0.25y + 0.30z = 0.21(2000)$$
$$z = 2y$$
Multiplying the first equation by –0.10:
$$-0.10x - 0.10y - 0.10z = -200$$
$$0.10x + 0.25y + 0.30z = 420$$
Adding yields $0.15y + 0.20z = 220$. Substituting $z = 2y$:
$$0.15y + 0.20(2y) = 220$$
$$0.15y + 0.40y = 220$$
$$0.55y = 220$$
$$y = 400$$
$$z = 2(400) = 800$$

Substituting into the first equation:
$$x + 400 + 800 = 2000$$
$$x + 1200 = 2000$$
$$x = 800$$

The final mixture contains 800 lb of 10%, 400 lb of 25%, and 800 lb of 30% nitrogen.

10. Sketching the graph:

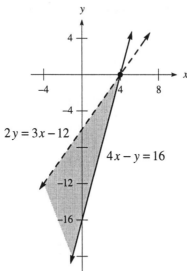

11. The system of inequalities is:
$$100X + 200Y \leq 10,000$$
$$4X + 3Y \leq 300$$
$$X \geq 0$$
$$Y \geq 0$$

Sketching the solution set:

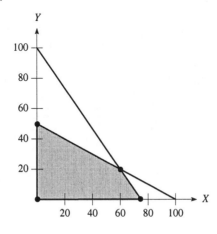

12.   The profit is $P = 60X + 50Y$. Evaluating P at these vertices:

$P(0,0) = 60(0) + 50(0) = 0$

$P(0,50) = 60(0) + 50(50) = 2500$

$P(75,0) = 60(75) + 50(0) = 4500$

$P(60,20) = 60(60) + 50(20) = 4600$

The maximum profit is \$4,600 if 60 of model X and 20 of model Y are ordered.

Chapter 10
Conic Sections and Nonlinear Systems

10.1 Conic Sections: Circles

1. The equation is $(x-3)^2 + (y+8)^2 = 25$.

3. The equation is $(x+3)^2 + (y-1)^2 = 1$. Converting to general form:

$$(x+3)^2 + (y-1)^2 = 1$$
$$x^2 + 6x + 9 + y^2 - 2y + 1 = 1$$
$$x^2 + y^2 + 6x - 2y + 9 = 0$$

5. The center is $(2,-3)$ and the radius is $\sqrt{81} = 9$.

7. The center is $\left(\frac{1}{2}, 0\right)$ and the radius is $\sqrt{18} = 3\sqrt{2}$.

9. First complete the square:

$$x^2 + y^2 - 6x - 8y + 9 = 0$$
$$\left(x^2 - 6x + 9\right) + \left(y^2 - 8y + 16\right) = -9 + 9 + 16$$
$$(x-3)^2 + (y-4)^2 = 16$$

The center is $(3,4)$ and the radius is $\sqrt{16} = 4$.

11. First complete the square:

$$x^2 + y^2 - 6x - 15 = 0$$
$$\left(x^2 - 6x + 9\right) + y^2 = 15 + 9$$
$$(x-3)^2 + y^2 = 24$$

The center is $(3,0)$ and the radius is $\sqrt{24} = 2\sqrt{6}$.

13. First complete the square:

$$4x^2 + 4y^2 - 4y - 47 = 0$$
$$x^2 + y^2 - y - \frac{47}{4} = 0$$
$$x^2 + \left(y^2 - y + \frac{1}{4}\right) = \frac{47}{4} + \frac{1}{4}$$
$$x^2 + \left(y - \frac{1}{2}\right)^2 = 12$$

The center is $\left(0, \frac{1}{2}\right)$ and the radius is $\sqrt{12} = 2\sqrt{3}$.

15. Since the slope from $P(1,2)$ to the center is 2, the tangent slope must be $-\frac{1}{2}$. Using the point-slope formula:

$$y-2=-\tfrac{1}{2}(x-1)$$
$$y-2=-\tfrac{1}{2}x+\tfrac{1}{2}$$
$$y=-\tfrac{1}{2}x+\tfrac{5}{2}$$

17. The slope from $P(2,5)$ to $(0,3)$ is $m=\dfrac{2}{2}=1$, so the tangent slope must be -1. Using the point-slope formula:

$$y-5=-1(x-2)$$
$$y-5=-x+2$$
$$y=-x+7$$

19. Sketching the solution set:

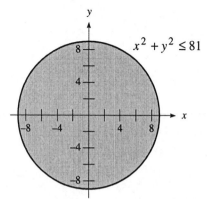

21. First complete the square:

$$x^2+y^2-8x-6y+9<0$$
$$\left(x^2-8x+16\right)+\left(y^2-6y+9\right)<-9+16+9$$
$$(x-4)^2+(y-3)^2<16$$

Sketching the solution set:

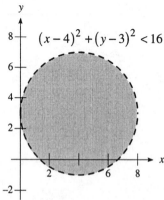

23. Using $\left(x_0,y_0\right)=(3,-2)$ and $A=3, B=-4, C=5$: $\ d=\dfrac{|3(3)-4(-2)+5|}{\sqrt{3^2+(-4)^2}}=\dfrac{22}{\sqrt{25}}=\dfrac{22}{5}$

25. The distance from the point $(0,0)$ to the line $2x - 3y - 12 = 0$ is the radius of the circle, therefore:

$$r = \frac{|2(0) - 3(0) - 12|}{\sqrt{2^2 + (-3)^2}} = \frac{12}{\sqrt{13}}$$

The equation is therefore $x^2 + y^2 = \frac{144}{13}$.

27. Using $(x_0, y_0) = (2, 4)$, $m = 3$ and $b = -4$: $d = \dfrac{|3(2) - 4 - 4|}{\sqrt{1 + 3^2}} = \dfrac{2}{\sqrt{10}} = \dfrac{2\sqrt{10}}{10} = \dfrac{\sqrt{10}}{5}$

29. Let $(x_0, y_0) = (x_0, mx_0 + b_1)$ be a point on the line $y = mx + b_1$. Using the distance formula:

$$d = \frac{|mx_0 + b_2 - (mx_0 + b_1)|}{\sqrt{1 + m^2}} = \frac{|b_2 - b_1|}{\sqrt{1 + m^2}} = \frac{|b_1 - b_2|}{\sqrt{1 + m^2}}$$

31. The center is $(0,0)$. Since the slope of the line drawn through the center of the circle is $\dfrac{y_0}{x_0}$, the slope

of the tangent line is $-\dfrac{x_0}{y_0}$. Using the point-slope formula:

$$y - y_0 = -\frac{x_0}{y_0}(x - x_0)$$

$$y_0 y - y_0^2 = -x_0 x + x_0^2$$

$$x_0 x + y_0 y = x_0^2 + y_0^2$$

$$x_0 x + y_0 y = r^2$$

Note that $r^2 = x_0^2 + y_0^2$, since (x_0, y_0) lies on the circle.

33. Graphing the circle:

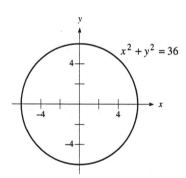

35. Graphing the circle:

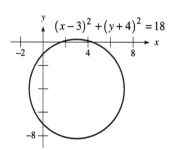

10.2 The Parabola

1.   Since $4p = 12$, $p = 3$. The focus is $(0,3)$ and the directrix is $y = -3$.

3.   Since $x = -8y^2$, $y^2 = -\frac{1}{8}x$. So $4p = -\frac{1}{8}$, thus $p = -\frac{1}{32}$. The focus is $\left(-\frac{1}{32}, 0\right)$ and the directrix is $x = \frac{1}{32}$.

5.   Converting to standard form:
$$6x^2 - 3y = 0$$
$$6x^2 = 3y$$
$$x^2 = \tfrac{1}{2}y$$
Since $4p = \frac{1}{2}$, $p = \frac{1}{8}$. The focus is $\left(0, \frac{1}{8}\right)$ and the directrix is $y = -\frac{1}{8}$.

7.   Converting to standard form:
$$4x^2 - 12y = 0$$
$$4x^2 = 12y$$
$$x^2 = 3y$$
Since $4p = 3$, $p = \frac{3}{4}$. The focus is $\left(0, \frac{3}{4}\right)$, the directrix is $y = -\frac{3}{4}$, and the axis is $x = 0$. Graphing the parabola:

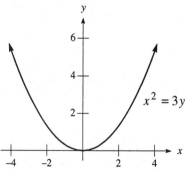

9.   Converting to standard form yields $y^2 = -x$. Since $4p = -1$, $p = -\frac{1}{4}$. The focus is $\left(-\frac{1}{4}, 0\right)$, the directrix is $x = \frac{1}{4}$, and the axis is $y = 0$. Graphing the parabola:

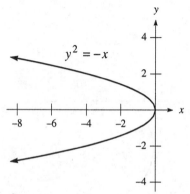

11.   Since $p = 4$, the equation is $x^2 = 16y$.

13.   Since $p = -3$, the equation is $x^2 = -12y$.

15. Since $p = 5$, the equation is $y^2 = 20x$.

17. Since $y = \frac{1}{12}x^2$, $x^2 = 12y$. So $4p = 12$, $p = 3$. The receiver should be placed 3 feet directly above

the vertex. To find the depth: $y = \frac{1}{12}(\pm 6)^2 = \frac{1}{12}(36) = 3$

The dish is 3 feet deep.

19. Since $y = \frac{1}{6}x^2$, $x^2 = 6y$. So $4p = 6$, $p = 1.5$. The mirror should be placed 1.5 feet = 18 inches

directly above the vertex. To find the depth: $y = \frac{1}{6}(\pm 2)^2 = \frac{1}{6}(4) = \frac{2}{3}$

The mirror is $\frac{2}{3}$ foot = 8 inches deep.

21. Call (0,0) the vertex of the parabola $x^2 = 4py$. Since (± 10,30) are points on the parabola:
$$x^2 = 4py$$
$$(\pm 10)^2 = 4p(30)$$
$$100 = 120p$$
$$p = \frac{5}{6}$$

So the equation of the parabola is $x^2 = \frac{10}{3}y$. Finding x when $y = 20$:
$$x^2 = \frac{10}{3}(20)$$
$$x^2 = \frac{200}{3}$$
$$x = \pm\sqrt{\frac{200}{3}} = \pm\frac{\sqrt{600}}{3} = \pm\frac{10\sqrt{6}}{3}$$

Thus the width is $2\left(\dfrac{10\sqrt{6}}{3}\right) = \dfrac{20\sqrt{6}}{3} \approx 16.3$ feet.

23. Let (0,0) be the vertex of the mirror $x^2 = 4py$. Since (± 2,2) are points on the parabola:
$$(\pm 2)^2 = 4p(2)$$
$$4 = 8p$$
$$p = \frac{1}{2}$$

The light source should be placed at the focus, which is $\frac{1}{2}$ inch above the vertex of the mirror.

25. Since $4p = 8$, $p = 2$. Since the vertex is (2,0), the focus is (2,2) and the directrix is $y = -2$.

27. First complete the square:
$$8x = y^2 - 2y - 15$$
$$8x + 15 = y^2 - 2y$$
$$y^2 - 2y + 1 = 8x + 16$$
$$(y-1)^2 = 8(x+2)$$

Since $4p = 8$, $p = 2$. Since the vertex is (–2,1), the focus is (0,1) and the directrix is $x = -4$.

29. First complete the square:
$$-5y = x^2 - 4x + 19$$
$$-5y - 19 = x^2 - 4x$$
$$x^2 - 4x + 4 = -5y - 15$$
$$(x-2)^2 = -5(y+3)$$

Since $4p = -5$, $p = -\frac{5}{4}$. Since the vertex is (2,–3), the focus is $\left(2, -\frac{17}{4}\right)$, and the directrix is

$y = -\frac{7}{4}$.

31.   First complete the square:
$$8y = x^2 + 10x + 33$$
$$8y - 33 = x^2 + 10x$$
$$x^2 + 10x + 25 = 8y - 8$$
$$(x+5)^2 = 8(y-1)$$
Since $4p = 8$, $p = 2$. The vertex is $(-5,1)$, the focus is $(-5,3)$, the directrix is $y = -1$, and the axis is $x = -5$. Graphing the parabola:

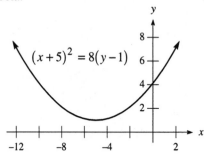

33.   First complete the square:
$$x = 3y^2 - 6y + 5$$
$$x - 5 = 3(y^2 - 2y)$$
$$3(y^2 - 2y + 1) = x - 5 + 2$$
$$3(y-1)^2 = x - 2$$
$$(y-1)^2 = \tfrac{1}{3}(x-2)$$
Since $4p = \tfrac{1}{3}$, $p = \tfrac{1}{12}$. The vertex is $(2,1)$, the focus is $\left(\tfrac{25}{12},1\right)$, the directrix is $x = \tfrac{23}{12}$, and the axis is $y = 1$. Graphing the parabola:

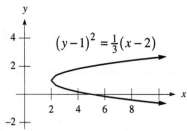

35.   The parabola is opening to the left with form $(y-4)^2 = 4p(x-5)$. Since $p = -3$, the equation is $(y-4)^2 = -12(x-5)$.

37.   The vertex is the midpoint of $(2,-3)$ and $(2,4)$, which is $\left(2,\tfrac{1}{2}\right)$. The parabola is opening down with form $(x-2)^2 = 4p\left(y-\tfrac{1}{2}\right)$. Since $p = -\tfrac{7}{2}$, the equation is $(x-2)^2 = -14\left(y-\tfrac{1}{2}\right)$.

39.   The parabola is opening to the right with form $(y-5)^2 = 4p(x-2)$. Since $p = 5$, the equation is $(y-5)^2 = 20(x-2)$.

41.   Graphing the solution set:

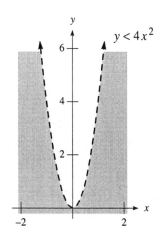

43.   First complete the square:

$$x \geq y^2 - 6y + 5$$
$$x - 5 \geq y^2 - 6y$$
$$x - 5 + 9 \geq y^2 - 6y + 9$$
$$x + 4 \geq (y - 3)^2$$

Graphing the solution set:

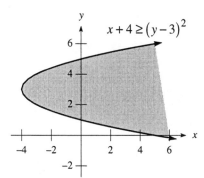

45.   Let (x, y) be a point on the parabola. Since the distance from any point on the parabola to the focus and the directrix must be equal:

$$\sqrt{(x-p)^2 + y^2} = x + p$$
$$(x-p)^2 + y^2 = (x+p)^2$$
$$x^2 - 2px + p^2 + y^2 = x^2 + 2px + p^2$$
$$-2px + y^2 = 2px$$
$$y^2 = 4px$$

47.   Using $(x_0, y_0) = (-4, 2)$ and $p = 2$, the tangent line is:

$$y = \frac{-4}{2 \cdot 2} x - 2$$
$$y = -x - 2$$

49. Finding when $y = p$:

$$x^2 = 4p \cdot p$$
$$x^2 = 4p^2$$
$$x = \pm 2p$$

The length of the latus rectum is therefore $2(2p) = 4p$.

51. Sketching the graph:

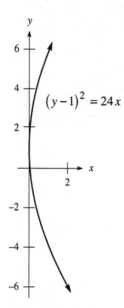

$$(y-1)^2 = 24x$$

53. First complete the square:

$$x^2 - 6x - 12y + 9 = 0$$
$$x^2 - 6x = 12y - 9$$
$$x^2 - 6x + 9 = 12y - 9 + 9$$
$$(x-3)^2 = 12y$$

Sketching the graph:

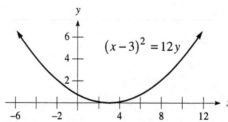

$$(x-3)^2 = 12y$$

55. Since Q lies on the y-axis, $x = 0$ and thus $y = -y_0$. The distance from F to Q is therefore $p + y_0$. Since P is y_0 units above the x-axis and $y = -p$ is p units below the x-axis, the distance from P to the directrix is $y_0 + p$. The distance from P to the focus is the same as the distance from P to the directrix, since P lies on the parabola. Since this distance is $y_0 + p$, $|FQ| = |FP|$. The triangle FPQ must thus be an isosceles triangle, so $\alpha = \theta$. Since $\theta = \beta$ by the construction that line q is parallel to the y-axis, so $\alpha = \beta$, as desired.

10.3 The Ellipse

1. Since $a^2 = 16$ and $b^2 = 9$, $a = 4$, $b = 3$, and $c = \sqrt{16-9} = \sqrt{7}$. The vertices are (4,0) and (–4,0), and the foci are $\left(\sqrt{7},0\right)$ and $\left(-\sqrt{7},0\right)$.

3. Since $a^2 = 18$ and $b^2 = 12$, $a = \sqrt{18} = 3\sqrt{2}$, $b = \sqrt{12} = 2\sqrt{3}$, and $c = \sqrt{18-12} = \sqrt{6}$. The vertices are $\left(0,3\sqrt{2}\right)$ and $\left(0,-3\sqrt{2}\right)$, and the foci are $\left(0,\sqrt{6}\right)$ and $\left(0,-\sqrt{6}\right)$.

5. Since $a^2 = 24$ and $b^2 = 9$, $a = \sqrt{24} = 2\sqrt{6}$, $b = \sqrt{9} = 3$, and $c = \sqrt{24-9} = \sqrt{15}$. The vertices are $\left(2\sqrt{6},0\right)$ and $\left(-2\sqrt{6},0\right)$, and the foci are $\left(\sqrt{15},0\right)$ and $\left(-\sqrt{15},0\right)$.

7. Dividing by 225, the standard form is $\dfrac{x^2}{9} + \dfrac{y^2}{25} = 1$. Since $a^2 = 25$ and $b^2 = 9$, $a = \sqrt{25} = 5$, $b = \sqrt{9} = 3$, and $c = \sqrt{25-9} = \sqrt{16} = 4$. The vertices are (0,5) and (0,–5), and the foci are (0,4) and (0,–4).

9. Dividing by 30, the standard form is $\dfrac{y^2}{30} + \dfrac{x^2}{1} = 1$. Since $a^2 = 30$ and $b^2 = 1$, $a = \sqrt{30}$, $b = 1$, and $c = \sqrt{30-1} = \sqrt{29}$. The vertices are $\left(0,\sqrt{30}\right)$ and $\left(0,-\sqrt{30}\right)$, and the foci are $\left(0,\sqrt{29}\right)$ and $\left(0,-\sqrt{29}\right)$.

11. Since $a^2 = 49$ and $b^2 = 9$, $a = 7$, $b = 3$, and $c = \sqrt{49-9} = \sqrt{40} = 2\sqrt{10}$. The vertices are (–7,0) and (7,0), and the foci are $\left(-2\sqrt{10},0\right)$ and $\left(2\sqrt{10},0\right)$. Graphing the ellipse:

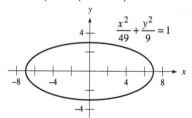

13. Dividing by 24, the standard form is $\dfrac{x^2}{2} + \dfrac{y^2}{24} = 1$. Since $a^2 = 24$ and $b^2 = 2$, $a = \sqrt{24} = 2\sqrt{6}$, $b = \sqrt{2}$, and $c = \sqrt{24-2} = \sqrt{22}$. The vertices are $\left(0,-2\sqrt{6}\right)$ and $\left(0,2\sqrt{6}\right)$, and the foci are $\left(0,-\sqrt{22}\right)$ and $\left(0,\sqrt{22}\right)$. Graphing the ellipse:

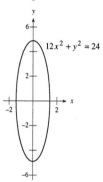

15. Dividing by 12, the standard form is $\dfrac{x^2}{4} + \dfrac{y^2}{3/2} = 1$. Since $a^2 = 4$ and $b^2 = \frac{3}{2}$, $a = \sqrt{4} = 2$,

$b = \sqrt{\frac{3}{2}} = \dfrac{\sqrt{6}}{2}$, and $c = \sqrt{4 - \frac{3}{2}} = \sqrt{\frac{5}{2}} = \dfrac{\sqrt{10}}{2}$. The vertices are $(-2,0)$ and $(2,0)$, and the foci are

$\left(-\dfrac{\sqrt{10}}{2}, 0\right)$ and $\left(\dfrac{\sqrt{10}}{2}, 0\right)$. Graphing the ellipse:

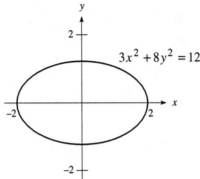

17. Since $a^2 = 36$ and $b^2 = 25$, $a = 6$, $b = 5$, and $c = \sqrt{36 - 25} = \sqrt{11}$. The eccentricity is

$e = \dfrac{c}{a} = \dfrac{\sqrt{11}}{6}$.

19. Dividing by 12, the standard form is $\dfrac{x^2}{3/2} + \dfrac{y^2}{4} = 1$. So $a^2 = 4$ and $b^2 = \frac{3}{2}$, thus $a = 2$,

$b = \sqrt{\frac{3}{2}} = \dfrac{\sqrt{6}}{2}$, and $c = \sqrt{4 - \frac{3}{2}} = \sqrt{\frac{5}{2}} = \dfrac{\sqrt{10}}{2}$. The eccentricity is $e = \dfrac{c}{a} = \dfrac{\sqrt{10}/2}{2} = \dfrac{\sqrt{10}}{4}$.

21. Sketching the solution set:

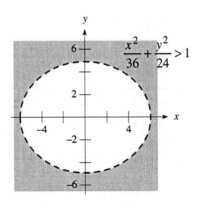

23. Dividing by 12, the standard form is $\dfrac{x^2}{4}+\dfrac{y^2}{3}\le 1$. Sketching the solution set:

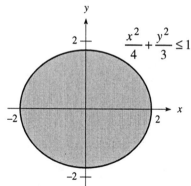

25. Given $c = 2$ and $a = 4$, find b:
$$2^2 = 4^2 - b^2$$
$$4 = 16 - b^2$$
$$b^2 = 12$$
$$b = 2\sqrt{3}$$
The equation is $\dfrac{x^2}{16}+\dfrac{y^2}{12}=1$.

27. Given $a = 4$ and $b = 2$, the equation is $\dfrac{x^2}{16}+\dfrac{y^2}{4}=1$.

29. Here $a = 40$ and $b = 20$, so $\dfrac{x^2}{40^2}+\dfrac{y^2}{20^2}=1$. Substituting $x = 30$:
$$\frac{30^2}{40^2}+\frac{y^2}{20^2}=1$$
$$\frac{y^2}{20^2}=1-\frac{30^2}{40^2}$$
$$y = 20\sqrt{1-\frac{30^2}{40^2}}=\frac{20\sqrt{7}}{4}=5\sqrt{7}\approx 13.23 \text{ feet}$$

31. Since $2a = 468972$, $a = 234486$. Therefore:
$$e = \frac{c}{a}$$
$$0.055 = \frac{c}{234486}$$
$$c \approx 12897$$
$$a - c = 234486 - 12897 = 221589$$
The distance is 221,589 miles.

33. The center is $(2,3)$. Since $a = 4$, the vertices are $(-2,3)$ and $(6,3)$. Finding the foci:
$$c = \sqrt{16-4}=\sqrt{12}=2\sqrt{3}$$
The foci are $\left(2-2\sqrt{3},3\right)$ and $\left(2+2\sqrt{3},3\right)$.

35. The center is $(-1,3)$. Since $a = \sqrt{18} = 3\sqrt{2}$, the vertices are $\left(-1,3-3\sqrt{2}\right)$ and $\left(-1,3+3\sqrt{2}\right)$.

Finding the foci: $c = \sqrt{18-12} = \sqrt{6}$

The foci are $\left(-1,3-\sqrt{6}\right)$ and $\left(-1,3+\sqrt{6}\right)$.

37. First complete the square:

$$4x^2 + y^2 - 16x - 6y + 21 = 0$$
$$4\left(x^2 - 4x + 4\right) + \left(y^2 - 6y + 9\right) = -21 + 16 + 9$$
$$4(x-2)^2 + (y-3)^2 = 4$$
$$\frac{(x-2)^2}{1} + \frac{(y-3)^2}{4} = 1$$

The center is $(2,3)$. Since $a = 2$, the vertices are $(2,1)$ and $(2,5)$. Finding the foci: $c = \sqrt{4-1} = \sqrt{3}$

The foci are $\left(2,3-\sqrt{3}\right)$ and $\left(2,3+\sqrt{3}\right)$.

39. First complete the square:

$$6x^2 + 5y^2 - 60x - 10y + 65 = 0$$
$$6\left(x^2 - 10x + 25\right) + 5\left(y^2 - 2y + 1\right) = -65 + +150 + 5$$
$$6(x-5)^2 + 5(y-1)^2 = 90$$
$$\frac{(x-5)^2}{15} + \frac{(y-1)^2}{18} = 1$$

The center is $(5,1)$. Since $a = \sqrt{18} = 3\sqrt{2}$, the vertices are $\left(5,1-3\sqrt{2}\right)$ and $\left(5,1+3\sqrt{2}\right)$. Finding

the foci: $c = \sqrt{18-15} = \sqrt{3}$

The foci are $\left(5,1-\sqrt{3}\right)$ and $\left(5,1+\sqrt{3}\right)$.

41. The center is $(2,-3)$ and the vertices are $(-1,-3)$ and $(5,-3)$. Since $c = \sqrt{9-8} = 1$, the foci are $(1,-3)$ and $(3,-3)$. Graphing the ellipse:

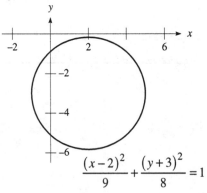

$$\frac{(x-2)^2}{9} + \frac{(y+3)^2}{8} = 1$$

43.   The center is (–2,1) and the vertices are $\left(-2,1-2\sqrt{6}\right)$ and $\left(-2,1+2\sqrt{6}\right)$. Since

$c = \sqrt{24-16} = \sqrt{8} = 2\sqrt{2}$, the foci are $\left(-2,1-2\sqrt{2}\right)$ and $\left(-2,1+2\sqrt{2}\right)$. Graphing the ellipse:

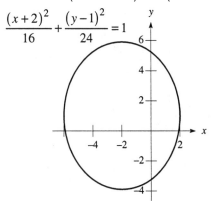

$$\frac{(x+2)^2}{16} + \frac{(y-1)^2}{24} = 1$$

45.   First complete the square:
$$16x^2 + 9y^2 - 32x + 54y = 47$$
$$16\left(x^2 - 2x + 1\right) + 9\left(y^2 + 6y + 9\right) = 47 + 16 + 81$$
$$16(x-1)^2 + 9(y+3)^2 = 144$$
$$\frac{(x-1)^2}{9} + \frac{(y+3)^2}{16} = 1$$

The center is (1,–3) and the vertices are (1,–7) and (1,1). Since $c = \sqrt{16-9} = \sqrt{7}$, the foci are $\left(1,-3-\sqrt{7}\right)$ and $\left(1,-3+\sqrt{7}\right)$. Graphing the ellipse:

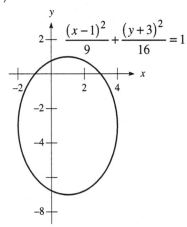

$$\frac{(x-1)^2}{9} + \frac{(y+3)^2}{16} = 1$$

47.   First complete the square:
$$x^2 + 4y^2 - 2x - 24y = -29$$
$$\left(x^2 - 2x + 1\right) + 4\left(y^2 - 6y + 9\right) = -29 + 1 + 36$$
$$(x-1)^2 + 4(y-3)^2 = 8$$
$$\frac{(x-1)^2}{8} + \frac{(y-3)^2}{2} = 1$$

The center is (1,3) and the vertices are $\left(1-2\sqrt{2},3\right)$ and $\left(1+2\sqrt{2},3\right)$. Since $c=\sqrt{8-2}=\sqrt{6}$, the foci are $\left(1-\sqrt{6},3\right)$ and $\left(1+\sqrt{6},3\right)$. Graphing the ellipse:

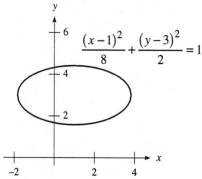

49. This is the graph of an ellipse. The center is (0,0) and the vertices are (–3,0) and (3,0). Since $c=\sqrt{9-4}=\sqrt{5}$, the foci are $\left(-\sqrt{5},0\right)$ and $\left(\sqrt{5},0\right)$. Graphing the ellipse:

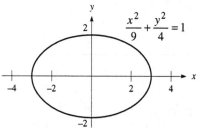

51. This is the graph of a circle. The center is (–2,1) and the radius is 4. Graphing the circle:

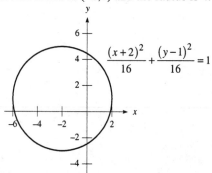

53. Since $x^2=9y$, $4p=9$ and thus $p=\frac{9}{4}$. This is the graph of a parabola. The vertex is (0,0), the focus is $\left(0,\frac{9}{4}\right)$, and the directrix is $y=-\frac{9}{4}$. Graphing the parabola:

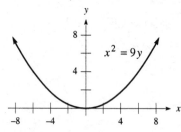

55. The center of the ellipse is (3,4), so $a = 3$ and $c = 2$. Finding b:

$$2^2 = 3^2 - b^2$$
$$4 = 9 - b^2$$
$$b^2 = 5$$

The equation of the ellipse is $\dfrac{(x-3)^2}{9} + \dfrac{(y-4)^2}{5} = 1$.

57. The center of the ellipse is (3,–5), $a = 2$, and $b = 1$, so its equation is $\dfrac{(x-3)^2}{4} + \dfrac{(y+5)^2}{1} = 1$.

59. Let (x,y) be a point on the ellipse. Using the definition where the sum of the distances from each focus is $2a$, we have:

$$\sqrt{x^2 + (y-c)^2} + \sqrt{x^2 + (y+c)^2} = 2a$$
$$\sqrt{x^2 + (y-c)^2} = 2a - \sqrt{x^2 + (y+c)^2}$$
$$x^2 + (y-c)^2 = 4a^2 - 4a\sqrt{x^2 + (y+c)^2} + x^2 + (y+c)^2$$
$$y^2 - 2cy + c^2 = 4a^2 - 4a\sqrt{x^2 + (y+c)^2} + y^2 + 2cy + c^2$$
$$-4cy - 4a^2 = -4a\sqrt{x^2 + (y+c)^2}$$
$$cy + a^2 = a\sqrt{x^2 + (y+c)^2}$$
$$\left(cy + a^2\right)^2 = a^2\left[x^2 + (y+c)^2\right]$$
$$c^2y^2 + 2a^2cy + a^4 = a^2x^2 + a^2y^2 + 2a^2cy + a^2c^2$$
$$a^4 - a^2c^2 = a^2x^2 + \left(a^2 - c^2\right)y^2$$
$$a^2\left(a^2 - c^2\right) = a^2x^2 + \left(a^2 - c^2\right)y^2$$
$$\frac{a^2x^2}{a^2\left(a^2 - c^2\right)} + \frac{\left(a^2 - c^2\right)y^2}{a^2\left(a^2 - c^2\right)} = 1$$
$$\frac{x^2}{a^2 - c^2} + \frac{y^2}{a^2} = 1$$
$$\frac{x^2}{b^2} + \frac{y^2}{a^2} = 1$$

Note the last step used the relationship $b^2 = a^2 - c^2$.

61. Using $(x_0, y_0) = (-2, \sqrt{3})$, $a = 4$, and $b = 2$, the tangent line is given by:

$$\frac{-2x}{4^2} + \frac{\sqrt{3}y}{2^2} = 1$$
$$-\frac{x}{8} + \frac{\sqrt{3}y}{4} = 1$$
$$-x + \left(2\sqrt{3}\right)y = 8$$

63. When $x = c$, the y-coordinate is given by:

$$\frac{c^2}{a^2} + \frac{y^2}{b^2} = 1$$

$$\frac{y^2}{b^2} = 1 - \frac{c^2}{a^2}$$

$$\frac{y^2}{b^2} = \frac{a^2 - c^2}{a^2}$$

$$\frac{y^2}{b^2} = \frac{b^2}{a^2}$$

$$y^2 = \frac{b^4}{a^2}$$

$$y = \frac{b^2}{a}$$

The length of the latus rectum is therefore $2y = \dfrac{2b^2}{a}$.

65. Graphing the ellipse:

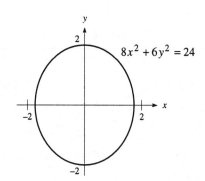

$$8x^2 + 6y^2 = 24$$

67. Graphing the ellipse:

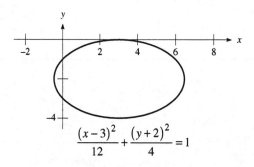

$$\frac{(x-3)^2}{12} + \frac{(y+2)^2}{4} = 1$$

10.4 The Hyperbola

1. The vertices are $(-3,0)$ and $(3,0)$. Since $c = \sqrt{9+16} = \sqrt{25} = 5$, the foci are $(-5,0)$ and $(5,0)$, and the asymptotes are $y = \pm\frac{4}{3}x$.

3. The vertices are $\left(0,-2\sqrt{3}\right)$ and $\left(0,2\sqrt{3}\right)$. Since $c = \sqrt{12+18} = \sqrt{30}$, the foci are $\left(0,-\sqrt{30}\right)$ and $\left(0,\sqrt{30}\right)$, and the asymptotes are $y = \pm\dfrac{\sqrt{12}}{\sqrt{18}}x = \pm\dfrac{\sqrt{6}}{3}x$.

5. The vertices are (–3,0) and (3,0). Since $c = \sqrt{9+18} = \sqrt{27} = 3\sqrt{3}$, the foci are $\left(-3\sqrt{3},0\right)$ and

$\left(3\sqrt{3},0\right)$, and the asymptotes are $y = \pm\dfrac{\sqrt{18}}{\sqrt{9}}x = \pm\sqrt{2}x$.

7. Writing the equation as $\dfrac{x^2}{9} - \dfrac{y^2}{25} = 1$, the vertices are (–3,0) and (3,0). Since $c = \sqrt{9+25} = \sqrt{34}$,

the foci are $\left(-\sqrt{34},0\right)$ and $\left(\sqrt{34},0\right)$, and the asymptotes are $y = \pm\frac{5}{3}x$.

9. Writing the equation as $\dfrac{y^2}{1} - \dfrac{x^2}{30} = 1$, the vertices are (0,–1) and (0,1). Since $c = \sqrt{1+30} = \sqrt{31}$, the

foci are $\left(0,-\sqrt{31}\right)$ and $\left(0,\sqrt{31}\right)$, and the asymptotes are $y = \pm\dfrac{1}{\sqrt{30}}x = \pm\dfrac{\sqrt{30}}{30}x$.

11. The vertices are (–3,0) and (3,0). Since $c = \sqrt{9+49} = \sqrt{58}$, the foci are $\left(-\sqrt{58},0\right)$ and $\left(\sqrt{58},0\right)$,

and the asymptotes are $y = \pm\frac{7}{3}x$. Graphing the hyperbola:

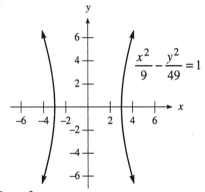

13. Writing the equation as $\dfrac{x^2}{3} - \dfrac{y^2}{24} = 1$, the vertices are $\left(-\sqrt{3},0\right)$ and $\left(\sqrt{3},0\right)$. Since

$c = \sqrt{3+24} = \sqrt{27} = 3\sqrt{3}$, the foci are $\left(-3\sqrt{3},0\right)$ and $\left(3\sqrt{3},0\right)$, and the asymptotes are

$y = \pm\dfrac{\sqrt{24}}{\sqrt{3}}x = \pm 2\sqrt{2}x$. Graphing the hyperbola:

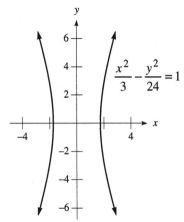

15. Writing the equation as $\dfrac{y^2}{8} - \dfrac{x^2}{3} = 1$, the vertices are $\left(0, -2\sqrt{2}\right)$ and $\left(0, 2\sqrt{2}\right)$. Since

$c = \sqrt{8+3} = \sqrt{11}$, the foci are $\left(0, -\sqrt{11}\right)$ and $\left(0, \sqrt{11}\right)$, and the asymptotes are

$y = \pm\dfrac{2\sqrt{2}}{\sqrt{3}}x = \pm\dfrac{2\sqrt{6}}{3}x$. Graphing the hyperbola:

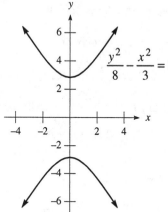

17. Assign $A(-50,0)$ and $B(50,0)$ as the positions of the coast guard stations, so $c = 50$. The difference between the distances of the boat is 196000 feet = 37.12 miles, so $2a = 37.12$, thus $a \approx 18.56$ miles. Finally, we find b:

$$18.56^2 + b^2 = 50^2$$
$$b^2 = 50^2 - 18.56^2 = 2155.53$$
$$b \approx 46.43$$

The equation of the hyperbola is $\dfrac{x^2}{18.56^2} - \dfrac{y^2}{46.43^2} = 1$, or $\dfrac{x^2}{344.5} - \dfrac{y^2}{2155.5} = 1$.

19. The hyperbola can be written as $\dfrac{x^2}{3} - \dfrac{y^2}{4} = 1$, so $a = \sqrt{3}$ and $b = 2$. Thus $c = \sqrt{3+4} = \sqrt{7}$. The

shortest distance to the comet is given by: $c - a = \sqrt{7} - \sqrt{3} \approx 0.9137$ AU

21. The center is $(1,3)$. Since $a = 2$ and $b = 2$, $c = \sqrt{4+4} = \sqrt{8} = 2\sqrt{2}$. The vertices are $(-1,3)$ and $(3,3)$,

and the foci are $\left(1 - 2\sqrt{2}, 3\right)$ and $\left(1 + 2\sqrt{2}, 3\right)$. Since $m = \pm\dfrac{2}{2} = \pm 1$, finding the asymptotes:

$$y - 3 = -1(x-1) \qquad\qquad y - 3 = 1(x-1)$$
$$y - 3 = -x + 1 \qquad\qquad y - 3 = x - 1$$
$$y = -x + 4 \qquad\qquad\quad y = x + 2$$

23. The center is $(-1,-3)$. Since $a = \sqrt{12} = 2\sqrt{3}$ and $b = \sqrt{18} = 3\sqrt{2}$, $c = \sqrt{12+18} = \sqrt{30}$. The vertices are $\left(-1,-3-2\sqrt{3}\right)$ and $\left(-1,-3+2\sqrt{3}\right)$, and the foci are $\left(-1,-3-\sqrt{30}\right)$ and $\left(-1,-3+\sqrt{30}\right)$. Since $m = \pm\dfrac{\sqrt{12}}{\sqrt{18}} = \pm\dfrac{\sqrt{6}}{3}$, finding the asymptotes:

$$y+3 = -\frac{\sqrt{6}}{3}(x+1)$$
$$y+3 = -\frac{\sqrt{6}}{3}x - \frac{\sqrt{6}}{3}$$
$$y = -\frac{\sqrt{6}}{3}x - \frac{\sqrt{6}+9}{3}$$

$$y+3 = \frac{\sqrt{6}}{3}(x+1)$$
$$y+3 = \frac{\sqrt{6}}{3}x + \frac{\sqrt{6}}{3}$$
$$y = \frac{\sqrt{6}}{3}x + \frac{\sqrt{6}-9}{3}$$

25. First complete the square:
$$x^2 - 4y^2 - 6x + 8y = 11$$
$$\left(x^2 - 6x + 9\right) - 4\left(y^2 - 2y + 1\right) = 11 + 9 - 4$$
$$(x-3)^2 - 4(y-1)^2 = 16$$
$$\frac{(x-3)^2}{16} - \frac{(y-1)^2}{4} = 1$$

The center is $(3,1)$. Since $a = 4$ and $b = 2$, $c = \sqrt{16+4} = \sqrt{20} = 2\sqrt{5}$. The vertices are $(-1,1)$ and $(7,1)$, and the foci are $\left(3-2\sqrt{5},1\right)$ and $\left(3+2\sqrt{5},1\right)$. Since $m = \pm\dfrac{2}{4} = \pm\dfrac{1}{2}$, finding the asymptotes:

$$y-1 = -\tfrac{1}{2}(x-3)$$
$$y-1 = -\tfrac{1}{2}x + \tfrac{3}{2}$$
$$y = -\tfrac{1}{2}x + \tfrac{5}{2}$$

$$y-1 = \tfrac{1}{2}(x-3)$$
$$y-1 = \tfrac{1}{2}x - \tfrac{3}{2}$$
$$y = \tfrac{1}{2}x - \tfrac{1}{2}$$

27. First complete the square:
$$2y^2 - 3x^2 + 4y + 6x = 49$$
$$2\left(y^2 + 2y + 1\right) - 3\left(x^2 - 2x + 1\right) = 49 + 2 - 3$$
$$2(y+1)^2 - 3(x-1)^2 = 48$$
$$\frac{(y+1)^2}{24} - \frac{(x-1)^2}{16} = 1$$

The center is $(1,-1)$. Since $a = \sqrt{24} = 2\sqrt{6}$ and $b = 4$, $c = \sqrt{24+16} = \sqrt{40} = 2\sqrt{10}$. The vertices are $\left(1,-1-2\sqrt{6}\right)$ and $\left(1,-1+2\sqrt{6}\right)$, and the foci are $\left(1,-1-2\sqrt{10}\right)$ and $\left(1,-1+2\sqrt{10}\right)$. Since $m = \pm\dfrac{\sqrt{24}}{\sqrt{16}} = \pm\dfrac{2\sqrt{6}}{4} = \pm\dfrac{\sqrt{6}}{2}$, finding the asymptotes:

$$y+1 = -\frac{\sqrt{6}}{2}(x-1)$$
$$y+1 = -\frac{\sqrt{6}}{2}x + \frac{\sqrt{6}}{2}$$
$$y = -\frac{\sqrt{6}}{2}x + \frac{\sqrt{6}-2}{2}$$

$$y+1 = \frac{\sqrt{6}}{2}(x-1)$$
$$y+1 = \frac{\sqrt{6}}{2}x - \frac{\sqrt{6}}{2}$$
$$y = \frac{\sqrt{6}}{2}x - \frac{\sqrt{6}+2}{2}$$

29. The center is $(-4,-3)$. Since $a = 3$ and $b = 4$, $c = \sqrt{9+16} = \sqrt{25} = 5$. The vertices are $(-7,-3)$ and $(-1,-3)$, and the foci are $(-9,-3)$ and $(1,-3)$. Since $m = \pm\frac{4}{3}$, finding the asymptotes:

$$y+3 = -\tfrac{4}{3}(x+4) \qquad\qquad\qquad y+3 = \tfrac{4}{3}(x+4)$$
$$y+3 = -\tfrac{4}{3}x - \tfrac{16}{3} \qquad\qquad\qquad y+3 = \tfrac{4}{3}x + \tfrac{16}{3}$$
$$y = -\tfrac{4}{3}x - \tfrac{25}{3} \qquad\qquad\qquad\quad y = \tfrac{4}{3}x + \tfrac{7}{3}$$

Graphing the hyperbola:

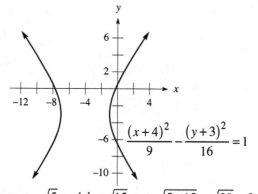

$$\frac{(x+4)^2}{9} - \frac{(y+3)^2}{16} = 1$$

31. The center is $(0,3)$. Since $a = \sqrt{5}$ and $b = \sqrt{15}$, $c = \sqrt{5+15} = \sqrt{20} = 2\sqrt{5}$. The vertices are $\left(-\sqrt{5},3\right)$ and $\left(\sqrt{5},3\right)$, and the foci are $\left(-2\sqrt{5},3\right)$ and $\left(2\sqrt{5},3\right)$. Since $m = \pm\dfrac{\sqrt{15}}{\sqrt{5}} = \pm\sqrt{3}$, finding the asymptotes:

$$y-3 = -\sqrt{3}x \qquad\qquad\qquad y-3 = \sqrt{3}x$$
$$y = -\sqrt{3}x + 3 \qquad\qquad\qquad y = \sqrt{3}x + 3$$

Graphing the hyperbola:

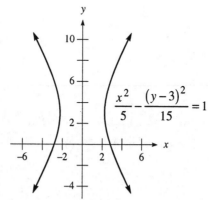

$$\frac{x^2}{5} - \frac{(y-3)^2}{15} = 1$$

33. First complete the square:

$$9x^2 - 16y^2 - 36x - 32y = 124$$
$$9\left(x^2 - 4x + 4\right) - 16\left(y^2 + 2y + 1\right) = 124 + 36 - 16$$
$$9(x-2)^2 - 16(y+1)^2 = 144$$
$$\frac{(x-2)^2}{16} - \frac{(y+1)^2}{9} = 1$$

The center is $(2,-1)$. Since $a = 4$ and $b = 3$, $c = \sqrt{16+9} = \sqrt{25} = 5$. The vertices are $(-2,-1)$ and $(6,-1)$, and the foci are $(-3,-1)$ and $(7,-1)$. Since $m = \pm\frac{3}{4}$, finding the asymptotes:

$$y+1 = -\frac{3}{4}(x-2)$$
$$y+1 = -\frac{3}{4}x + \frac{3}{2}$$
$$y = -\frac{3}{4}x + \frac{1}{2}$$

$$y+1 = \frac{3}{4}(x-2)$$
$$y+1 = \frac{3}{4}x - \frac{3}{2}$$
$$y = \frac{3}{4}x - \frac{5}{2}$$

Graphing the hyperbola:

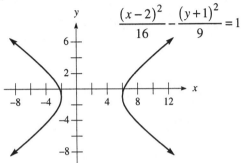

$$\frac{(x-2)^2}{16} - \frac{(y+1)^2}{9} = 1$$

35. First complete the square:

$$25y^2 - 36x^2 - 150y + 288x - 1251 = 0$$
$$25\left(y^2 - 6y + 9\right) - 36\left(x^2 - 8x + 16\right) = 1251 + 225 - 576$$
$$25(y-3)^2 - 36(x-4)^2 = 900$$
$$\frac{(y-3)^2}{36} - \frac{(x-4)^2}{25} = 1$$

The center is $(4,3)$. Since $a = 6$ and $b = 5$, $c = \sqrt{36+25} = \sqrt{61}$. The vertices are $(4,-3)$ and $(4,9)$, and the foci are $\left(4, 3-\sqrt{61}\right)$ and $\left(4, 3+\sqrt{61}\right)$. Since $m = \pm\frac{6}{5}$, finding the asymptotes:

$$y-3 = -\frac{6}{5}(x-4)$$
$$y-3 = -\frac{6}{5}x + \frac{24}{5}$$
$$y = -\frac{6}{5}x + \frac{39}{5}$$

$$y-3 = \frac{6}{5}(x-4)$$
$$y-3 = \frac{6}{5}x - \frac{24}{5}$$
$$y = \frac{6}{5}x - \frac{9}{5}$$

Graphing the hyperbola:

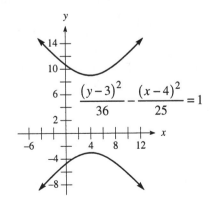

$$\frac{(y-3)^2}{36} - \frac{(x-4)^2}{25} = 1$$

37. The center is $(6,-1)$, so $a = 4$ and $c = 6$. Finding b:

$$4^2 + b^2 = 6^2$$
$$16 + b^2 = 36$$
$$b^2 = 20$$

The equation is $\dfrac{(x-6)^2}{16} - \dfrac{(y+1)^2}{20} = 1$.

39. The graph is an ellipse. Since $c = \sqrt{12 - 4} = \sqrt{8} = 2\sqrt{2}$, the vertices are $\left(-2\sqrt{3},0\right)$ and $\left(2\sqrt{3},0\right)$, and the foci are $\left(-2\sqrt{2},0\right)$ and $\left(2\sqrt{2},0\right)$. Sketching the graph:

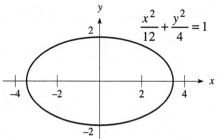

41. The graph is a hyperbola. The center is $(-2,1)$. Since $c = \sqrt{9 + 25} = \sqrt{34}$, the vertices are $(-5,1)$ and $(1,1)$, and the foci are $\left(-2 - \sqrt{34},1\right)$ and $\left(-2 + \sqrt{34},1\right)$. The asymptotes have slopes of $m = \pm\frac{5}{3}$, so their equations are:

$$y - 1 = -\tfrac{5}{3}(x+2) \qquad\qquad y - 1 = \tfrac{5}{3}(x+2)$$
$$y - 1 = -\tfrac{5}{3}x - \tfrac{10}{3} \qquad\qquad y - 1 = \tfrac{5}{3}x + \tfrac{10}{3}$$
$$y = -\tfrac{5}{3}x - \tfrac{7}{3} \qquad\qquad y = \tfrac{5}{3}x + \tfrac{13}{3}$$

Sketching the graph:

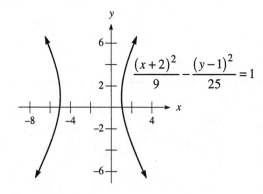

43. Simplifying the equation:
$$x^2 - 2y - 12 = 0$$
$$x^2 = 2y + 12$$
$$x^2 = 2(y + 6)$$

The graph is a parabola. Since $4p = 2$, $p = \frac{1}{2}$. The vertex is $(0, -6)$, the focus is $\left(0, -\frac{11}{2}\right)$, and the directrix is $y = -\frac{13}{2}$. Sketching the graph:

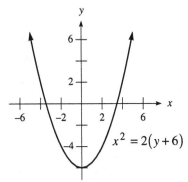

45. Let (x, y) be a point on the hyperbola. Since the difference of distances from this point to each of the foci is $2a$, we have:

$$\sqrt{(x+c)^2 + y^2} - \sqrt{(x-c)^2 + y^2} = 2a$$
$$\sqrt{(x+c)^2 + y^2} = 2a + \sqrt{(x-c)^2 + y^2}$$
$$(x+c)^2 + y^2 = 4a^2 + 4a\sqrt{(x-c)^2 + y^2} + (x-c)^2 + y^2$$
$$x^2 + 2cx + c^2 + y^2 = 4a^2 + 4a\sqrt{(x-c)^2 + y^2} + x^2 + 2cx + c^2 + y^2$$
$$4cx - 4a^2 = 4a\sqrt{(x-c)^2 + y^2}$$
$$cx - a^2 = a\sqrt{(x-c)^2 + y^2}$$
$$\left(cx - a^2\right)^2 = a^2\left[(x-c)^2 + y^2\right]$$
$$c^2 x^2 - 2a^2 cx + a^4 = a^2 x^2 - 2a^2 cx + a^2 c^2 + a^2 y^2$$
$$\left(c^2 - a^2\right)x^2 - a^2 y^2 = a^2 c^2 - a^4$$
$$\left(c^2 - a^2\right)x^2 - a^2 y^2 = a^2\left(c^2 - a^2\right)$$
$$\frac{\left(c^2 - a^2\right)x^2}{a^2\left(c^2 - a^2\right)} - \frac{a^2 y^2}{a^2\left(c^2 - a^2\right)} = 1$$
$$\frac{x^2}{a^2} - \frac{y^2}{c^2 - a^2} = 1$$
$$\frac{x^2}{a^2} - \frac{y^2}{b^2} = 1$$

Note the last step used the relationship $b^2 = c^2 - a^2$.

47.   Using $(x_0, y_0) = (4, 0)$, $a = 4$, $b = 2$, the tangent line is:

$$\frac{4x}{4^2} - \frac{0y}{2^2} = 1$$

$$\frac{x}{4} = 1$$

$$x = 4$$

49.   Substituting $x = c$:

$$\frac{c^2}{a^2} - \frac{y^2}{b^2} = 1$$

$$\frac{y^2}{b^2} = \frac{c^2}{a^2} - 1$$

$$\frac{y^2}{b^2} = \frac{c^2 - a^2}{a^2}$$

$$\frac{y^2}{b^2} = \frac{b^2}{a^2}$$

$$y^2 = \frac{b^4}{a^2}$$

$$y = \frac{b^2}{a}$$

51.   Graphing the hyperbola:

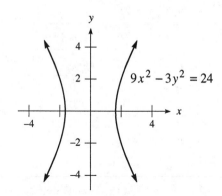

$9x^2 - 3y^2 = 24$

53.   Graphing the hyperbola:

$$\frac{(x+3)^2}{8} - \frac{(y-2)^2}{2} = 1$$

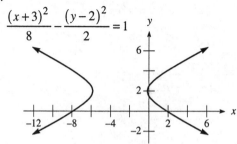

55. **a.** Solving for y:

$$\frac{x^2}{a^2} - \frac{y^2}{b^2} = 1$$

$$\frac{y^2}{b^2} = \frac{x^2}{a^2} - 1$$

$$y^2 = \frac{b^2\left(x^2 - a^2\right)}{a^2}$$

$$y = \pm\frac{b}{a}\sqrt{x^2 - a^2}$$

b. Factoring out x^2: $\;y = \pm\dfrac{b}{a}\sqrt{x^2 - a^2} = \pm\dfrac{b}{a}\sqrt{x^2\left(1 - \dfrac{a^2}{x^2}\right)} = \pm\dfrac{b}{a}x\sqrt{1 - \dfrac{a^2}{x^2}}$

c. As $x \to \infty$ or $x \to -\infty$, the radical term approaches 1.

d. As $x \to \infty$ or $x \to -\infty$, $y \to \pm\dfrac{b}{a}$. That is, as $x \to \infty$ or $x \to -\infty$, the y-coordinates of the hyperbola approach the asymptotes.

57. The graph is a pair of lines $x = \pm 2y$:

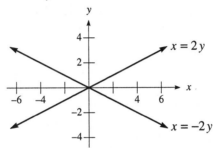

59. Sketching the inequality:

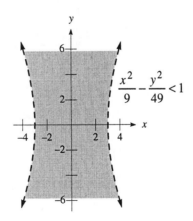

61.   Sketching the inequality:

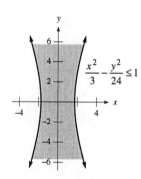

$$\frac{x^2}{3} - \frac{y^2}{24} \leq 1$$

10.5 Identifying Conic Sections: Degenerate Forms

1.   Since A and C have the same sign but $A \neq C$, the graph will be an ellipse (or its degenerate form).
3.   Since A and C are different signs, the graph will be a hyperbola (or its degenerate form).
5.   Since $A = C$, the graph will be a circle (or its degenerate form).
7.   Since A and C are different signs, the graph will be a hyperbola (or its degenerate form).
9.   Since A and C have the same sign but $A \neq C$, the graph will be an ellipse (or its degenerate form).
11.   Writing the equation as $y^2 = -16x$, the graph is a parabola:

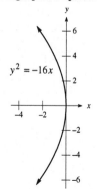

13.   Completing the square:
$$x^2 - 10x + 25 = 0$$
$$(x - 5)^2 = 0$$
$$x = 5$$
The graph is a line:

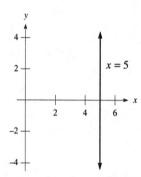

15.   Completing the square:

$$2x^2 + y^2 - 8x - 2y - 7 = 0$$
$$2\left(x^2 - 4x + 4\right) + \left(y^2 - 2y + 1\right) = 7 + 8 + 1$$
$$2(x-2)^2 + (y-1)^2 = 16$$
$$\frac{(x-2)^2}{8} + \frac{(y-1)^2}{16} = 1$$

The graph is an ellipse:

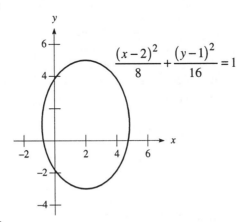

17.   Completing the square:

$$25x^2 + y^2 - 50x - 6y - 16 = 0$$
$$25\left(x^2 - 2x + 1\right) + \left(y^2 - 6y + 9\right) = 16 + 25 + 9$$
$$25(x-1)^2 + (y-3)^2 = 50$$
$$\frac{(x-1)^2}{2} + \frac{(y-3)^2}{50} = 1$$

The graph is an ellipse:

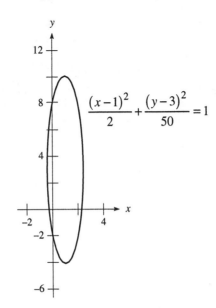

19. Simplifying:
$$25x^2 + 9y^2 - 225 = 0$$
$$25x^2 + 9y^2 = 225$$
$$\frac{x^2}{9} + \frac{y^2}{25} = 1$$

The graph is an ellipse:

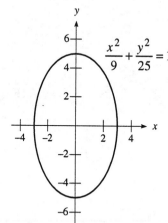

21. Writing the equation as $x^2 - y^2 = 18$, the graph is a hyperbola:

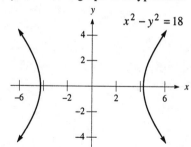

23. Simplifying the equation:
$$y^2 - 6y - 16 = 0$$
$$(y-8)(y+2) = 0$$
$$y = -2, 8$$

The graph is two lines:

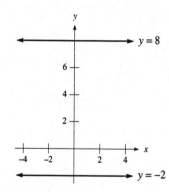

25. Simplifying the equation:

$$-10x^2 + y = 0$$
$$10x^2 = y$$
$$x^2 = \tfrac{1}{10}y$$

The graph is a parabola:

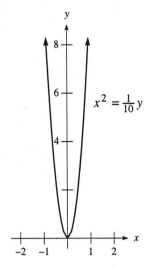

27. Completing the square:

$$x^2 + y^2 + 2x - 6y - 5 = 0$$
$$\left(x^2 + 2x + 1\right) + \left(y^2 - 6y + 9\right) = 5 + 1 + 9$$
$$(x+1)^2 + (y-3)^2 = 15$$

The graph is a circle:

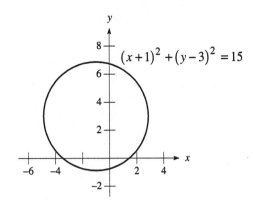

29.   Completing the square:
$$x^2 + y^2 - 4x - 2y + 3 = 0$$
$$\left(x^2 - 4x + 4\right) + \left(y^2 - 2y + 1\right) = -3 + 4 + 1$$
$$(x-2)^2 + (y-1)^2 = 2$$
The graph is a circle:

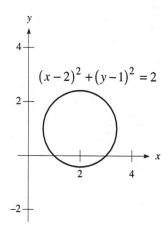

31.   Completing the square:
$$9x^2 - y^2 - 18x - 4y - 139 = 0$$
$$9\left(x^2 - 2x + 1\right) - \left(y^2 + 4y + 4\right) = 139 + 9 - 4$$
$$9(x-1)^2 - (y+2)^2 = 144$$
$$\frac{(x-1)^2}{16} - \frac{(y+2)^2}{144} = 1$$
The graph is a hyperbola:

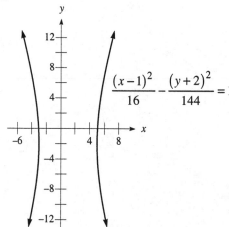

33.   The equation is $x^2 + y^2 = -36$, which has no solution. There is no graph.

35. Completing the square:
$$9x^2 + 25y^2 + 18x + 9 = 0$$
$$9\left(x^2 + 2x + 1\right) + 25y^2 = -9 + 9$$
$$9(x+1)^2 + 25y^2 = 0$$
The only solution to this equation is the point $(-1,0)$. The graph is a point:

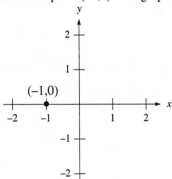

37. Simplifying the equation:
$$16y^2 - x^2 = 0$$
$$16y^2 = x^2$$
$$4y = \pm x$$
$$y = \pm \tfrac{1}{4}x$$
The graph is two lines:

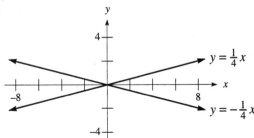

39. Completing the square:
$$x^2 - 2x + 12y - 47 = 0$$
$$\left(x^2 - 2x + 1\right) = -12y + 47 + 1$$
$$(x-1)^2 = -12(y-4)$$
The graph is a parabola:

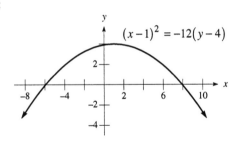

41. a.   If $AC < 0$, the graph must be a hyperbola (or its degenerate form).

b.   If $AC > 0$, the graph is either an ellipse (if $A \neq C$) or a circle (if $A = C$), or their degenerate forms.

c.   If $AC = 0$, the graph must be a parabola (or its degenerate form).

10.6 Translations and Rotations of Coordinate Axes

1.   Since $h = 2$ and $k = -4$, $x' = x - 2$ and $y' = y + 4$. If $(x, y) = (3, 5)$, $(x', y') = (1', 9')$.

3.   Since $h = 2$ and $k = -4$, $x = x' + 2$ and $y = y' - 4$. If $(x', y') = (2', 5')$, $(x, y) = (4, 1)$.

5.   Since $h = 2$ and $k = -4$, $x' = x - 2$ and $y' = y + 4$. If $(x, y) = (-4, 4)$, $(x', y') = (-6', 8')$.

7.   Since $h = 2$ and $k = -4$, $x = x' + 2$ and $y = y' - 4$. If $(x', y') = (0', 0')$, $(x, y) = (2, -4)$.

9.   The translation is $x' = x - 3$, $y' = y + 4$.

11.   The translation is $x' = x$, $y' = y - 3$.

13.   First complete the square:

$$3x^2 + 4y^2 - 6x + 56y - 187 = 0$$

$$3\left(x^2 - 2x + 1\right) + 4\left(y^2 + 14y + 49\right) = 187 + 3 + 196$$

$$3(x - 1)^2 + 4(y + 7)^2 = 386$$

The translation is $x' = x - 1$, $y' = y + 7$.

15.   First complete the square:

$$x^2 - 4y^2 - 10x + 16y + 1 = 0$$

$$\left(x^2 - 10x + 25\right) - 4\left(y^2 - 4y + 4\right) = -1 + 25 - 16$$

$$(x - 5)^2 - 4(y - 2)^2 = 8$$

The translation is $x' = x - 5$, $y' = y - 2$.

17.   Using the rotation equations:

$$x' = 2\cos 45° + 5\sin 45° = 2\left(\frac{\sqrt{2}}{2}\right) + 5\left(\frac{\sqrt{2}}{2}\right) = \frac{7\sqrt{2}}{2}$$

$$y' = -2\sin 45° + 5\cos 45° = -2\left(\frac{\sqrt{2}}{2}\right) + 5\left(\frac{\sqrt{2}}{2}\right) = \frac{3\sqrt{2}}{2}$$

The new point is $(x', y') = \left(\dfrac{7\sqrt{2}}{2}, \dfrac{3\sqrt{2}}{2}\right)'$.

19.   Using the rotation equations:

$$x = 2\cos 60° - (-1)\sin 60° = 2\left(\frac{1}{2}\right) + 1\left(\frac{\sqrt{3}}{2}\right) = \frac{\sqrt{3} + 2}{2}$$

$$y = 2\sin 60° + (-1)\cos 60° = 2\left(\frac{\sqrt{3}}{2}\right) - 1\left(\frac{1}{2}\right) = \frac{2\sqrt{3} - 1}{2}$$

The point is $(x, y) = \left(\dfrac{\sqrt{3} + 2}{2}, \dfrac{2\sqrt{3} - 1}{2}\right)$.

21. Using the rotation equations:

$$x' = 2\cos 60° + (-1)\sin 60° = 2\left(\frac{1}{2}\right) - 1\left(\frac{\sqrt{3}}{2}\right) = \frac{2-\sqrt{3}}{2}$$

$$y' = -2\sin 60° + (-1)\cos 60° = -2\left(\frac{\sqrt{3}}{2}\right) - 1\left(\frac{1}{2}\right) = \frac{-2\sqrt{3}-1}{2}$$

The point is $(x', y') = \left(\dfrac{2-\sqrt{3}}{2}, \dfrac{-2\sqrt{3}-1}{2}\right)'$.

23. Using the rotation equations:

$$x' = -2\cos 45° + 5\sin 45° = -2\left(\frac{\sqrt{2}}{2}\right) + 5\left(\frac{\sqrt{2}}{2}\right) = \frac{3\sqrt{2}}{2}$$

$$y' = -(-2)\sin 45° + 5\cos 45° = 2\left(\frac{\sqrt{2}}{2}\right) + 5\left(\frac{\sqrt{2}}{2}\right) = \frac{7\sqrt{2}}{2}$$

The point is $(x', y') = \left(\dfrac{3\sqrt{2}}{2}, \dfrac{7\sqrt{2}}{2}\right)'$.

25. The rotation equations are:

$$x = x'\cos 60° - y'\sin 60° = \frac{1}{2}x' - \frac{\sqrt{3}}{2}y'$$

$$y = x'\sin 60° + y'\cos 60° = \frac{\sqrt{3}}{2}x' + \frac{1}{2}y'$$

Now converting the equation:

$$2x^2 = y$$

$$2\left(\frac{1}{2}x' - \frac{\sqrt{3}}{2}y'\right)^2 = \frac{\sqrt{3}}{2}x' + \frac{1}{2}y'$$

$$2\left(\frac{1}{4}x'^2 - \frac{\sqrt{3}}{2}x'y' + \frac{3}{4}y'^2\right) = \frac{\sqrt{3}}{2}x' + \frac{1}{2}y'$$

$$\frac{1}{2}x'^2 - \sqrt{3}x'y' + \frac{3}{2}y'^2 - \frac{\sqrt{3}}{2}x' - \frac{1}{2}y' = 0$$

$$x'^2 - 2\sqrt{3}x'y' + 3y'^2 - \sqrt{3}x' - y' = 0$$

27.   The rotation equations are:

$$x = x'\cos 30° - y'\sin 30° = \frac{\sqrt{3}}{2}x' - \frac{1}{2}y'$$

$$y = x'\sin 30° + y'\cos 30° = \frac{1}{2}x' + \frac{\sqrt{3}}{2}y'$$

Now converting the equation:

$$2x^2 - 4y^2 = 6$$

$$x^2 - 2y^2 = 3$$

$$\left(\frac{\sqrt{3}}{2}x' - \frac{1}{2}y'\right)^2 - 2\left(\frac{1}{2}x' + \frac{\sqrt{3}}{2}y'\right)^2 = 3$$

$$\frac{3}{4}x'^2 - \frac{\sqrt{3}}{2}x'y' + \frac{1}{4}y'^2 - \frac{1}{2}x'^2 - \sqrt{3}x'y' - \frac{3}{2}y'^2 = 8$$

$$3x'^2 - 2\sqrt{3}x'y' + y'^2 - 2x'^2 - 4\sqrt{3}x'y' - 6y'^2 = 32$$

$$x'^2 - 6\sqrt{3}x'y' - 5y'^2 - 32 = 0$$

29.   The rotation equations are:

$$x = x'\cos\frac{\pi}{4} - y'\sin\frac{\pi}{4} = \frac{1}{\sqrt{2}}x' - \frac{1}{\sqrt{2}}y'$$

$$y = x'\sin\frac{\pi}{4} + y'\cos\frac{\pi}{4} = \frac{1}{\sqrt{2}}x' + \frac{1}{\sqrt{2}}y'$$

Now converting the equation:

$$x^2 + y^2 = 4$$

$$\left(\frac{1}{\sqrt{2}}x' - \frac{1}{\sqrt{2}}y'\right)^2 + \left(\frac{1}{\sqrt{2}}x' + \frac{1}{\sqrt{2}}y'\right)^2 = 4$$

$$\frac{1}{2}x'^2 - x'y' + \frac{1}{2}y'^2 + \frac{1}{2}x'^2 + x'y' + \frac{1}{2}y'^2 = 4$$

$$x'^2 + y'^2 = 4$$

31.   The rotation equations are:

$$x' = x\cos 45° + y\sin 45° = \frac{1}{\sqrt{2}}x + \frac{1}{\sqrt{2}}y$$

$$y' = -x\sin 45° + y\cos 45° = -\frac{1}{\sqrt{2}}x + \frac{1}{\sqrt{2}}y$$

Now converting the equation:

$$3x'y' = 4$$

$$3\left(\frac{1}{\sqrt{2}}x + \frac{1}{\sqrt{2}}y\right)\left(-\frac{1}{\sqrt{2}}x + \frac{1}{\sqrt{2}}y\right) = 4$$

$$3\left(-\frac{1}{2}x^2 + \frac{1}{2}y^2\right) = 4$$

$$\frac{3}{2}y^2 - \frac{3}{2}x^2 = 4$$

$$3y^2 - 3x^2 = 8$$

33. The rotation equations are:

$$x' = x\cos 30° + y\sin 30° = \frac{\sqrt{3}}{2}x + \frac{1}{2}y$$

$$y' = -x\sin 30° + y\cos 30° = -\frac{1}{2}x + \frac{\sqrt{3}}{2}y$$

Now converting the equation:

$$x'^2 - 2x'y' = 8$$

$$\left(\frac{\sqrt{3}}{2}x + \frac{1}{2}y\right)^2 - 2\left(\frac{\sqrt{3}}{2}x + \frac{1}{2}y\right)\left(-\frac{1}{2}x + \frac{\sqrt{3}}{2}y\right) = 8$$

$$\frac{3}{4}x^2 + \frac{\sqrt{3}}{2}xy + \frac{1}{4}y^2 + \frac{\sqrt{3}}{2}x^2 + \frac{1}{2}xy - \frac{3}{2}xy - \frac{\sqrt{3}}{2}y^2 = 8$$

$$3x^2 + 2\sqrt{3}xy + y^2 + 2\sqrt{3}x^2 - 4xy - 2\sqrt{3}y^2 = 32$$

$$\left(3 + 2\sqrt{3}\right)x^2 + \left(2\sqrt{3} - 4\right)xy + \left(1 - 2\sqrt{3}\right)y^2 = 32$$

35. Finding the rotation angle:

$$\cot 2\phi = \frac{2-3}{\sqrt{3}} = \frac{-1}{\sqrt{3}}$$

$$2\phi = 60°$$

$$\phi = 30°$$

37. Finding the rotation angle:

$$\cot 2\phi = \frac{2-1}{1} = 1$$

$$2\phi = 45°$$

$$\phi = 22.5°$$

39. Finding the rotation angle:

$$\cot 2\phi = \frac{0-0}{1} = 0$$

$$2\phi = 90°$$

$$\phi = 45°$$

The rotation equations are:

$$x = x'\cos 45° - y'\sin 45° = \frac{1}{\sqrt{2}}x' - \frac{1}{\sqrt{2}}y'$$

$$y = x'\sin 45° + y'\cos 45° = \frac{1}{\sqrt{2}}x' + \frac{1}{\sqrt{2}}y'$$

Now converting the equation:

$$xy = 8$$

$$\left(\frac{1}{\sqrt{2}}x' - \frac{1}{\sqrt{2}}y'\right)\left(\frac{1}{\sqrt{2}}x' + \frac{1}{\sqrt{2}}y'\right) = 8$$

$$\frac{1}{2}x'^2 - \frac{1}{2}y'^2 = 8$$

$$x'^2 - y'^2 = 16$$

Rotating 45°, graph the equation:

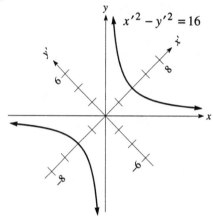

$$x'^2 - y'^2 = 16$$

41. Finding the rotation angle:

$$\cot 2\phi = \frac{1-11}{-10\sqrt{3}} = \frac{1}{\sqrt{3}}$$

$$2\phi = 60°$$

$$\phi = 30°$$

The rotation equations are:

$$x = x'\cos 30° - y'\sin 30° = \frac{\sqrt{3}}{2}x' - \frac{1}{2}y'$$

$$y = x'\sin 30° + y'\cos 30° = \frac{1}{2}x' + \frac{\sqrt{3}}{2}y'$$

Now converting the equation:

$$x^2 - 10\sqrt{3}xy + 11y^2 - 16 = 0$$

$$\left(\frac{\sqrt{3}}{2}x' - \frac{1}{2}y'\right)^2 - 10\sqrt{3}\left(\frac{\sqrt{3}}{2}x' - \frac{1}{2}y'\right)\left(\frac{1}{2}x' + \frac{\sqrt{3}}{2}y'\right) + 11\left(\frac{1}{2}x' + \frac{\sqrt{3}}{2}y'\right)^2 - 16 = 0$$

$$\frac{3}{4}x'^2 - \frac{\sqrt{3}}{2}x'y' + \frac{1}{4}y'^2 - \frac{15}{2}x'^2 - 5\sqrt{3}x'y' + \frac{15}{2}y'^2 + \frac{11}{4}x'^2 + \frac{11\sqrt{3}}{2}x'y' + \frac{33}{4}y'^2 = 16$$

$$-4x'^2 + 16y' = 16$$

$$\frac{y'^2}{1} - \frac{x'^2}{4} = 1$$

Rotating 45°, graph the equation:

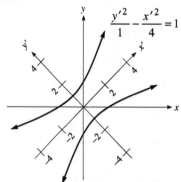

$$\frac{y'^2}{1} - \frac{x'^2}{4} = 1$$

43. Finding the rotation angle:

$$\cot 2\phi = \frac{23 - 5}{18\sqrt{3}} = \frac{1}{\sqrt{3}}$$

$$2\phi = 60°$$

$$\phi = 30°$$

The rotation equations are:

$$x = x'\cos 30° - y'\sin 30° = \frac{\sqrt{3}}{2}x' - \frac{1}{2}y'$$

$$y = x'\sin 30° + y'\cos 30° = \frac{1}{2}x' + \frac{\sqrt{3}}{2}y'$$

Now converting the equation:

$$23x^2 + 18\sqrt{3}xy + 5y^2 - 256 = 0$$

$$23\left(\frac{\sqrt{3}}{2}x' - \frac{1}{2}y'\right)^2 + 18\sqrt{3}\left(\frac{\sqrt{3}}{2}x' - \frac{1}{2}y'\right)\left(\frac{1}{2}x' + \frac{\sqrt{3}}{2}y'\right) + 5\left(\frac{1}{2}x' + \frac{\sqrt{3}}{2}y'\right)^2 = 256$$

$$\tfrac{69}{4}x'^2 - \frac{23\sqrt{3}}{2}x'y' + \tfrac{23}{4}y'^2 + \tfrac{27}{2}x'^2 + 9\sqrt{3}x'y' - \tfrac{27}{2}y'^2 + \tfrac{5}{4}x'^2 + \frac{5\sqrt{3}}{2}x'y' + \tfrac{15}{4}y'^2 = 256$$

$$69x'^2 - 46\sqrt{3}x'y' + 23y'^2 + 54x'^2 + 36\sqrt{3}x'y' - 54y'^2 + 5x'^2 + 10\sqrt{3}x'y' + 15y'^2 = 1024$$

$$128x'^2 - 16y'^2 = 1024$$

$$\frac{x'^2}{8} - \frac{y'^2}{64} = 1$$

Rotating 30°, graph the equation:

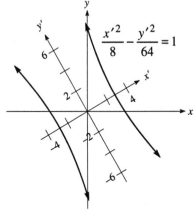

45. Taking an alternate approach:

$$x^2 + 2xy + y^2 = 1$$

$$(x + y)^2 = 1$$

$$x + y = 1 \qquad \text{or} \qquad x + y = -1$$

$$y = -x + 1 \qquad\qquad\qquad y = -x - 1$$

The graph consists of two lines:

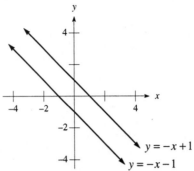

$y = -x + 1$

$y = -x - 1$

47 .   Finding the rotation angle:

$$\cot 2\phi = \frac{31 - 21}{10\sqrt{3}} = \frac{1}{\sqrt{3}}$$
$$2\phi = 60°$$
$$\phi = 30°$$

The rotation equations are:

$$x = x' \cos 30° - y' \sin 30° = \frac{\sqrt{3}}{2} x' - \frac{1}{2} y'$$

$$y = x' \sin 30° + y' \cos 30° = \frac{1}{2} x' + \frac{\sqrt{3}}{2} y'$$

Now converting the equation:

$$31x^2 + 10\sqrt{3}xy + 21y^2 + 48x - 48\sqrt{3}y = 0$$

$$\tfrac{93}{4} x'^2 - \tfrac{31\sqrt{3}}{2} x'y' + \tfrac{31}{4} y'^2 + \tfrac{15}{2} x'^2 + 5\sqrt{3}x'y' - \tfrac{15}{2} y'^2 + \tfrac{21}{4} x'^2 + \tfrac{21\sqrt{3}}{2} x'y' + \tfrac{63}{4} y'^2 - 96y' = 0$$

$$144x'^2 + 64y'^2 - 384y' = 0$$

$$144x'^2 + 64\left(y'^2 - 6y' + 9\right) = 576$$

$$144x'^2 + 64(y' - 3)^2 = 576$$

$$\frac{x'^2}{4} + \frac{(y' - 3)^2}{9} = 1$$

Rotating 30°, graph the equation:

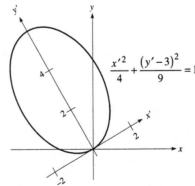

$$\frac{x'^2}{4} + \frac{(y' - 3)^2}{9} = 1$$

49 .   Since $B^2 - 4AC = (-7)^2 - 4(5)(3) = -11 < 0$, the graph is an ellipse (or its degenerate form).

51. Since $B^2 - 4AC = (-7)^2 - 4(2)(0) = 49 > 0$, the graph is a hyperbola (or its degenerate form).

53. We will use Cramer's rule. First find the required determinants:

$$D = \begin{vmatrix} \cos\phi & \sin\phi \\ -\sin\phi & \cos\phi \end{vmatrix} = \cos^2\phi + \sin^2\phi = 1$$

$$D_x = \begin{vmatrix} x' & \sin\phi \\ y' & \cos\phi \end{vmatrix} = x'\cos\phi - y'\sin\phi$$

$$D_y = \begin{vmatrix} \cos\phi & x' \\ -\sin\phi & y' \end{vmatrix} = y'\cos\phi + x'\sin\phi$$

Using Cramer's rule:

$$x = \frac{D_x}{D} = x'\cos\phi - y'\sin\phi \qquad\qquad y = \frac{D_y}{D} = x'\sin\phi + y'\cos\phi$$

55. The quadratic formula directly results in the answer.

57. Graphing the equation:

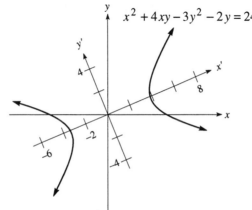

$$x^2 + 4xy - 3y^2 - 2y = 24$$

10.7 Nonlinear Systems of Equations and Inequalities

1. Solving the second equation for y yields $y = 2 - x$. Substituting:

$$x^2 + (2-x)^2 = 10$$
$$x^2 + 4 - 4x + x^2 = 10$$
$$2x^2 - 4x - 6 = 0$$
$$x^2 - 2x - 3 = 0$$
$$(x-3)(x+1) = 0$$
$$x = 3, -1$$
$$y = -1, 3$$

The solutions are $(3,-1)$ and $(-1,3)$.

3. Solving the second equation for y yields $y = 2x - 6$. Substituting:

$$4x^2 + (2x - 6)^2 = 36$$
$$4x^2 + 4x^2 - 24x + 36 = 36$$
$$8x^2 - 24x = 0$$
$$8x(x - 3) = 0$$
$$x = 0, 3$$
$$y = -6, 0$$

The solutions are $(0,-6)$ and $(3,0)$.

5. Solving the second equation for y yields $y = 3x - 1$. Substituting:

$$x^2 - (3x - 1)^2 = 16$$
$$x^2 - 9x^2 + 6x - 1 = 16$$
$$-8x^2 + 6x - 17 = 0$$
$$8x^2 - 6x + 17 = 0$$

Using the quadratic formula: $x = \dfrac{6 \pm \sqrt{36 - 544}}{16} = \dfrac{6 \pm \sqrt{-508}}{16}$

There are no real solutions.

7. Multiplying the first equation by -1:

$$-x^2 + y^2 = -8$$
$$x^2 + 2y^2 = 11$$

Adding yields:

$$3y^2 = 3$$
$$y^2 = 1$$
$$y = \pm 1$$

Substituting:

$$x^2 - 1 = 8$$
$$x^2 = 9$$
$$x = \pm 3$$

The solutions are $(-3,1)$, $(-3,-1)$, $(3,1)$, and $(3,-1)$.

9. Multiplying the second equation by -1:

$$x^2 + y^2 = 1$$
$$-\dfrac{x^2}{4} - y^2 = -1$$

Adding yields:

$$\dfrac{3x^2}{4} = 0$$
$$x = 0$$

Substituting:

$$0^2 + y^2 = 1$$
$$y = \pm 1$$

The solutions are $(0,1)$ and $(0,-1)$.

11. Solving the second equation for y yields $y = x - 2$. Substituting into the first equation:

$$x - 2 = x^2 - 2x - 4$$
$$x^2 - 3x - 2 = 0$$

Using the quadratic formula: $x = \dfrac{3 \pm \sqrt{9+8}}{2} = \dfrac{3 \pm \sqrt{17}}{2}$

Substituting $x = \dfrac{3 - \sqrt{17}}{2}$: $y = \dfrac{3 - \sqrt{17}}{2} - 2 = \dfrac{-1 - \sqrt{17}}{2}$

Substituting $x = \dfrac{3 + \sqrt{17}}{2}$: $y = \dfrac{3 + \sqrt{17}}{2} - 2 = \dfrac{-1 + \sqrt{17}}{2}$

The solutions are $\left(\dfrac{3 - \sqrt{17}}{2}, \dfrac{-1 - \sqrt{17}}{2} \right)$ and $\left(\dfrac{3 + \sqrt{17}}{2}, \dfrac{-1 + \sqrt{17}}{2} \right)$.

13. Substituting into the second equation:

$$2x + \sqrt{x} = 10$$
$$\sqrt{x} = 10 - 2x$$
$$x = 100 - 40x + 4x^2$$
$$4x^2 - 41x + 100 = 0$$
$$(4x - 25)(x - 4) = 0$$
$$x = \tfrac{25}{4}, 4$$

If $x = \tfrac{25}{4}$, $y = \tfrac{5}{2}$, now checking: $2\left(\tfrac{25}{4}\right) + \tfrac{5}{2} = \tfrac{25}{2} + \tfrac{5}{2} = 15 \neq 10$ (doesn't check)

If x = 4, y = 2, now checking: $2(4) + 2 = 8 + 2 = 10$

The solution is (4,2).

15. Substituting into the first equation:

$$(x - 2)^2 + \left(\sqrt{3 - x}\right)^2 = 13$$
$$x^2 - 4x + 4 + 3 - x = 13$$
$$x^2 - 5x - 6 = 0$$
$$(x - 6)(x + 1) = 0$$
$$x = -1, 6 \quad \text{(not possible)}$$
$$y = 2$$

The solution is (–1,2).

17. Substituting for $(x - 3)^2$:

$$y + y^2 = 12$$
$$y^2 + y - 12 = 0$$
$$(y + 4)(y - 3) = 0$$
$$y = 3, -4 \quad \text{(impossible)}$$

When $y = 3$:

$$(x - 3)^2 = 3$$
$$x - 3 = \pm\sqrt{3}$$
$$x = 3 \pm \sqrt{3}$$

The solutions are $\left(3 - \sqrt{3}, 3\right)$ and $\left(3 + \sqrt{3}, 3\right)$.

19.   Solving the second equation for x yields $x = 7 - 3y$. Substituting into the first equation:

$$(7 - 3y)y = 2$$
$$7y - 3y^2 = 2$$
$$3y^2 - 7y + 2 = 0$$
$$(3y - 1)(y - 2) = 0$$
$$y = \tfrac{1}{3}, 2$$
$$x = 6, 1$$

The solutions are $\left(6, \tfrac{1}{3}\right)$ and $(1,2)$.

21.   Substituting:

$$\log_2(x + 3) - 1 = \log_2(x - 1)$$
$$\log_2(x + 3) - \log_2(x - 1) = 1$$
$$\log_2 \frac{x + 3}{x - 1} = 1$$
$$\frac{x + 3}{x - 1} = 2$$
$$x + 3 = 2x - 2$$
$$x = 5$$
$$y = \log_2 4 = 2$$

The solution is $(5,2)$.

23.   Multiplying the first equation by -1:

$$-x^2 - y^2 = -1$$
$$x^2 + (y - 1)^2 = 1$$

Adding yields:

$$(y - 1)^2 - y^2 = 0$$
$$y^2 - 2y + 1 - y^2 = 0$$
$$-2y + 1 = 0$$
$$-2y = -1$$
$$y = \tfrac{1}{2}$$

Substituting $y = \tfrac{1}{2}$:

$$x^2 + \tfrac{1}{4} = 1$$
$$x^2 = \tfrac{3}{4}$$
$$x = \pm \frac{\sqrt{3}}{2}$$

The solutions are $\left(-\dfrac{\sqrt{3}}{2},\dfrac{1}{2}\right)$ and $\left(\dfrac{\sqrt{3}}{2},\dfrac{1}{2}\right)$. Sketching the graphs:

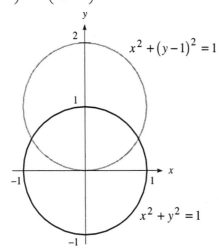

25. Substituting:
$$2x^2 = x^2 - 1$$
$$x^2 = -1$$
There are no real solutions. Sketching the graphs:

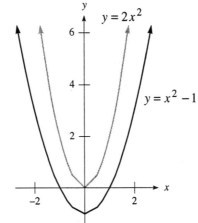

27. Solving the second equation for y yields $y = x - 3$. Substituting into the first equation:
$$x^2 + (x-3)^2 = 9$$
$$x^2 + x^2 - 6x + 9 = 9$$
$$2x^2 - 6x = 0$$
$$2x(x-3) = 0$$
$$x = 0, 3$$
$$y = -3, 0$$

The solutions are $(0,-3)$ and $(3,0)$. Sketching the graphs:

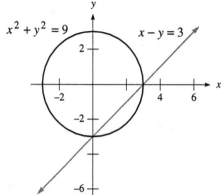

29.   Substituting into the second equation:

$$2x + 3(x^2 - 6x) + 13 = 0$$
$$2x + 3x^2 - 18x + 13 = 0$$
$$3x^2 - 16x + 13 = 0$$
$$(3x - 13)(x - 1) = 0$$
$$x = 1, \tfrac{13}{3}$$

When $x = 1$: $y = (1)^2 - 6(1) = -5$

When $x = \tfrac{13}{3}$: $y = \left(\tfrac{13}{3}\right)^2 - 6\left(\tfrac{13}{3}\right) = \tfrac{169}{9} - 26 = -\tfrac{65}{9}$

The solutions are $(1,-5)$ and $\left(\tfrac{13}{3}, -\tfrac{65}{9}\right)$. Sketching the graphs:

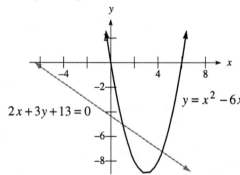

31.   Solving the second equation for x^2 yields $x^2 = 2y + 1$. Substituting into the first equation:

$$2y + 1 + y^2 = 9$$
$$y^2 + 2y - 8 = 0$$
$$(y + 4)(y - 2) = 0$$
$$y = 2, -4 \text{ (impossible)}$$

When $y = 2$:
$$x^2 = 2(2) + 1$$
$$x^2 = 5$$
$$x = \pm\sqrt{5}$$

The solutions are $\left(-\sqrt{5},2\right)$ and $\left(\sqrt{5},2\right)$. Sketching the graphs:

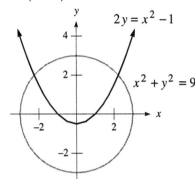

33. Solving the second equation for y yields $y = 2 - x$. Substituting into the first equation:

$$x^2 + (2-x)^2 = 16$$
$$x^2 + 4 - 4x + x^2 = 16$$
$$2x^2 - 4x - 12 = 0$$
$$x^2 - 2x - 6 = 0$$

Using the quadratic formula: $x = \dfrac{2 \pm \sqrt{4 - 4(-6)}}{2} = \dfrac{2 \pm \sqrt{28}}{2} = \dfrac{2 \pm 2\sqrt{7}}{2} = 1 \pm \sqrt{7}$

When $x = 1 - \sqrt{7}$: $y = 2 - \left(1 - \sqrt{7}\right) = 1 + \sqrt{7}$

When $x = 1 + \sqrt{7}$: $y = 2 - \left(1 + \sqrt{7}\right) = 1 - \sqrt{7}$

The solutions are $\left(1 - \sqrt{7}, 1 + \sqrt{7}\right)$ and $\left(1 + \sqrt{7}, 1 - \sqrt{7}\right)$. Sketching the graphs:

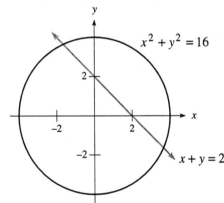

35. Using the hint, the system is:

$$8u - 4v = 3$$
$$6u + 8v = 5$$

Multiplying the first equation by 2:

$$16u - 8v = 6$$
$$6u + 8v = 5$$

Adding yields $22u = 11$, so $u = \frac{1}{2}$. Substituting into the second equation:

$$6\left(\tfrac{1}{2}\right) + 8v = 5$$
$$3 + 8v = 5$$
$$8v = 2$$
$$v = \tfrac{1}{4}$$

Since $u = \dfrac{1}{x^2}$ and $v = \dfrac{1}{y^2}$, $x^2 = 2$ and $y^2 = 4$, so $x = \pm\sqrt{2}$ and $y = \pm 2$. The solutions are $\left(-\sqrt{2}, 2\right)$, $\left(\sqrt{2}, 2\right)$, $\left(-\sqrt{2}, -2\right)$, and $\left(\sqrt{2}, -2\right)$.

37. Substituting into the second equation:

$$y = y^2 - 12$$
$$y^2 - y - 12 = 0$$
$$(y - 4)(y + 3) = 0$$
$$y = 4 \quad (y = -3 \text{ is impossible})$$
$$5^x = 4$$
$$x = \log_5 4 \approx 0.9$$

The solution is $\left(\log_5 4, 4\right) \approx (0.9, 4)$. The two curves intersect at one point.

39. Let w and l represent the width and length, respectively. We have the system of equations:

$$wl = 35$$
$$2w + 2l = 27$$

Solving the first equation for l yields $l = \dfrac{35}{w}$. Substituting into the second equation:

$$2w + 2\left(\frac{35}{w}\right) = 27$$
$$2w^2 + 70 = 27w$$
$$2w^2 - 27w + 70 = 0$$
$$(2w - 7)(w - 10) = 0$$
$$w = \tfrac{7}{2}, 10$$

If $w = \tfrac{7}{2}$, $l = \dfrac{35}{7/2} = 10$ and if $w = 10$, $l = \dfrac{35}{10} = \tfrac{7}{2}$. The dimensions are 3.5 cm by 10 cm.

41. Let b and h represent the base and height, respectively. We have the system of equations:

$$b^2 + h^2 = 25^2$$
$$\tfrac{1}{2}bh = 84$$

Solving the second equation for h yields $h = \dfrac{168}{b}$. Substituting into the first equation:

$$b^2 + \left(\frac{168}{b}\right)^2 = 25^2$$
$$b^2 + \frac{28224}{b^2} = 625$$
$$b^4 + 28224 = 625b^2$$
$$b^4 - 625b^2 + 28224 = 0$$
$$\left(b^2 - 49\right)\left(b^2 - 576\right) = 0$$
$$(b+7)(b-7)(b+24)(b-24) = 0$$
$$b = 7, 24 \qquad \left(b = -7, -24 \text{ are impossible}\right)$$
$$h = 24, 7$$

The dimensions are 7 cm by 24 cm.

43. **a.** Using substitution:

$$x^2 - 2x - 3 = K$$
$$x^2 - 2x - 3 - K = 0$$

In order for only one solution to occur, the discriminant must equal 0. Finding the discriminant:

$$(-2)^2 - 4(1)(-3 - K) = 0$$
$$4 + 12 + 4K = 0$$
$$16 + 4K = 0$$
$$4K = -16$$
$$K = -4$$

b. For no real solutions to exist, the discriminant must be negative:

$$16 + 4K < 0$$
$$4K < -16$$
$$K < -4$$

45. Multiplying the first equation by -1:

$$-x^2 - y^2 = -16$$
$$4x^2 + y^2 = 64$$

Adding the two equations:

$$3x^2 = 48$$
$$x^2 = 16$$
$$x = \pm 4$$
$$y = 0$$

The points are $(-4, 0)$ and $(4, 0)$.

47. Let x represent the length of fence parallel to the barn and y represent the length of fence perpendicular to the barn. The system of equations is:
$$x + 2y = 180$$
$$xy = 4000$$
Solving the first equation for x yields $x = 180 - 2y$. Substituting:
$$(180 - 2y)y = 4000$$
$$180y - 2y^2 = 4000$$
$$y^2 - 90y + 2000 = 0$$
$$(y - 50)(y - 40) = 0$$
$$y = 40, 50$$
$$x = 100, 80$$
Dimensions of either 40 ft by 100 ft, or 50 ft by 80 ft, will meet the required conditions.

49. Sketching the solution set:

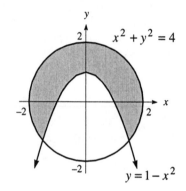

51. Sketching the solution set:

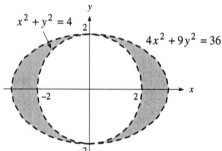

53. Sketching the solution set:

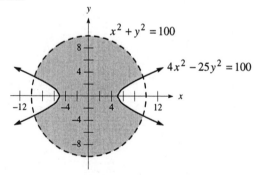

55. Using substitution:
$$x^2 = 3x^2 - 2$$
$$-2x^2 = -2$$
$$x^2 = 1$$
$$x = \pm 1$$
$$y = 1$$
The solutions are $(-1,1)$ and $(1,1)$. These solutions are verified using a graphing calculator.

57. Solving the second equation for x yields $x = y + 3$. Substituting into the first equation:
$$(y+3)^2 + y^2 = 16$$
$$y^2 + 6y + 9 + y^2 = 16$$
$$2y^2 + 6y - 7 = 0$$
Using the quadratic formula: $y = \dfrac{-6 \pm \sqrt{6^2 - 4(2)(-7)}}{2(2)} = \dfrac{-6 \pm \sqrt{92}}{4} = \dfrac{-6 \pm 2\sqrt{23}}{4} = \dfrac{-3 \pm \sqrt{23}}{2}$

If $y = \dfrac{-3 - \sqrt{23}}{2}$, $x = \dfrac{-3 - \sqrt{23}}{2} + 3 = \dfrac{3 - \sqrt{23}}{2}$. If $y = \dfrac{-3 + \sqrt{23}}{2}$, $x = \dfrac{-3 + \sqrt{23}}{2} + 3 = \dfrac{3 + \sqrt{23}}{2}$.

The solutions are $\left(\dfrac{3 - \sqrt{23}}{2}, \dfrac{-3 - \sqrt{23}}{2} \right) \approx (-0.9, -3.9)$ and $\left(\dfrac{3 + \sqrt{23}}{2}, \dfrac{-3 + \sqrt{23}}{2} \right) \approx (3.9, 0.9)$.

These solutions are verified using a graphing calculator.

59. The maximum degree of such an equation would be four, so there are a maximum of four solutions to this equation. Graphically, if two conic sections are intersected, the maximum number of solutions is four.

Chapter 10 Review Exercises

1. The center is $(3,-4)$ and the radius is $\sqrt{18} = 3\sqrt{2}$.

2. The center is $(0,1)$ and the radius is $\sqrt{12} = 2\sqrt{3}$.

3. Completing the square:
$$x^2 + y^2 - 8x - 8y - 4 = 0$$
$$\left(x^2 - 8x + 16\right) + \left(y^2 - 8y + 16\right) = 4 + 16 + 16$$
$$(x-4)^2 + (y-4)^2 = 36$$
The center is $(4,4)$ and the radius is $\sqrt{36} = 6$.

4. Completing the square:
$$x^2 + y^2 - 10x + 6y + 15 = 0$$
$$\left(x^2 - 10x + 25\right) + \left(y^2 + 6y + 9\right) = -15 + 25 + 9$$
$$(x-5)^2 + (y+3)^2 = 19$$
The center is $(5,-3)$ and the radius is $\sqrt{19}$.

5.   Since the slope from $P(1,-3)$ to the center is $\dfrac{-3}{1} = -3$, the tangent slope must be $\frac{1}{3}$. Using the
point-slope formula:
$$y + 3 = \tfrac{1}{3}(x-1)$$
$$y + 3 = \tfrac{1}{3}x - \tfrac{1}{3}$$
$$y = \tfrac{1}{3}x - \tfrac{10}{3}$$

6.   The slope from $P(2,5)$ to $(0,5)$ is 0, so the tangent line must be vertical. Its equation is thus $x = 2$.

7.   Converting to standard form:
$$8x - y^2 = 0$$
$$y^2 = 8x$$
Since $4p = 8$, $p = 2$. The focus is $(2,0)$, the directrix is $x = -2$, and the axis is $y = 0$. Graphing the
parabola:

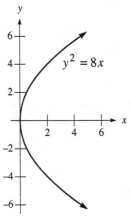

8.   Converting to standard form:
$$y + 5x^2 = 0$$
$$5x^2 = -y$$
$$x^2 = -\tfrac{1}{5}y$$
Since $4p = -\tfrac{1}{5}$, $p = -\tfrac{1}{20}$. The focus is $\left(0, -\tfrac{1}{20}\right)$, the directrix is $y = \tfrac{1}{20}$, and the axis is $x = 0$.
Graphing the parabola:

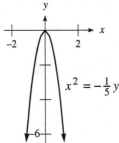

9. Since $4p = 4$, $p = 1$. The vertex is (4,0), the focus is (4,1), the directrix is $y = -1$, and the axis is $x = 4$. Graphing the parabola:

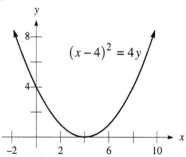

10. Since $4p = -16$, $p = -4$. The vertex is (0,2), the focus is (–4,2), the directrix is $x = 4$, and the axis is $y = 2$. Graphing the parabola:

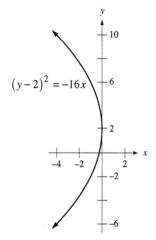

11. First complete the square:
$$y^2 - 10y + 21 = 4x$$
$$y^2 - 10y + 25 = 4x - 21 + 25$$
$$(y-5)^2 = 4(x+1)$$

Since $4p = 4$, $p = 1$. The vertex is (–1,5), the focus is (0,5), the directrix is $x = -2$, and the axis is $y = 5$. Graphing the parabola:

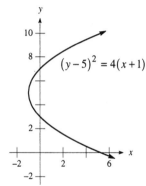

12.   First complete the square:

$$x^2 - 4x - 2 = 6y$$
$$x^2 - 4x + 4 = 6y + 2 + 4$$
$$(x-2)^2 = 6(y+1)$$

Since $4p = 6$, $p = \frac{3}{2}$. The vertex is $(2,-1)$, the focus is $\left(2,\frac{1}{2}\right)$, the directrix is $y = -\frac{5}{2}$, and the axis is $x = 2$. Graphing the parabola:

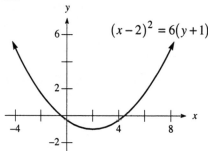

13.   Since $p = -3$, the equation is $y^2 = -12x$.

14.   Since $p = -3$, the equation is $(x-1)^2 = -12(y+1)$.

15.   Since $y = \frac{1}{8}x^2$, $x^2 = 8y$. So $4p = 8$, $p = 2$. The light source should be placed 2 feet directly above the vertex. To find the depth: $y = \frac{1}{8}(\pm 3)^2 = \frac{9}{8}$

The mirror is $\frac{9}{8} = 1\frac{1}{8}$ feet deep.

16.   Call $(0,0)$ the vertex of the parabola $x^2 = 4py$. Since $(\pm 15, 60)$ are points on the parabola:

$$(\pm 15)^2 = 4p(60)$$
$$225 = 240p$$
$$p = \frac{15}{16}$$

So the equation of the parabola is $x^2 = \frac{15}{4}y$. Finding x when $y = 50$:

$$x^2 = \frac{15}{4}(50) = \frac{375}{2}$$
$$x = \frac{5\sqrt{15}}{\sqrt{2}} = \frac{5\sqrt{30}}{2}$$

The width is $2\left(\frac{5\sqrt{30}}{2}\right) = 5\sqrt{30} \approx 27.39$ feet.

17.   Since $a = 6$ and $b = 3$, $c = \sqrt{36-9} = \sqrt{27} = 3\sqrt{3}$. The vertices are $(0,-6)$ and $(0,6)$, and the foci are $\left(0,-3\sqrt{3}\right)$ and $\left(0,3\sqrt{3}\right)$. Graphing the ellipse:

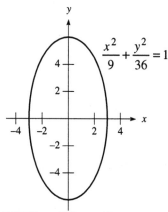

$$\frac{x^2}{9} + \frac{y^2}{36} = 1$$

18.   Since $a = 9$ and $b = \sqrt{6}$, $c = \sqrt{81-6} = \sqrt{75} = 5\sqrt{3}$. The vertices are $(-9,0)$ and $(9,0)$, and the foci are $\left(-5\sqrt{3},0\right)$ and $\left(5\sqrt{3},0\right)$. Graphing the ellipse:

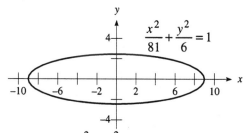

$$\frac{x^2}{81} + \frac{y^2}{6} = 1$$

19.   Dividing by 48, the standard form is $\dfrac{x^2}{2} + \dfrac{y^2}{24} = 1$. Since $a = \sqrt{24} = 2\sqrt{6}$ and $b = \sqrt{2}$, $c = \sqrt{24-2} = \sqrt{22}$. The vertices are $\left(0,-2\sqrt{6}\right)$ and $\left(0,2\sqrt{6}\right)$, and the foci are $\left(0,-\sqrt{22}\right)$ and $\left(0,\sqrt{22}\right)$. Graphing the ellipse:

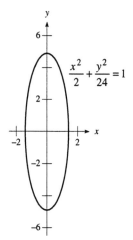

$$\frac{x^2}{2} + \frac{y^2}{24} = 1$$

20.   Dividing by 40, the standard form is $\dfrac{y^2}{2}+\dfrac{x^2}{4}=1$. Since $a=2$ and $b=\sqrt{2}$, $c=\sqrt{4-2}=\sqrt{2}$. The vertices are $(-2,0)$ and $(2,0)$, and the foci are $\left(-\sqrt{2},0\right)$ and $\left(\sqrt{2},0\right)$. Graphing the ellipse:

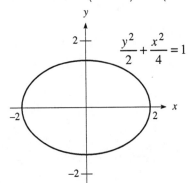

21.   Since $a=\sqrt{6}$ and $b=2$, $c=\sqrt{6-4}=\sqrt{2}$. The center is $(2,-4)$, the vertices are $\left(2-\sqrt{6},-4\right)$ and $\left(2+\sqrt{6},-4\right)$, and the foci are $\left(2-\sqrt{2},-4\right)$ and $\left(2+\sqrt{2},-4\right)$. Graphing the ellipse:

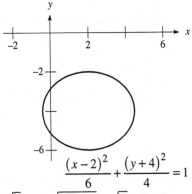

22.   Since $a=4$ and $b=\sqrt{12}=2\sqrt{3}$, $c=\sqrt{16-12}=\sqrt{4}=2$. The center is $(0,-5)$, the vertices are $(-4,-5)$ and $(4,-5)$, and the foci are $(-2,-5)$ and $(2,-5)$. Graphing the ellipse:

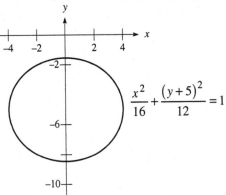

23.   Completing the square:
$$x^2 + 4y^2 - 6x + 8y = 3$$
$$\left(x^2 - 6x + 9\right) + 4\left(y^2 + 2y + 1\right) = 3 + 9 + 4$$
$$(x-3)^2 + 4(y+1)^2 = 16$$
$$\frac{(x-3)^2}{16} + \frac{(y+1)^2}{4} = 1$$

Since $a = 4$ and $b = 2$, $c = \sqrt{16-4} = \sqrt{12} = 2\sqrt{3}$. The center is $(3,-1)$, the vertices are $(-1,-1)$ and $(7,-1)$, and the foci are $\left(3 - 2\sqrt{3}, -1\right)$ and $\left(3 + 2\sqrt{3}, -1\right)$. Graphing the ellipse:

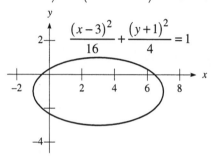

24.   Completing the square:
$$6x^2 + 5y^2 + 12x - 20y - 4 = 0$$
$$6\left(x^2 + 2x + 1\right) + 5\left(y^2 - 4y + 4\right) = 4 + 6 + 20$$
$$6(x+1)^2 + 5(y-2)^2 = 30$$
$$\frac{(x+1)^2}{5} + \frac{(y-2)^2}{6} = 1$$

Since $a = \sqrt{6}$ and $b = \sqrt{5}$, $c = \sqrt{6-5} = 1$. The center is $(-1,2)$, the vertices are $\left(-1, 2 - \sqrt{6}\right)$ and $\left(-1, 2 + \sqrt{6}\right)$, and the foci are $(-1,1)$ and $(-1,3)$. Graphing the ellipse:

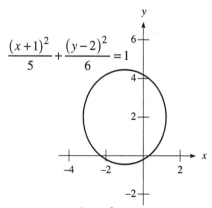

25.   Here $a = 4$ and $b = 3$, so the equation is $\dfrac{x^2}{9} + \dfrac{y^2}{16} = 1$.

26. The center is (5,5), so $a = 5$ and $c = 3$. Thus $b = \sqrt{25-9} = \sqrt{16} = 4$, so the equation is
$$\frac{(x-5)^2}{25} + \frac{(y-5)^2}{16} = 1.$$

27. Since $a = 7$ and $b = 2$, $c = \sqrt{49+4} = \sqrt{53}$. The vertices are (–7,0) and (7,0), the foci are $\left(-\sqrt{53},0\right)$ and $\left(\sqrt{53},0\right)$, and the asymptotes are $y = \pm\frac{2}{7}x$. Graphing the hyperbola:

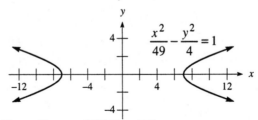

28. Since $a = 9$ and $b = \sqrt{8} = 2\sqrt{2}$, $c = \sqrt{81+8} = \sqrt{89}$. The vertices are (0,–9) and (0,9), the foci are $\left(0,-\sqrt{89}\right)$ and $\left(0,\sqrt{89}\right)$, and the asymptotes are $y = \pm\frac{9}{2\sqrt{2}}x = \pm\frac{9\sqrt{2}}{4}x$. Graphing the hyperbola:

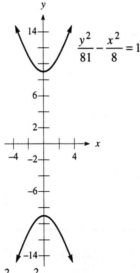

29. Dividing by 24, the equation is $\frac{x^2}{24} - \frac{y^2}{3} = 1$. Since $a = \sqrt{24} = 2\sqrt{6}$ and $b = \sqrt{3}$, $c = \sqrt{24+3} = \sqrt{27} = 3\sqrt{3}$. The vertices are $\left(-2\sqrt{6},0\right)$ and $\left(2\sqrt{6},0\right)$, the foci are $\left(-3\sqrt{3},0\right)$ and $\left(3\sqrt{3},0\right)$, and the asymptotes are $y = \pm\frac{\sqrt{3}}{2\sqrt{6}}x = \pm\frac{\sqrt{2}}{4}x$. Graphing the hyperbola:

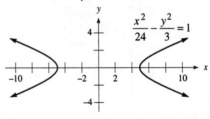

30. Dividing by 40, the equation is $\dfrac{y^2}{4} - \dfrac{x^2}{8} = 1$. Since $a = 2$ and $b = \sqrt{8} = 2\sqrt{2}$,

$c = \sqrt{4+8} = \sqrt{12} = 2\sqrt{3}$. The vertices are $(0,-2)$ and $(0,2)$, the foci are $\left(0,-2\sqrt{3}\right)$ and $\left(0,2\sqrt{3}\right)$,

and the asymptotes are $y = \pm\dfrac{2}{2\sqrt{2}}x = \pm\dfrac{\sqrt{2}}{2}x$. Graphing the hyperbola:

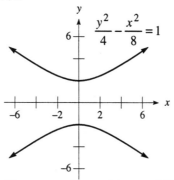

31. The center is $(1,2)$. Since $a = \sqrt{2}$ and $b = \sqrt{8} = 2\sqrt{2}$, $c = \sqrt{2+8} = \sqrt{10}$. The vertices are

$\left(1, 2 - \sqrt{2}\right)$ and $\left(1, 2 + \sqrt{2}\right)$, and the foci are $\left(1, 2 - \sqrt{10}\right)$ and $\left(1, 2 + \sqrt{10}\right)$. The slopes of the

asymptotes are $m = \pm\dfrac{\sqrt{2}}{2\sqrt{2}} = \pm\frac{1}{2}$, so the asymptotes are:

$$y - 2 = -\tfrac{1}{2}(x-1) \qquad\qquad y - 2 = \tfrac{1}{2}(x-1)$$
$$y - 2 = -\tfrac{1}{2}x + \tfrac{1}{2} \qquad\qquad y - 2 = \tfrac{1}{2}x - \tfrac{1}{2}$$
$$y = -\tfrac{1}{2}x + \tfrac{5}{2} \qquad\qquad y = \tfrac{1}{2}x + \tfrac{3}{2}$$

Graphing the hyperbola:

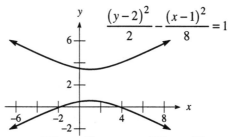

32. The center is $(-3,0)$. Since $a = \sqrt{12} = 2\sqrt{3}$ and $b = \sqrt{24} = 2\sqrt{6}$, $c = \sqrt{12+24} = \sqrt{36} = 6$. The

vertices are $\left(-3 - 2\sqrt{3}, 0\right)$ and $\left(-3 + 2\sqrt{3}, 0\right)$, and the foci are $(-9,0)$ and $(3,0)$. The slopes of the

asymptotes are $m = \pm\dfrac{2\sqrt{6}}{2\sqrt{3}} = \pm\sqrt{2}$, so the asymptotes are:

$$y - 0 = -\sqrt{2}(x+3) \qquad\qquad y - 0 = \sqrt{2}(x+3)$$
$$y = -\sqrt{2}x - 3\sqrt{2} \qquad\qquad y = \sqrt{2}x + 3\sqrt{2}$$

Graphing the hyperbola:

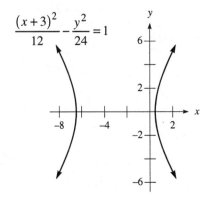

$$\frac{(x+3)^2}{12} - \frac{y^2}{24} = 1$$

33. Completing the square:
$$x^2 - 4y^2 - 6x - 8y = 11$$
$$\left(x^2 - 6x + 9\right) - 4\left(y^2 + 2y + 1\right) = 11 + 9 - 4$$
$$(x-3)^2 - 4(y+1)^2 = 16$$
$$\frac{(x-3)^2}{16} + \frac{(y+1)^2}{4} = 1$$

The center is $(3,-1)$. Since $a = 4$ and $b = 2$, $c = \sqrt{16+4} = \sqrt{20} = 2\sqrt{5}$. The vertices are $(-1,-1)$ and $(7,-1)$, and the foci are $\left(3 - 2\sqrt{5}, -1\right)$ and $\left(3 + 2\sqrt{5}, -1\right)$. The slopes of the asymptotes are $m = \pm\frac{2}{4} = \pm\frac{1}{2}$, so the asymptotes are:

$$y + 1 = -\tfrac{1}{2}(x-3) \qquad\qquad y + 1 = \tfrac{1}{2}(x-3)$$
$$y + 1 = -\tfrac{1}{2}x + \tfrac{3}{2} \qquad\qquad y + 1 = \tfrac{1}{2}x - \tfrac{3}{2}$$
$$y = -\tfrac{1}{2}x + \tfrac{1}{2} \qquad\qquad y = \tfrac{1}{2}x - \tfrac{5}{2}$$

Graphing the hyperbola:

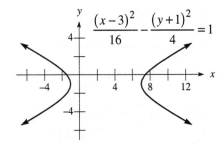

$$\frac{(x-3)^2}{16} - \frac{(y+1)^2}{4} = 1$$

34. Completing the square:
$$5y^2 - 6x^2 - 20y - 12x - 16 = 0$$
$$5\left(y^2 - 4y + 4\right) - 6\left(x^2 + 2x + 1\right) = 16 + 20 - 6$$
$$5(y-2)^2 - 6(x+1)^2 = 30$$
$$\frac{(y-2)^2}{6} - \frac{(x+1)^2}{5} = 1$$

The center is $(-1,2)$. Since $a = \sqrt{6}$ and $b = \sqrt{5}$, $c = \sqrt{6+5} = \sqrt{11}$. The vertices are $\left(-1, 2 - \sqrt{6}\right)$ and $\left(-1, 2 + \sqrt{6}\right)$, and the foci are $\left(-1, 2 - \sqrt{11}\right)$ and $\left(-1, 2 + \sqrt{11}\right)$. The slopes of the asymptotes are $m = \pm\dfrac{\sqrt{6}}{\sqrt{5}} = \pm\dfrac{\sqrt{30}}{5}$, so the asymptotes are:

$$y - 2 = -\frac{\sqrt{30}}{5}(x+1)$$
$$y - 2 = -\frac{\sqrt{30}}{5}x - \frac{\sqrt{30}}{5}$$
$$y = -\frac{\sqrt{30}}{5}x + \frac{10 - \sqrt{30}}{5}$$

$$y - 2 = \frac{\sqrt{30}}{5}(x+1)$$
$$y - 2 = \frac{\sqrt{30}}{5}x + \frac{\sqrt{30}}{5}$$
$$y = \frac{\sqrt{30}}{5}x + \frac{10 + \sqrt{30}}{5}$$

Graphing the hyperbola:

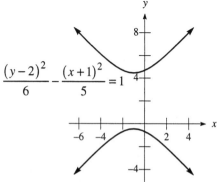

$$\frac{(y-2)^2}{6} - \frac{(x+1)^2}{5} = 1$$

35. Since A and C are the same sign but $A \neq C$, the graph will be an ellipse (or its degenerate form).

36. Since $A = 0$, the graph will be a parabola (or its degenerate form).

37. Since A and C are different signs, the graph will be a hyperbola (or its degenerate form).

38. Since $A = C$, the graph will be a circle (or its degenerate form).

39. The graph is a hyperbola:

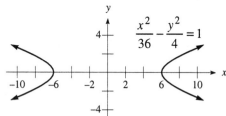

$$\frac{x^2}{36} - \frac{y^2}{4} = 1$$

40. The graph is an ellipse:

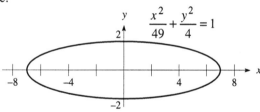

$$\frac{x^2}{49} + \frac{y^2}{4} = 1$$

41.   The graph is a line:

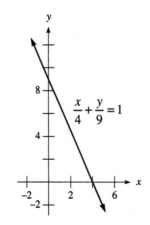

$$\frac{x}{4}+\frac{y}{9}=1$$

42.   The graph is an ellipse:

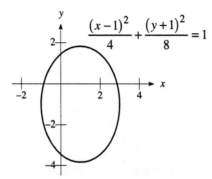

$$\frac{(x-1)^2}{4}+\frac{(y+1)^2}{8}=1$$

43.   The graph is a circle:

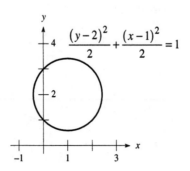

$$\frac{(y-2)^2}{2}+\frac{(x-1)^2}{2}=1$$

44.   The graph is a hyperbola:

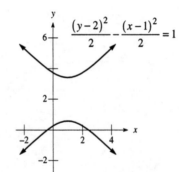

$$\frac{(y-2)^2}{2}-\frac{(x-1)^2}{2}=1$$

45. Since $x^2 = -64y$, the graph is a parabola:

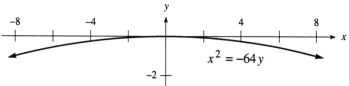

$$x^2 = -64y$$

46. Completing the square:
$$x^2 + 4y^2 - 2x + 24y + 21 = 0$$
$$\left(x^2 - 2x + 1\right) + 4\left(y^2 + 6y + 9\right) = -21 + 1 + 36$$
$$(x-1)^2 + 4(y+3)^2 = 16$$
$$\frac{(x-1)^2}{16} + \frac{(y+3)^2}{4} = 1$$

The graph is an ellipse:

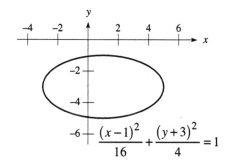

$$\frac{(x-1)^2}{16} + \frac{(y+3)^2}{4} = 1$$

47. Completing the square:
$$x^2 - 8x + 16 = 0$$
$$(x-4)^2 = 0$$
$$x = 4$$

The graph is a line:

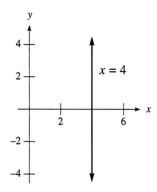

$$x = 4$$

48. Factoring:
$$3y^2 - 3y - 18 = 0$$
$$3\left(y^2 - y - 6\right) = 0$$
$$3(y-3)(y+2) = 0$$
$$y = -2, 3$$

The graph is two lines:

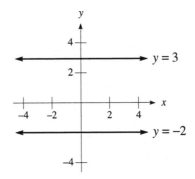

49. Completing the square:
$$x^2 + 2y^2 - 2x - 8y - 7 = 0$$
$$\left(x^2 - 2x + 1\right) + 2\left(y^2 - 4y + 4\right) = 7 + 1 + 8$$
$$(x-1)^2 + 2(y-2)^2 = 16$$
$$\frac{(x-1)^2}{16} + \frac{(y-2)^2}{8} = 1$$

The graph is an ellipse:

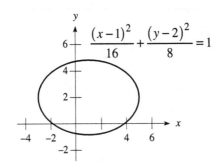

50. Completing the square:
$$x^2 + 4x - 5y + 24 = 0$$
$$x^2 + 4x + 4 = 5y - 24 + 4$$
$$(x+2)^2 = 5(y-4)$$

The graph is a parabola:

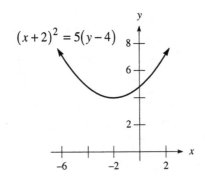

51. Completing the square:
$$x^2 + 25y^2 - 6x - 50y + 34 = 0$$
$$\left(x^2 - 6x + 9\right) + 25\left(y^2 - 2y + 1\right) = -34 + 9 + 25$$
$$(x-3)^2 + 25(y-1)^2 = 0$$

The graph is the point $(3,1)$:

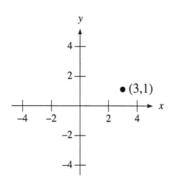

52. Completing the square:
$$3x^2 - y^2 - 6x - 6y - 6 = 0$$
$$3\left(x^2 - 2x + 1\right) - \left(y^2 + 6y + 9\right) = 6 + 3 - 9$$
$$3(x-1)^2 - (y+3)^2 = 0$$
$$(y+3)^2 = 3(x-1)^2$$
$$y + 3 = \pm\sqrt{3}(x-1)$$

The graph is two lines:

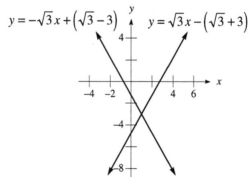

53. Since $h = 2$ and $k = 5$, $x = x' + 2$ and $y = y' + 5$. If $(x',y') = (2',5')$, $(x,y) = (4,10)$.

54. Since $h = 2$ and $k = 5$, $x' = x - 2$ and $y' = y - 5$. If $(x,y) = (3,-4)$, $(x',y') = (1',-9')$.

55. Using the rotation equations:
$$x' = 3\cos 30° + 5\sin 30° = 3\left(\frac{\sqrt{3}}{2}\right) + 5\left(\frac{1}{2}\right) = \frac{3\sqrt{3} + 5}{2}$$
$$y' = -3\sin 30° + 5\cos 30° = -3\left(\frac{1}{2}\right) + 5\left(\frac{\sqrt{3}}{2}\right) = \frac{-3 + 5\sqrt{3}}{2}$$

The point is $(x',y') = \left(\dfrac{3\sqrt{3} + 5}{2}, \dfrac{-3 + 5\sqrt{3}}{2}\right)'$.

56.   Using the rotation equations:

$$x' = -2\cos\frac{\pi}{4} + 3\sin\frac{\pi}{4} = -2\left(\frac{\sqrt{2}}{2}\right) + 3\left(\frac{\sqrt{2}}{2}\right) = \frac{\sqrt{2}}{2}$$

$$y' = -(-2)\sin\frac{\pi}{4} + 3\cos\frac{\pi}{4} = 2\left(\frac{\sqrt{2}}{2}\right) + 3\left(\frac{\sqrt{2}}{2}\right) = \frac{5\sqrt{2}}{2}$$

The point is $(x', y') = \left(\frac{\sqrt{2}}{2}, \frac{5\sqrt{2}}{2}\right)'$.

57.   The rotation equations are:

$$x = x'\cos 30° - y'\sin 30° = \frac{\sqrt{3}}{2}x' - \frac{1}{2}y'$$

$$y = x'\sin 30° + y'\cos 30° = \frac{1}{2}x' + \frac{\sqrt{3}}{2}y'$$

Now converting the equation:

$$x^2 - xy - 1 = 0$$

$$\left(\frac{\sqrt{3}}{2}x' - \frac{1}{2}y'\right)^2 - \left(\frac{\sqrt{3}}{2}x' - \frac{1}{2}y'\right)\left(\frac{1}{2}x' + \frac{\sqrt{3}}{2}y'\right) = 1$$

$$\frac{3}{4}x'^2 - \frac{\sqrt{3}}{2}x'y' + \frac{1}{4}y'^2 - \frac{\sqrt{3}}{4}x'^2 - \frac{1}{2}x'y' + \frac{\sqrt{3}}{4}y'^2 = 1$$

$$3x'^2 - 2\sqrt{3}x'y' + y'^2 - \sqrt{3}x'^2 - 2x'y' + \sqrt{3}y'^2 = 4$$

$$\left(3 - \sqrt{3}\right)x'^2 - \left(2\sqrt{3} + 2\right)x'y' + \left(1 + \sqrt{3}\right)y'^2 = 4$$

58.   The rotation equations are:

$$x' = x\cos 45° + y\sin 45° = \frac{1}{\sqrt{2}}x + \frac{1}{\sqrt{2}}y$$

$$y' = -x\sin 45° + y\cos 45° = -\frac{1}{\sqrt{2}}x + \frac{1}{\sqrt{2}}y$$

Now converting the equation:

$$3y'^2 = -16x'$$

$$3\left(-\frac{1}{\sqrt{2}}x + \frac{1}{\sqrt{2}}y\right)^2 = -16\left(\frac{1}{\sqrt{2}}x + \frac{1}{\sqrt{2}}y\right)$$

$$3\left(\tfrac{1}{2}x^2 - xy + \tfrac{1}{2}y^2\right) = -8\sqrt{2}x - 8\sqrt{2}y$$

$$\tfrac{3}{2}x^2 - 3xy + \tfrac{3}{2}y^2 = -8\sqrt{2}x - 8\sqrt{2}y$$

$$3x^2 - 6xy + 3y^2 = -16\sqrt{2}x - 16\sqrt{2}y$$

$$3x^2 - 6xy + 3y^2 + 16\sqrt{2}x + 16\sqrt{2}y = 0$$

59.   Finding the rotation angle:

$$\cot 2\phi = \frac{0 - 0}{1} = 0$$

$$2\phi = 90°$$

$$\phi = 45°$$

The rotation equations are:

$$x = x'\cos 45° - y'\sin 45° = \frac{1}{\sqrt{2}}x' - \frac{1}{\sqrt{2}}y'$$

$$y = x'\sin 45° + y'\cos 45° = \frac{1}{\sqrt{2}}x' + \frac{1}{\sqrt{2}}y'$$

Now converting the equation:

$$xy = 10$$

$$\left(\frac{1}{\sqrt{2}}x' - \frac{1}{\sqrt{2}}y'\right)\left(\frac{1}{\sqrt{2}}x' + \frac{1}{\sqrt{2}}y'\right) = 10$$

$$\tfrac{1}{2}x'^2 - \tfrac{1}{2}y'^2 = 10$$

$$x'^2 - y'^2 = 20$$

$$\frac{x'^2}{20} - \frac{y'^2}{20} = 1$$

Rotating 45°, graph the equation:

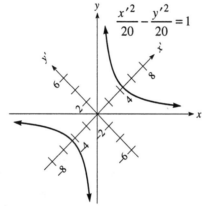

60. Finding the rotation angle:

$$\cot 2\phi = \frac{2-2}{-2} = 0$$

$$2\phi = 90°$$

$$\phi = 45°$$

The rotation equations are:

$$x = x'\cos 45° - y'\sin 45° = \frac{1}{\sqrt{2}}x' - \frac{1}{\sqrt{2}}y'$$

$$y = x'\sin 45° + y'\cos 45° = \frac{1}{\sqrt{2}}x' + \frac{1}{\sqrt{2}}y'$$

Now converting the equation:

$$2x^2 - 2xy + 2y^2 = 9$$

$$2\left(\frac{1}{\sqrt{2}}x' - \frac{1}{\sqrt{2}}y'\right)^2 - 2\left(\frac{1}{\sqrt{2}}x' - \frac{1}{\sqrt{2}}y'\right)\left(\frac{1}{\sqrt{2}}x' + \frac{1}{\sqrt{2}}y'\right) + 2\left(\frac{1}{\sqrt{2}}x' + \frac{1}{\sqrt{2}}y'\right)^2 = 9$$

$$x'^2 - 2x'y' + y'^2 - x'^2 + y'^2 + x'^2 + 2x'y' + y'^2 = 9$$

$$x'^2 + 3y'^2 = 9$$

$$\frac{x'^2}{9} + \frac{y'^2}{3} = 1$$

Rotating 45°, graph the equation:

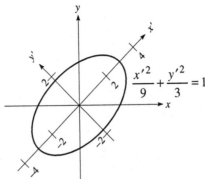

$$\frac{x'^2}{9} + \frac{y'^2}{3} = 1$$

61. Finding the rotation angle:

$$\cot 2\phi = \frac{11-1}{10\sqrt{3}} = \frac{1}{\sqrt{3}}$$

$$2\phi = 60°$$

$$\phi = 30°$$

The rotation equations are:

$$x = x'\cos 30° - y'\sin 30° = \frac{\sqrt{3}}{2}x' - \frac{1}{2}y'$$

$$y = x'\sin 30° + y'\cos 30° = \frac{1}{2}x' + \frac{\sqrt{3}}{2}y'$$

Now converting the equation:

$$11x^2 + 10\sqrt{3}xy + y^2 = 16$$

$$11\left(\frac{\sqrt{3}}{2}x' - \frac{1}{2}y'\right)^2 + 10\sqrt{3}\left(\frac{\sqrt{3}}{2}x' - \frac{1}{2}y'\right)\left(\frac{1}{2}x' + \frac{\sqrt{3}}{2}y'\right) + \left(\frac{1}{2}x' + \frac{\sqrt{3}}{2}y'\right)^2 = 16$$

$$\frac{33}{4}x'^2 - \frac{11\sqrt{3}}{2}x'y' + \frac{11}{4}y'^2 + \frac{15}{2}x'^2 + 5\sqrt{3}x'y' - \frac{15}{2}y'^2 + \frac{1}{4}x'^2 + \frac{\sqrt{3}}{2}x'y' + \frac{3}{4}y'^2 = 16$$

$$16x'^2 - 4y'^2 = 16$$

$$\frac{x'^2}{1} - \frac{y'^2}{4} = 1$$

Rotating 30°, graph the equation:

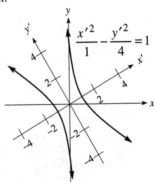

$$\frac{x'^2}{1} - \frac{y'^2}{4} = 1$$

62.   Finding the rotation angle:
$$\cot 2\phi = \frac{1-1}{1} = 0$$
$$2\phi = 90°$$
$$\phi = 45°$$

The rotation equations are:
$$x = x'\cos 45° - y'\sin 45° = \frac{1}{\sqrt{2}}x' - \frac{1}{\sqrt{2}}y'$$
$$y = x'\sin 45° + y'\cos 45° = \frac{1}{\sqrt{2}}x' + \frac{1}{\sqrt{2}}y'$$

Now converting the equation:
$$x^2 + xy + y^2 + \sqrt{2}x - \sqrt{2}y = 4$$
$$\tfrac{1}{2}x'^2 - x'y' + \tfrac{1}{2}y'^2 + \tfrac{1}{2}x'^2 - \tfrac{1}{2}y'^2 + \tfrac{1}{2}x'^2 + x'y' + \tfrac{1}{2}y'^2 + x' - y' - x' - y' = 4$$
$$\tfrac{3}{2}x'^2 + \tfrac{1}{2}y'^2 - 2y' = 4$$
$$3x'^2 + y'^2 - 4y' = 8$$
$$3x'^2 + \left(y'^2 - 4y' + 4\right) = 8 + 4$$
$$3x'^2 + \left(y' - 2\right)^2 = 12$$
$$\frac{x'^2}{4} + \frac{\left(y' - 2\right)^2}{12} = 1$$

Rotating 45°, graph the equation:
$$\frac{x'^2}{4} + \frac{\left(y' - 2\right)^2}{12} = 1$$

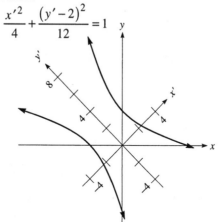

63.   Substituting into the first equation:
$$x^2 - (x-2)^2 = 8$$
$$x^2 - \left(x^2 - 4x + 4\right) = 8$$
$$4x - 4 = 8$$
$$4x = 12$$
$$x = 3$$
$$y = 1$$

The solution is $(3, 1)$.

64. Substituting for x^2:
$$y = 1 - \left(y^2 - 5\right)$$
$$y = 1 - y^2 + 5$$
$$y^2 + y - 6 = 0$$
$$(y+3)(y-2) = 0$$
$$y = -3, 2$$

When $y = -3$:
$$1 - x^2 = -3$$
$$x^2 = 4$$
$$x = \pm 2$$

When $y = 2$:
$$1 - x^2 = 2$$
$$x^2 = -1 \quad \text{(impossible)}$$

The solutions are $(-2,-3)$ and $(2,-3)$.

65. Solving the second equation for y yields $y = 10 - x$. Substituting into the first equation:
$$x^2 - (10-x)^2 = 40$$
$$x^2 - \left(100 - 20x + x^2\right) = 40$$
$$-100 + 20x = 40$$
$$20x = 140$$
$$x = 7$$
$$y = 10 - 7 = 3$$

The solution is $(7,3)$.

66. Solving the second equation for y yields $y = 3x - 10$. Substituting into the first equation:
$$2x^2 + (3x-10)^2 = 19$$
$$2x^2 + 9x^2 - 60x + 100 = 19$$
$$11x^2 - 60x + 81 = 0$$
$$(11x - 27)(x - 3) = 0$$
$$x = \tfrac{27}{11}, 3$$

When $x = \tfrac{27}{11}$: $y = 3\left(\tfrac{27}{11}\right) - 10 = -\tfrac{29}{11}$
When $x = 3$: $y = 3(3) - 10 = -1$

The solutions are $\left(\tfrac{27}{11}, -\tfrac{29}{11}\right)$ and $(3,-1)$.

67. Multiplying the second equation by 4:
$$x^2 + y^2 = 4$$
$$-x^2 + y^2 = 4$$

Adding the two equations:
$$2y^2 = 8$$
$$y^2 = 4$$
$$y = \pm 2$$
$$x = 0$$

The solutions are $(0,-2)$ and $(0,2)$.

68. Multiplying the first equation by 4 and the second equation by -3:
$$8x^2 + 12y^2 = 32$$
$$-15x^2 - 12y^2 = -21$$
Adding the two equations:
$$-7x^2 = 11$$
$$x^2 = -\tfrac{11}{7}$$
There is no solution to this system.

69. Solving the second equation for x^2 yields $x^2 = y + 1$. Substituting into the first equation:
$$(y+1) + y^2 = 13$$
$$y^2 + y - 12 = 0$$
$$(y+4)(y-3) = 0$$
$$y = -4, 3$$
When $y = -4$:
$$x^2 - 1 = -4$$
$$x^2 = -3 \quad \text{(impossible)}$$
When $y = 3$:
$$x^2 - 1 = 3$$
$$x^2 = 4$$
$$x = \pm 2$$
$$y = 3$$
The solutions are $(-2,3)$ and $(2,3)$.

70. Multiplying the second equation by -1:
$$x^2 + 2y^2 = 6$$
$$-x^2 + y^2 = -3$$
Adding the two equations:
$$3y^2 = 3$$
$$y^2 = 1$$
$$y = \pm 1$$
When $y = \pm 1$:
$$x^2 - 1 = 3$$
$$x^2 = 4$$
$$x = \pm 2$$
The solutions are $(-2,1)$, $(-2,-1)$, $(2,-1)$, and $(2,1)$.

71. Let b and h represent the two legs. The system of equations is:
$$b^2 + h^2 = 4^2$$
$$\tfrac{1}{2}bh = 4$$

Solving the second equation for h yields $h = \dfrac{8}{b}$. Substituting into the first equation:

$$b^2 + \left(\frac{8}{b}\right)^2 = 16$$
$$b^2 + \frac{64}{b^2} = 16$$
$$b^4 + 64 = 16b^2$$
$$b^4 - 16b^2 + 64 = 0$$
$$\left(b^2 - 8\right)^2 = 0$$
$$b^2 = 8$$
$$b = 2\sqrt{2} \qquad \left(b = -2\sqrt{2} \text{ is impossible}\right)$$
$$h = \frac{8}{2\sqrt{2}} = 2\sqrt{2}$$

The legs are both $2\sqrt{2} \approx 2.83$ cm.

72. Let r represent the radius and l represent the length. The system of equations is:
$$\pi r^2 l = 18\pi$$
$$2\pi r l = 21\pi$$

Solving the first equation for l yields $l = \dfrac{18}{r^2}$. Substituting into the second equation:

$$2\pi r \cdot \frac{18}{r^2} = 21\pi$$
$$\frac{36}{r} = 21$$
$$r = \tfrac{12}{7} = 1\tfrac{5}{7}$$
$$l = \frac{18}{\left(\frac{12}{7}\right)^2} = \tfrac{49}{8} = 6\tfrac{1}{8}$$

The radius is $1\tfrac{5}{7}$ cm and the length is $6\tfrac{1}{8}$ cm.

73. Sketching the solution set:

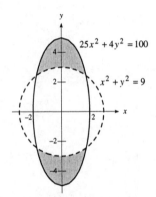

74. Sketching the solution set:

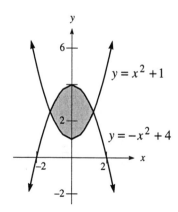

<u>Chapter 10 Practice Test</u>

1. Converting the equation to standard form:
$$4x^2 + 4y = 0$$
$$4x^2 = -4y$$
$$x^2 = -y$$
Since $4p = -1$, $p = -\frac{1}{4}$. This is a parabola with vertex $(0,0)$, focus $\left(0, -\frac{1}{4}\right)$, and directrix $y = \frac{1}{4}$.
Sketching the parabola:

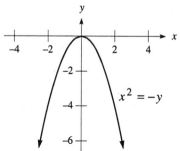

2. Dividing by 4, the equation is $x^2 + y^2 = 9$. The center is $(0,0)$ and the radius is 3. Sketching the circle:

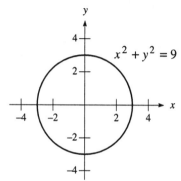

3.   Since $a = 7$ and $b = 5$, $c = \sqrt{49 - 25} = \sqrt{24} = 2\sqrt{6}$. The vertices are $(0,-7)$ and $(0,7)$, and the foci are $\left(0,-2\sqrt{6}\right)$ and $\left(0,2\sqrt{6}\right)$. Sketching the ellipse:

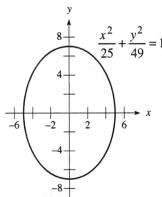

4.   Since $a = 5$ and $b = 7$, $c = \sqrt{25 + 49} = \sqrt{74}$. The vertices are $(-5,0)$ and $(5,0)$, the foci are $\left(-\sqrt{74},0\right)$ and $\left(\sqrt{74},0\right)$, and the asymptotes are $y = \pm\frac{7}{5}x$. Sketching the hyperbola:

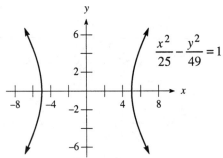

5.   The center is $(3,-5)$ and the radius is $\sqrt{5}$. Sketching the circle:

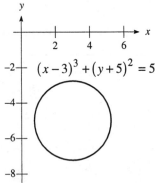

6.   Converting the equation to standard form:
$$2(x-3)^2 + 8y + 24 = 0$$
$$2(x-3)^2 = -8(y+3)$$
$$(x-3)^2 = -4(y+3)$$

Since $4p = -4$, $p = -1$. This is a parabola with vertex $(3,-3)$, focus $(3,-4)$, and directrix $y = -2$. Sketching the graph:

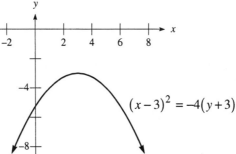

$$(x-3)^2 = -4(y+3)$$

7.   The center is $(3,2)$. Since $a = \sqrt{12} = 2\sqrt{3}$ and $b = 2$, $c = \sqrt{12-4} = \sqrt{8} = 2\sqrt{2}$. This is an ellipse with vertices $\left(3, 2 - 2\sqrt{3}\right)$ and $\left(3, 2 + 2\sqrt{3}\right)$, and foci $\left(3, 2 - 2\sqrt{2}\right)$ and $\left(3, 2 + 2\sqrt{2}\right)$. Sketching the ellipse:

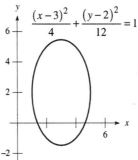

$$\frac{(x-3)^2}{4} + \frac{(y-2)^2}{12} = 1$$

8.   The center is $(0,-2)$. Since $a = 5$ and $b = \sqrt{18} = 3\sqrt{2}$, $c = \sqrt{25+18} = \sqrt{43}$. This is a hyperbola with vertices $(0,-7)$ and $(0,3)$, foci $\left(0, -2 - \sqrt{43}\right)$ and $\left(0, -2 + \sqrt{43}\right)$, and asymptote slopes

$m = \pm\dfrac{5}{3\sqrt{2}} = \pm\dfrac{5\sqrt{2}}{6}$. Finding the asymptotes:

$$y + 2 = -\frac{5\sqrt{2}}{6}x \qquad\qquad y + 2 = \frac{5\sqrt{2}}{6}x$$

$$y = -\frac{5\sqrt{2}}{6}x - 2 \qquad\qquad y = \frac{5\sqrt{2}}{6}x - 2$$

Sketching the hyperbola:

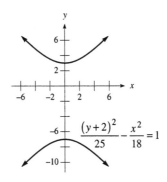

$$\frac{(y+2)^2}{25} - \frac{x^2}{18} = 1$$

9.   Completing the square:

$$4x^2 + 5y^2 - 24x + 30y + 61 = 0$$
$$4\left(x^2 - 6x + 9\right) + 5\left(y^2 + 6y + 9\right) = -61 + 36 + 45$$
$$4(x-3)^2 + 5(y+3)^2 = 20$$
$$\frac{(x-3)^2}{5} + \frac{(y+3)^2}{4} = 1$$

The center is (3,–3). Since $a = \sqrt{5}$ and $b = 2$, $c = \sqrt{5-4} = 1$. This is an ellipse with vertices $\left(3 - \sqrt{5}, -3\right)$ and $\left(3 + \sqrt{5}, -3\right)$, and foci (2,–3) and (4,–3). Graphing the ellipse:

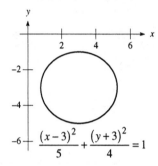

$$\frac{(x-3)^2}{5} + \frac{(y+3)^2}{4} = 1$$

10.   Completing the square:

$$3x^2 - 4y^2 - 6x = 45$$
$$3\left(x^2 - 2x + 1\right) - 4y^2 = 45 + 3$$
$$3(x-1)^2 - 4y^2 = 48$$
$$\frac{(x-1)^2}{16} - \frac{y^2}{12} = 1$$

The center is (1,0). Since $a = 4$ and $b = \sqrt{12} = 2\sqrt{3}$, $c = \sqrt{16+12} = \sqrt{28} = 2\sqrt{7}$. This is a hyperbola with vertices (–3,0) and (5,0), foci $\left(1 - 2\sqrt{7}, 0\right)$ and $\left(1 + 2\sqrt{7}, 0\right)$, and asymptote slopes $m = \pm\frac{2\sqrt{3}}{4} = \pm\frac{\sqrt{3}}{2}$. Finding the asymptotes:

$$y = -\frac{\sqrt{3}}{2}(x-1) \qquad\qquad y = \frac{\sqrt{3}}{2}(x-1)$$
$$y = -\frac{\sqrt{3}}{2}x + \frac{\sqrt{3}}{2} \qquad\qquad y = \frac{\sqrt{3}}{2}x - \frac{\sqrt{3}}{2}$$

Graphing the hyperbola:

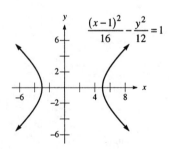

$$\frac{(x-1)^2}{16} - \frac{y^2}{12} = 1$$

11. Since $p = 2$, the equation is $y^2 = 8x$.

12. The center is $(1,1)$. Since $a = 4$ and $c = 2$, $b = \sqrt{16-4} = \sqrt{12} = 2\sqrt{3}$. The equation is
$$\frac{(x-1)^2}{16} + \frac{(y-1)^2}{12} = 1.$$

13. **a.** The rotation equations are:
$$x' = 2\cos 45° - 1\sin 45° = 2\left(\frac{\sqrt{2}}{2}\right) - 1\left(\frac{\sqrt{2}}{2}\right) = \frac{\sqrt{2}}{2}$$
$$y' = -2\sin 45° - 1\cos 45° = -2\left(\frac{\sqrt{2}}{2}\right) - 1\left(\frac{\sqrt{2}}{2}\right) = -\frac{3\sqrt{2}}{2}$$

The point is $(x', y') = \left(\frac{\sqrt{2}}{2}, -\frac{3\sqrt{2}}{2}\right)'$.

b. The rotation equations are:
$$x = x'\cos 45° - y'\sin 45° = \frac{1}{\sqrt{2}}x' - \frac{1}{\sqrt{2}}y'$$
$$y = x'\sin 45° + y'\cos 45° = \frac{1}{\sqrt{2}}x' + \frac{1}{\sqrt{2}}y'$$

Now converting the equation:
$$2y^2 - \sqrt{2}x = 1$$
$$2\left(\frac{1}{\sqrt{2}}x' + \frac{1}{\sqrt{2}}y'\right)^2 - \sqrt{2}\left(\frac{1}{\sqrt{2}}x' - \frac{1}{\sqrt{2}}y'\right) = 1$$
$$2\left(\tfrac{1}{2}x'^2 + x'y' + \tfrac{1}{2}y'^2\right) - x' + y' = 1$$
$$x'^2 + 2x'y' + y'^2 - x' + y' - 1 = 0$$

14. Finding the rotation angle:
$$\cot 2\phi = \frac{1-0}{\sqrt{3}} = \frac{1}{\sqrt{3}}$$
$$2\phi = 60°$$
$$\phi = 30°$$

The rotation equations are:
$$x = x'\cos 30° - y'\sin 30° = \frac{\sqrt{3}}{2}x' - \frac{1}{2}y'$$
$$y = x'\sin 30° + y'\cos 30° = \frac{1}{2}x' + \frac{\sqrt{3}}{2}y'$$

Now converting the equation:

$$x^2 + \sqrt{3}xy = 6$$

$$\left(\frac{\sqrt{3}}{2}x' - \frac{1}{2}y'\right)^2 + \sqrt{3}\left(\frac{\sqrt{3}}{2}x' - \frac{1}{2}y'\right)\left(\frac{1}{2}x' + \frac{\sqrt{3}}{2}y'\right) = 6$$

$$\frac{3}{4}x'^2 - \frac{\sqrt{3}}{2}x'y' + \frac{1}{4}y'^2 + \frac{3}{4}x'^2 + \frac{\sqrt{3}}{2}x'y' - \frac{3}{4}y'^2 = 6$$

$$\tfrac{3}{2}x'^2 - \tfrac{1}{2}y'^2 = 6$$

$$3x'^2 - y'^2 = 12$$

$$\frac{x'^2}{4} - \frac{y'^2}{12} = 1$$

Rotating 30°, graph the hyperbola:

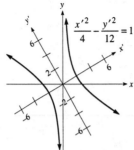

15. Multiplying the second equation by 2:
$$x^2 + 2y^2 = 3$$
$$4x^2 - 2y^2 = 2$$
Adding yields:
$$5x^2 = 5$$
$$x^2 = 1$$
$$x = \pm 1$$
Substituting into the first equation:
$$1 + 2y^2 = 3$$
$$2y^2 = 2$$
$$y^2 = 1$$
$$y = \pm 1$$
The solutions are $(-1,-1)$, $(-1,1)$, $(1,-1)$, and $(1,1)$.

16. Sketching the solution set:

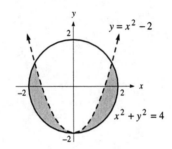

Chapter 11
Sequences, Series, and Related Topics

11.1 Sequences

1. Finding the required terms:
$$a_1 = 5(1) - 3 = 2 \qquad a_2 = 5(2) - 3 = 7$$
$$a_3 = 5(3) - 3 = 12 \qquad a_4 = 5(4) - 3 = 17$$
$$a_8 = 5(8) - 3 = 37$$

3. Finding the required terms:
$$b_1 = 3(1) + 2 = 5 \qquad b_2 = 3(2) + 2 = 8$$
$$b_3 = 3(3) + 2 = 11 \qquad b_4 = 3(4) + 2 = 14$$
$$b_8 = 3(8) + 2 = 26$$

5. Finding the required terms:
$$c_1 = 3^1 = 3 \qquad c_2 = 3^2 = 9$$
$$c_3 = 3^3 = 27 \qquad c_4 = 3^4 = 81$$
$$c_8 = 3^8 = 6561$$

7. Finding the required terms:
$$x_1 = \frac{1+1}{1} = 2 \qquad x_2 = \frac{2+1}{2} = \tfrac{3}{2}$$
$$x_3 = \frac{3+1}{3} = \tfrac{4}{3} \qquad x_4 = \frac{4+1}{4} = \tfrac{5}{4}$$
$$x_8 = \frac{8+1}{8} = \tfrac{9}{8}$$

9. Finding the required terms:
$$x_1 = \frac{(-1)^{1+1}}{1+2} = \tfrac{1}{3} \qquad x_2 = \frac{(-1)^{2+1}}{2+2} = -\tfrac{1}{4}$$
$$x_3 = \frac{(-1)^{3+1}}{3+2} = \tfrac{1}{5} \qquad x_4 = \frac{(-1)^{4+1}}{4+2} = -\tfrac{1}{6}$$
$$x_8 = \frac{(-1)^{8+1}}{8+2} = -\tfrac{1}{10}$$

11. Finding the required terms:

$$a_1 = \frac{(-1)^1 \cdot 1}{1+1} = -\frac{1}{2} \qquad a_2 = \frac{(-1)^2 \cdot 2}{2+1} = \frac{2}{3}$$

$$a_3 = \frac{(-1)^3 \cdot 3}{3+1} = -\frac{3}{4} \qquad a_4 = \frac{(-1)^4 \cdot 4}{4+1} = \frac{4}{5}$$

$$a_8 = \frac{(-1)^8 \cdot 8}{8+1} = \frac{8}{9}$$

13. Finding the required terms:

$$a_1 = 2^1 + |1-3| = 4 \qquad a_2 = 2^2 + |2-3| = 5$$

$$a_3 = 2^3 + |3-3| = 8 \qquad a_4 = 2^4 + |4-3| = 17$$

$$a_8 = 2^8 + |8-3| = 261$$

15. Finding the required terms:

$$b_1 = 1 + (-0.1)^1 = 0.9 \qquad b_2 = 1 + (-0.1)^2 = 1.01$$

$$b_3 = 1 + (-0.1)^3 = 0.999 \qquad b_4 = 1 + (-0.1)^4 = 1.0001$$

$$b_8 = 1 + (-0.1)^8 = 1.00000001$$

17. Finding the required terms:

$$x_1 = \frac{2^1}{1^2} = 2 \qquad x_2 = \frac{2^2}{2^2} = 1$$

$$x_3 = \frac{2^3}{3^2} = \frac{8}{9} \qquad x_4 = \frac{2^4}{4^2} = 1$$

$$x_8 = \frac{2^8}{8^2} = 4$$

19. Finding the required terms:

$$a_1 = \sin\frac{\pi}{2} = 1 \qquad a_2 = \sin\frac{2\pi}{2} = 0$$

$$a_3 = \sin\frac{3\pi}{2} = -1 \qquad a_4 = \sin\frac{4\pi}{2} = 0$$

$$a_8 = \sin\frac{8\pi}{2} = 0$$

21. Finding the required terms:

$$c_1 = \left(1 + \frac{1}{1}\right)^1 = 2 \qquad c_2 = \left(1 + \frac{1}{2}\right)^2 = \frac{9}{4}$$

$$c_3 = \left(1 + \frac{1}{3}\right)^3 = \frac{64}{27} \qquad c_4 = \left(1 + \frac{1}{4}\right)^4 = \frac{625}{256}$$

$$c_8 = \left(1 + \frac{1}{8}\right)^8 = \frac{43046721}{16777216}$$

23. Finding the required terms:

$$a_1 = \sqrt{1^2 + 1} = \sqrt{2} \qquad a_2 = \sqrt{2^2 + 1} = \sqrt{5}$$

$$a_3 = \sqrt{3^2 + 1} = \sqrt{10} \qquad a_4 = \sqrt{4^2 + 1} = \sqrt{17}$$

$$a_8 = \sqrt{8^2 + 1} = \sqrt{65}$$

25. Finding the required terms:

$$a_1 = \frac{1}{1!} = 1 \qquad\qquad a_2 = \frac{1}{2!} = \tfrac{1}{2}$$

$$a_3 = \frac{1}{3!} = \tfrac{1}{6} \qquad\qquad a_4 = \frac{1}{4!} = \tfrac{1}{24}$$

$$a_8 = \frac{1}{8!} = \tfrac{1}{40320}$$

27. Finding the required terms:

$$c_1 = \frac{1!}{1^1} = 1 \qquad\qquad c_2 = \frac{2!}{2^2} = \tfrac{1}{2}$$

$$c_3 = \frac{3!}{3^3} = \tfrac{2}{9} \qquad\qquad c_4 = \frac{4!}{4^4} = \tfrac{3}{32}$$

$$c_8 = \frac{8!}{8^8} = \tfrac{315}{131072}$$

29. Finding the required terms:

$$t_1 = \frac{3!}{3^1 \cdot 1!} = 2 \qquad\qquad t_2 = \frac{6!}{3^2 \cdot 2!} = 40$$

$$t_3 = \frac{9!}{3^3 \cdot 3!} = 2240 \qquad\qquad t_4 = \frac{12!}{3^4 \cdot 4!} = 246400$$

$$t_8 = \frac{24!}{3^8 \cdot 8!} \approx 2.35 \times 10^{15}$$

31. Finding the first six terms:

$$a_1 = 3 \qquad\qquad a_2 = 2(3) - 1 = 5$$

$$a_3 = 2(5) - 1 = 9 \qquad\qquad a_4 = 2(9) - 1 = 17$$

$$a_5 = 2(17) - 1 = 33 \qquad\qquad a_6 = 2(33) - 1 = 65$$

33. Finding the first six terms:

$$a_1 = -1 \qquad\qquad a_2 = (1+1)^2 = 4$$

$$a_3 = (1-4)^2 = 9 \qquad\qquad a_4 = (1-9)^2 = 64$$

$$a_5 = (1-64)^2 = 3,969 \qquad\qquad a_6 = (1-3969)^2 = 15,745,024$$

35. Finding the first six terms:

$$a_1 = 1 \qquad\qquad a_2 = 2^1 = 2$$

$$a_3 = 2^2 = 4 \qquad\qquad a_4 = 2^4 = 16$$

$$a_5 = 2^{16} = 65,536 \qquad\qquad a_6 = 2^{65,536}$$

37. Finding the first six terms:

$$a_1 = 2 \qquad\qquad a_2 = \frac{2}{2} = 1$$

$$a_3 = \frac{3}{1} = 3 \qquad\qquad a_4 = \tfrac{4}{3}$$

$$a_5 = \frac{5}{4/3} = \tfrac{15}{4} \qquad\qquad a_6 = \frac{6}{15/4} = \tfrac{8}{5}$$

39. Finding the first six terms:

$$a_1 = 0 \qquad\qquad a_2 = 1$$

$$a_3 = \tfrac{1}{2}(1+0) = \tfrac{1}{2} \qquad\qquad a_4 = \tfrac{1}{2}\left(\tfrac{1}{2}+1\right) = \tfrac{3}{4}$$

$$a_5 = \tfrac{1}{2}\left(\tfrac{3}{4}+\tfrac{1}{2}\right) = \tfrac{5}{8} \qquad\qquad a_6 = \tfrac{1}{2}\left(\tfrac{5}{8}+\tfrac{3}{4}\right) = \tfrac{11}{16}$$

41. Finding the first six terms:

$$a_1 = -1 \qquad\qquad a_2 = 1$$
$$a_3 = (-1)(1) = -1 \qquad\qquad a_4 = (1)(-1) = -1$$
$$a_5 = (-1)(-1) = 1 \qquad\qquad a_6 = (-1)(1) = -1$$

43. The first five terms are:

$$y_1 = 2^0 = 1, y_2 = 2^1 = 2, y_3 = 2^2 = 4, y_4 = 2^3 = 8, y_5 = 2^4 = 16$$

Graphing these terms:

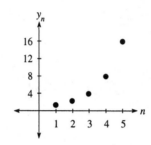

45. The first five terms are:

$$y_1 = (-1)^1 = -1, y_2 = (-1)^2 = 1, y_3 = (-1)^3 = -1, y_4 = (-1)^4 = 1, y_5 = (-1)^5 = -1$$

Graphing these terms:

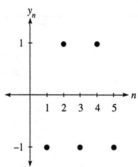

47. Remember that c_n represents the **total** number of calls made. The first five terms are:

$$c_1 = 100, c_2 = 100 + 120 = 220, c_3 = 220 + 140 = 360, c_4 = 360 + 160 = 520, c_5 = 520 + 180 = 700$$

Graphing these terms:

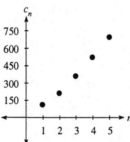

49. A formula is $a_n = 5n$.

51. A formula is $a_n = 2n - 1$.

53. A formula is $a_n = \dfrac{2n-1}{2n}$.

55. A formula is $a_n = \dfrac{(-1)^{n+1}}{n+1}$.

57. A formula is $a_n = (-1)^n \cdot \dfrac{4}{10^n}$.

59. The first three terms are given by:
$$a_1 = 24,500 + 650 = \$25,150$$
$$a_2 = 25,150 + 650 = \$25,800$$
$$a_3 = 25,800 + 650 = \$26,450$$

61. The first three terms are given by:
$$a_1 = 60,000 + 0.02(60,000) = 61,200$$
$$a_2 = 61,200 + 0.02(61,200) = 62,424$$
$$a_3 = 62,424 + 0.02(62,424) \approx 63,672$$

63. The first five terms are given by:
$$a_1 = 1800 + 0.035(1800) = \$1863$$
$$a_2 = 1863 + 0.035(1863) \approx \$1928.21$$
$$a_3 = 1928.21 + 0.035(1928.21) \approx \$1995.70$$
$$a_4 = 1995.70 + 0.035(1995.70) \approx \$2065.55$$
$$a_5 = 2065.55 + 0.035(2065.55) \approx \$2137.84$$

65. The first five terms are given by:
$$a_1 = 2.3 - 0.0135(2.3) \approx 2.27 \text{ billion}$$
$$a_2 = 2.27 - 0.0135(2.27) \approx 2.24 \text{ billion}$$
$$a_3 = 2.24 - 0.0135(2.24) \approx 2.21 \text{ billion}$$
$$a_4 = 2.21 - 0.0135(2.21) \approx 2.18 \text{ billion}$$
$$a_5 = 2.18 - 0.0135(2.18) \approx 2.15 \text{ billion}$$

67. The first five terms are given by:
$$a_1 = \tfrac{1}{2}(60) = 30 \text{ feet}$$
$$a_2 = \tfrac{1}{2}(30) = 15 \text{ feet}$$
$$a_3 = \tfrac{1}{2}(15) = 7.5 \text{ feet}$$
$$a_4 = \tfrac{1}{2}(7.5) = 3.75 \text{ feet}$$
$$a_5 = \tfrac{1}{2}(3.75) = 1.875 \text{ feet}$$

69. The first four terms are given by:
$$a_1 = 200 + 0.01(3000) = \$230$$
$$a_2 = 200 + 0.01(2770) = \$227.70$$
$$a_3 = 200 + 0.01(2542.30) \approx \$225.42$$
$$a_4 = 200 + 0.01(2316.88) \approx \$223.17$$

71.   **a.**   The first eight terms are 1, 1, 2, 3, 5, 8, 13, 21.

     b.   Computing the values:

$$F_1 + F_2 + F_3 + F_4 + F_5 = 1 + 1 + 2 + 3 + 5 = 12$$
$$F_7 - 1 = 13 - 1 = 12$$
$$F_1 + F_2 + F_3 + F_4 + F_5 + F_6 + F_7 = 1 + 1 + 2 + 3 + 5 + 8 + 13 = 33$$
$$F_9 - 1 = 34 - 1 = 33$$
$$F_1 + F_2 + \ldots + F_{10} = 1 + 1 + 2 + 3 + 5 + 8 + 13 + 21 + 34 + 55 = 143$$
$$F_{12} - 1 = 144 - 1 = 143$$

Note that $F_1 + F_2 + \ldots + F_n = F_{n+2} - 1$.

     c.   Computing the values:

$$F_1^2 + F_2^2 + F_3^2 + F_4^2 = 1^2 + 1^2 + 2^2 + 3^2 = 15$$
$$F_4 F_5 = 3 \cdot 5 = 15$$
$$F_1^2 + F_2^2 + F_3^2 + F_4^2 + F_5^2 + F_6^2 = 1^2 + 1^2 + 2^2 + 3^2 + 5^2 + 8^2 = 104$$
$$F_6 F_7 = 8 \cdot 13 = 104$$
$$F_1^2 + F_2^2 + \ldots + F_9^2 = 1^2 + 1^2 + 2^2 + 3^2 + 5^2 + 8^2 + 13^2 + 21^2 + 34^2 = 1870$$
$$F_9 F_{10} = 34 \cdot 55 = 1870$$

Note that $F_1^2 + F_2^2 + \ldots + F_n^2 = F_n F_{n+1}$.

73.   **a.**   Simplifying: $\dfrac{(n+1)!}{n!} = \dfrac{(n+1)n!}{n!} = n+1$

     b.   Simplifying: $\dfrac{(2n+3)!}{(2n)!} = \dfrac{(2n+3)(2n+2)(2n+1)(2n)!}{(2n)!} = (2n+3)(2n+2)(2n+1)$

     c.   Simplifying: $\dfrac{(2n-1)!}{(2n+1)!} = \dfrac{(2n-1)!}{(2n+1)(2n)(2n-1)!} = \dfrac{1}{(2n+1)(2n)} = \dfrac{1}{4n^2 + 2n}$

11.2 Series and Sigma Notation

1.   Finding the sum: $S_5 = 3 + 7 + 11 + 15 + 19 = 55$

3.   Finding the sum: $S_6 = -2 + 0 + 2 + 4 + 6 + 8 = 18$

5.   Finding the sum: $S_7 = 10 + 6 + 2 - 2 - 6 - 10 - 14 = -14$

7.   Finding the sum: $S_6 = 1 + 2 + 4 + 8 + 16 + 32 = 63$

9.   Since $a_1 = 3(1) + 1 = 4$ and $a_2 = 3(2) + 1 = 7$: $S_2 = 4 + 7 = 11$

11.   Since $b_1 = 5(1) - 1 = 4$, $b_2 = 5(2) - 1 = 9$, $b_3 = 5(3) - 1 = 14$, and $b_4 = 5(4) - 1 = 19$:
$$S_4 = 4 + 9 + 14 + 19 = 46$$

13.   Since $x_1 = 2^1 = 2$, $x_2 = 2^2 = 4$, $x_3 = 2^3 = 8$, $x_4 = 2^4 = 16$, and $x_5 = 2^5 = 32$:
$$S_5 = 2 + 4 + 8 + 16 + 32 = 62$$

15.   Since $x_1 = 2^1 + 1 = 3$, $x_2 = 2^2 + 1 = 5$, $x_3 = 2^3 + 1 = 9$, $x_4 = 2^4 + 1 = 17$, $x_5 = 2^5 + 1 = 33$, and
$x_6 = 2^6 + 1 = 65$: $S_6 = 3 + 5 + 9 + 17 + 33 + 65 = 132$

17.   Since $x_1 = 3^{1-2} = \frac{1}{3}$, $x_2 = 3^{2-2} = 0$, $x_3 = 3^{3-2} = 3$, $x_4 = 3^{4-2} = 9$, and $x_5 = 3^{5-2} = 27$:
$$S_5 = \tfrac{1}{3} + 1 + 3 + 9 + 27 = \tfrac{121}{3}$$

19. Since $a_1 = \dfrac{1+1}{1} = 2$, $a_2 = \dfrac{2+1}{2} = \dfrac{3}{2}$, $a_3 = \dfrac{3+1}{3} = \dfrac{4}{3}$, and $a_4 = \dfrac{4+1}{4} = \dfrac{5}{4}$:

$S_4 = 2 + \dfrac{3}{2} + \dfrac{4}{3} + \dfrac{5}{4} = \dfrac{73}{12}$

21. Since $b_1 = \dfrac{(-1)^2}{1+2} = \dfrac{1}{3}$, $b_2 = \dfrac{(-1)^3}{2+2} = -\dfrac{1}{4}$, $b_3 = \dfrac{(-1)^4}{3+2} = \dfrac{1}{5}$, $b_4 = \dfrac{(-1)^5}{4+2} = -\dfrac{1}{6}$, and

$b_5 = \dfrac{(-1)^6}{5+2} = \dfrac{1}{7}$: $S_5 = \dfrac{1}{3} - \dfrac{1}{4} + \dfrac{1}{5} - \dfrac{1}{6} + \dfrac{1}{7} = \dfrac{109}{420}$

23. Since $r_1 = \dfrac{1}{1!} = 1$, $r_2 = \dfrac{1}{2!} = \dfrac{1}{2}$, $r_3 = \dfrac{1}{3!} = \dfrac{1}{6}$, and $r_4 = \dfrac{1}{4!} = \dfrac{1}{24}$: $S_4 = 1 + \dfrac{1}{2} + \dfrac{1}{6} + \dfrac{1}{24} = \dfrac{41}{24}$

25. Since $a_1 = 1 - \dfrac{1}{2} = \dfrac{1}{2}$, $a_2 = \dfrac{1}{2} - \dfrac{1}{3} = \dfrac{1}{6}$, $a_3 = \dfrac{1}{3} - \dfrac{1}{4} = \dfrac{1}{12}$, $a_4 = \dfrac{1}{4} - \dfrac{1}{5} = \dfrac{1}{20}$, $a_5 = \dfrac{1}{5} - \dfrac{1}{6} = \dfrac{1}{30}$,

$a_6 = \dfrac{1}{6} - \dfrac{1}{7} = \dfrac{1}{42}$, and $a_7 = \dfrac{1}{7} - \dfrac{1}{8} = \dfrac{1}{56}$:

$S_7 = \left(1 - \dfrac{1}{2}\right) + \left(\dfrac{1}{2} - \dfrac{1}{3}\right) + \left(\dfrac{1}{3} - \dfrac{1}{4}\right) + \left(\dfrac{1}{4} - \dfrac{1}{5}\right) + \left(\dfrac{1}{5} - \dfrac{1}{6}\right) + \left(\dfrac{1}{6} - \dfrac{1}{7}\right) + \left(\dfrac{1}{7} - \dfrac{1}{8}\right) = 1 - \dfrac{1}{8} = \dfrac{7}{8}$

27. Computing the sum: $\displaystyle\sum_{n=1}^{5} n = 1 + 2 + 3 + 4 + 5 = 15$

29. Computing the sum: $\displaystyle\sum_{i=1}^{6} 5(i-1) = 0 + 5 + 10 + 15 + 20 + 25 = 75$

31. Computing the sum: $\displaystyle\sum_{k=2}^{6} k^2 = 4 + 9 + 16 + 25 + 36 = 90$

33. Computing the sum: $\displaystyle\sum_{m=3}^{7} \left(2m^2 - 5\right) = 13 + 27 + 45 + 67 + 93 = 245$

35. Computing the sum: $\displaystyle\sum_{n=2}^{4} \left(n^2 - 3n + 1\right) = -1 + 1 + 5 = 5$

37. Computing the sum: $\displaystyle\sum_{j=1}^{5} \dfrac{1}{j+1} = \dfrac{1}{2} + \dfrac{1}{3} + \dfrac{1}{4} + \dfrac{1}{5} + \dfrac{1}{6} = \dfrac{29}{20}$

39. Computing the sum:

$\displaystyle\sum_{n=1}^{5} \left(\sqrt{n+1} - \sqrt{n}\right) = \left(\sqrt{2} - 1\right) + \left(\sqrt{3} - \sqrt{2}\right) + \left(2 - \sqrt{3}\right) + \left(\sqrt{5} - 2\right) + \left(\sqrt{6} - \sqrt{5}\right) = \sqrt{6} - 1$

41. Computing the sum: $\displaystyle\sum_{n=1}^{5} 3 = 3 + 3 + 3 + 3 + 3 = 15$

43. The sum is $S_5 = \displaystyle\sum_{k=1}^{5} 4k$.

45. The sum is $S_7 = \displaystyle\sum_{k=1}^{7} \left(2k^2\right)$.

47. The sum is $\displaystyle\sum_{k=1}^{5} \left(4k - 2\right)$.

49. The sum is $\displaystyle\sum_{n=1}^{26}(2n-1)$.

51. The sum is $\displaystyle\sum_{k=1}^{99}\frac{1}{k(k+1)}$.

53. Since $a_1=3$, $a_2=9$, $a_3=21$, $a_4=45$, $a_5=93$, and $a_6=189$:
$$S_6=3+9+21+45+93+189=360$$

55. Since $a_1=-2$, $a_2=1$, $a_3=4$, $a_4=25$, $a_5=676$, and $a_6=458329$:
$$S_6=-2+1+4+25+676+458329=459033$$

57. Since $a_1=2$, $a_2=3$, $a_3=6$, $a_4=18$, $a_5=108$, and $a_6=1944$:
$$S_6=2+3+6+18+108+1944=2081$$

59. Writing out the sum:
$$\sum_{n=1}^{8}\log_3\frac{n+1}{n}=\log_3 2+\log_3\tfrac{3}{2}+\log_3\tfrac{4}{3}+\log_3\tfrac{5}{4}+\log_3\tfrac{6}{5}+\log_3\tfrac{7}{6}+\log_3\tfrac{8}{7}+\log_3\tfrac{9}{8}$$
$$=\log_3\left(2\cdot\tfrac{3}{2}\cdot\tfrac{4}{3}\cdot\tfrac{5}{4}\cdot\tfrac{6}{5}\cdot\tfrac{7}{6}\cdot\tfrac{8}{7}\cdot\tfrac{9}{8}\right)$$
$$=\log_3 9$$
$$=2$$

61. Finding the function values:
$$f(1)=1+1^2+1^3+1^4+1^5=1+1+1+1+1=5$$
$$f(-2)=-2+(-2)^2+(-2)^3+(-2)^4+(-2)^5=-2+4-8+16-32=-22$$
$$f\left(\tfrac{1}{2}\right)=\tfrac{1}{2}+\left(\tfrac{1}{2}\right)^2+\left(\tfrac{1}{2}\right)^3+\left(\tfrac{1}{2}\right)^4+\left(\tfrac{1}{2}\right)^5=\tfrac{1}{2}+\tfrac{1}{4}+\tfrac{1}{8}+\tfrac{1}{16}+\tfrac{1}{32}=\tfrac{31}{32}$$

63. **a.** Finding the function values:
$$F(1)=\frac{1^0}{0!}+\frac{1^1}{1!}+\frac{1^2}{2!}+\frac{1^3}{3!}+\frac{1^4}{4!}+\frac{1^5}{5!}+\frac{1^6}{6!}$$
$$=1+1+\tfrac{1}{2}+\tfrac{1}{6}+\tfrac{1}{24}+\tfrac{1}{120}+\tfrac{1}{720}$$
$$=\tfrac{1957}{720}$$
$$\approx 2.7181$$
$$F(2)=\frac{2^0}{0!}+\frac{2^1}{1!}+\frac{2^2}{2!}+\frac{2^3}{3!}+\frac{2^4}{4!}+\frac{2^5}{5!}+\frac{2^6}{6!}$$
$$=1+2+2+\tfrac{4}{3}+\tfrac{2}{3}+\tfrac{4}{15}+\tfrac{4}{45}$$
$$=\tfrac{331}{45}$$
$$\approx 7.3556$$
$$F(-1)=\frac{(-1)^0}{0!}+\frac{(-1)^1}{1!}+\frac{(-1)^2}{2!}+\frac{(-1)^3}{3!}+\frac{(-1)^4}{4!}+\frac{(-1)^5}{5!}+\frac{(-1)^6}{6!}$$
$$=1-1+\tfrac{1}{2}-\tfrac{1}{6}+\tfrac{1}{24}-\tfrac{1}{120}+\tfrac{1}{720}$$
$$=\tfrac{53}{144}$$
$$\approx 0.3681$$

b. Finding the required values:
$$e\approx 2.1783,\; e^2\approx 7.3891,\; e^{-1}\approx 0.3679$$
The values are very close to each other.

65. Expanding the sum: $\displaystyle\sum_{k=1}^{N} c = c + c + \ldots + c = Nc$

67. In sigma notation, the polynomial is $p(x) = \displaystyle\sum_{k=0}^{n} a_k x^k$.

11.3 Arithmetic Sequences and Series

1. Since the common difference is 3, this is an arithmetic sequence.
3. Since $14 - 10 = 4$ and $20 - 14 = 6$, there is no common difference. This is not an arithmetic sequence.
5. Since the common difference is -2, this is an arithmetic sequence.
7. Using the formula $a_n = a_1 + (n-1)d$:
$$a_5 = a_1 + 4d$$
$$23 = 3 + 4d$$
$$4d = 20$$
$$d = 5$$
9. Using the formula $a_k - a_j = (k - j)d$:
$$a_{20} - a_5 = 15d$$
$$10 = 15d$$
$$d = \tfrac{2}{3}$$
11. Here $d = 4$.
13. Using the formula $a_k - a_j = (k - j)d$:
$$a_8 - a_4 = 4d$$
$$10 - 8 = 4d$$
$$4d = 2$$
$$d = \tfrac{1}{2}$$
Using the formula $a_n = a_1 + (n-1)d$:
$$a_4 = a_1 + 3d$$
$$8 = a_1 + 3 \cdot \tfrac{1}{2}$$
$$8 = a_1 + \tfrac{3}{2}$$
$$a_1 = \tfrac{13}{2}$$
15. Using the formula $a_n = a_1 + (n-1)d$:
$$a_5 = a_1 + 4d$$
$$1 = a_1 + 4\left(\tfrac{2}{5}\right)$$
$$1 = a_1 + \tfrac{8}{5}$$
$$a_1 = -\tfrac{3}{5}$$
Using the same formula: $a_{20} = a_1 + 19d = -\tfrac{3}{5} + 19\left(\tfrac{2}{5}\right) = 7$
17. Here $a_1 = 2$ and $d = 3$. Using the formula $a_n = a_1 + (n-1)d$: $a_{18} = a_1 + 17d = 2 + 17(3) = 53$
19. Here $a_1 = \tfrac{1}{2}$ and $d = \tfrac{1}{3}$. Using the formula $a_n = a_1 + (n-1)d$: $a_{10} = a_1 + 9d = \tfrac{1}{2} + 9\left(\tfrac{1}{3}\right) = \tfrac{7}{2}$

21. Using the formula $a_k - a_j = (k-j)d$:

$$a_4 - a_3 = d$$
$$d = -\frac{5}{6} + \frac{1}{3} = -\frac{1}{2}$$

Using the formula $a_n = a_1 + (n-1)d$:

$$a_3 = a_1 + 2d$$
$$-\frac{1}{3} = a_1 + 2\left(-\frac{1}{2}\right)$$
$$-\frac{1}{3} = a_1 - 1$$
$$a_1 = \frac{2}{3}$$

Using the same formula: $a_9 = a_1 + 8d = \frac{2}{3} + 8\left(-\frac{1}{2}\right) = -\frac{10}{3}$

23. Using the formula $a_k - a_j = (k-j)d$:

$$a_7 - a_4 = 3d$$
$$-\frac{5}{4} = 3d$$
$$d = -\frac{5}{12}$$

Now find a_1:

$$a_{10} = a_1 + 9d$$
$$11 = a_1 + 9\left(-\frac{5}{12}\right)$$
$$11 = a_1 - \frac{15}{4}$$
$$a_1 = \frac{59}{4}$$

25. Using $a_1 = 25250$, $d = 250$, and $n = 7$: $a_7 = a_1 + 6d = 25250 + 6(250) = \$26,750$

27. Using $a_1 = 50$, $d = 2$, and $n = 20$: $a_{20} = a_1 + 19d = 50 + 19(2) = 88$ pounds

29. Using $a_1 = 16$ and $d = 32$:

$$a_5 = a_1 + 4d = 16 + 4(32) = 144 \text{ feet}$$
$$a_6 = a_1 + 5d = 16 + 5(32) = 176 \text{ feet}$$

31. Using the formula $S_n = \frac{n}{2}\left[2a_1 + (n-1)d\right]$: $S_6 = \frac{6}{2}[2 \cdot 4 + 5 \cdot 5] = 99$

33. Using $a_k - a_j = (k-j)d$:

$$b_7 - b_4 = 3d$$
$$17 - 2 = 3d$$
$$3d = 15$$
$$d = 5$$

Now using $a_n = a_1 + (n-1)d$:

$$b_4 = b_1 + 3d$$
$$2 = b_1 + 3 \cdot 5$$
$$2 = b_1 + 15$$
$$b_1 = -13$$

Now using the formula $S_n = \frac{n}{2}\left[2a_1 + (n-1)d\right]$: $S_7 = \frac{7}{2}[2(-13) + 6 \cdot 5] = 14$

35. Using $a_n = a_1 + (n-1)d$:

$$a_3 = a_1 + 2d$$
$$2 = a_1 + 2 \cdot \tfrac{1}{2}$$
$$2 = a_1 + 1$$
$$a_1 = 1$$

Now using the formula $S_n = \dfrac{n}{2}\left[2a_1 + (n-1)d\right]$: $\ S_7 = \dfrac{7}{2}\left[2 \cdot 1 + 6 \cdot \tfrac{1}{2}\right] = \tfrac{35}{2}$

37. Here $a_1 = 3$ and $d = -5$, so: $\ S_6 = \dfrac{6}{2}\left[2 \cdot 3 + 5 \cdot (-5)\right] = -57$

39. Here $a_1 = -4$ and $d = 2$, so: $\ S_{10} = \dfrac{10}{2}\left[2 \cdot (-4) + 9 \cdot 2\right] = 50$

41. Here $x_1 = \tfrac{1}{4}$ and $x_1 = \tfrac{1}{4}, d = \tfrac{1}{3} - \tfrac{1}{4} = \tfrac{1}{12}$, so: $\ S_8 = \dfrac{8}{2}\left[2 \cdot \tfrac{1}{4} + 7 \cdot \tfrac{1}{12}\right] = 4 \cdot \tfrac{13}{12} = \tfrac{13}{3}$

43. Here $a_1 = 4$ and $d = 4$, so: $\ S_8 = \dfrac{8}{2}\left[2 \cdot 4 + 7 \cdot 4\right] = 144$

45. Here $a_1 = 10$ and $d = 10$, so: $\ S_{10} = \dfrac{10}{2}\left[2 \cdot 10 + 9 \cdot 10\right] = 550$

47. Here $a_1 = 7$ and $d = 3$, so: $\ S_7 = \dfrac{7}{2}\left[2 \cdot 7 + 6 \cdot 3\right] = 112$

49. Here $a_1 = 1$ and $d = 1$. Finding the difference:

$$S_{70} - S_{39} = \frac{70}{2}\left[2 \cdot 1 + 69 \cdot 1\right] - \frac{39}{2}\left[2 \cdot 1 + 38 \cdot 1\right] = \frac{70}{2}(71) - \frac{39}{2}(40) = 1705$$

51. Here $a_1 = \dfrac{1}{\pi}$ and $d = \dfrac{2}{\pi}$, so: $\ S_8 = \dfrac{8}{2}\left[2 \cdot \dfrac{1}{\pi} + 7 \cdot \dfrac{2}{\pi}\right] = 4 \cdot \dfrac{16}{\pi} = \dfrac{64}{\pi}$

53. Here $a_1 = 40$ and $d = -2$, so: $\ S_{12} = \dfrac{12}{2}\left[2 \cdot 40 + 11 \cdot (-2)\right] = 348$ logs

55. Here $a_1 = 10000$ and $d = -500$, so:

$$a_{18} = 10000 + 17(-500) = 1500$$
$$S_{18} = \frac{18}{2}\left[2 \cdot 10000 + 17 \cdot (-500)\right] = 103{,}500$$

The eighteenth prize is \$1500, and the total value of all prizes is \$103,500.

57. a. Using $a_1 = 1$, $d = 2$, and $S_n = 12 \cdot 48 = 576$:

$$S_n = \frac{n}{2}\left[2a_1 + (n-1)d\right]$$
$$576 = \frac{n}{2}\left[2 + 2(n-2)\right]$$
$$576 = \frac{n}{2}(2n)$$
$$n^2 = 576$$
$$n = 24$$

Yes, it is possible to construct the pyramid.

b. The pyramid will be 24 boxes high.

c. Finding a_{24}: $\ a_{24} = a_1 + 23d = 1 + 23 \cdot 2 = 47$

The bottom row must have 47 boxes in it.

59. Using $a_1 = 1$ and $d = 2$: $\ S_{100} = \dfrac{100}{2}\left[2 \cdot 1 + 99 \cdot 2\right] = 10{,}000$

61. Using $a_1 = -3$ and $d = 10$: $S_{25} = \dfrac{25}{2}[2\cdot(-3)+24\cdot10] = 2925$

63. Using the formula $S_n = \dfrac{n}{2}\left(a_1 + a_n\right)$:

$$S_{20} = \frac{20}{2}\left(a_1 + a_{20}\right)$$
$$910 = 10\left(a_1 + 95\right)$$
$$91 = a_1 + 95$$
$$a_1 = -4$$

65. Using the formula $S_n = \dfrac{n}{2}\left(a_1 + a_n\right)$:

$$S_{15} = \frac{15}{2}\left(a_1 + a_{15}\right)$$
$$390 = \frac{15}{2}\left(a_1 + 56\right)$$
$$780 = 15a_1 + 840$$
$$15a_1 = -60$$
$$a_1 = -4$$

Now finding d:
$$a_{15} = a_1 + 14d$$
$$56 = -4 + 14d$$
$$14d = 60$$
$$d = \tfrac{30}{7}$$

Now finding S_5: $S_5 = \dfrac{5}{2}\left[2(-4)+4\left(\tfrac{30}{7}\right)\right] = \tfrac{5}{2}\cdot\tfrac{64}{7} = \tfrac{160}{7}$

67. Using the formula $S_n = \dfrac{n}{2}\left[2a_1 + (n-1)d\right]$:

$$S_{22} = \frac{22}{2}\left[2a_1 + 21d\right]$$
$$2431 = 11[2(-5)+21d]$$
$$221 = -10 + 21d$$
$$21d = 231$$
$$d = 11$$

69. No. If $k < j$, the formula is still valid.

11.4 Geometric Sequences and Series

1. This sequence is arithmetic, with $d = 3$.

3. This sequence is geometric, with $r = \tfrac{1}{4}$.

5. This sequence is neither arithmetic nor geometric.

7. This sequence is geometric, with $r = \tfrac{4}{7}$.

9. This sequence is geometric with $r = 5$. Using $a_n = a_1 r^{n-1}$: $a_7 = a_1 r^6 = 1\cdot5^6 = 15625$

11. This sequence is geometric with $r = -\tfrac{2}{3}$. Using $a_n = a_1 r^{n-1}$: $a_5 = a_1 r^4 = \tfrac{1}{2}\cdot\left(-\tfrac{2}{3}\right)^4 = \tfrac{8}{81}$

13. This sequence is arithmetic with $d = 7$. Using $a_n = a_1 + (n-1)d$: $a_{10} = 1 + 9 \cdot 7 = 64$

15. This sequence is geometric with $r = \frac{1}{3}$. Using $a_n = a_1 r^{n-1}$: $a_6 = a_1 r^5 = 9 \cdot \left(\frac{1}{3}\right)^5 = \frac{1}{27}$

17. This sequence is geometric with $r = \frac{1}{2}$. Using $a_n = a_1 r^{n-1}$: $a_9 = a_1 r^8 = 2 \cdot \left(\frac{1}{2}\right)^8 = \frac{1}{128}$

19. This sequence is geometric with $r = \frac{1}{4}$. Using $a_n = a_1 r^{n-1}$: $a_9 = a_1 r^8 = 8 \cdot \left(\frac{1}{4}\right)^8 = \frac{1}{8192}$

21. This sequence is geometric with $r = \sqrt{5}$. Using $a_n = a_1 r^{n-1}$: $a_7 = a_1 r^6 = \sqrt{5} \cdot \left(\sqrt{5}\right)^6 = 125\sqrt{5}$

23. Using $a_n = a_1 r^{n-1}$: $a_4 = a_1 r^3 = 10 \cdot 2^3 = 80$

25. Using $a_n = a_1 r^{n-1}$:

$$a_6 = a_1 r^5 \qquad\qquad a_3 = a_1 r^2$$
$$216 = a_1 r^5 \qquad\qquad 1 = a_1 r^2$$

Dividing:

$$\frac{a_1 r^5}{a_1 r^2} = 216$$
$$r^3 = 216$$
$$r = 6$$

Thus:

$$1 = a_1 \cdot 6^2$$
$$a_1 = \frac{1}{36}$$

27. Using $a_n = a_1 r^{n-1}$:

$$a_2 = a_1 r^1 \qquad\qquad a_5 = a_1 r^4$$
$$\frac{1}{\sqrt{3}} = a_1 r \qquad\qquad -\frac{1}{9} = a_1 r^4$$

Dividing:

$$\frac{a_1 r^4}{a_1 r} = \frac{-\frac{1}{9}}{\frac{1}{\sqrt{3}}}$$
$$r^3 = -\frac{\sqrt{3}}{9} = -\frac{1}{3^{3/2}}$$
$$r = -\frac{1}{\sqrt{3}}$$

Thus:

$$\frac{1}{\sqrt{3}} = a_1 \cdot \left(-\frac{1}{\sqrt{3}}\right)$$
$$a_1 = -1$$

Now finding a_8: $a_8 = a_1 r^7 = (-1) \cdot \left(-\frac{1}{\sqrt{3}}\right)^7 = \frac{1}{27\sqrt{3}} = \frac{\sqrt{3}}{81}$

29. Finding r:
$$a_4 = a_1 r^3$$
$$343 = 1r^3$$
$$r = 7$$
Therefore: $a_5 = a_1 r^4 = 1 \cdot 7^4 = 2401$

31. Using $a_k - a_j = (k - j)d$:
$$a_4 - a_1 = 3d$$
$$-40 - 5 = 3d$$
$$3d = -45$$
$$d = -15$$
Now using $a_n = a_1 + (n-1)d$: $a_8 = a_1 + 7d = 5 + 7(-15) = -100$

33. Using $a_n = a_1 r^{n-1}$:
$$a_5 = a_1 r^4$$
$$\tfrac{1}{81} = 1 \cdot r^4$$
$$r = \tfrac{1}{3}$$
Therefore: $a_7 = a_1 r^6 = 1 \cdot \left(\tfrac{1}{3}\right)^6 = \tfrac{1}{729}$

35. Using $a_1 = 28000$ and $r = \tfrac{9}{10}$: $a_5 = a_1 r^4 = 28000\left(\tfrac{9}{10}\right)^4 \approx \$18{,}370.80$

37. Using $a_1 = 29000$ and $r = \tfrac{4}{5}$: $a_4 = a_1 r^3 = 29000\left(\tfrac{4}{5}\right)^3 = \$14{,}848$

39. Using $a_1 = 100{,}000$ and $r = 1.025$: $a_{11} = a_1 r^{10} = 100{,}000(1.025)^{10} \approx 128{,}008$

41. Using $a_1 = 10{,}000$ and $r = 1.05$: $a_6 = a_1 r^5 = 10{,}000(1.05)^5 \approx \$12{,}762.82$

43. Using $a_1 = 5{,}000$ and $r = 1 + \dfrac{0.06}{2} = 1.03$: $a_{13} = a_1 r^{12} = 5{,}000(1.03)^{12} \approx \$7{,}128.80$

45. Using $a_1 = 20$ and $r = \tfrac{1}{2}$: $a_6 = a_1 r^5 = 20\left(\tfrac{1}{2}\right)^5 = 0.625$ feet

47. Using $a_1 = 4$ and $r = 3$: $S_5 = \dfrac{a_1\left(1 - r^5\right)}{1 - r} = \dfrac{4\left(1 - 3^5\right)}{1 - 3} = 484$

49. Using $a_1 = \tfrac{2}{3}$ and $r = -\tfrac{3}{5}$: $S_6 = \dfrac{a_1\left(1 - r^6\right)}{1 - r} = \dfrac{\tfrac{2}{3}\left(1 - \left(-\tfrac{3}{5}\right)^6\right)}{1 - \left(-\tfrac{3}{5}\right)} = \dfrac{\tfrac{2}{3}\left(1 - \tfrac{729}{15625}\right)}{\tfrac{8}{5}} = \tfrac{3724}{9375} \approx 0.3972$

51. Using $a_1 = 2$ and $d = 4$: $S_8 = \dfrac{8}{2}\left[2a_1 + 7d\right] = 4(2 \cdot 2 + 7 \cdot 4) = 128$

53. Using $a_1 = -\tfrac{3}{4}$ and $r = -\tfrac{1}{3}$: $S_5 = \dfrac{a_1\left(1 - r^5\right)}{1 - r} = \dfrac{-\tfrac{3}{4}\left[1 - \left(-\tfrac{1}{3}\right)^5\right]}{1 - \left(-\tfrac{1}{3}\right)} = \dfrac{-\tfrac{3}{4}\left(1 + \tfrac{1}{243}\right)}{\tfrac{4}{3}} = -\tfrac{61}{108} \approx -0.5648$

55. Using $a_1 = \tfrac{1}{4}$ and $r = 2$: $S_{10} = \dfrac{a_1\left(1 - r^{10}\right)}{1 - r} = \dfrac{\tfrac{1}{4}\left[1 - 2^{10}\right]}{1 - 2} = \tfrac{1023}{4} = 255.75$

57. Using $a_1 = 2$ and $r = 2$: $S_7 = \dfrac{a_1\left(1 - r^7\right)}{1 - r} = \dfrac{2\left[1 - 2^7\right]}{1 - 2} = 254$

59. Using $a_1 = \frac{1}{3}$ and $r = -\frac{1}{3}$: $S_8 = \dfrac{a_1\left(1-r^8\right)}{1-r} = \dfrac{\frac{1}{3}\left[1-\left(-\frac{1}{3}\right)^8\right]}{1-\left(-\frac{1}{3}\right)} = \frac{1640}{6561} \approx 0.2500$

61. Using $a_1 = 3$ and $d = 2$: $S_7 = \dfrac{7}{2}\left[2a_1 + 6d\right] = \dfrac{7}{2}\left[2\cdot 3 + 6\cdot 2\right] = 63$

63. The sum is $1 + \frac{1}{10} + \frac{1}{100} + \frac{1}{1,000} + \frac{1}{10,000} + \frac{1}{100,000} = 1.11111$.

65. The sum is $\frac{1}{100} + \frac{1}{1000} + \ldots + \frac{1}{10^9} = 0.011111111$.

67. The total distance is given by: $60 + 2\left(\frac{1}{3}\cdot 60\right) + 2\left(\frac{1}{9}\cdot 60\right) + 2\left(\frac{1}{27}\cdot 60\right) + 2\left(\frac{1}{81}\cdot 60\right) \approx 119.26$ feet

69. **a.** Since the number of letters sent follows the sequence $5, 5^2, 5^3, \ldots$, this forms a geometric sequence.

b. $5^4 = 625$ letters

c. Finding S_8: $S_8 = \dfrac{a_1\left(1-r^8\right)}{1-r} = \dfrac{5\left(1-5^8\right)}{1-5} = 488,280$ letters

71. Using $a_1 = 2$ and $r = \frac{1}{2}$: $S = \dfrac{a_1}{1-r} = \dfrac{2}{1-\frac{1}{2}} = 4$

73. Using $a_1 = 3$ and $r = \frac{1}{3}$: $S = \dfrac{a_1}{1-r} = \dfrac{3}{1-\frac{1}{3}} = \frac{9}{2}$

75. Using $a_1 = \frac{1}{5}$ and $r = \frac{1}{2}$: $S = \dfrac{a_1}{1-r} = \dfrac{\frac{1}{5}}{1-\frac{1}{2}} = \frac{2}{5}$

77. Using $a_1 = 5$ and $r = \frac{1}{3}$: $S = \dfrac{a_1}{1-r} = \dfrac{5}{1-\frac{1}{3}} = \frac{15}{2}$

79. Since $r = \frac{4}{3} > 1$, the sum does not exist.

81. The series is $\frac{47}{100} + \frac{47}{10000} + \frac{47}{1000000} + \ldots$. Using $a_1 = \frac{47}{100}$ and $r = \frac{1}{100}$: $S = \dfrac{a_1}{1-r} = \dfrac{\frac{47}{100}}{1-\frac{1}{100}} = \frac{47}{99}$

83. The series is $1.0 + \left(\frac{681}{10000} + \frac{681}{10^7} + \ldots\right)$. Using $a_1 = \frac{681}{10000}$ and $r = \frac{1}{1000}$:

$$1.0 + S = 1.0 + \dfrac{a_1}{1-r} = 1 + \dfrac{\frac{681}{10000}}{1-\frac{1}{1000}} = 1 + \frac{681}{9990} = \frac{10671}{9990}$$

85. The total distance can be written as: $15 + 2\left(10 + \frac{20}{3} + \frac{40}{9} + \ldots\right) = 15 + 2S$

To find S, let $a_1 = 10$ and $r = \frac{2}{3}$: $S = \dfrac{a_1}{1-r} = \dfrac{10}{1-\frac{2}{3}} = 30$

So the total distance is: $15 + 2(30) = 75$ feet

87. Multiplying: $(1-r)\left(1 + r + r^2 + \ldots + r^{n-1}\right) = 1 + r + r^2 + \ldots + r^{n-1} - r - r^2 - r^3 - \ldots - r^n = 1 - r^n$

89. **a.** Using $a_1 = 20$ and $r = 5$: $S_{10} = \dfrac{20\left(1-5^{10}\right)}{1-5} = 48828120$

b. Using $a_1 = 6$ and $r = 2$: $S_{10} = \dfrac{6\left(1-2^{10}\right)}{1-2} = 6138$

91.   **a.**   The infinite sum is: $S = \dfrac{\frac{2}{5}}{1 - \frac{2}{5}} = \frac{2}{3}$

$S_{10} \approx 0.666597,\ S_{20} \approx 0.666668,\ S_{30} \approx 0.666667$

b.   The infinite sum is: $S = \dfrac{\frac{3}{16}}{1 - \frac{3}{4}} = \frac{3}{4}$

$S_{10} \approx 0.707765,\ S_{20} \approx 0.747622,\ S_{30} \approx 0.749866$

c.   The infinite sum is: $S = \dfrac{\frac{125}{6}}{1 - \frac{5}{6}} = 125$

$S_{10} \approx 104.811802,\ S_{20} \approx 121.739493,\ S_{30} \approx 124.47341$

Note in all three cases how S_n approaches the value of S as n becomes larger.

11.5 Mathematical Induction

1.   Since $1 = \dfrac{1(1+1)}{2}$, P_1 is true. Assume P_n is true:

$$1 + 2 + 3 + \ldots + n = \frac{n(n+1)}{2}$$

$$1 + 2 + 3 + \ldots + n + n + 1 = \frac{n(n+1)}{2} + (n+1)$$

$$1 + 2 + 3 + \ldots + n + n + 1 = \frac{n(n+1) + 2(n+1)}{2}$$

$$1 + 2 + 3 + \ldots + n + n + 1 = \frac{(n+1)(n+2)}{2}$$

This last statement is P_{n+1}, so the induction is complete.

3.   Since $3 = 1(2+1)$, T_1 is true. Assume T_n is true:

$$3 + 7 + 11 + \ldots + (4n-1) = n(2n+1)$$
$$3 + 7 + 11 + \ldots + (4n-1) + (4n+3) = n(2n+1) + (4n+3)$$
$$3 + 7 + 11 + \ldots + (4n-1) + (4n+3) = 2n^2 + 5n + 3$$
$$3 + 7 + 11 + \ldots + (4n-1) + (4n+3) = (n+1)(2n+3)$$
$$3 + 7 + 11 + \ldots + (4n-1) + (4n+3) = (n+1)\left[2(n+1)+1\right]$$

This last statement is T_{n+1}, so the induction is complete.

5. Using $n = 1$, $2 = \dfrac{1 \cdot 4}{2}$, P_1 is true. Assume P_n is true:

$$2 + 5 + 8 + \ldots + (3n - 1) = \frac{n(3n + 1)}{2}$$

$$2 + 5 + 8 + \ldots + (3n - 1) + (3n + 2) = \frac{n(3n + 1)}{2} + (3n + 2)$$

$$2 + 5 + 8 + \ldots + (3n - 1) + (3n + 2) = \tfrac{1}{2}\left(3n^2 + n + 6n + 4\right)$$

$$2 + 5 + 8 + \ldots + (3n - 1) + (3n + 2) = \tfrac{1}{2}\left(3n^2 + 7n + 4\right)$$

$$2 + 5 + 8 + \ldots + (3n - 1) + (3n + 2) = \tfrac{1}{2}(n + 1)(3n + 4)$$

$$2 + 5 + 8 + \ldots + (3n - 1) + (3n + 2) = \frac{(n + 1)\left[3(n + 1) + 1\right]}{2}$$

This last statement is P_{n+1}, so the induction to complete.

7. Since $2 = \dfrac{1 \cdot 4}{2}$, R_1 is true. Assume R_n is true:

$$2 + 9 + 16 + \ldots + (7n - 5) = \frac{n(7n - 3)}{2}$$

$$2 + 9 + 16 + \ldots + (7n - 5) + (7n + 2) = \frac{n(7n - 3)}{2} + (7n + 2)$$

$$2 + 9 + 16 + \ldots + (7n - 5) + (7n + 2) = \tfrac{1}{2}\left(7n^2 - 3n + 14n + 4\right)$$

$$2 + 9 + 16 + \ldots + (7n - 5) + (7n + 2) = \tfrac{1}{2}\left(7n^2 + 11n + 4\right)$$

$$2 + 9 + 16 + \ldots + (7n - 5) + (7n + 2) = \tfrac{1}{2}(n + 1)(7n + 4)$$

$$2 + 9 + 16 + \ldots + (7n - 5) + (7n + 2) = \frac{(n + 1)\left[7(n + 1) - 3\right]}{2}$$

This last statement is R_{n+1}, so the induction is complete.

9. Since $2^2 = \dfrac{2 \cdot 2 \cdot 3}{3}$, P_1 is true. Assume P_n is true:

$$2^2 + 4^2 + 6^2 + \ldots + (2n)^2 = \frac{2n(n + 1)(2n + 1)}{3}$$

$$2^2 + 4^2 + 6^2 + \ldots + (2n)^2 + (2n + 2)^2 = \frac{2n(n + 1)(2n + 1)}{3} + (2n + 2)^2$$

$$2^2 + 4^2 + 6^2 + \ldots + (2n)^2 + (2n + 2)^2 = \frac{2n(n + 1)(2n + 1)}{3} + \frac{4(n + 1)^2}{1}$$

$$2^2 + 4^2 + 6^2 + \ldots + (2n)^2 + (2n + 2)^2 = \frac{2(n + 1)\left[n(2n + 1) + 6(n + 1)\right]}{3}$$

$$2^2 + 4^2 + 6^2 + \ldots + (2n)^2 + (2n + 2)^2 = \frac{2(n + 1)\left(2n^2 + 7n + 6\right)}{3}$$

$$2^2 + 4^2 + 6^2 + \ldots + (2n)^2 + (2n + 2)^2 = \frac{2(n + 1)(n + 2)(2n + 3)}{3}$$

$$2^2 + 4^2 + 6^2 + \ldots + (2n)^2 + (2n + 2)^2 = \frac{2(n + 1)(n + 2)\left[2(n + 1) + 1\right]}{3}$$

This last statement is P_{n+1}, so the induction is complete.

11. Since $2 = 2^2 - 2$, P_1 is true. Assume P_n is true:
$$2 + 2^2 + 2^3 + ... + 2^n = 2^{n+1} - 2$$
$$2 + 2^2 + 2^3 + ... + 2^n + 2^{n+1} = 2^{n+1} - 2 + 2^{n+1}$$
$$2 + 2^2 + 2^3 + ... + 2^n + 2^{n+1} = 2 \cdot 2^{n+1} - 2$$
$$2 + 2^2 + 2^3 + ... + 2^n + 2^{n+1} = 2^{n+2} - 2$$
This last statement is P_{n+1}, so the induction is complete.

13. Since $\dfrac{1}{1 \cdot 2} = \dfrac{1}{1+1}$, T_1 is true. Assume T_n is true:

$$\frac{1}{1 \cdot 2} + \frac{1}{2 \cdot 3} + \frac{1}{3 \cdot 4} + ... + \frac{1}{n(n+1)} = \frac{n}{n+1}$$

$$\frac{1}{1 \cdot 2} + \frac{1}{2 \cdot 3} + \frac{1}{3 \cdot 4} + ... + \frac{1}{n(n+1)} + \frac{1}{(n+1)(n+2)} = \frac{n}{n+1} + \frac{1}{(n+1)(n+2)}$$

$$\frac{1}{1 \cdot 2} + \frac{1}{2 \cdot 3} + \frac{1}{3 \cdot 4} + ... + \frac{1}{n(n+1)} + \frac{1}{(n+1)(n+2)} = \frac{n(n+2)+1}{(n+1)(n+2)}$$

$$\frac{1}{1 \cdot 2} + \frac{1}{2 \cdot 3} + \frac{1}{3 \cdot 4} + ... + \frac{1}{n(n+1)} + \frac{1}{(n+1)(n+2)} = \frac{n^2 + 2n + 1}{(n+1)(n+2)}$$

$$\frac{1}{1 \cdot 2} + \frac{1}{2 \cdot 3} + \frac{1}{3 \cdot 4} + ... + \frac{1}{n(n+1)} + \frac{1}{(n+1)(n+2)} = \frac{(n+1)^2}{(n+1)(n+2)}$$

$$\frac{1}{1 \cdot 2} + \frac{1}{2 \cdot 3} + \frac{1}{3 \cdot 4} + ... + \frac{1}{n(n+1)} + \frac{1}{(n+1)(n+2)} = \frac{n+1}{n+2}$$

This last statement is T_{n+1}, so the induction is complete.

15. Since $1 \cdot 2 = \dfrac{1 \cdot 2 \cdot 3}{3}$, S_1 is true. Assume S_n is true:

$$1 \cdot 2 + 2 \cdot 3 + 3 \cdot 4 + ... + n(n+1) = \frac{n(n+1)(n+2)}{3}$$

$$1 \cdot 2 + 2 \cdot 3 + 3 \cdot 4 + ... + n(n+1) + (n+1)(n+2) = \frac{n(n+1)(n+2)}{3} + (n+1)(n+2)$$

$$1 \cdot 2 + 2 \cdot 3 + 3 \cdot 4 + ... + n(n+1) + (n+1)(n+2) = \frac{n(n+1)(n+2) + 3(n+1)(n+2)}{3}$$

$$1 \cdot 2 + 2 \cdot 3 + 3 \cdot 4 + ... + n(n+1) + (n+1)(n+2) = \frac{(n+1)(n+2)(n+3)}{3}$$

This last statement is S_{n+1}, so the induction is complete.

17. Since $1 + \dfrac{1}{1} = 1 + 1$, T_1 is true. Assume T_n is true:

$$\left(1 + \frac{1}{1}\right)\left(1 + \frac{1}{2}\right)\left(1 + \frac{1}{3}\right) \cdots \left(1 + \frac{1}{n}\right) = n + 1$$

$$\left(1 + \frac{1}{1}\right)\left(1 + \frac{1}{2}\right)\left(1 + \frac{1}{3}\right) \cdots \left(1 + \frac{1}{n}\right)\left(1 + \frac{1}{n+1}\right) = (n+1)\left(1 + \frac{1}{n+1}\right)$$

$$\left(1 + \frac{1}{1}\right)\left(1 + \frac{1}{2}\right)\left(1 + \frac{1}{3}\right) \cdots \left(1 + \frac{1}{n}\right)\left(1 + \frac{1}{n+1}\right) = (n+1) + 1$$

$$\left(1 + \frac{1}{1}\right)\left(1 + \frac{1}{2}\right)\left(1 + \frac{1}{3}\right) \cdots \left(1 + \frac{1}{n}\right)\left(1 + \frac{1}{n+1}\right) = n + 2$$

This last statement is T_{n+1}, so the induction is complete.

19. Since $a = 1(a+0)$, S_1 is true. Assume S_n is true:

$$a+(a+d)+(a+2d)+...+[a+(n-1)d] = n\left(a+\frac{n-1}{2}d\right)$$

$$a+(a+d)+(a+2d)+...+[a+(n-1)d]+[a+nd] = n\left(a+\frac{n-1}{2}d\right)+(a+nd)$$

$$a+(a+d)+(a+2d)+...+[a+(n-1)d]+[a+nd] = an+\frac{n(n-1)}{2}d+a+nd$$

$$a+(a+d)+(a+2d)+...+[a+(n-1)d]+[a+nd] = a(n+1)+\frac{nd}{2}(n-1+2)$$

$$a+(a+d)+(a+2d)+...+[a+(n-1)d]+[a+nd] = a(n+1)+\frac{n(n+1)d}{2}$$

$$a+(a+d)+(a+2d)+...+[a+(n-1)d]+[a+nd] = (n+1)\left[a+\frac{n}{2}d\right]$$

This last statement is S_{n+1}, so the induction is complete.

21. For $n = 1$: $1^3 - 1 + 6 = 6$, which is divisible by 3

Let P_n be the statement that $n^3 - n + 6$ is divisible by 3. Thus $n^3 - n + 6 = 3k$ for some constant k.
Now consider P_{n+1}:

$$
\begin{aligned}
(n+1)^3 - (n+1) + 6 &= n^3 + 3n^2 + 3n + 1 - n - 1 + 6 \\
&= n^3 + 3n^2 + 2n + 6 \\
&= \left(n^3 - n + 6\right) + 3n^2 + 3n \\
&= 3k + 3\left(n^2 + n\right) \\
&= 3\left(k + n^2 + n\right)
\end{aligned}
$$

Thus P_{n+1} is true and the induction is complete. Alternatively, note that:

$$
\begin{aligned}
(n+1)^3 - (n+1) + 6 &= (n+1)\left[(n+1)^2 - 1\right] + 6 \\
&= (n+1)\left(n^2 + 2n + 1 - 1\right) + 6 \\
&= (n+1)\left(n^2 + 2n\right) + 6 \\
&= n(n+1)(n+2) + 6
\end{aligned}
$$

But note that $n(n+1)(n+2)$ must be divisible by 3, since it represents the product of three consecutive integers. Since 6 is divisible by 3 also, the expression $(n+1)^3 - (n+1) + 6$ must be divisible by 3.

23. For $n = 1$: $x^1 - 1 = x - 1$, so P_1 is true.

Let P_n denote the statement that $x - 1$ is a factor of $x^n - 1$. Thus $x^n - 1 = (x-1)Q(x)$ for some polynomial $Q(x)$. Now consider P_{n+1}:

$$x^{n+1} - 1 = x^{n+1} - x + x - 1 = x\left(x^n - 1\right) + (x-1) = x(x-1)Q(x) + (x-1) = (x-1)[xQ(x)+1]$$

Thus $x - 1$ is a factor of $x^{n+1} - 1$, and P_{n+1} is true. An alternate approach might be to use long division.

25. For $n = 1$: $x^1 - y^1 = x - y$, so P_1 is true.

Let P_n denote the statement that $x - y$ is a factor of $x^n - y^n$. Thus $x^n - y^n = (x - y)Q(x)$ for some polynomial $Q(x)$. Now consider P_{n+1}:

$$\begin{aligned} x^{n+1} - y^{n+1} &= x^{n+1} - xy^n + xy^n - y^{n+1} \\ &= x\left(x^n - y^n\right) + y^n (x - y) \\ &= x(x - y)Q(x) + y^n (x - y) \\ &= (x - y)\left[xQ(x) + y^n\right] \end{aligned}$$

Thus $x - y$ is a factor of $x^{n+1} - y^{n+1}$, and P_{n+1} is true. An alternate approach might be to use long division.

27. For $n = 1$: $9^1 - 1 = 8$, which is divisible by 4

Let P_n denote the statement that $9^n - 1$ is divisible by 4, so $9^n - 1 = 4k$ for some constant k. Now consider P_{n+1}: $9^{n+1} - 1 = 9^{n+1} - 9 + 9 - 1 = 9\left(9^n - 1\right) + 8 = 9 \cdot 4k + 8 = 4(9k + 2)$

Thus 4 is a factor of $9^{n+1} - 1$, and P_{n+1} is true.

29. For $n = 1$: $\left[r(\cos\theta + i\sin\theta)\right]^1 = r(\cos\theta + i\sin\theta)$, so P_1 is true

Let P_n denote the statement $\left[r(\cos\theta + i\sin\theta)\right]^n = r^n (\cos n\theta + i\sin n\theta)$. Now consider P_{n+1}:

$$\begin{aligned} \left[r(\cos\theta + i\sin\theta)\right]^{n+1} &= r(\cos\theta + i\sin\theta) \cdot \left[r(\cos\theta + i\sin\theta)\right]^n \\ &= r(\cos\theta + i\sin\theta) \cdot r^n (\cos n\theta + i\sin n\theta) \\ &= r^{n+1}\left[\cos\theta \cos n\theta - \sin\theta\sin n\theta + i(\sin\theta\cos n\theta + \cos\theta\sin n\theta)\right] \\ &= r^{n+1}\left[\cos(n+1)\theta + i\sin(n+1)\theta\right] \end{aligned}$$

This last statement is P_{n+1}, so the induction is complete.

31. For $n = 3$: $3^2 > 2 \cdot 3 + 1$, so P_3 is true

Assume P_n: $n^2 > 2n + 1$ is true. Now consider P_{n+1}:

$$(n+1)^2 = n^2 + 2n + 1 > (2n+1) + (2n+1) = 4n + 2 = (2n+3) + (2n-1) > 2n + 3 = 2(n+1) + 1$$

Thus P_{n+1} is true and the induction is complete.

33. For $n = 5$: $\left(\frac{7}{5}\right)^5 = 5.37824 > 5$, so P_5 is true

Assume P_n: $\left(\frac{7}{5}\right)^n > n$ is true. Now consider P_{n+1}:

$$\left(\frac{7}{5}\right)^{n+1} = \left(\frac{7}{5}\right)\left(\frac{7}{5}\right)^n > \frac{7}{5}n = n + \frac{2}{5}n \geq n + \frac{2}{5}(5) > n + 1$$

Thus P_{n+1} is true and the induction is complete.

35. For $n = 11$: $11^2 = 121 > 10(11)$, so P_{11} is true

Assume P_n: $n^2 > 10n$ is true. Now consider P_{n+1}:

$$(n+1)^2 = n^2 + 2n + 1 > 10n + 2n + 1 > 10n + 2(10) + 1 = 10n + 21 > 10n + 10 = 10(n+1)$$

Thus P_{n+1} is true and the induction is complete. Another approach is to consider:

$$y = x^2 - 10x = \left(x^2 - 10x + 25\right) - 25 = (x - 5)^2 - 25$$

For $x > 5$, $y > 0$, thus $x^2 - 10x > 0$ for all $x > 5$. Thus $n^2 - 10n > 0$, so $n^2 > 10n$.

37. For $n = 4$: $4! = 24$ and $2^4 = 16$, so P_4 is true

Assume P_n: $n! > 2^n$ is true. Now consider P_{n+1}: $(n+1)! = (n+1)n! > (n+1)2^n > 2 \cdot 2^n = 2^{n+1}$

Thus P_{n+1} is true and the induction is complete.

39. For $n = 1$: $|a_1| \le |a_1|$, so P_1 is true

Assume P_n: $|a_1 + a_2 + \ldots + a_n| \le |a_1| + |a_2| + \ldots + |a_n|$ is true. Now consider P_{n+1}:

$$|a_1 + a_2 + \ldots + a_n + a_{n+1}| \le |a_1 + a_2 + \ldots + a_n| + |a_{n+1}| \le |a_1| + |a_2| + \ldots + |a_n| + |a_{n+1}|$$

Thus P_{n+1} is true and the induction is complete.

41. Note that $\left(\frac{4}{3}\right)^8 \approx 9.98$ and $\frac{4}{3}(8) \approx 10.67$, so $\left(\frac{4}{3}\right)^8 < \frac{4}{3}(8)$. Also note that $\left(\frac{4}{3}\right)^9 \approx 13.32$ and

$\frac{4}{3}(9) = 12$, so $\left(\frac{4}{3}\right)^9 > \frac{4}{3}(9)$. Thus $N = 9$ is the smallest natural number such that $\left(\frac{4}{3}\right)^N > \frac{4}{3}N$.

Assume P_n: $\left(\frac{4}{3}\right)^n > \frac{4}{3}n$ is true. Now consider P_{n+1}:

$$\left(\frac{4}{3}\right)^{n+1} = \frac{4}{3} \cdot \left(\frac{4}{3}\right)^n > \frac{4}{3}\left(\frac{4}{3}n\right) = \frac{4}{3}\left(n + \frac{1}{3}n\right) > \frac{4}{3}\left(n + \frac{1}{3} \cdot 9\right) = \frac{4}{3}(n+3) > \frac{4}{3}(n+1)$$

Thus P_{n+1} is true and the induction is complete.

43. Note that $4^3 = 64$ and $(4+5)^2 = 81$, so $4^3 < (4+5)^2$. Also note that $5^3 = 125$ and

$(5+5)^2 = 100$, so $5^3 > (5+5)^2$. Thus $N = 5$ is the smallest natural number such that $n^3 > (n+5)^2$.

Assume P_n: $n^3 > (n+5)^2$ is true. Now consider P_{n+1}:

$$\begin{aligned}
(n+1)^3 &= n^3 + 3n^2 + 3n + 1 \\
&> (n+5)^2 + 3n^2 + 3n + 1 \\
&= n^2 + 10n + 25 + 3n^2 + 3n + 1 \\
&= 4n^2 + 13n + 26 \\
&= \left(n^2 + 12n + 36\right) + \left(3n^2 + n - 10\right) \\
&> (n+6)^2 + \left[3 \cdot 5^2 + 5 - 10\right] \\
&= (n+6)^2 + 70 \\
&> (n+6)^2
\end{aligned}$$

Thus P_{n+1} is true and the induction is complete.

45. Note that $2^4 = 16$ and $4^2 = 16$, so $2^4 = 4^2$. Also note that $2^5 = 32$ and $5^2 = 25$, so $2^5 > 5^2$.

Thus $N = 5$ is the smallest natural number such that $2^n > n^2$. Assume P_n: $2^n > n^2$ is true. Now

consider P_{n+1}: $2^{n+1} = 2 \cdot 2^n > 2 \cdot n^2 = n^2 + n^2 > n^2 + 2n + 1 = (n+1)^2$

Thus P_{n+1} is true and the induction is complete. Note that we used the inequality $n^2 > 2n + 1$,

which was proved in Exercise 31.

47. It can be concluded that $P_1, P_2, \ldots, P_{49}$ are true.

49. It can be concluded that P_n is true for $n \ge 10$.

51. It can only be concluded that P_{10} is not true.

53. It can be concluded that P_n is true for odd natural numbers n.

55. It can be concluded that P_n is true for all natural numbers n.

57.   It can be concluded that $P_1, P_3, P_9, P_{27}, \ldots$ are true, so P_n is true for odd powers of 3.

59.   This is true. For $n = 1$, $5^1 + 3 = 8$, which is divisible by 4. Assume $5^n + 3 = 4k$, for some natural number k. Form P_{n+1}: $5^{n+1} + 3 = 5 \cdot 5^n + 3 = 5(4k - 3) + 3 = 20k - 12 = 4(5k - 3)$

Thus $5^{n+1} + 3$ is divisible by 4, so P_{n+1} is true and the induction is complete.

11.6 Permutations and Combinations

1.   The value is: $P(8,3) = 8 \cdot 7 \cdot 6 = 336$

This represents the number of ways to choose 3 objects out of 8, where order matters.

3.   The value is: $\binom{12}{7} = \dfrac{12 \cdot 11 \cdot 10 \cdot 9 \cdot 8}{5 \cdot 4 \cdot 3 \cdot 2 \cdot 1} = 792$

This represents the number of ways to choose 7 objects out of 12, where order does not matter.

5.   The value is: $P(6,5) = 6 \cdot 5 \cdot 4 \cdot 3 \cdot 2 = 720$

This represents the number of ways to choose 5 objects out of 6, where order matters.

7.   The value is: $\binom{18}{12} = \dfrac{18 \cdot 17 \cdot 16 \cdot 15 \cdot 14 \cdot 13}{6 \cdot 5 \cdot 4 \cdot 3 \cdot 2 \cdot 1} = 18,564$

This represents the number of ways to choose 12 objects out of 18, where order does not matter.

9.   The value is: $\dfrac{P(8,4)}{P(4,4)} = \dfrac{8 \cdot 7 \cdot 6 \cdot 5}{4 \cdot 3 \cdot 2 \cdot 1} = 70$

11.   The equation is:
$$\frac{n(n-1)}{2} = 105$$
$$n(n-1) = 210$$
$$n^2 - n - 210 = 0$$
$$(n-15)(n+14) = 0$$
$$n = 15 \qquad (n = -14 \text{ is impossible})$$

13.   The number of outfits is: $5 \cdot 8 \cdot 3 = 120$

15.   The number of types is: $8 \cdot 5 \cdot 4 = 160$

17.   The number of arrangements is: $8! = 40,320$

19.   The number of playing orders is: $15! \approx 3 \times 10^{12}$ (3 trillion!)

21.   The number of rings is: $26 \cdot 26 = 676$

23.   **a.**   The number of plates is: $26 \cdot 26 \cdot 26 \cdot 26 \cdot 9 \cdot 9 \cdot 9 = 333,135,504$

   b.   The number of plates is: $26 \cdot 26 \cdot 26 \cdot 26 \cdot 5 \cdot 9 \cdot 9 = 185,075,280$

   c.   The number of plates is: $21 \cdot 26 \cdot 26 \cdot 26 \cdot 4 \cdot 9 \cdot 9 = 119,587,104$

25.   For 4 choices, the number of ways is $4^{20} \approx 1.1 \times 10^{12}$.

For 5 choices, the number or ways is $5^{20} \approx 9.5 \times 10^{13}$.

27.   The number of sets is: $C(12,10) = \dfrac{12!}{2!10!} = 66$

29. **a.** The number of choices is: $C(7,5) = \dfrac{7 \cdot 6}{2!} = 21$

b. The number of choices is: $C(9,4) = \dfrac{9!}{5!4!} = 126$

c. The number of choices is: $C(16,9) = \dfrac{16!}{9!7!} = 11,440$

d. The number of choices is: $C(7,3) \cdot C(9,6) = \dfrac{7!}{3!4!} \cdot \dfrac{9!}{6!3!} = 2940$

31. The number of ways is: $C(25,6) \cdot C(6,2) = \dfrac{25!}{6!19!} \cdot \dfrac{6!}{2!4!} = 2,656,500$

33. In order for the series to go the full seven games, only 3 games can be won by either team during the first six games. The number of ways is: $2 \cdot C(6,3) = 2 \cdot \dfrac{6!}{3!3!} = 40$

35. **a.** The number of combinations is: $5 \cdot 5 \cdot 5 = 125$
b. The number of combinations is: $5 \cdot 5 \cdot 2 = 50$

37. **a.** The number of ways is: $C(13,5) = \dfrac{13!}{5!8!} = 1287$

b. The number of ways is: $C(4,1) \cdot C(13,5) = 4 \cdot \dfrac{13!}{5!8!} = 5148$

c. The number of ways is: $C(4,3) \cdot C(48,2) = 4 \cdot \dfrac{48!}{46!2!} = 4152$

d. The number of ways is: $P(13,2) \cdot C(4,3) \cdot C(4,2) = 13 \cdot 12 \cdot 4 \cdot \dfrac{4!}{2!2!} = 3744$

39. **a.** The number of lines are: $C(10,2) = \dfrac{10!}{2!8!} = 45$

b. The number of triangles are: $C(10,3) = \dfrac{10!}{3!7!} = 120$

41. **a.** The number of arrangements is: $8! = 40,320$

b. The number of arrangements is: $\dfrac{12!}{2!2!} = 119,750,400$

c. The number of arrangements is: $\dfrac{12!}{4!2!} = 9,979,200$

43. The number of routes is: $3 \cdot 5 \cdot 2 = 30$

45. Working from the left-side:

$$\binom{n}{r} + \binom{n}{r-1} = \dfrac{n!}{r!(n-r)!} + \dfrac{n!}{(r-1)!(n-r+1)!}$$

$$= \dfrac{n!(n-r+1) + n!r}{r!(n-r+1)!}$$

$$= \dfrac{n!(n-r+1+r)}{r!(n-r+1)!}$$

$$= \dfrac{n!(n+1)}{r!(n-r+1)!}$$

$$= \dfrac{(n+1)!}{r!(n-r+1)!}$$

$$= \binom{n+1}{r}$$

47. A permutation is counting with order, while a combination is counting without order. The algebraic connection is $P(n,k) = k!C(n,k)$, or $C(n,k) = \dfrac{P(n,k)}{k!}$.

11.7 The Binomial Theorem

1. Using the binomial theorem:

$$(x-4)^3 = \binom{3}{0}x^3 + \binom{3}{1}x^2(-4)^1 + \binom{3}{2}x^1(-4)^2 + \binom{3}{3}(-4)^3 = x^3 - 12x^2 + 48x - 64$$

3. Using the binomial theorem:

$$(a+2)^7$$

$$= \binom{7}{0}a^7 + \binom{7}{1}a^6(2)^1 + \binom{7}{2}a^5(2)^2 + \binom{7}{3}a^4(2)^3 + \binom{7}{4}a^3(2)^4$$

$$+ \binom{7}{5}a^2(2)^5 + \binom{7}{6}a(2)^6 + \binom{7}{7}(2)^7$$

$$= a^7 + 14a^6 + 84a^5 + 280a^4 + 560a^3 + 672a^2 + 448a + 128$$

5. Using the binomial theorem:

$$(2x+y)^6$$

$$= \binom{6}{0}(2x)^6 + \binom{6}{1}(2x)^5 y + \binom{6}{2}(2x)^4 y^2 + \binom{6}{3}(2x)^3 y^3$$

$$+ \binom{6}{4}(2x)^2 y^4 + \binom{6}{5}(2x)y^5 + \binom{6}{6}y^6$$

$$= 64x^6 + 192x^5 y + 240x^4 y^2 + 160x^3 y^3 + 60x^2 y^4 + 12xy^5 + y^6$$

7. Using the binomial theorem:

$$(t-3r)^5$$

$$= \binom{5}{0}t^5 + \binom{5}{1}t^4(-3r) + \binom{5}{2}t^3(-3r)^2 + \binom{5}{3}t^2(-3r)^3 + \binom{5}{4}t(-3r)^4 + \binom{5}{5}(-3r)^5$$

$$= t^5 - 15t^4 r + 90t^3 r^2 - 270t^2 r^3 + 405tr^4 - 243r^5$$

9. Using the binomial theorem:

$$\left(w^2 + 1\right)^8$$

$$= \binom{8}{0}\left(w^2\right)^8 + \binom{8}{1}\left(w^2\right)^7(1) + \binom{8}{2}\left(w^2\right)^6(1)^2 + \binom{8}{3}\left(w^2\right)^5(1)^3 + \binom{8}{4}\left(w^2\right)^4(1)^4$$

$$+ \binom{8}{5}\left(w^2\right)^3(1)^5 + \binom{8}{6}\left(w^2\right)^2(1)^6 + \binom{8}{7}\left(w^2\right)(1)^7 + \binom{8}{8}(1)^8$$

$$= w^{16} + 8w^{14} + 28w^{12} + 56w^{10} + 70w^8 + 56w^6 + 28w^4 + 8w^2 + 1$$

11. Using the binomial theorem:

$$\left(x^3 - y^3\right)^4$$

$$= \binom{4}{0}\left(x^3\right)^4 + \binom{4}{1}\left(x^3\right)^3\left(-y^3\right) + \binom{4}{2}\left(x^3\right)^2\left(-y^3\right)^2 + \binom{4}{3}\left(x^3\right)\left(-y^3\right)^3 + \binom{4}{4}\left(-y^3\right)^4$$

$$= x^{12} - 4x^9 y^3 + 6x^6 y^6 - 4x^3 y^9 + y^{12}$$

13. Using the binomial theorem:

$$\left(a^2 + 2b\right)^9$$

$$= \binom{9}{0}\left(a^2\right)^9 + \binom{9}{1}\left(a^2\right)^8 (2b) + \binom{9}{2}\left(a^2\right)^7 (2b)^2 + \binom{9}{3}\left(a^2\right)^6 (2b)^3 + \binom{9}{4}\left(a^2\right)^5 (2b)^4$$

$$+ \binom{9}{5}\left(a^2\right)^4 (2b)^5 + \binom{9}{6}\left(a^2\right)^3 (2b)^6 + \binom{9}{7}\left(a^2\right)^2 (2b)^7 + \binom{9}{8}\left(a^2\right)(2b)^8 + \binom{9}{9}(2b)^9$$

$$= a^{18} + 18a^{16}b + 144a^{14}b^2 + 672a^{12}b^3 + 2016a^{10}b^4 + 4032a^8b^5 + 5376a^6b^6$$

$$+ 4608a^4b^7 + 2304a^2b^8 + 512b^9$$

15. Using the binomial theorem:

$$\left(x - \tfrac{1}{2}\right)^4 = \binom{4}{0}x^4 + \binom{4}{1}x^3\left(-\tfrac{1}{2}\right) + \binom{4}{2}x^2\left(-\tfrac{1}{2}\right)^2 + \binom{4}{3}x\left(-\tfrac{1}{2}\right)^3 + \binom{4}{4}\left(-\tfrac{1}{2}\right)^4$$

$$= x^4 - 2x^3 + \tfrac{3}{2}x^2 - \tfrac{1}{2}x + \tfrac{1}{16}$$

17. Using the binomial theorem:

$$\left(\tfrac{x}{3} + 4\right)^3 = \binom{3}{0}\left(\tfrac{x}{3}\right)^3 + \binom{3}{1}\left(\tfrac{x}{3}\right)^2 (4) + \binom{3}{2}\left(\tfrac{x}{3}\right)(4)^2 + \binom{3}{3}(4)^3 = \tfrac{1}{27}x^3 + \tfrac{4}{3}x^2 + 16x + 64$$

19. Using the binomial theorem:

$$\left(x - \sqrt{x}\right)^6$$

$$= \binom{6}{0}x^6 + \binom{6}{1}x^5\left(-\sqrt{x}\right) + \binom{6}{2}x^4\left(-\sqrt{x}\right)^2 + \binom{6}{3}x^3\left(-\sqrt{x}\right)^3$$

$$+ \binom{6}{4}x^2\left(-\sqrt{x}\right)^4 + \binom{6}{5}x\left(-\sqrt{x}\right)^5 + \binom{6}{6}\left(-\sqrt{x}\right)^6$$

$$= x^6 - 6x^5\sqrt{x} + 15x^5 - 20x^4\sqrt{x} + 15x^4 - 6x^3\sqrt{x} + x^3$$

21. Using the binomial theorem:

$$\left(\sqrt{x} + \sqrt{y}\right)^6$$

$$= \binom{6}{0}\left(\sqrt{x}\right)^6 + \binom{6}{1}\left(\sqrt{x}\right)^5\left(\sqrt{y}\right) + \binom{6}{2}\left(\sqrt{x}\right)^4\left(\sqrt{y}\right)^2 + \binom{6}{3}\left(\sqrt{x}\right)^3\left(\sqrt{y}\right)^3$$

$$+ \binom{6}{4}\left(\sqrt{x}\right)^2\left(\sqrt{y}\right)^4 + \binom{6}{5}\left(\sqrt{x}\right)\left(\sqrt{y}\right)^5 + \binom{6}{6}\left(\sqrt{y}\right)^6$$

$$= x^3 + 6x^2\sqrt{xy} + 15x^2y + 20xy\sqrt{xy} + 15xy^2 + 6y^2\sqrt{xy} + y^3$$

23. Using the binomial theorem:

$$\left(x - \frac{1}{x}\right)^5$$

$$= \binom{5}{0}x^5 + \binom{5}{1}x^4\left(-\frac{1}{x}\right) + \binom{5}{2}x^3\left(-\frac{1}{x}\right)^2 + \binom{5}{3}x^2\left(-\frac{1}{x}\right)^3 + \binom{5}{4}x\left(-\frac{1}{x}\right)^4 + \binom{5}{5}\left(-\frac{1}{x}\right)^5$$

$$= x^5 - 5x^3 + 10x - \frac{10}{x} + \frac{5}{x^3} - \frac{1}{x^5}$$

25. Using the binomial theorem:

$$\left(\frac{x}{y}+\frac{y}{x}\right)^4 = \binom{4}{0}\left(\frac{x}{y}\right)^4 + \binom{4}{1}\left(\frac{x}{y}\right)^3\left(\frac{y}{x}\right) + \binom{4}{2}\left(\frac{x}{y}\right)^2\left(\frac{y}{x}\right)^2 + \binom{4}{3}\left(\frac{x}{y}\right)\left(\frac{y}{x}\right)^3 + \binom{4}{4}\left(\frac{y}{x}\right)^4$$

$$= \frac{x^4}{y^4} + \frac{4x^2}{y^2} + 6 + \frac{4y^2}{x^2} + \frac{y^4}{x^4}$$

27. Using the binomial theorem:

$$\left(x^{-1}-y^2\right)^5$$

$$= \binom{5}{0}\left(x^{-1}\right)^5 + \binom{5}{1}\left(x^{-1}\right)^4\left(-y^2\right) + \binom{5}{2}\left(x^{-1}\right)^3\left(-y^2\right)^2$$

$$+ \binom{5}{3}\left(x^{-1}\right)^2\left(-y^2\right)^3 + \binom{5}{4}\left(x^{-1}\right)\left(-y^2\right)^4 + \binom{5}{5}\left(-y^2\right)^5$$

$$= x^{-5} - 5x^{-4}y^2 + 10x^{-3}y^4 - 10x^{-2}y^6 + 5x^{-1}y^8 - y^{10}$$

29. Using the binomial theorem:

$$\left(2r^{-1}+s^{-2}\right)^6$$

$$= \binom{6}{0}\left(2r^{-1}\right)^6 + \binom{6}{1}\left(2r^{-1}\right)^5\left(s^{-2}\right) + \binom{6}{2}\left(2r^{-1}\right)^4\left(s^{-2}\right)^2 + \binom{6}{3}\left(2r^{-1}\right)^3\left(s^{-2}\right)^3$$

$$+ \binom{6}{4}\left(2r^{-1}\right)^2\left(s^{-2}\right)^4 + \binom{6}{5}\left(2r^{-1}\right)\left(s^{-2}\right)^5 + \binom{6}{6}\left(s^{-2}\right)^6$$

$$= 64r^{-6} + 192r^{-5}s^{-2} + 240r^{-4}s^{-4} + 160r^{-3}s^{-6} + 60r^{-2}s^{-8} + 12r^{-1}s^{-10} + s^{-12}$$

31. The coefficient is: $\binom{10}{4}(5)^4 = 131,250$

33. The coefficient is: $\binom{7}{2}(-3)^2 = 189$

35. The coefficient is: $\binom{12}{2}(2)^{10}(1)^2 = 67584$

37. The coefficient is: $\binom{8}{6}(1)^6(-1)^2 = 28$

39. Using the binomial theorem:

$$[(a+b)+c]^4$$

$$= \binom{4}{0}(a+b)^4 + \binom{4}{1}(a+b)^3 c + \binom{4}{2}(a+b)^2 c^2 + \binom{4}{3}(a+b)c^3 + \binom{4}{4}c^4$$

$$= \left(a^4 + 4a^3b + 6a^2b^2 + 4ab^3 + b^4\right) + 4c\left(a^3 + 3a^2b + 3ab^2 + b^3\right)$$

$$+ 6\left(a^2 + 2ab + b^2\right)c^2 + 4(a+b)c^3 + c^4$$

$$= a^4 + 4a^3b + 6a^2b^2 + 4ab^3 + b^4 + 4a^3c + 12a^2bc + 12ab^2c + 4b^3c$$

$$+ 6a^2c^2 + 12abc^2 + 6b^2c^2 + 4ac^3 + 4bc^3 + c^4$$

41. Simplifying the expression: $\binom{9}{4}\binom{4}{2} = 126 \cdot 6 = 756$

43. Simplifying the expression: $\dfrac{(n+1)!}{n!} = \dfrac{(n+1)\cdot n!}{n!} = n+1$

45. Yes, this is true.

47. Using the binomial theorem:

$$(1+i)^5 = \binom{5}{0}(1)^5 + \binom{5}{1}(1)^4 i + \binom{5}{2}(1)^3 i^2 + \binom{5}{3}(1)^2 i^3 + \binom{5}{4}(1)i^4 + \binom{5}{5}i^5$$
$$= 1 + 5i + 10i^2 + 10i^3 + 5i^4 + i^5$$
$$= 1 + 5i - 10 - 10i + 5 + i$$
$$= -4 - 4i$$

49. Using the binomial theorem:

$$\left(2+\sqrt{-9}\right)^3 = (2+3i)^3$$
$$= \binom{3}{0}(2)^3 + \binom{3}{1}(2)^2(3i) + \binom{3}{2}(2)(3i)^2 + \binom{3}{3}(3i)^3$$
$$= 8 + 36i + 54i^2 + 27i^3$$
$$= 8 + 36i - 54 - 27i$$
$$= -46 + 9i$$

51. It would seem $\left(n^2\right)!$ would be larger than $(n!)^2$. Using $n = 5$:

$$\left(5^2\right)! = 25! \approx 1.55 \times 10^{25}$$
$$(5!)^2 = 120^2 = 14400$$

Chapter 11 Review Exercises

1. Finding the required terms:

$$a_1 = 3(1) - 5 = -2 \qquad\qquad a_2 = 3(2) - 5 = 1$$
$$a_3 = 3(3) - 5 = 4 \qquad\qquad a_4 = 3(4) - 5 = 7$$
$$a_{12} = 3(12) - 5 = 31$$

2. Finding the required terms:

$$b_1 = 7(1) + 1 = 8 \qquad\qquad b_2 = 7(2) + 1 = 15$$
$$b_3 = 7(3) + 1 = 22 \qquad\qquad b_4 = 7(4) + 1 = 29$$
$$b_{12} = 7(12) + 1 = 85$$

3. Finding the required terms:

$$x_1 = 5(1)^2 = 5 \qquad\qquad x_2 = 5(2)^2 = 20$$
$$x_3 = 5(3)^2 = 45 \qquad\qquad x_4 = 5(4)^2 = 80$$
$$x_{12} = 5(12)^2 = 720$$

4. Finding the required terms:

$$y_1 = 4(1) = 4 \qquad\qquad y_2 = 4(2) = 8$$
$$y_3 = 4(3) = 12 \qquad\qquad y_4 = 4(4) = 16$$
$$y_{12} = 4(12) = 48$$

5.   Finding the required terms:

$$a_1 = \frac{(-1)^1}{1+2} = -\frac{1}{3} \qquad\qquad a_2 = \frac{(-1)^2}{2+2} = \frac{1}{4}$$

$$a_3 = \frac{(-1)^3}{3+2} = -\frac{1}{5} \qquad\qquad a_4 = \frac{(-1)^4}{4+2} = \frac{1}{6}$$

$$a_{12} = \frac{(-1)^{12}}{12+2} = \frac{1}{14}$$

6.   Finding the required terms:

$$b_1 = \frac{(-1)^{1+2}}{1+1} = -\frac{1}{2} \qquad\qquad b_2 = \frac{(-1)^{2+2}}{2+1} = \frac{1}{3}$$

$$b_3 = \frac{(-1)^{3+2}}{3+1} = -\frac{1}{4} \qquad\qquad b_4 = \frac{(-1)^{4+2}}{4+1} = \frac{1}{5}$$

$$b_{12} = \frac{(-1)^{12+2}}{12+1} = \frac{1}{13}$$

7.   Finding the required terms:

$$x_1 = 4+(-1)^1 = 3 \qquad\qquad x_2 = 4+(-1)^2 = 5$$

$$x_3 = 4+(-1)^3 = 3 \qquad\qquad x_4 = 4+(-1)^4 = 5$$

$$x_{12} = 4+(-1)^{12} = 5$$

8.   Finding the required terms:

$$y_1 = 2^1 +|1-5| = 6 \qquad\qquad y_2 = 2^2 +|2-5| = 7$$

$$y_3 = 2^3 +|3-5| = 10 \qquad\qquad y_4 = 2^4 +|4-5| = 17$$

$$y_{12} = 2^{12} +|12-5| = 4103$$

9.   A formula is $a_n = 5n - 7$.

10.   A formula is $a_n = 3 \cdot 5^{n-1}$.

11.   A formula is $a_n = \dfrac{(-1)^{n+1}}{3n}$.

12.   A formula is $a_n = \dfrac{1}{2^n}$.

13.   His salary is $p_n = 250 + 8(n-1) = 8n + 242$.

14.   Over the next four years, the population is:

$$a_1 = 22,500$$
$$a_2 = 22500 + 0.03(22500) = 23,175$$
$$a_3 = 23175 + 0.03(23175) \approx 23,870$$
$$a_4 = 23,870 + 0.03(23,870) \approx 24,586$$
$$a_5 = 24,586 + 0.03(24,586) \approx 25,323$$

15.   In 5 years, the bond will be worth: $5000(1.05)^5 \approx \$6,381.41$

16.   In 5 years, the bond will be worth: $5000(1.06)^5 \approx \$6,691.13$

17.   Using $a_1 = 3$ and $d = 3$: $S_7 = \frac{7}{2}\left[2a_1 + 6d\right] = \frac{7}{2}[2 \cdot 3 + 6 \cdot 3] = 84$

18.   Using $a_1 = 3$ and $d = 4$: $S_5 = \frac{5}{2}\left[2a_1 + 4d\right] = \frac{5}{2}[2 \cdot 3 + 4 \cdot 4] = 55$

19. Using $a_1 = 2$ and $r = 2$: $S_6 = \dfrac{a_1\left(1 - r^6\right)}{1 - r} = \dfrac{2\left(1 - 2^6\right)}{1 - 2} = 126$

20. Using $a_1 = 3$ and $r = 3$: $S_8 = \dfrac{a_1\left(1 - r^8\right)}{1 - r} = \dfrac{3\left(1 - 3^8\right)}{1 - 3} = 9840$

21. Using $a_1 = 3$ and $d = 5$: $S_6 = \dfrac{6}{2}\left[2a_1 + 5d\right] = 3\left[2 \cdot 3 + 5 \cdot 5\right] = 93$

22. Using $b_1 = 7$ and $d = 3$: $S_8 = \dfrac{8}{2}\left[2b_1 + 7d\right] = 4\left[2 \cdot 7 + 7 \cdot 3\right] = 140$

23. Computing the sum: $\displaystyle\sum_{m=1}^{6} 2m = 2 + 4 + 6 + 8 + 10 + 12 = 42$

24. Computing the sum: $\displaystyle\sum_{j=4}^{9} 4j = 16 + 20 + 24 + 28 + 32 + 36 = 156$

25. Computing the sum: $\displaystyle\sum_{n=1}^{5} (n-2)^2 = 1 + 0 + 1 + 4 + 9 = 15$

26. Computing the sum: $\displaystyle\sum_{k=1}^{4} k^3 = 1 + 8 + 27 + 64 = 100$

27. Computing the sum: $\displaystyle\sum_{i=1}^{7} 5i^2 = 5 + 20 + 45 + 80 + 125 + 180 + 245 = 700$

28. Computing the sum: $\displaystyle\sum_{k=2}^{4} \left(k^2 - 6k + 1\right) = -7 - 8 - 7 = -22$

29. Since $d = 6$, the next two terms are $a_3 = 13 + 6 = 19$ and $a_4 = 19 + 6 = 25$. Finding a_{10}:
$$a_{10} = a_1 + 9d = 7 + 9 \cdot 6 = 61$$

30. Since $d = -5$, the next two terms are $a_5 = -1 + (-5) = -6$ and $a_6 = -6 + (-5) = -11$. Finding a_1:
$$a_3 = a_1 + 2d$$
$$4 = a_1 + 2(-5)$$
$$4 = a_1 - 10$$
$$a_1 = 14$$
Now finding a_{10}: $a_{10} = a_1 + 9d = 14 + 9 \cdot (-5) = -31$

31. Since $d = 5$, the next two terms are $a_5 = 21 + 5 = 26$ and $a_6 = 26 + 5 = 31$. Finding a_{10}:
$$a_{10} = a_1 + 9d = 6 + 9 \cdot 5 = 51$$

32. Since $d = 2$, the next two terms are $a_5 = 2 + 2 = 4$ and $a_6 = 4 + 2 = 6$. Finding a_{10}:
$$a_{10} = a_1 + 9d = -4 + 9 \cdot 2 = 14$$

33. Since $d = -2$, the next two terms are $a_5 = -4 - 2 = -6$ and $a_6 = -6 - 2 = -8$. Finding a_{10}:
$$a_{10} = a_1 + 9d = 2 + 9 \cdot (-2) = -16$$

34. Since $d = -\frac{1}{3}$, the next two terms are $a_4 = -\frac{1}{3} - \frac{1}{3} = -\frac{2}{3}$ and $a_5 = -\frac{2}{3} - \frac{1}{3} = -1$. Finding a_{10}:
$$a_{10} = a_1 + 9d = \tfrac{1}{3} + 9 \cdot (-\tfrac{1}{3}) = -\tfrac{8}{3}$$

35. Using $a_k - a_j = (k-j)d$:

$$x_8 - x_4 = 4d$$
$$22 - 10 = 4d$$
$$4d = 12$$
$$d = 3$$

Finding x_1:

$$x_4 = x_1 + 3d$$
$$10 = x_1 + 3 \cdot 3$$
$$10 = x_1 + 9$$
$$x_1 = 1$$

36. Using $a_k - a_j = (k-j)d$:

$$y_6 - y_4 = 2d$$
$$-13 + 9 = 2d$$
$$2d = -4$$
$$d = -2$$

Finding y_1:

$$y_4 = y_1 + 3d$$
$$-9 = y_1 + 3 \cdot (-2)$$
$$-9 = y_1 - 6$$
$$y_1 = -3$$

Now finding y_2: $y_2 = y_1 + d = -3 + (-2) = -5$

37. Using $a_1 = 16$ and $d = 32$: $a_8 = a_1 + 7d = 16 + 7(32) = 240$ feet

38. Finding a_{10}: $a_{10} = a_1 + 9d = 16 + 9(32) = 304$ feet

39. Using $a_1 = 8$ and $d = 6$: $S_{12} = \frac{12}{2}\left[2a_1 + 11d\right] = 6[2 \cdot 8 + 11 \cdot 6] = 492$

40. Using $b_1 = -9$ and $d = -2$: $S_9 = \frac{9}{2}\left[2b_1 + 8d\right] = \frac{9}{2}[2 \cdot (-9) + 8 \cdot (-2)] = -153$

41. Using $a_1 = 4$ and $d = 7$: $S_7 = \frac{7}{2}\left[2a_1 + 6d\right] = \frac{7}{2}[2 \cdot 4 + 6 \cdot 7] = 175$

42. Using $a_1 = -6$ and $d = 4$: $S_8 = \frac{8}{2}\left[2a_1 + 7d\right] = 4[2 \cdot (-6) + 7 \cdot 4] = 64$

43. Using $a_1 = 4$ and $d = 4$: $\displaystyle\sum_{i=1}^{40} 4i = S_{40} = \frac{40}{2}\left[2a_1 + 39d\right] = 20[2 \cdot 4 + 39 \cdot 4] = 3280$

44. Using $a_1 = 2$ and $d = 2$: $\displaystyle\sum_{i=10}^{20} 2i = S_{20} - S_9 = \frac{20}{2}[2 \cdot 2 + 19 \cdot 2] - \frac{9}{2}[2 \cdot 2 + 8 \cdot 2] = 420 - 90 = 330$

45. Using $a_1 = 1$ and $d = 2$: $S_{25} = \frac{25}{2}[2 \cdot 1 + 24 \cdot 2] = 625$

46. Using $a_1 = 2$ and $d = 2$: $S_{25} = \frac{25}{2}[2 \cdot 2 + 24 \cdot 2] = 650$

47. First find d:
$$a_9 - a_5 = 4d$$
$$8 - 5 = 4d$$
$$4d = 3$$
$$d = \tfrac{3}{4}$$

Now find a_1:
$$a_5 = a_1 + 4d$$
$$5 = a_1 + 4 \cdot \tfrac{3}{4}$$
$$5 = a_1 + 3$$
$$a_1 = 2$$

Now finding the sum: $S_{12} = \dfrac{12}{2}\left[2 \cdot 2 + 11 \cdot \tfrac{3}{4}\right] = 73.5$

48. First find d:
$$a_7 - a_2 = 5d$$
$$5 - 2 = 5d$$
$$5d = 3$$
$$d = \tfrac{3}{5}$$

Now find a_1:
$$a_2 = a_1 + d$$
$$2 = a_1 + \tfrac{3}{5}$$
$$a_1 = \tfrac{7}{5}$$

Now finding the sum: $S_{10} = \dfrac{10}{2}\left[2 \cdot \tfrac{7}{5} + 9 \cdot \tfrac{3}{5}\right] = 41$

49. Using $a_1 = 2$ and $r = -3$: $a_5 = a_1 r^4 = 2(-3)^4 = 162$

50. Using $a_1 = 1$ and $r = 4$: $a_6 = a_1 r^5 = 1 \cdot 4^5 = 1024$

51. Using $a_1 = 1$ and $r = \tfrac{1}{2}$: $a_7 = a_1 r^6 = 1 \cdot \left(\tfrac{1}{2}\right)^6 = \tfrac{1}{64}$

52. Using $a_1 = 4$ and $r = -\tfrac{1}{3}$: $a_8 = a_1 r^7 = 4 \cdot \left(-\tfrac{1}{3}\right)^7 = -\tfrac{4}{2187}$

53. Using $a_1 = 4$ and $r = 3$: $a_5 = a_1 r^4 = 4 \cdot 3^4 = 324$

54. First find a_1:
$$a_2 = a_1 r$$
$$4 = a_1 \cdot \tfrac{1}{2}$$
$$a_1 = 8$$

Using $a_1 = 8$ and $r = \tfrac{1}{2}$: $a_6 = a_1 r^5 = 8 \cdot \left(\tfrac{1}{2}\right)^5 = \tfrac{1}{4}$

55. Using $a_1 = 3$ and $r = 2$: $a_{10} = a_1 r^9 = 3 \cdot 2^9 = \$1536$

56. Finding S_{12}: $S_{12} = \dfrac{a_1\left(1 - r^{12}\right)}{1 - r} = \dfrac{3\left(1 - 2^{12}\right)}{1 - 2} = \$12,285$

57. Finding the sum with $a_1 = 5$ and $r = \tfrac{1}{5}$: $S = \dfrac{a_1}{1 - r} = \dfrac{5}{1 - \tfrac{1}{5}} = \tfrac{25}{4}$

58. Finding the sum with $a_1 = 1$ and $r = -\frac{1}{3}$: $S = \dfrac{a_1}{1-r} = \dfrac{1}{1+\frac{1}{3}} = \dfrac{3}{4}$

59. Since $r = 2 > 1$, the sum does not exist.

60. Since $r = \frac{7}{3} > 1$, the sum does not exist.

61. The series is $\frac{39}{100} + \frac{39}{10000} + \frac{39}{10^6} + \dots$. Using $a_1 = \frac{39}{100}$ and $r = \frac{1}{100}$: $S = \dfrac{a_1}{1-r} = \dfrac{\frac{39}{100}}{1-\frac{1}{100}} = \dfrac{39}{99} = \dfrac{13}{33}$

62. The series is $\frac{21}{10} + \left(\frac{745}{10000} + \frac{745}{10^7} + \dots \right)$. Using $a_1 = \frac{745}{10000}$ and $r = \frac{1}{1000}$:

$$\frac{21}{10} + S = \frac{21}{10} + \frac{\frac{745}{10000}}{1 - \frac{1}{1000}} = \frac{21}{10} + \frac{745}{9990} = \frac{21724}{9990}$$

63. The total distance traveled is: $80 + 2\left[60 + \frac{3}{4} \cdot 60 + \left(\frac{3}{4}\right)^2 \cdot 60 + \dots \right] = 80 + 2S$

Using $a_1 = 60$ and $r = \frac{3}{4}$: $S = \dfrac{a_1}{1-r} = \dfrac{60}{1 - \frac{3}{4}} = 240$

The total distance is therefore: $80 + 2(240) = 560$ feet

64. For $n = 2$: $1 \cdot 2 = \dfrac{(2-1) \cdot 2 \cdot (2+1)}{3} = 2$, so P_1 is true

Assume P_n is true:

$$1 \cdot 2 + 2 \cdot 3 + 3 \cdot 4 + \dots + (n-1)n = \frac{(n-1)n(n+1)}{3}$$

$$1 \cdot 2 + 2 \cdot 3 + 3 \cdot 4 + \dots + (n-1)n + n(n+1) = \frac{(n-1)n(n+1)}{3} + n(n+1)$$

$$1 \cdot 2 + 2 \cdot 3 + 3 \cdot 4 + \dots + (n-1)n + n(n+1) = \frac{n(n+1)}{3}[n-1+3]$$

$$1 \cdot 2 + 2 \cdot 3 + 3 \cdot 4 + \dots + (n-1)n + n(n+1) = \frac{n(n+1)(n+2)}{3}$$

So P_{n+1} is true and the induction is complete.

65. For $n = 4$: $2^4 = 16$ and $4! = 24$, so $2^4 < 4!$

Assume $P_n : 2^n < n!$ is true. Form P_{n+1}: $2^{n+1} = 2 \cdot 2^n < 2 \cdot n! < (n+1) \cdot n! = (n+1)!$

So P_{n+1} is true and the induction is complete.

66. For $n = 6$: $6! = 720$ and $6^3 = 216$, so $6! > 6^3$

Assume $P_n : n! > n^3$ is true. Form P_{n+1}: $(n+1)! = (n+1) \cdot n! > (n+1) \cdot n^3 = n^4 + n^3$

Now $(n+1)^3 = n^3 + 3n^2 + 3n + 1$, so we must show $n^4 > 3n^2 + 3n + 1$, which is true whenever $n \geq 3$ (this can be verified using a graphing calculator). Therefore:

$$(n+1)! > n^3 + n^4 > n^3 + 3n^2 + 3n + 1 = (n+1)^3$$

So P_{n+1} is true and the induction is complete.

67. The number of finishes is: $P(12,3) = 12 \cdot 11 \cdot 10 = 1320$

68. The number of picks is: $C(8,3) \cdot C(10,3) = 56 \cdot 120 = 6720$

69. The number of arrangements is: $9! = 362,880$

70. The number of arrangements is: $\dfrac{8!}{3!} = 6720$

71. The number of arrangements is: $\dfrac{7!}{3!2!} = 420$

72. **a.** The number of combinations is: $9 \cdot 9 \cdot 9 = 729$

    b. The number of combinations is: $9 \cdot 9 \cdot 5 = 405$

    c. The number of combinations is: $4 \cdot 9 \cdot 9 - 1 = 323$

       Note that we excluded the combination 500.

73. **a.** The number of ways is: $C(13,5) = 1287$

    b. Note that there are 10 possible straights, starting with A...5 through 10...A. The number of ways is: $10 \cdot 4^5 = 10,240$

    c. The number of ways is: $4 \cdot 10 = 40$

74. The family can have either 0, 1, 2, 3, 4, or 5 girls, so there are 6 combinations.

75. The number of sequences is $2^5 = 32$.

76. Using the binomial theorem:

$$(2x-y)^5$$

$$= \binom{5}{0}(2x)^5 + \binom{5}{1}(2x)^4(-y) + \binom{5}{2}(2x)^3(-y)^2 + \binom{5}{3}(2x)^2(-y)^3$$

$$+ \binom{5}{4}(2x)(-y)^4 + \binom{5}{5}(-y)^5$$

$$= 32x^5 - 80x^4y + 80x^3y^2 - 40x^2y^3 + 10xy^4 - y^5$$

77. Using the binomial theorem:

$$\left(\frac{x}{3}+\frac{y}{4}\right)^3 = \binom{3}{0}\left(\frac{x}{3}\right)^3 + \binom{3}{1}\left(\frac{x}{3}\right)^2\left(\frac{y}{4}\right) + \binom{3}{2}\left(\frac{x}{3}\right)\left(\frac{y}{4}\right)^2 + \binom{3}{3}\left(\frac{y}{4}\right)^3$$

$$= \frac{x^3}{27} + \frac{x^2y}{12} + \frac{xy^2}{16} + \frac{y^3}{64}$$

78. Using the binomial theorem:

$$\left(x^{-2}+y^{-1}\right)^6$$

$$= \binom{6}{0}\left(x^{-2}\right)^6 + \binom{6}{1}\left(x^{-2}\right)^5\left(y^{-1}\right) + \binom{6}{2}\left(x^{-2}\right)^4\left(y^{-1}\right)^2 + \binom{6}{3}\left(x^{-2}\right)^3\left(y^{-1}\right)^3$$

$$+ \binom{6}{4}\left(x^{-2}\right)^2\left(y^{-1}\right)^4 + \binom{6}{5}\left(x^{-2}\right)\left(y^{-1}\right)^5 + \binom{6}{6}\left(y^{-1}\right)^6$$

$$= x^{-12} + 6x^{-10}y^{-1} + 15x^{-8}y^{-2} + 20x^{-6}y^{-3} + 15x^{-4}y^{-4} + 6x^{-2}y^{-5} + y^{-6}$$

79. The coefficient is: $\binom{8}{4} \cdot 1^4 \cdot 5^4 = 43750$

80. The coefficient is: $\binom{10}{3} \cdot 1^7 \cdot (-3)^3 = -3240$

Chapter 11 Practice Test

1. **a.** Finding the required terms:
$$x_1 = 1^2 - 3(1) = -2 \qquad\qquad x_2 = 2^2 - 3(2) = -2$$
$$x_3 = 3^2 - 3(3) = 0 \qquad\qquad x_8 = 8^2 - 3(8) = 40$$

   b. Finding the required terms:
$$y_1 = 5(1)^3 - 2 = 3 \qquad\qquad y_2 = 5(2)^3 - 2 = 38$$
$$y_3 = 5(3)^3 - 2 = 133 \qquad\qquad y_8 = 5(8)^3 - 2 = 2558$$

2. Computing the sum: $\displaystyle\sum_{k=4}^{7}\left(3k^2 - 5k + 1\right) = 29 + 51 + 79 + 113 = 272$

3. Using $a_1 = -4$ and $d = 6$: $a_{112} = a_1 + 111d = -4 + 111(6) = 662$

4. Using $a_1 = 3$ and $r = -2$: $a_6 = a_1 r^5 = 3(-2)^5 = -96$

5. Using $a_1 = 4$ and $d = 4$: $S_{20} = \dfrac{20}{2}\left[2a_1 + 19d\right] = 10[2 \cdot 4 + 19 \cdot 4] = 840$

6. Using $a_1 = 3$ and $r = 3$: $S_5 = \dfrac{a_1\left(1 - r^5\right)}{1 - r} = \dfrac{3\left(1 - 3^5\right)}{1 - 3} = 363$

7. **a.** Using $a_1 = 5$ and $d = 4$: $a_7 = a_1 + 6d = 5 + 6 \cdot 4 = 29$

   b. Using $a_1 = 1$ and $r = 5$: $a_7 = a_1 r^6 = 1 \cdot 5^6 = 15625$

   c. Using $a_1 = \frac{1}{3}$ and $r = 6$: $a_5 = a_1 r^4 = \frac{1}{3} \cdot 6^4 = 432$

   d. Finding d:
$$a_5 - a_3 = 2d$$
$$2 - 1 = 2d$$
$$d = \tfrac{1}{2}$$

   Now find a_1:
$$a_3 = a_1 + 2d$$
$$1 = a_1 + 2 \cdot \tfrac{1}{2}$$
$$a_1 = 0$$

   Using $a_1 = 0$ and $d = \frac{1}{2}$: $S_{10} = \dfrac{10}{2}\left[2a_1 + 9d\right] = 5\left[2 \cdot 0 + 9 \cdot \tfrac{1}{2}\right] = \tfrac{45}{2}$

8. Using $a_1 = 10$ and $d = 10$: $S_{20} = \dfrac{20}{2}\left[2a_1 + 19d\right] = 10[2 \cdot 10 + 19 \cdot 10] = 2100$

9. Using $a_1 = 8$ and $r = \frac{3}{4}$: $S = \dfrac{a_1}{1 - r} = \dfrac{8}{1 - \frac{3}{4}} = 32$ feet

10. For $n = 1$: $x - 2$ is a factor of $x - 2$, which is true

Assume P_n: $x - 2$ is a factor of $x^n - 2^n$. So $x^n - 2^n = (x - 2) \cdot Q(x)$, for some polynomial $Q(x)$.

Forming P_{n+1}:

$$x^{n+1} - 2^{n+1} = x^{n+1} - x \cdot 2^n + x \cdot 2^n - 2^{n+1}$$
$$= x(x^n - 2^n) + 2^n (x - 2)$$
$$= x \cdot (x - 2)Q(x) + 2^n (x - 2)$$
$$= (x - 2)(xQ(x) + 2^n)$$

Thus $x - 2$ is a factor of $x^{n+1} - 2^{n+1}$, so P_{n+1} is true and the induction is complete.

11. **a.** The number of ways is: $C(23, 6) = \dfrac{23!}{6!17!} = 100,947$

   b. The number of ways is: $C(10, 2) \cdot C(8, 2) \cdot C(5, 2) = 12,600$

12. The number of arrangements is: $\dfrac{11!}{5!2!2!} = 83,160$

13. Using the binomial theorem:

$$\left(2x^2 - 3y\right)^8$$

$$= \binom{8}{0}\left(2x^2\right)^8 + \binom{8}{1}\left(2x^2\right)^7(-3y) + \binom{8}{2}\left(2x^2\right)^6(-3y)^2 + \binom{8}{3}\left(2x^2\right)^5(-3y)^3$$

$$+ \binom{8}{4}\left(2x^2\right)^4(-3y)^4 + \binom{8}{5}\left(2x^2\right)^3(-3y)^5 + \binom{8}{6}\left(2x^2\right)^2(-3y)^6$$

$$+ \binom{8}{7}\left(2x^2\right)(-3y)^7 + \binom{8}{8}(-3y)^8$$

$$= 256x^{16} - 3072x^{14}y + 16128x^{12}y^2 - 48384x^{10}y^3 + 90720x^8y^4 - 108864x^6y^5$$
$$+ 81648x^4y^6 - 34992x^2y^7 + 6561y^8$$